Cell Mech

From Single Scale-Based Models to Multiscale Modeling

CHAPMAN & HALL/CRC

Mathematical and Computational Biology Series

Aims and scope:

This series aims to capture new developments and summarize what is known over the entire spectrum of mathematical and computational biology and medicine. It seeks to encourage the integration of mathematical, statistical, and computational methods into biology by publishing a broad range of textbooks, reference works, and handbooks. The titles included in the series are meant to appeal to students, researchers, and professionals in the mathematical, statistical and computational sciences, fundamental biology and bioengineering, as well as interdisciplinary researchers involved in the field. The inclusion of concrete examples and applications, and programming techniques and examples, is highly encouraged.

Series Editors

N. F. Britton
Department of Mathematical Sciences
University of Bath

Xihong Lin
Department of Biostatistics
Harvard University

Hershel M. Safer

Mona Singh
Department of Computer Science
Princeton University

Anna Tramontano
Department of Biochemical Sciences
University of Rome La Sapienza

Proposals for the series should be submitted to one of the series editors above or directly to:
CRC Press, Taylor & Francis Group
4th, Floor, Albert House
1-4 Singer Street
London EC2A 4BQ
UK

Published Titles

Algorithms in Bioinformatics: A Practical Introduction
Wing-Kin Sung

Bioinformatics: A Practical Approach
Shui Qing Ye

Biological Sequence Analysis Using the SeqAn C++ Library
Andreas Gogol-Döring and Knut Reinert

Cancer Modelling and Simulation
Luigi Preziosi

Cell Mechanics: From Single Scale-Based Models to Multiscale Modeling
Arnaud Chauvière, Luigi Preziosi, and Claude Verdier

Combinatorial Pattern Matching Algorithms in Computational Biology Using Perl and R
Gabriel Valiente

Computational Biology: A Statistical Mechanics Perspective
Ralf Blossey

Computational Neuroscience: A Comprehensive Approach
Jianfeng Feng

Data Analysis Tools for DNA Microarrays
Sorin Draghici

Differential Equations and Mathematical Biology, Second Edition
D.S. Jones, M.J. Plank, and B.D. Sleeman

Engineering Genetic Circuits
Chris J. Myers

Exactly Solvable Models of Biological Invasion
Sergei V. Petrovskii and Bai-Lian Li

Gene Expression Studies Using Affymetrix Microarrays
Hinrich Göhlmann and Willem Talloen

Glycome Informatics: Methods and Applications
Kiyoko F. Aoki-Kinoshita

Handbook of Hidden Markov Models in Bioinformatics
Martin Gollery

Introduction to Bioinformatics
Anna Tramontano

An Introduction to Systems Biology: Design Principles of Biological Circuits
Uri Alon

Kinetic Modelling in Systems Biology
Oleg Demin and Igor Goryanin

Knowledge Discovery in Proteomics
Igor Jurisica and Dennis Wigle

Meta-analysis and Combining Information in Genetics and Genomics
Rudy Guerra and Darlene R. Goldstein

Modeling and Simulation of Capsules and Biological Cells
C. Pozrikidis

Niche Modeling: Predictions from Statistical Distributions
David Stockwell

Normal Mode Analysis: Theory and Applications to Biological and Chemical Systems
Qiang Cui and Ivet Bahar

Optimal Control Applied to Biological Models
Suzanne Lenhart and John T. Workman

Pattern Discovery in Bioinformatics: Theory & Algorithms
Laxmi Parida

Python for Bioinformatics
Sebastian Bassi

Spatial Ecology
Stephen Cantrell, Chris Cosner, and Shigui Ruan

Spatiotemporal Patterns in Ecology and Epidemiology: Theory, Models, and Simulation
Horst Malchow, Sergei V. Petrovskii, and Ezio Venturino

Stochastic Modelling for Systems Biology
Darren J. Wilkinson

Structural Bioinformatics: An Algorithmic Approach
Forbes J. Burkowski

The Ten Most Wanted Solutions in Protein Bioinformatics
Anna Tramontano

Chapman & Hall/CRC Mathematical and Computational Biology Series

CELL MECHANICS

FROM SINGLE SCALE-BASED MODELS TO MULTISCALE MODELING

EDITED BY

ARNAUD CHAUVIÈRE,
LUIGI PREZIOSI,
AND CLAUDE VERDIER

CRC Press is an imprint of the
Taylor & Francis Group, an **informa** business
A CHAPMAN & HALL BOOK

CRC Press
Taylor & Francis Group
6000 Broken Sound Parkway NW, Suite 300
Boca Raton, FL 33487-2742

First issued in paperback 2019

© 2010 by Taylor and Francis Group, LLC
CRC Press is an imprint of Taylor & Francis Group, an Informa business

No claim to original U.S. Government works

ISBN-13: 978-0-8493-3816-8 (hbk)
ISBN-13: 978-0-367-39061-7 (pbk)

This book contains information obtained from authentic and highly regarded sources. Reasonable efforts have been made to publish reliable data and information, but the author and publisher cannot assume responsibility for the validity of all materials or the consequences of their use. The authors and publishers have attempted to trace the copyright holders of all material reproduced in this publication and apologize to copyright holders if permission to publish in this form has not been obtained. If any copyright material has not been acknowledged please write and let us know so we may rectify in any future reprint.

Except as permitted under U.S. Copyright Law, no part of this book may be reprinted, reproduced, transmitted, or utilized in any form by any electronic, mechanical, or other means, now known or hereafter invented, including photocopying, microfilming, and recording, or in any information storage or retrieval system, without written permission from the publishers.

For permission to photocopy or use material electronically from this work, please access www.copyright.com (http://www.copyright.com/) or contact the Copyright Clearance Center, Inc. (CCC), 222 Rosewood Drive, Danvers, MA 01923, 978-750-8400. CCC is a not-for-profit organization that provides licenses and registration for a variety of users. For organizations that have been granted a photocopy license by the CCC, a separate system of payment has been arranged.

Trademark Notice: Product or corporate names may be trademarks or registered trademarks, and are used only for identification and explanation without intent to infringe.

Visit the Taylor & Francis Web site at
http://www.taylorandfrancis.com

and the CRC Press Web site at
http://www.crcpress.com

Contents

III Mechanical Effects of Environment on Cell Behavior

IV From Cellular to Multicellular Models

Preface

This book is the result of years of collaboration among different teams interacting in the framework of a European Research and Training Network on "Modeling, Mathematical Methods and Computer Simulation of Tumour Growth and Therapy." Throughout this network, the collaboration of groups has brought new insight and motivated research leading to mechanical, mathematical, physical, and biological approaches all related to the simulation of the behavior of cells and, in particular, of tumor cells.

While looking at the phenomena from different points of view, it soon became clear that multiscale problems are ubiquitous and fundamental in cell mechanics as illustrated in Figure P.1. For instance, such problems naturally emerge from the description of situations involving cell adhesion, migration, invasion, intravasation and extravasation, mitosis, cell–cell interactions, etc. Modeling the role of mechano-transduction in the behavior of the cell, its fate (duplication, survival, or death) or even its differentiation, or how the internal dynamics of actin and microtubules determine the migratory characteristics of the cells leads also to a combination of intrinsically coupled scales. These aspects are, of course, not only of major interest to describe the growth of tumors, but also in embryogenesis, in the description of physiological and pathological states of all tissues, and nowadays in tissue engineering, tissue remediation, and regeneration. It is our motivation here to present non-exhaustive tools that will help graduate students, researchers, and professors in their research and teaching activities.

The book is a multidisciplinary effort that contains reviews or articles of interest to students and researchers from different disciplines, such as the ones cited above. Some of the collected works were presented in a European Workshop on "Multiscale Approaches in Cell Mechanics" (January 7–10, 2008), which took place in Autrans (France) and was organized by Arnaud Chauvière (then at the Technische Universität, Dresden, Germany) and Claude Verdier (CNRS, Grenoble, France). Several other works from experts in the field are also included in order to provide a more general overview of the subject. At the moment, it appears that solving such multiscale problems is still a difficult task to achieve.

Throughout this textbook, various approaches and methods are presented in order to highlight how phenomena happening at various scales can be modeled and coupled to account for multiscale aspects. It is also our purpose here to show the existing tools available, and by default to bring the need to carry on

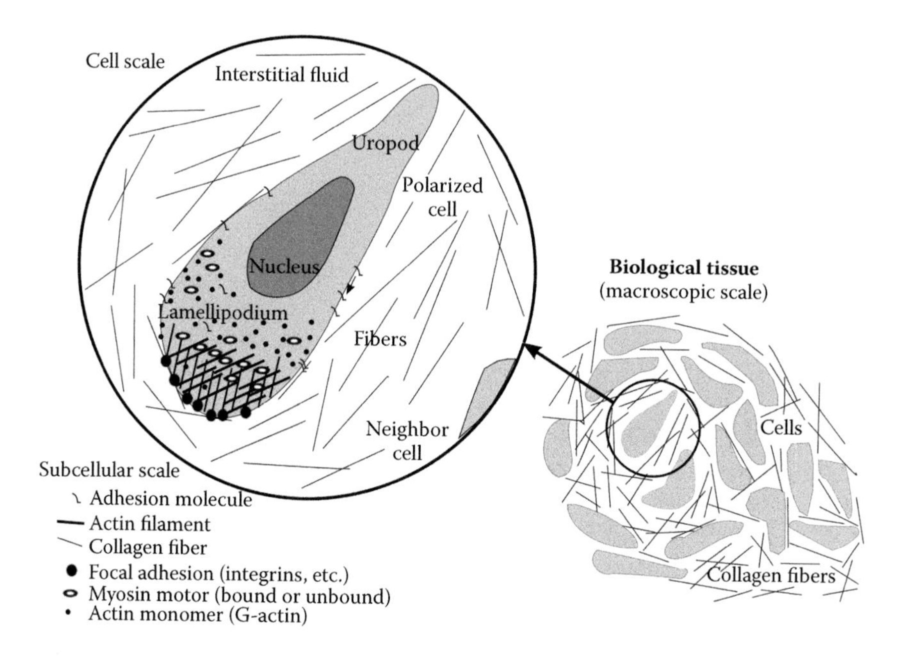

FIGURE P.1 Multiscale view of cell mechanics. A biological tissue consists of several components including cells and the extracellular matrix. A zoom allows one to identify a single cell embedded in the extracellular matrix and surrounded by neighboring cells. Details illustrate the configuration at the cellular scale. At the microscopic scale, various elements interact to create cell motion (actin filaments, adhesion molecules within focal adhesions, etc.). At the nanoscale, cell sub-elements combine to create actin filaments generating forces through the acto-myosin complex.

new approaches. In the spirit of the workshop mentioned above, efforts have been made to propose, in some chapters, both experimental and theoretical aspects of the same topic. In other chapters, experiments and theories are discussed separately in order to provide a deeper insight into biological aspects in the first case, or complex modeling in the second case.

The book is divided into four parts. In Part I, some subcellular and cellular aspects and their links are presented. Microrheology (Chapter 1) is shown to be a powerful tool for the investigation of cell mechanical properties. Links between length scales are properly identified in a unified concept. Chapter 2 presents multiphysics and multiscale approaches to investigate intracellular mechanisms involved in cell motility. Minimal principles are used to explain how actin polymerization can take place and trigger forces to generate motion. This approach is very promising for further investigations and understanding of cell motion. Chapter 3 discusses how subcellular effects involving certain genes can play a role in inducing cell motility in the context of cancer. The multiscale approach is explained within the concept of sophisticated experiments.

Part II considers cell migration. The basic principles proposed earlier[1] are revisited through three different models, respectively, based on physical (Chapter 4), mathematical (Chapter 5), and computational (Chapter 6) approaches. All of them introduce and develop tools to account for the description of the complex interplay of cell adhesion molecules and the dynamic evolution of the cell cytoskeleton. They underline the ubiquity of multiscale aspects when one aims to understand the complex interaction among molecules at the nanometer scale, and their combination to create motion at the cellular scale (i.e., micron scale).

Part III is devoted to the study of cell interactions with the environment, in particular the role of external mechanical forces and their effects on cell behavior. Chapter 7 considers a model approach to see how a flow field (involving shear stresses) can affect the dynamic behavior of a cell (a bead here) in close contact with a wall. Such movement is very important when considering the rolling motion of leukocytes on the vessel wall, but may also be relevant when cancer cells interact with the endothelium of blood vessels to intravasate or extravasate. Chapters 8 and 9 consider the role of a substrate on the adhesion of a cell and the exerted traction forces. In particular, the role of substrate rigidity is very important and affects mechano-transduction (Chapter 8) as well as the development of stable or unstable focal adhesions. Another related aspect concerns the corresponding forces exerted by cells on a rigid substrate (Chapter 9), which depend on its rigidity. Such features are important when considering many cells in their natural environment—for example, when in contact with a soft tissue. Mechano-transduction is also reported in Chapter 10 in an attempt to see how external mechanical forces can lead to calcium fluxes and then to biochemical responses at the cell level.

Part IV presents models of multicellular systems. A substantial review of innovative models developed in the context of cancer has been presented earlier[2] but some recent developments are now available and can be considered complementary. This is the aim of Part IV: to propose such new approaches. Chapter 11 investigates the collective amoeboid motion of cells within the extracellular matrix (i.e., collagen in real tissue). This problem is particularly important when looking at cancer cell migration.[3] The next three chapters present applications of population models to developmental biology and cancer (Chapter 12), embryogenesis (Chapter 13), and cancer invasion (Chapter 14).

[1]D.A. Lauffenburger and A.F. Horwitz (1996). Cell migration: a physically integrated molecular process. *Cell* 84:359–369.

[2]*Cancer Modelling and Simulation*, Chapman & Hall/CRC Mathematical and Medicine Series, Ed. Luigi Preziosi, 2003.

[3]P. Friedl and K. Wolf (2003). Tumour-cell invasion and migration: diversity and escape mechanisms. *Nature* 3:362–374.

Finally, Chapter 15 presents a new approach based on Delaunay object dynamics to study the behavior of highly motile cells involved in certain tissues.

Of course, there is still a large gap between the acceptance of multiscale models and their effectiveness in predicting accurate pathologies, and their import toward real medical applications. Nevertheless, we have tried here to collect new methods to apply to cells and tissues through a multiscale concept.

The editors wish to thank all the authors who contributed to this book. Special thanks are also given to all participants of the Marie Curie RTN Network (2004–2008) for bringing into play such challenging scientific efforts. We are also indebted to Sunil Nair at Taylor & Francis for his helpful advice.

Arnaud Chauvière
Luigi Preziosi
Claude Verdier

Editors

Arnaud Chauvière studied physics at Pierre et Marie Curie University in Paris and received his Ph.D. in theoretical fluid mechanics in 2001. He has held several postdoctoral positions in the LMM (Theoretical Mechanics Lab) at Pierre et Marie Curie University, the Politecnico di Torino, the Centre of Mathematical Medicine and Biology of the University of Nottingham, and the Technische Universität Dresden. He is now Assistant Professor of Health Informatics at the Health Science Center of the University of Texas in Houston. His principal research area is cell migration modeling and its application in the context of tumor growth.

Luigi Preziosi graduated in mathematics from the University of Naples. He received his Ph.D.s in mechanics and in mathematics, studying at the University of Minnesota and at the University of Naples, respectively. He is now a full professor of mathematical physics at the Engineering Faculty of the Politecnico di Torino. Preziosi is the author of more than 100 publications in the international literature in several fields of mathematical physics and applied mathematics. For the past ten years he has focused his research on cancer modeling. Related to this last subject, he edited, among others, the CRC volume *Cancer Modelling and Simulation* and, together with F. Mollica and K.R. Rajagopal, a book on *Modeling of Biological Materials.*

Claude Verdier graduated with an engineering degree from Ecole Centrale Paris in 1986 before moving to the University of Minnesota where he received his Ph.D. in 1990, working on viscoelastic fluids. He joined the CNRS (Centre National de la Recherche Scientifique) in 1992 where he undertook research in various areas, including polymer adhesion, ultrasonic characterization, drop deformation, and biorheology. He is now Research Director at CNRS, working at LSP (Physics Lab) in Grenoble. Author of more than fifty publications in his research areas, he currently focuses on experimental and theoretical biomechanics, and he has been involved in two large-scale European projects. He is a member of the Comité National and the Conseil de l'Institut ST2I at CNRS.

Contributors

Richard J. Adams
Department of Physiology, Development, and Neuroscience
University of Cambridge
Cambridge, United Kingdom

Wolfgang Alt
Group of Theoretical Biology
University of Bonn
Bonn, Germany

Davide Ambrosi
Dipartimento di Matematica
Politecnico di Milano
Milano, Italy

Atef Asnacios
Laboratoire Matière et Systèmes Complexes
Université Paris Diderot
Paris, France

Tilo Beyer
Institute for Molecular and Clinical Immunology
Otto von Guericke-University
Magdeburg, Germany

Guy B. Blanchard
Department of Physiology, Development, and Neuroscience
University of Cambridge
Cambridge, United Kingdom

Martin Bock
Group of Theoretical Biology
University of Bonn
Bonn, Germany

Julien Browaeys
Laboratoire Matière et Systèmes Complexes
Université Paris Diderot
Paris, France

Denis Caillerie
Laboratoire 3S–R
Domaine Universitaire
Grenoble, France

Arnaud Chauvière
School of Health Information Sciences
University of Texas Health Science Center
Houston, Texas, United States

Dirk Drasdo
INRIA
Le Chesnay, France

Alain Duperray
Institut Albert Bonniot
Université Joseph Fourier
La Tronche, France

Sophie Féréol
INSERM
Biomécanique Cellulaire et Respiratoire
Faculté de Médecine
Créteil, France

Redouane Fodil
INSERM
Biomécanique Cellulaire et Respiratoire
Faculté de Médecine
Créteil, France

François Gallet
Laboratoire Matière et Systèmes Complexes
Université Paris Diderot
Paris, France

Alf Gerisch
Institut für Mathematik
Martin Luther Universität
Halle-Wittenberg, Germany

Sylvie Hénon
Laboratoire Matière et Systèmes Complexes
Université Paris Diderot
Paris, France

Daniel Isabey
INSERM
Biomécanique Cellulaire et Respiratoire
Faculté de Médecine
Créteil, France

Mourad Ismail
Laboratoire de Spectrométrie Physique
Saint Martin d'Hères, France

Nick Jagiella
INRIA
Le Chesnay, France

Oliver E. Jensen
School of Mathematical Sciences
University of Nottingham
Nottingham, United Kingdom

Karin John
Laboratoire de Spectrométrie Physique
Saint Martin d'Hères, France

Alexandre Kabla
Engineering Department Cambridge University
Cambridge, United Kingdom

Tae-Jin Kim
Neuroscience Program and the Beckman Institute for Advanced Science and Technology
University of Illinois
Urbana, Illinois, United States

Valérie M. Laurent
Laboratoire de Spectrométrie Physique
Saint Martin d'Hères, France

Bruno Louis
INSERM
Biomécanique Cellulaire et Respiratoire
Faculté de Médecine
Créteil, France

L. Mahadevan
School of Engineering and Applied Sciences
Harvard University
Cambridge, Massachusetts United States

Michael Meyer-Hermann
Frankfurt Institute for Advanced Studies
Frankfurt am Main, Germany

Chaouqi Misbah
Laboratoire de Spectrométrie Physique
Saint Martin d'Hères, France

Christoph Möhl
Institute of Bio- and Nanosystems
Research Centre
Jülich GmbH
Jülich, Germany

Dietmar Ölz
Faculty of Mathematics
Vienna University
Wolfgang Pauli Institute
Vienna, Austria

Kevin J. Painter
Department of Mathematics
Maxwell Institute for Mathematical Sciences
School of Mathematical and Computer Sciences
Heriot-Watt University
Edinburgh, United Kingdom

Rémy Pedeux
Institut Albert Bonniot
Université Joseph Fourier
La Tronche, France

Gabriel Pelle
INSERM
Biomécanique Cellulaire et Respiratoire
Faculté de Médecine and Hôpital Henri Mondor
Créteil, France

Valentina Peschetola
Laboratoire de Spectrométrie Physique
Saint Martin d'Hères, France

Philippe Peyla
Laboratoire de Spectrométrie Physique
Saint Martin d'Hères, France

Emmanuelle Planus
Centre de Recherche
Equipe DySAD, IAB Site Santé
Grenoble, France

Luigi Preziosi
Dipartimento di Matematica
Politecnico di Torino
Torino, Italy

Jacques Prost
Physico-Chimie Institut Curie
Paris, France

Ignacio Ramis-Conde
INRIA
Le Chesnay, France

Annie Raoult
Laboratoire MAP5
Université Paris Descartes
Paris, France

Sylvain Reboux
Institute for Theoretical Computer Science
Zürich, Switzerland

Giles Richardson
School of Mathematics
University of Southampton
Southampton, United Kingdom

Christian Schmeiser
Faculty of Mathematics
Vienna University
Wolfgang Pauli Institute
Vienna, Austria
and Johann Radon Institute for Computational and Applied Mathematics
Linz, Austria

Angélique Stéphanou
Institut de l'Ingénierie et de l'Information de Santé
Faculté de Médecine de Grenoble
La Tronche, France

Claude Verdier
Laboratoire de Spectrométrie Physique
Saint Martin d'Hères, France

Irene E. Vignon-Clementel
INRIA
Le Chesnay, France

Yingxiao Wang
Department of Bioengineering and Department of Molecular and Integrative Physiology
University of Illinois
Urbana, Illinois, United States

William Weens
INRIA
Le Chesnay, France

Damien Ythier
Institut Albert Bonniot
Université Joseph Fourier
La Tronche, France

Part I

From Subcellular to Cellular Properties

Chapter 1. Microrheology of Living Cells at Different Time and Length Scales

Microrheology is a powerful tool to investigate the mechanical properties of living cells, which are essential to many biological functions such as cellular adhesion, migration, and division. After a synthetic presentation of rheology basics for complex systems, we describe the optical tweezers and the uniaxial stretching techniques, which allow us to probe the microrheology of individual cells at different length scales and over a wide time range. The creep function $J(t)$ and viscoelastic modulus $G_e(\omega)$ are retrieved from both experiments and always exhibit weak power law behaviors as functions of time and frequency, respectively. The exponent $\alpha \approx 0.2$ of this power law appears very robust and shows almost no dependence on the cell type, nor on the typical length scale of the experiment. On the contrary, the typical rigidity of the cells, characterized for instance by the viscoelastic modulus G_0 at 1 Hz, varies over almost two orders of magnitude from one cell type to another. We propose an interpretation of the observed power law rheology, based on a semi-phenomenological model involving a scale-free structure of the cellular network and a broad and dense distribution of relaxation times in the system. This model leads to power law mechanical responses and accurately predicts the normal distribution of the exponent α and the log-normal distribution of G_0, as experimentally observed.

Chapter 2. Actin-Based Propulsion: Intriguing Interplay between Material Properties and Growth Processes

Eukaryotic cells and intracellular pathogens such as bacteria or viruses utilize the actin polymerization machinery to propel themselves forward. Thereby, the onset of motion and choice of direction may be the result of a spontaneous symmetry breaking or might be triggered by external signals and preexisting asymmetries, for example, through a previous septation in bacteria. Although very complex, a key feature of cellular motility is the ability of actin to form dense polymeric networks, whose microstructures are tightly regulated by the cell. These polar actin networks produce the forces necessary for propulsion but may also be at the origin of a spontaneous symmetry breaking. Understanding the exact role of actin dynamics in cell motility requires multiscale approaches that capture at the same time the polymer network structure and dynamics on the scale of a few nanometers and the macroscopic distribution of elastic stresses on the scale of the whole cell. In this chapter we review a selection of theories on how mechanical material properties and growth processes interact to induce the onset of actin-based motion.

Chapter 3. Cancer: Cell Motility and Tumor Suppressor Genes

Cell migration plays a critical role in cancer pathologies because metastasis is the ultimate stage of the disease that leads to the death of the individual. During tumorigenesis and especially the metastasis process, transformed cells need to gain functions to migrate and invade new tissues. Recently, it has been shown that a new family of tumor suppressor genes, the ING genes, may be involved in cell migration and motility. This chapter describes the importance of cell motility and invasion in the tumor process and highlights the role that tumor suppressor genes may play, especially the ING genes. Some of the methodologies used to investigate these processes are also described.

Chapter 1

Microrheology of Living Cells at Different Time and Length Scales

Atef Asnacios, Sylvie Hénon, Julien Browaeys, and François Gallet

Contents

1.1 Introduction

To perform their functions, living cells must adapt to external stresses and to varying mechanical properties of their environment. Thus, rheological properties (i.e., stress–strain relationships) are key features of living cells. Actually, mechanics play a major role in many biological processes such as cell crawling, wound healing, protein regulation, and even apoptosis [19]. Conversely, several pathologies, like metastasis, asthma, and sickle cell anemia, involve alteration of the mechanical properties of a given cell type. All these processes are mainly controlled by the structure and mechanical properties of the cytoskeletal network. This network is a dynamical assembly of macromolecules, principally made of actin filaments, intermediate filaments, and microtubules, and interacting with a variety of associated proteins, crosslinkers, and molecular motors.

The mechanical properties of the cytoskeletal network are therefore the subject of many experimental studies, made possible by the development of numerous quantitative micromanipulation techniques, such as micropipettes [37], cell poking [32], shear flow cytometry [4,12], atomic force microscopy (AFM) [1,35,36], microplates [9,11,39], optical tweezers [3,24,47], optical stretchers [16,45], magnetic tweezers [5,44], magnetic twisters [13,24], and particle tracking [23,42,46]. These techniques are complementary in the sense that they probe the behavior of the intracellular medium at different length scales and time scales, and that they implement stresses and strains in different geometries and with different orders of magnitude.

Most recent results in microrheology of the intracellular medium have demonstrated that it is a complex viscoelastic medium that cannot be simply modeled by associating a finite, small number of elastic and viscous elements. Indeed, the viscoelastic complex modulus of the cell medium exhibits a weak power law behavior over a wide frequency range [3,14]. Similarly the creep function behaves as a power law of elapsed time [11,26]. This clearly indicates that there is a broad and dense distribution of dissipation times in the cell, and that the mechanisms responsible for the storage of elastic energy and its dissipation are strongly correlated. Such behavior, characteristic of structural damping, is found in other complex viscoelastic systems, like colloids, gels, pastes, and more generally the class of so-called "soft glassy materials" [6,43].

Beyond the formal analogy that can be made between the behaviors of the intracellular medium and of such soft glassy materials a detailed interpretation of structural damping in terms of elementary elastic and dissipative mechanisms was still to be built. This is one of the objectives of this chapter, in which we propose a global description of the mechanics of the cytoskeletal network, consistent with the data gathered from numerous experiments and based on the existence of a quasi self-similar structure in the cytoskeletal network. This description allows the comparison of the results obtained with different microrheological techniques and/or under various experimental conditions, and offers a unifying frame to interpret the experiments performed at the nano- and microscale levels and at the scale of the whole cell.

1.2 Elements of Rheology

The purpose of rheology is to derive the relationships between the mechanical stress σ applied to a viscoelastic material and the induced strain ϵ. This response can be expressed either as a function of time elapsed after the application of a step stress (creep function $J(t)$) or as a function of angular velocity ω for an oscillating stress (viscoelastic modulus $G(\omega)$). The two descriptions are equivalent, and we recall here the translation rules from one to the other.

We restrict ourselves to the linear response regime, corresponding to the conditions of the experiments performed on living cells, and described in the next section. Then the linear response theory indicates that the strain $\epsilon(t)$ generated by a time-dependent stress $\sigma(t)$ initiated at $t = 0$ is given by:

$$\epsilon(t) = J(t)\sigma(t=0) + \int_0^t J(t-t')\dot{\sigma}(t')\,dt' \quad (1.1)$$

This defines the creep function $J(t)$ of the material. Indeed the strain generated by a step stress σ_0 applied at $t = 0$ is proportional to the creep function, according to $\epsilon(t) = \sigma_0 J(t)$.

Taking the Laplace transform of the above equation leads to:

$$\tilde{\epsilon}(s) = s\tilde{J}(s)\tilde{\sigma}(s) \quad (1.2)$$

where the Laplace transform of a function f is defined as: $LT[f(t)] = \tilde{f}(s) = \int_0^{+\infty} e^{-st} f(t)dt$. Notice that the usual definition of the compliance J^* is related to the creep function $J(t)$ through $J^*(s) = s\tilde{J}(s)$.

On the other hand, for an oscillating stress $\sigma(t) = \sigma(\omega)\exp(j\omega t)$, the induced strain will be $\epsilon(t) = \epsilon(\omega)\exp(j\omega t)$, where the phase shift between the $\sigma(t)$ and $\epsilon(t)$ is the argument of the complex number $\epsilon(\omega)$.

The viscoelastic complex modulus is defined as $G(\omega) = \frac{\sigma(\omega)}{\epsilon(\omega)}$. Its real component $G'(\omega)$ (storage modulus) represents the elastic part of the response, and its imaginary component $G''(\omega)$ (loss modulus) the dissipative one. $G(\omega)$ generalizes at a non-zero frequency the usual definition of the elastic constants for a solid material because of the limit that $\omega \to 0$, $G(\omega)$ reduces either to the Young modulus E or to the shear modulus μ of the material, according to the stress geometry. In the following, we refer to the modulus $G_e(\omega)$ associated with E.

To relate $G_e(\omega)$ to $\tilde{J}(s)$, we recall that the Fourier transform $FT[f(t)] = \hat{f}(\omega) = \int_{-\infty}^{+\infty} e^{-j\omega t} f(t)dt$ and the Laplace transform $\tilde{f}(s)$ are linked through: $\hat{f}(\omega) = \tilde{f}(s = j\omega)$. Then one can rewrite Equation (1.2) as $\hat{\epsilon}(\omega) = j\omega\hat{J}(\omega)\hat{\sigma}(\omega)$, and consequently:

$$G_e(\omega) = \frac{1}{j\omega\tilde{J}(j\omega)} = \frac{1}{j\omega\hat{J}(\omega)} \quad (1.3)$$

This equation establishes the correspondence between $G_e(\omega)$ and $J(t)$ in the most general case.

In Section 1.3, we will show that the creep function of the intracellular medium behaves as a power law of time: $J(t) = A_0(\frac{t}{t_0})^\alpha$. Here, t_0 is an arbitrary reference time, which will be chosen for convenience equal to 1 s. The exponent α may take any value between 0 (elastic solid) and 1 (viscous Newtonian fluid). Thus, α is representative of the balance between the elastic and dissipative character of the considered viscoelastic medium.

In this particular case, the Laplace transform of $J(t)$ is equal to:

$$\tilde{J}(s) = \frac{A_0\Gamma(1+\alpha)}{s(st_0)^\alpha} \quad (1.4)$$

where $\Gamma(1+\alpha) = \int_0^{+\infty} e^{-x} x^{\alpha}\, dx$ is the gamma Euler function.

Following Equation (1.3) the corresponding viscoelastic complex modulus then takes the form:

$$G_e(\omega) = |G_e| e^{j\delta} = \frac{(j\omega t_0)^{\alpha}}{A_0 \Gamma(1+\alpha)} \tag{1.5}$$

The amplitude of G_e is worth:

$$|G_e| = \frac{\omega^{\alpha} t_0^{\alpha}}{A_0 \Gamma(1+\alpha)} \tag{1.6}$$

and the phase shift

$$\delta = \frac{\alpha\pi}{2} \tag{1.7}$$

is independent of the angular velocity ω.

Hence, the amplitude of the viscoelastic modulus behaves as a power law of frequency $|G_e|(f) = G_0(\frac{f}{f_0})^{\alpha}$, with:

$$G_0 = \frac{(2\pi t_0 f_0)^{\alpha}}{A_0 \Gamma(1+\alpha)} \tag{1.8}$$

where f_0 is an arbitrary reference frequency, which will be chosen for convenience equal to 1 Hz. In the following we use Equations (1.4) to (1.8) to compare the data $J(t)$ and $G_e(\omega)$ respectively obtained from creep experiments and oscillating stress experiments.

1.3 Experimental Set-Ups and Protocols

1.3.1 Optical Tweezers Experiments

The set-up used for optical tweezers experiments has been described elsewhere in detail [2,3,18] and is schematically represented in Figure 1.1. Briefly, the experimental chamber is mounted on a piezoelectric stage, fixed on the stage of an inverted microscope, enclosed in a 37°C thermalization box. Optical tweezers are created by focusing an infrared laser beam (Nd:YAG laser, 600 mW maximum power) through the oil immersion objective of the microscope (×100, 1.25 numerical aperture). Silica microbeads, about 3.5 μm in diameter, are used as handles to apply forces to cells with optical tweezers. For a specific binding to integrins, the beads are coated with a polypeptide containing the arginine–glycine–aspartic (RGD) sequence (PepTite-2000, Telios Pharmaceuticals, San Diego, California). The applied force F is approximately proportional to the laser power and to the bead trap distance. It is independently pre-calibrated [3,25], and its maximal value is about 200 pN with the set-up described here. The cellular deformation is related to the displacement x of the bead with respect to the cell.

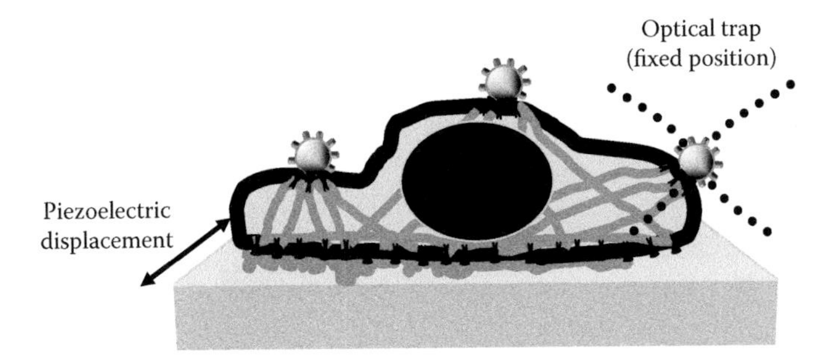

FIGURE 1.1 Principle of the microrheological measurements using the optical tweezers set-up. A silica microbead, specifically bound to given membrane receptors, is trapped in the focused laser beam, and is used as a handle to apply a force to the cytoskeleton. The experimental chamber is actuated by a piezoelectric stage, while the optical trap is kept at a fixed position.

Two types of experiments are performed with optical tweezers. In the first one, performed to measure the viscoelastic modulus of a cell as a function of frequency [3], the optical trap is kept at a fixed position, while a Labview program generates a sequence of successive sinusoidal signals at given frequencies (from 0.05 to 50 Hz) that control the piezoelectric stage motion. The Labview program triggers image acquisition, performed with a fast CCD camera. The bead position is measured on the images, from which one derives the force applied to the bead and the cell deformation.

In the second type of experiments, performed to measure the creep function of a cell [18], the optical trap is switched at initial time ($t = 0$) from the center of the bead to a position that is kept constant, while a Labview program controls the piezoelectric stage motion to keep the bead at a fixed position (measured on a quadrant photodiode), ensuring that the force exerted by the trap on the bead is constant. The cell deformation is inferred from the piezo stage displacement.

The assessment of the creep function or viscoelastic modulus from the F–x measurements takes into account the actual immersion of the bead into the cell [24].

1.3.2 Uniaxial Stretcher Experiments

The uniaxial stretching rheometer (USR, [10,11]) can essentially perform the same oscillatory and creep experiments as described above for optical tweezers, but strain is here applied to the whole cell. In this set-up, a single cell can be stretched or compressed uniaxially between two parallel microplates: one rigid, the other flexible. The flexible microplate is used as a nano-Newton force sensor of calibrated stiffness, the force being simply proportional to the plate deflection. An original design of the microplates allows us to achieve an

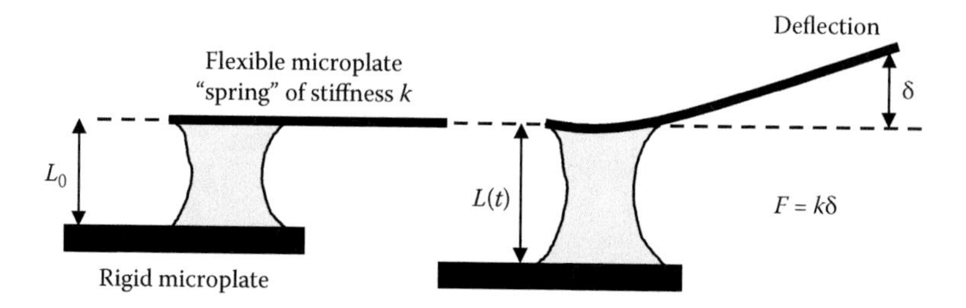

FIGURE 1.2 Diagram of the uniaxial stretching rheometer. A single cell is attached by specific or nonspecific binding to a rigid glass microplate (bottom) and a flexible microplate (top). The stiffness of the flexible plate is calibrated so that one simultaneously measures the force applied to the cell and its deformation.

efficient feedback control of either strain or stress applied to the cell. Controlling the flexible plate deflection with a typical precision of less than 200 nm, we are able to apply stresses ranging from a few Pascals to thousands of Pascals with a precision better than 2%.

The rheometer is composed of two arms fixed symmetrically on each side of the optical axis of a Leica DMIRB inverted microscope (Leica Microsystems, France) (Figure 1.2). Each arm bears a stainless steel (rod-like) microplate holder mounted on an M-UTR46A precision manual rotation stage (Micro-Controle, France). The latter is fixed by a pivot pin on a piezoelectric stage $100 \times 100 \times 100$ μm travel NanoCube XYZ (Polytech-PI, France). The rheometer is enclosed in a Plexiglas box and maintained at 37 ± 0.2°C by an Air-Therm heater controller (World Precision Instruments, United Kingdom). To ensure efficient vibration isolation, the whole set-up lies on a TS-150 active antivibration table (HWL Scientific Instruments Gmbh, Germany), itself supported by a stiff optical table (Newport, France).

The microplates with a cell stretched between them are visualized under bright light illumination with a Plan Fluotar L 63×/0.70 objective and a Micromax digital CCD camera (Princeton Instruments, Roper Scientific, France). The control of the flexible plate deflection is achieved by direct imaging of the plate tip on a photosensitive detector mounted on the second phototube of the microscope.

In creep experiments, the flexible plate deflection is kept constant (constant applied stress) through a PID correction applied to the rigid plate. Thus, the rigid plate displacement (correction command) is a direct measurement of the cell strain, and consequently, of the creep function.

In oscillatory experiments, the flexible plate basis is displaced sinusoidally (f ranging from 0.02 to 10 Hz) and the plate tip position is recorded. The ratio of plate tip to basis displacements and their phase shift allow one to determine the storage ($G'(f)$) and the loss ($G''(f)$) dynamic moduli.

1.4 Results and Discussion

Typical recordings of the creep function $J(t)$ are shown in Figure 1.3, as obtained for two single cells from the C2 myogenic line, either with optical tweezers or with the stretching rheometer. Over three time decades ($0.1 < t < 100$ s), $J(t)$ is remarkably well fitted by a power law: $J(t) = A_0(\frac{t}{t_0})^{\alpha}$. For these two particular cells: α and A_0 are equal to 0.25 and 0.017 Pa^{-1} (stretching rheometer) and to 0.18 and 0.0058 Pa^{-1} (optical tweezers). As already emphasized in [11], any attempt to adjust $J(t)$ by the sum of a few exponential functions in the full time range leads to much poorer agreement.

There is an exact equivalence between a power law behavior of $J(t)$ as a function of time t and a power law behavior of the complex viscoelastic modulus $G_e(\omega)$ as a function of the frequency $f = \omega/(2\pi)$ (see Section 1.2). Using Equation (1.8) we derive values of G_0 (value of $|G_e|$ at 1 Hz) from the measured values of A_0 and α: for the two cells of Figure 1.3, $G_0 = 10^3$ Pa (stretching rheometer) and 260 Pa (optical tweezers).

Similarly, Figure 1.4 shows the values of the modulus $|G_e|$ and of the phase δ of the viscoelastic modulus G_e, as a function of the frequency f, measured for

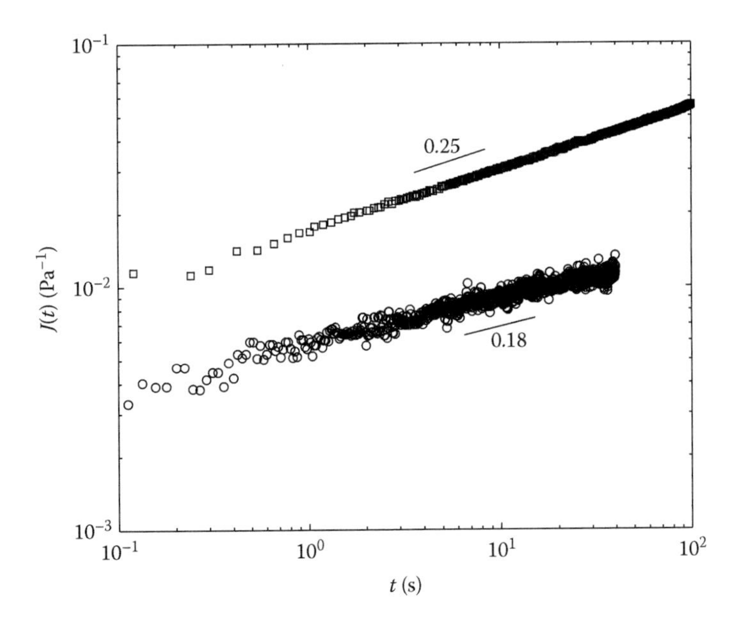

FIGURE 1.3 Plot of the creep function $J(t)$, measured for two single C2 myoblast cells, one using the stretching rheometer (□), the other using the optical tweezers (○). Over three decades in time, $J(t)$ is very well fitted by a power law: $J(t) = A_0(t/t_0)^{\alpha}$.

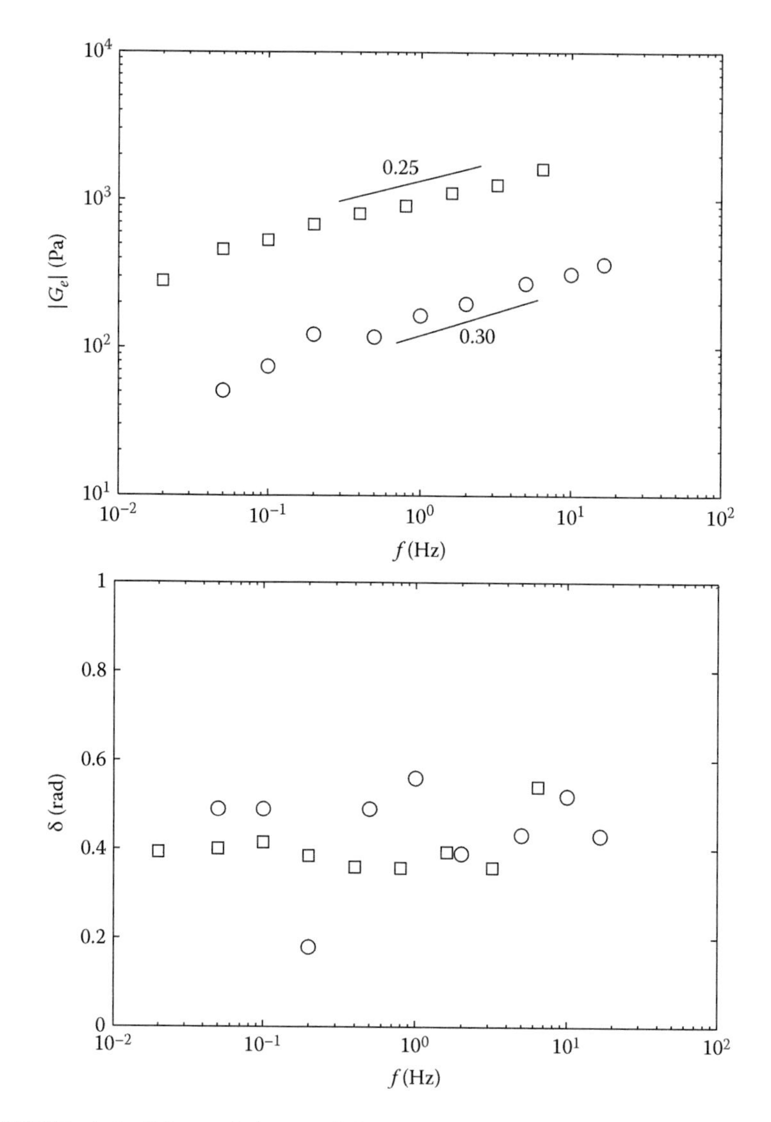

FIGURE 1.4 Plots of the modulus $|G_e|$ and the phase δ of the complex viscoelastic coefficient $G_e(\omega)$, measured for two single C2 myoblast cells, one using the stretching rheometer (□), the other using the optical tweezers (○). $|G_e|$ behaves as a power law of the frequency f over three frequency decades. The phase shift δ remains roughly constant in the studied frequency range.

two single C2 cells, either with optical tweezers or with the uniaxial stretcher. $|G_e|$ behaves as a power law of f over three frequency decades. For these two particular cells, the exponent α and the modulus G_0 of $|G_e|$ at 1 Hz are found equal to 0.30 and 155 Pa (optical tweezers) and 0.28 and 979 Pa (uniaxial stretcher), respectively. Moreover, the measured phase shift δ remains constant within a good approximation. Its average value is equal to 0.45 rad (resp. 0.40), very close to the theoretically expected value $\alpha\pi/2 = 0.47$ (resp. 0.44) (see Equation (1.7)).

A remarkable feature is that all the tested cells belonging to the same C2 population exhibit a similar power law rheology behavior, but with different values of α and G_0. Figures 1.5(a) and 1.5(b) show the statistical distributions and cumulative distributions of α in linear scale and G_0 in log scale, as measured on a set of 43 different C2 cells with the uniaxial stretcher (creep experiments, glutaraldehyde coating). Both cumulative distributions are well fitted by an error function, which is the cumulative function of a Gaussian [2,11]. This means that the values of α are normally distributed while the values of G_0 are log-normally distributed. From the fits, the best estimates for the mean value $\langle\alpha\rangle$ and the median value G_0^M are $\langle\alpha\rangle = 0.24 \pm 0.01$ and $G_0^M = 640^{+80}_{-70}$ Pa. For a whole set of cells tested either with optical tweezers or with the uniaxial stretcher, measuring either the creep function or the viscoelastic modulus, the α distribution always follows a normal law, while the distribution of G_0 always appears log-normal.

It is remarkable that the four types of experiments lead to the same power law behavior, with approximately the same exponent $\langle\alpha\rangle \approx 0.24$ (within experimental error).

As reported in Table 1.1, we have performed microrheological experiments on several other cell types, using either uniaxial stretching or optical tweezers, and in various coating conditions. The individual cell behavior appears strikingly independent of the cell type and of the experimental conditions. When performing a creep experiment, $J(t)$ is accurately adjusted by a power law function of time t. Similarly, in oscillating force experiments, the viscoelastic modulus $G_e(\omega)$ behaves as a power law of the excitation frequency. As seen in Table 1.1, the average exponent $\langle\alpha\rangle$ of the power law always remains in the range 0.15 to 0.30, whatever the cell type and function: this holds for premuscular cells (C2 myoblasts), epithelial cells (alveolar A549 and MDCK), fibroblasts (primary and L929), and macrophages (primary). Although the number of cells tested may not always be high enough to yield an accurate statistic, the prefactor G_0 of the complex modulus at 1 Hz seems to follow a log-normal distribution. Contrary to what is observed for the exponent α, the median value G_0^M of G_0 appears to depend on the cell type and on the experimental conditions.

These results are confirmed by numerous experimental results performed on other cell types with different techniques. For instance, magnetic twisting cytometry (MTC) was used to probe human airway smooth muscle (HASM) cells [13,26], mouse embryonic carcinoma cells (F9), human bronchial epithelial cells, mouse macrophages (J744.A) [14], monkey kidney epithelial cells (TC7), and 3T3 fibroblasts [17,29]. Similarly, the same epithelial alveolar cells (A549) as studied above were mechanically probed by atomic force microscopy (AFM) [1] and by MTC [40]. Optical tweezers were also used to apply forces on endogenous granules in human neutrophils [47], while chains of magnetic

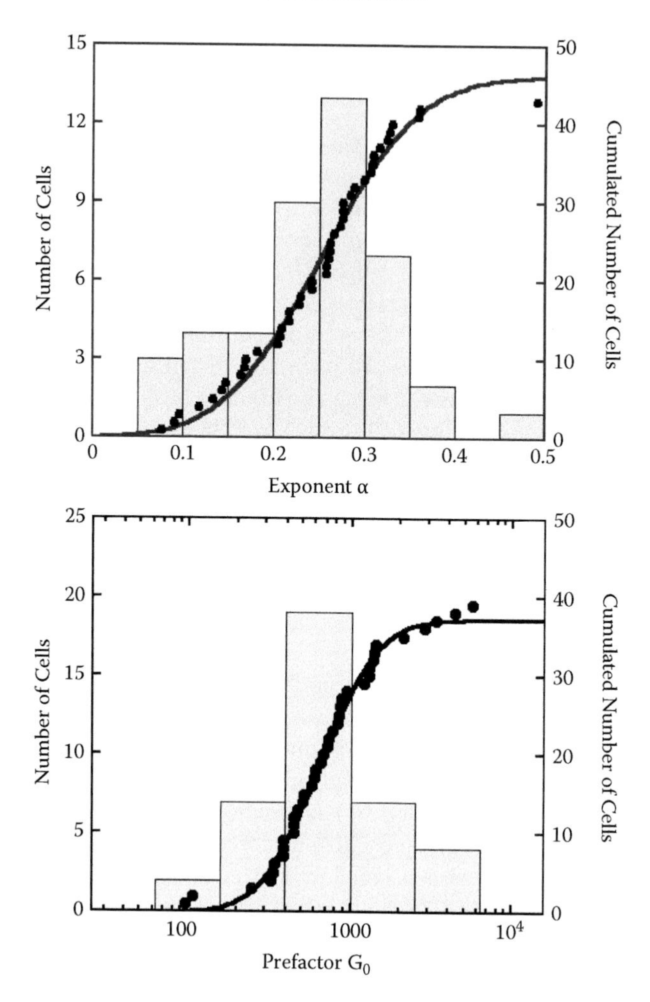

FIGURE 1.5 Histograms of the distributions of the exponents α and of the prefactor G_0, measured with the stretching rheometer on a set of 43 cells (C2 myoblasts). The cumulative distributions are also plotted (black dots). They are correctly fitted by an error function $1 + \mathrm{erf}[(x - \langle x \rangle)/a]$ (lines). This indicates that the distribution of the exponents α is normal, and that the distribution of the prefactors G_0 is log-normal. The best estimates are $\langle \alpha \rangle = 0.242 \pm 0.013$ for the average exponent and $G_0^M = 640^{+80}_{-70}$ Pa for the median value of the prefactor. Adapted from M. Ballard, N. Desprat, D. Icard, S. Fereol, A. Asnacios, J. Browaeys, S. Henon, and F. Gallet (2006). Power laws in microrheology experiments on living cells: comparative analysis and modeling. *Phys. Rev.* E 74:021911.

TABLE 1.1 Exponents and Prefactors Obtained in Different Microrheology Experiments

Cell Type	$\langle \alpha \rangle$	σ_α (Std Dev)	Δ_α (Std Err)	G_0^M (Pa) (Median)	$\langle G_0 \rangle$ (Pa) (Mean)	No. Cells	Technique/ Experiment	Coating
Myoblast C2	0.242	0.082	0.013	640 (+80/−70)	1060 ± 190	43	Uniaxial stretch / creep	Glutaraldehyde
Myoblast C2	0.29	0.07	0.02	850 (+75/−65)		12	Uniaxial stretch / creep	Cadherin
Myoblast C2	0.208	0.098	0.021	310 (+130/−100)	570 ± 150	22	Optical tweezers / oscillating	RGD
Myoblast C2C12	0.264	0.085	0.02	115 (+21/−18)	123 ± 33	29	Optical tweezers / creep	RGD
Alveolar epithelial A549	0.219	0.067	0.014	420 (+80/−70)	705 ± 155	23	Optical tweezers / oscillating	RGD
Alveolar epithelial A549	0.181	0.06	0.014	80 (+25/−20)	160 ± 40	19	Optical tweezers / oscillating	Anti-ICAM-1
Alveolar epithelial A549	0.29	0.09	0.02	305 (+45/−40)	355 ± 65	14	Optical tweezers / creep	RGD
Macrophage	0.25	0.11	0.04	2210 (+390/−330)		8	Uniaxial stretch / creep	Glutaraldehyde
Macrophage	0.20	0.08	0.03	1910 (+600/−460)		7	Uniaxial stretch / creep	Fibronectin
Fibroblast L929	0.15	0.06	0.02	65 (+25/−20)	170 ± 90	10	Optical tweezers oscillating	RGD
Fibroblast primary cell	0.26	0.07	0.02	750 (+380/−250)		13	Uniaxial stretch / creep	Glutaraldehyde
Canine kidney MDCK	0.18	0.10	0.03	1950 (+950/−600)	3660 ± 1150	10	Optical tweezers / creep	RGD

endosomes embedded in the cytoplasm of HeLa cells were submitted to an oscillating magnetic field [44].

The most prominent feature emerging from the experiments is the robustness of the power law behavior, independent of cell types and experimental conditions. The average values $\langle \alpha \rangle$ of the exponent remain very close to 0.2 when probing the cytoskeleton in the cortical region, and take slightly higher values (between 0.3 and 0.4) in the deep intracellular medium [17]. This indicates that the mechanical structure of the network is closer to an elastic solid in the vicinity of the cell membrane. In the same time, the mean prefactor G_0 may vary by nearly two orders of magnitude (roughly between 10^2 Pa and 10^4 Pa) from one cell type to the other, or according to the experimental conditions. An extensive discussion of the possible origins of such variations is beyond the scope of this chapter and can be found for instance in [2]. Of course, the cell type and the cell function are most likely to determine the structure and the density of the cytoskeletal network, but these are not the only parameters that may influence the rigidity measurements. Indeed, the measured stiffness is not homogeneous within a single cell and depends on the cell activity [20,28,44,47]. Also, the nature, structure and maturity of the contact through which the stress is applied are known to significantly affect the local stiffness measurements [8,18,34].

1.5 Statistical Self-Similar Model for Cell Rheology

Complex viscoelastic systems showing a power law rheological behavior are well documented in the literature. Colloidal systems close to the sol/gel transition [33], or "soft glassy materials" (foams, pastes, emulsions, and slurries [6,43]), exhibit such a behavior. Due to their structural complexity, the dynamics of those systems cannot be described by a small, finite number of relaxation times. The mechanical dissipation must take into account multiscale dynamical processes, meaning that their response to an external mechanical stress involves a broad and dense distribution of relaxation times.

A framework of the mechanics of soft glassy materials has been developed recently [7,38]. Structural disorder, metastability, and rearrangements are taken into account in order to describe generic out-of-equilibrium and disordered systems. This model might appear as a good candidate to describe the cell medium. However, at the present stage, the analogy between the general description of soft glassy materials and the cytoskeleton network dynamics remains quite formal because the elementary biophysical and biochemical mechanisms that govern the cytoskeleton rearrangements are not precisely described.

Other authors have developed a more phenomenological approach, where the cytoskeletal network is seen as a polarized liquid crystal, and its dynamics are coupled to the activity of molecular motors [21]. Quantitative predictions about the microrheological behavior of the cellular medium have yet to be obtained from this model.

We propose another description, intermediate between a formal approach and a phenomenological structural model. It stems from the observation that a power law creep function (or viscoelastic modulus, see Section 1.2) translates into a continuous population of relaxation times that also follows a power law.

1.5.1 Power Law Distribution of Relaxation Times

If a viscoelastic material has a single relaxation time, its creep function $J(t)$ is proportional to $1 - e^{-t/\tau}$. We choose to consider the time derivative of the creep function $\dot{J}(t)$ in order to avoid mathematical convergence problems. This last quantity is therefore proportional to $e^{-t/\tau}$. Considering a distribution of relaxation times $P(\tau)$, one has:

$$\dot{J}(t) = \int P(\tau) \exp(-\frac{t}{\tau})\, d\tau \tag{1.9}$$

Let $P(\tau) = Bt^{\alpha-2}$ represent the density of relaxation times assumed to follow a power law. Substituting this expression in Equation (1.9), one gets:

$$\dot{J}(t) = B\,\Gamma(\alpha - 1)\, t^{\alpha-1} \tag{1.10}$$

which yields upon integration:

$$J(t) = \frac{B\,\Gamma(\alpha - 1)}{\alpha} t^{\alpha} = A_0 \left(\frac{t}{t_0}\right)^{\alpha} \tag{1.11}$$

where Γ is the Euler function.

1.5.2 Self-Similar Time and Length Scales

The structure of the cytoskeleton may explain why $P(\tau)$ follows a power law. It is made of many interconnected units of different length scales, from actin individual filaments to actin bundles and stress fibers. Their size continuously spreads from the nanometer scale to the scale of the whole cell. We describe the mechanical response of each unit, labeled by the index i, by a response time τ_i. Given the cytoskeleton structure, it is reasonable to assume that the characteristic response times τ_i are widely and densely distributed. The elementary creep function $j_i(t)$ associated with each unit i is such that $\frac{dj_i}{dt} = \exp(-\frac{t}{\tau_i})$.

The description in terms of passive units may appear oversimplified because it does not seem to take into account the active molecular mechanisms related to the molecular motor activity or to the filaments remodeling. However, a precise description of the elastic and dissipation processes (such as fiber tension, cytoskeleton remodeling, molecular motors activity, and passive viscosity) at the molecular level may not be necessary to derive macroscopic mechanical behaviors.

The choice of a model where the units are associated in series (Figure 1.6a) is based on convenience. The choice of the dual representation in which elements are placed in parallel would have yielded the same results. Those two

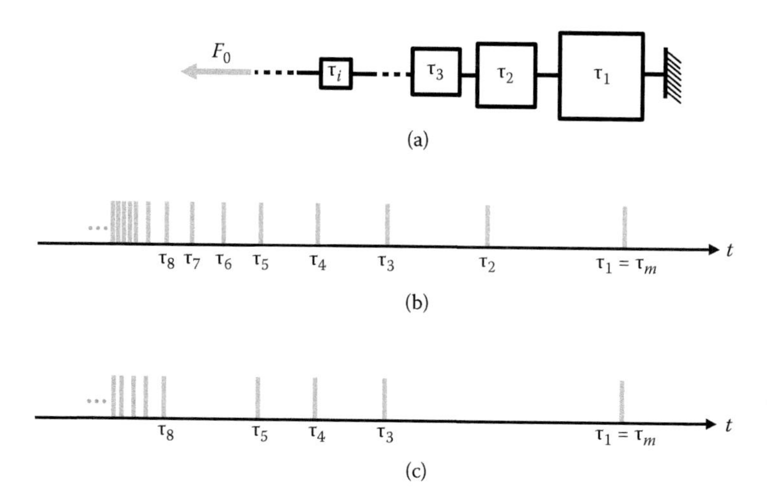

FIGURE 1.6 The cytoskeleton network is modeled as an infinite assembly of elementary units labeled by the index i, each of them showing a relaxation time τ_i (a). In the ideal case, the relaxation times τ_i are assumed to be exactly distributed according to a power law. This distribution P_0 is represented in (b). To take into account natural dispersion, some units are randomly removed from the ideal distribution. An example of the resulting distribution P_k is represented in (c).

representations are equivalent because any given model can be reduced to an equivalent series or parallel model [41]. Besides, it should be clear that it is not because we consider viscoelastic elements in series that the complex filament network of the cytoplasm is organized in such a way.

To explain why the distribution $P(\tau_i)$ of relaxation times τ_i is a power law, we assume that the cytoskeletal structure in the cell is close to a self-similar one. This assumption is especially supported by fluorescent images of the actin cytoskeleton, showing similarities between the large stress fibers structures at the scale of the cell and the structures of individual filaments at the nanometer scale. The number of units having a given size l is taken proportional to $l^{-\xi}$, where $\xi > 0$ represents the fractal dimension of the network. Concerning the dependence of the response time τ_i with the size l_i of elementary units, it is reasonable to assume that it is depicted by a simple scaling: $\tau_i \propto {l_i}^{\beta}$. As a consequence the distribution of times τ_i in the cell will itself be a power law of τ_i:

$$P(\tau_i) \propto \tau_i^{\alpha-2} \tag{1.12}$$

with $\alpha = 1 - \xi/\beta$. Actually, such a power law distribution is a commonly used assumption in several models of complex viscoelastic solids [31]. This multiscale coupling between elasticity and dissipation processes is a main characteristic of structural damping.

In our model, the exponent α of the power law is related to the fractal dimension ξ of the network and to the exponent β characterizing the dependence of the response time τ on the scale l. Lacking more information about ξ and β, it is not possible at this stage to make a quantitative prediction for α.

1.5.3 Discretization

The discretization of the population of relaxation times has no fundamental effect on the resulting creep function. Indeed we show that the resulting creep function $J_0(t) = \Sigma j_i(\tau)$ is still a power law of time, provided the number of units is high enough.

Let us assume that the relaxation times τ_i of elementary units are exactly distributed along the time axis according to $\tau_i = \tau_m i^{(-1/(1-\alpha))}$, with $0 < \alpha < 1$. The label i varies from 1 to ∞, so that τ_m represents the largest relaxation time in the system. Figure 1.6b shows a schematic drawing of this distribution P_0. This form reduces in the continuum limit to a power law distribution $P(\tau) = \frac{di}{d\tau} \propto \tau^{\alpha-2}$. In this limit, one recovers an exact power law for the corresponding creep function $J_0(t)$ as obtained by integrating:

$$\begin{aligned} \dot{J}_0(t) &= \sum_i \frac{dj_i}{dt} \cong \int_1^{\infty} \exp\left(-\frac{t}{\tau_i}\right) di \\ &= \int_0^{\tau_m} \exp\left(-\frac{t}{\tau}\right) (1-\alpha) \left(\frac{\tau}{\tau_m}\right)^{\alpha-2} \frac{d\tau}{\tau_m} = B_0\, t^{\alpha-1} \end{aligned} \tag{1.13}$$

To evaluate the influence of the discrete character of the τ_i distribution, we have performed numerical simulations in which we calculated the value of $\dot{J}_0(\theta)$ as a function of the reduced time $\theta = t/\tau_m$. The resulting function, shown in Figure 1.7 (upper curve) for a typical value $\alpha = 0.20$ of the exponent, was obtained by summing 10^5 elementary units ($i = 1$ to 10^5). This covers six orders of magnitude for the reduced response times τ_i/τ_m. As expected, $\dot{J}_0(\theta)$ is perfectly adjusted by a power law of exponent $\alpha - 1 = -0.8$, in the range $10^{-6} < \theta < 1$. An increased number of elementary units would only extend the range of validity of the power law at the smallest times: it is therefore unnecessary.

1.5.4 Randomization

To build a more realistic picture of the cytoskeleton dynamics, we assume now that only a proportion p of the elementary units are actually present in a given cell. Indeed, the response times τ_i are very unlikely to follow the smooth distribution P_0 in a real cell. One has to take into account the dispersion of results actually observed from one cell to the other. A given cell should then be represented by its actual distribution P_k of time constants, constructed by selecting, with a given probability p, a random set of relaxation times τ_i from the distribution P_0.

An example of P_k distribution is schematically represented in Figure 1.6c. Under these assumptions, the creep function $J_k(t)$ of the k cell will be calculated from $\dot{J}_k(t) = \sum_i p_i \frac{dj_i}{dt}$, where p_i is a Bernoulli random variable equal to 1 with a probability p and equal to 0 with a probability $1-p$. The underlying

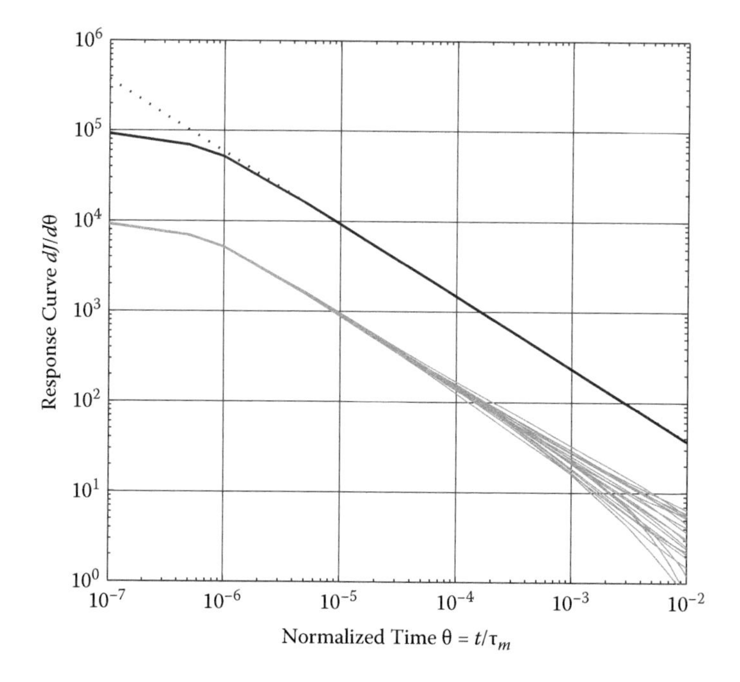

FIGURE 1.7 Plot of the time derivative of the creep function $dJ/d\theta$, numerically calculated for the ideal power-law distribution P_0 (top curve), and for 20 more realistic distributions P_k (bottom curves) randomly extracted from P_0 as explained in the text. The time scale is normalized by the largest relaxation time in the cell τ_m. The simulations involve 10^5 elementary units, and the reference exponent in the distribution P_0 is equal to 0.20. As expected, the response function $dJ/d\theta$ corresponding to the distribution P_0 exactly merges into a power law (dashed curve) of exponent $\alpha - 1 = -0.8$, in the range $10^{-6} < \theta < 1$. For the distributions P_k, the response functions $dJ/d\theta$ are well adjusted by power laws in the range $10^{-6} < \theta < 10^{-2}$, with a distribution of exponents α_k close to α. Adapted from M. Ballard, N. Desprat, D. Icard, S. Fereol, A. Asnacios, J. Browaeys, S. Henon, and F. Gallet (2006). Power laws in microrheology experiments on living cells: comparative analysis and modeling. *Phys. Rev.* E 74:021911.

response times τ_i remain distributed according to $\tau_i = \tau_m i^{(-1/(1-\alpha))}$. For each distribution P_k, we then compute the new creep function $J_k(t)$ and analyze the distribution of $J_k(t)$ over different P_k.

A set of numerical calculations of $dJ/d\theta$ versus the reduced time $\theta = t/\tau_m$ are shown in Figure 1.7 (lower curves) for twenty different realizations of P_k distributions ($k = 1$ to 20), with the same probability $p = 0.1$. These curves show that, at least in the range $10^{-6} < \theta < 10^{-2}$, all the $\dot{J}_k(\theta)$ functions behave roughly as power laws of θ and exhibit approximately the same exponent $\alpha - 1 \simeq -0.8$ as $\dot{J}_0(\theta)$.

1.5.5 Analysis of Results

Each curve $\dot{J}_k(\theta)$ is fitted in the same range by a power law $\dot{J}_k(\theta) = b_k\theta^{\alpha_k - 1}$, leading to an exponent α_k and a prefactor b_k for each realization k. For 500 realizations of P_k, the histograms of α_k and b_k are presented in Figures 1.8a and 1.8b. The distribution of α_k is a Gaussian centered on the value $\alpha = 0.20$, and the distribution of b_k is log-normal.

The parameter α_k corresponds to the scaling exponent α of the creep function for a unique realization of the simulation. Using the relation $J_k(\theta) = \frac{b_k}{\alpha_k}\theta^{\alpha_k}$ together with Equation (1.8), we can calculate the prefactor G_{0k} corresponding to the viscoelastic modulus at $\omega/(2\pi) = 1$ Hz of a unique simulated cell (k):

$$G_{0k} = \frac{\alpha_k \tau_m^{\alpha_k - 1}(2\pi)^{\alpha_k}}{\Gamma(1+\alpha_k)} \frac{1}{b_k} \tag{1.14}$$

As long as the exponents α_k remain close to the averaged value α, a log-normal distribution for b_k corresponds to a log-normal distribution for $G_{0k} \propto 1/b_k$. So far the model predictions are consistent with experimental data, in which α and G_0, experimentally measured on a given set of cells, respectively follow a normal and log-normal distribution (Figure 1.5).

Actually, one parameter has been adjusted. The width of the distributions calculated in the simulations depends on the drawing probability p. The particular value $p = 0.1$ has been chosen to match the standard deviations of experimental data.

Another important feature emerging from these simulations is that the exponent α_k and prefactor b_k of the power law $\dot{J}_k(\theta)$ are not independent parameters. The quantities $\ln(b_k)$ and α_k appear strongly correlated through a linear relationship [2]. Numerically, the slope $s = \frac{d(\ln(b_k))}{d\alpha_k}$ is found equal to 9.8 for the choice of drawing probability $p = 0.1$. Numerical tests indicate that $s \approx 10 \pm 0.5$ is almost independent of the choice of p in a wide range: $0.01 < p < 0.8$. It is noteworthy that the linear relationship between $\ln(b_k)$ and α_k translates into an (approximate) linear relationship between $\ln(G_{0k})$ and α_k, such as:

$$\left.\frac{d(\ln(G_{0k}))}{d\alpha_k}\right|_{\alpha_k=\langle\alpha\rangle} = -\left.\frac{d(\ln(b_k))}{d\alpha_k}\right|_{\alpha_k=\langle\alpha\rangle} + \ln(2\pi\tau_m) + \frac{1}{\langle\alpha\rangle} - \psi(\langle\alpha\rangle + 1), \tag{1.15}$$

where $\psi(\alpha) = \frac{d(\ln(\Gamma(\alpha))}{d\alpha}$ is the digamma function [15].

To step further into the comparison between the model and the data, one may focus on the correlations between exponents α_k and prefactors G_{0k}. Figure 1.9 gathers the experimental data of $\ln(G_0)$ versus α for all the C2 cells, as determined either in optical tweezers or uniaxial stretching experiments. Despite a noticeable dispersion of the results, $\ln(G_0)$ appears to be an increasing function of the exponent α. This is consistent with the model, which predicts a correlation between $\ln(G_{0k})$ and α_k.

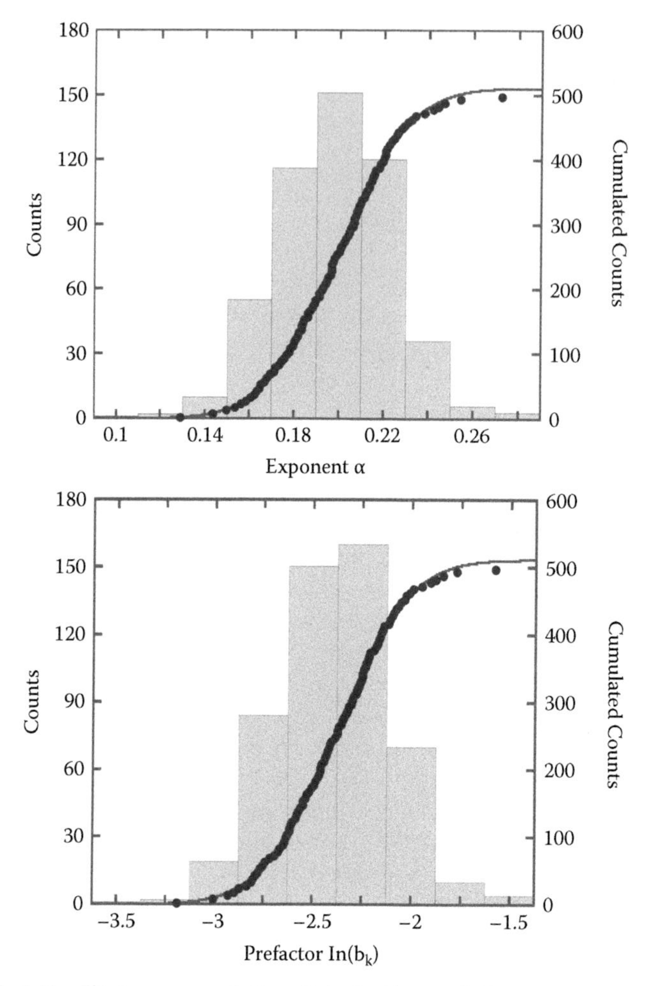

FIGURE 1.8 Histograms of the distributions of the exponents α_k and of the logarithm of the prefactor $\ln(b_k)$. These quantities are measured from the numerically calculated curves $dJ/d\theta$ for 500 different realizations of P_k distributions. The cumulated distributions and their best fits by erf functions are also represented. They show that, according to this model, α_k and b_k are, respectively, normally and log-normally distributed. Adapted from M. Ballard, N. Desprat, D. Icard, S. Fereol, A. Asnacios, J. Browaeys, S. Henon, and F. Gallet (2006). Power laws in microrheology experiments on living cells: comparative analysis and modeling. *Phys. Rev.* E 74:021911.

Fitting a linear relationship between the experimental measurements of $\ln(G_0)$ and α in Figure 1.9, we obtain a slope $\frac{d(\ln(G_{0k}))}{d\alpha_k} = 5.2$. It is possible to make this value consistent with the predicted value $s = 10$ by adjusting the only unknown parameter in our model, τ_m, which represents the longest relaxation time in the cell. This adjustment leads to $\tau_m \approx 3200$ s for the C2 cell type.

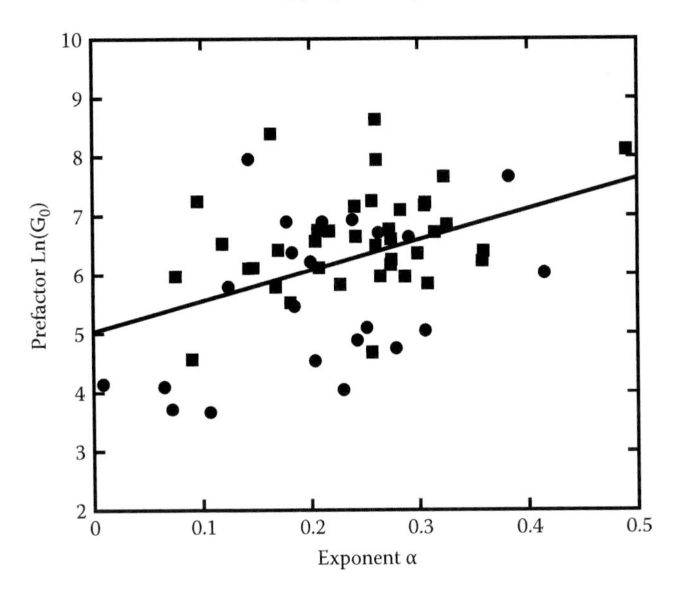

FIGURE 1.9 Plot of the experimental values of the prefactors $\ln(G_0)$ versus exponents α, measured for C2 myoblasts. The data from optical tweezers experiments (●) and stretching rheometer experiments (■) are plotted together. Despite the dispersion of the results, $\ln(G_0)$ appears to be an increasing function of the exponent α. This is consistent with the prediction of the model. Adapted from M. Ballard, N. Desprat, D. Icard, S. Fereol, A. Asnacios, J. Browaeys, S. Henon, and F. Gallet (2006). Power laws in microrheology experiments on living cells: comparative analysis and modeling. *Phys. Rev.* E 74:021911.

Since $\theta = t/\tau_m$, the reduced time range $10^{-6} < \theta < 10^{-2}$ then corresponds to a real-time range of $0.003 < t < 30$ s, which exactly matches the experimental range of our measurements. This reinforces the validity of our approach.

From the same analysis performed on our data on alveolar epithelial cells A549 we infer that the highest response time is $\tau_m \approx 300$ s for this type of cell. Other experiments [14] yield a typical time $\tau_m \approx 3200$ s. It is remarkable that these estimates of the longest time responses in different cell types, derived from different experiments, are roughly consistent with each other and lie in the range of 5 min to 1 hr. Moreover, we emphasize that their common order of magnitude is quite reasonable, as far as it effectively corresponds to a typical relaxation time at the scale of the whole cell. Indeed, an independent rough evaluation of τ_m may be obtained by dividing a typical value of the cytoplasm viscosity at long time scale (~10 kPa.s) [32,39] by a typical Young modulus (~10^2 Pa) measured at the cell scale in quasi-static experiments. Beyond this time range, some macroscopic remodeling processes (treadmilling, signalization cascades) are known to take place and to interfere with the cell mechanical properties.

1.6 Summary and Conclusions

This study makes a parallel analysis of the results obtained by two different microrheological experiments on single living cells. It allows us to bring out some striking common features in the mechanical properties of the cytoskeleton network.

First, as demonstrated by the data shown here and by numerous works in the literature, the mechanical response function presents a quasi-universal power law behavior, whatever the experimental technique and the cellular type: the viscoelastic complex modulus of the cell is a weak power law function of the exciting mechanical frequency f. Correlatively, the creep function of the intracellular medium is a power law of elapsed time, with the same exponent $\alpha \approx 0.15$–0.35. This demonstrates that there is no characteristic dissipation time in the cellular response, or more precisely that these relaxation times are broadly distributed over a wide time interval, extending at least from 0.01 to 100 s. This is a characteristic of structural damping, where the mechanisms responsible for the storage of elastic energy and its dissipation are strongly correlated.

Moreover, the prefactor of the response function, which represents the value of the elastic modulus at a given reference frequency, varies by almost two orders of magnitude, in the range 100 to 10^5 Pa, according to the cellular type and to the experimental conditions. Interestingly, for a set of cells of a given type, probed in the same experiment, the prefactor distribution is found log-normal. A full discussion about the possible influence of several physical or biological parameters, such as temperature, probe size, nature of the mechanical receptor, and cell inhomogeneities, on the average value of this prefactor can be found elsewhere [2].

The semiphenomenological model presented here is able to accurately predict the mechanical response of a living cell submitted to a controlled stress, over a wide range of time scales. The mechanical behavior of the cell is modeled by the association of a large number of elementary units, which account for the different scales in the cytoskeletal network. Assuming that the structure of the network is self-similar, and that the dissipative time constants of the units are distributed according to a power law, one recovers all the features of the macroscopic behavior observed for different cell types. This approach quantitatively accounts for the power law responses measured in different rheological experiments, and also for the normal and log-normal distributions retrieved for the exponents and prefactors. The largest relaxation time in the cell, which is the only adjustable parameter of the model, is consistent with other independent estimates. A further step will consist of interpreting the dissipative elements in terms of elementary biological mechanisms, such as molecular motor activity and crosslinker dynamics, which play a crucial role in cytoskeleton remodeling.

Acknowledgments

We thank Martial Balland, Nicolas Desprat, Delphine Icard-Arcizet, Sophie Féréol, and Axel Guiroy for performing the experiments; and Mireille Lambert for providing C2 cells and cadherin protocols. This work was partially supported by grants from Université Paris 7 (Bonus Qualité Recherche) from the French Ministère de la Recherche (ACI Jeunes chercheurs), from the French Centre National de la Recherche Scientifique (Programme Physique et Chimie du Vivant), and from the Association pour la Recherche sur le Cancer (subvention libre n°3115).

References

[1] J. Alcaraz, L. Buscemi, M. Grabulosa, X. Trepat, B. Fabry, R. Farré, and D. Navajas (2003). Microrheology of human lung epithelial cells measured by atomic force microscopy. *Biophys. J.* 84:2071–2079.

[2] M. Balland, N. Desprat, D. Icard, S. Féréol, A. Asnacios, J. Browaeys, S. Hénon, and F. Gallet (2006). Power laws in microrheology experiments on living cells: comparative analysis and modeling. *Phys. Rev. E* 74:021911.

[3] M. Balland, A. Richert, and F. Gallet (2005). The dissipative contribution of myosin II in the cytoskeleton dynamics of fibroblasts. *Eur. Biophys. J.* 34:255–261.

[4] K.A. Barbee, P.F. Davies, and R. Lal (1994). Shear stress-induced reorganization of the surface topography of living endothelial cells imaged by atomic force microscopy. *Circ. Res.* 74:163–171.

[5] A.R. Bausch, W. Möller, and E. Sackmann (1999). Measurement of local viscoelasticity and forces in living cells by magnetic tweezers. *Biophys. J.* 76:573–579.

[6] D. Bonn, P. Coussot, H.T. Huynh, F. Bertrand, and G. Debrégeas (2002). Rheology of soft glassy materials. *Europhys. Lett.* 59:786–792.

[7] J.P. Bouchaud (1992). Weak ergodicity breaking and aging in disordered systems. *J. Phys. I France* 2:1705–1713.

[8] P. Bursac, G. Lenormand, B. Fabry, M. Oliver, D.A. Weitz, V. Viasnoff, J.P. Butler, and J.J. Fredberg (2005). Cytoskeletal remodelling and slow dynamics in the living cell. *Nat. Mater.* 4:557–561.

[9] N. Caille, O. Thoumine, Y. Tardy, and J.-J. Meister (2002). Contribution of the nucleus to the mechanical properties of endothelial cells. *J. Biomech.* 35:177–187.

[10] N. Desprat, A. Guiroy, and A. Asnacios (2006). Microplates-based rheometer for a single living cell. *Rev. Sci. Inst.* 77:055111.

[11] N. Desprat, A. Richert, J. Simeon, and A. Asnacios (2005). Creep function of a single living cell. *Biophys. J.* 88:2224–2233.

[12] C.F. Dewey Jr., S.R. Bussolari, M.A. Gimbrone Jr., and P.F. Davies (1981). The dynamic response of vascular endothelial cells to fluid shear stress. *J. Biomech. Eng.* 103:177–188.

[13] B. Fabry, G.N. Maksym, J.P. Butler, M. Glogauer, D. Navajas, and J.J. Fredberg (2001). Scaling the microrheology of living cells. *Phys. Rev. Lett.* 87:148102.

[14] B. Fabry, G.N. Maksym, J.P. Butler, M. Glogauer, D. Navajas, N.A. Taback, E.J. Millet, and J.J. Fredberg (2003). Time scale and other invariants of integrative mechanical behavior in living cells. *Phys. Rev. E* 68:041914.

[15] I.S. Gradshteyn, I.M. Ryzhik, A. Jeffrey, and D. Zwillinger (1980). *Table of Integrals, Series and Products*, Academic Press, New York, Chapter 8.36, p. 943.

[16] J. Guck, R. Ananthakrishnan, H. Mahmood, T.J. Moon, C.C. Cunningham, and J. Käs (2001). The optical stretcher: a novel laser tool to micromanipulate cells. *Biophys. J.* 81:767–784.

[17] B.D. Hoffman, G. Massiera, K.M. Van Citters, and J.C. Crocker (2006). The consensus mechanics of cultured mammalian cells. *Proc. Natl. Acad. Sci. USA* 103:10259–10264.

[18] D. Icard-Arcizet, O. Cardoso, A. Richert, and S. Hénon (2008). Cell stiffening in response to external stress is correlated to actin recruitment. *Biophys. J.* 94:2906–2913.

[19] P.A. Janmey (1998). The cytoskeleton and cell signaling: component localization and mechanical coupling. *Physiol. Rev.* 78:763–781.

[20] T.P. Kole, Y. Tseng, I. Jiang, J.L. Katz, and D. Wirtz (2005). Intracellular mechanics of migrating fibroblasts. *Mol. Biol. Cell* 16: 328–338.

[21] K. Kruse, J.F. Joanny, F. Jülicher, J. Prost, and K. Sekimoto (2004). Asters, vortices and rotating spirals in active gels of polar filaments. *Phys. Rev. Lett.* 92:078101.

[22] L. Landau and E. Lifchitz (1959). *Theory of Elasticity*, Pergamon Press, London.

[23] A.W.C. Lau, B.D. Hoffmann, A. Davies, J.C. Crocker, and T.C. Lubensky (2003). Microrheology, stress fluctuations, and active behavior of living cells. *Phys. Rev. Lett.* 91:198101.

[24] V. Laurent, S. Hénon, E. Planus, R. Fodil, M. Balland, D. Isabey, and F. Gallet (2002). Assessment of mechanical properties of adherent living cells by bead micromanipulation: comparison of magnetic twisting cytometry vs optical tweezers. *J. Biomech. Eng.* 124:408–421.

[25] G. Lenormand, S. Hénon, A. Richert, J. Siméon, and F. Gallet (2001). Direct measurement of the area expansion and shear moduli of the human red blood cell membrane skeleton. *Biophys. J.* 81:43–58.

[26] G. Lenormand, E. Millet, B. Fabry, J.P. Butler, and J.J. Fredberg (2004). Linearity and time-scale invariance of the creep function in living cells. *J. Royal Soc. Lond. Interface* 1:91–97.

[27] C.-M. Lo and J. Ferrier (1999). Electrically measuring viscoelastic parameters of adherent cell layers under controlled magnetic forces. *Eur. Biophys. J.* 28:112–118.

[28] G.N. Maksym, B. Fabry, J.P. Butler, D. Navajas, D.J. Tschumberlin, J.D. Laporte, and J.J. Fredberg (2000). Mechanical properties of cultured human airway smooth muscle cells from 0.05 to 0.4 Hz. *J. Appl. Physiol.* 89:1619–1632.

[29] G. Massiera, K.M. Van Citters, P.L. Biancaniello, and J.C. Crocker (2007). Mechanics of single cells: rheology, time dependance, and fluctuations. *Biophys. J.* 93:3703–3713.

[30] J. Ohayon, P. Tracqui, R. Fodil, S. Féreol, V.M. Laurent, E. Planus, and D. Isabey (2004). Analysis of nonlinear responses of adherent epithelial cells probed by magnetic bead twisting: a finite element model based on a homogenization approach. *J. Biomech. Eng.* 126:685–698.

[31] A. Oustaloup (1994). *La Dérivation non Entière*, Hermès ed., Paris.

[32] N.O. Petersen, W.B. McConnaughey, and E.L. Elson (1982). Dependence of locally measured cellular deformability on position on the cell, temperature, and cytochalasin B. *Proc. Natl Acad. Sci. USA* 79: 5327–5331.

[33] A. Ponton, S. Warlus, and P. Griesmar (2002). Rheological study of the sol-gel transition in silica alkoxydes. *J. Coll. Interf. Sci.* 249:209–216.

[34] M. Puig-de-Morales, E.J. Millet, B. Fabry, D. Navajas, N. Wang, J.P. Butler, and J.J. Fredberg (2004). Cytoskeletal mechanics in adherent human airway smooth muscle cells: probe specificity and scaling of protein-protein dynamics. *Am. J. Physiol. Cell Physiol.* 287:C643–C654.

[35] C. Rotsch, F. Braet, E. Wisse, and M. Radmacher (1997). AFM imaging and elasticity measurements on living rat liver macrophages. *Cell. Biol. Intl.* 21:685–696.

[36] M. Sato, K. Nagayama, N. Kataoka, M. Sasaki, and K. Hane (2000). Local mechanical properties measured by atomic force microscopy for

cultured bovine endothelial cells exposed to shear stress. *J. Biomech.* 33:127–135.

[37] M. Sato, D.P. Theret, L. Wheeler, N. Ohshima, and R.M. Nerem (1990). Application of the micropipette technique to the measurement of culture porcine aortic endothelial cell viscoelastic properties. *J. Biomech. Eng.* 112:263–268.

[38] P. Sollich (1998). Rheological constitutive equation for a model of soft glassy materials. *Phys. Rev. E* 58:738–759.

[39] O. Thoumine and A. Ott (1997). Time scale dependent viscoelastic and contractile regimes in fibroblasts probed by microplate manipulation. *J. Cell Sci.* 110:2109–2116.

[40] X. Trepat, M. Grabulosa, L. Buscemi, F. Rico, R. Farré, and D. Navajas (2005). Thrombin and histamine induce stiffening of alveolar epithelial cells. *J. Appl. Physiol.* 98:1567–1574.

[41] N.W. Tschoegl (1989). *The Phenomenological Theory of Linear Viscoelastic Behavior*, Springer, Berlin.

[42] Y. Tseng, T.P. Kole, and D. Wirtz (2002). Micromechanical mapping of live cells by multiple-particle-tracking microrheology. *Biophys. J.* 83:3162–3176.

[43] E.R. Weeks, J.C. Crocker, A.C. Levitt, A. Schofiled, and D.A. Weitz (2000). Three-dimensional imaging of structural relaxation near the colloidal glass transition. *Science* 287:627–631.

[44] C. Wilhelm, F. Gazeau, and J.-C. Bacri (2003). Rotational magnetic endosome microrheology: viscoelastic architecture inside living cells. *Phys. Rev. E* 67:061908.

[45] F. Wottawah, S. Schinkinger, B. Lincoln, R. Ananthakrishnan, M. Romeyke, J. Guck, and J. Käs (2005). Optical rheology of biological cells. *Phys. Rev. Lett.* 94:098103.

[46] S. Yamada, D. Wirtz, and S.C. Kuo (2000). Mechanics of living cells measured by laser tracking microrheology. *Biophys. J.* 78:1736–1747.

[47] M. Yanai, J.P. Butler, T. Suzuki, H. Sasaki, and H. Higushi (2004). Regional rheological differences in locomoting neutrophils. *Am. J. Physiol. Cell Physiol.* 287:C603–C611.

[48] F. Ziemann, J. Rädler, and E. Sackmann (1994). Local measurements of viscoelastic moduli of entangled actin networks using an oscillating magnetic bead micro-rheometer. *Biophys. J.* 66:2210–2216.

Chapter 2

Actin-Based Propulsion: Intriguing Interplay between Material Properties and Growth Processes

Karin John, Denis Caillerie, Philippe Peyla, Mourad Ismail, Annie Raoult, Jacques Prost, and Chaouqi Misbah

Contents

2.1 Introduction

Most living cells are able to perform a directed motion, either by swimming in a liquid environment, by crawling on a solid support or by squeezing through a three-dimensional matrix of fibers.

The speed of swimming bacteria can reach up to 100 $\mu m\,s^{-1}$, whereas eukaryotic cell crawling can be as fast as 1 $\mu m\,s^{-1}$ (Ref. [17] and references therein). Given their size and speed, the motion of single cells is governed by viscous forces, not inertia; that is, the Reynolds number $Re \ll 1$.

Unicellular organisms move in search of a food or light source. Intracellular pathogens like bacteria or viruses spread by exiting their host cell and entering a neighboring cell. Other simple organisms like the slime mold *Dictyostelium discoideum* migrate under unfavorable conditions (e.g., starvation) toward an aggregation center to form a multicellular organism.

Typically in a multicellular organism, not all cells are motile all the time but they can be mobilized by the appropriate stimuli. For example, the ability to move plays a crucial role during embryonic development, in wound healing, and in the immune response. Additionally, cellular motility is a prerequisite for metastasis formation during cancer development.

The biological realizations to produce a propulsive force are diverse. Most swimming cells (e.g., sperm cells or the bacterium *Escherichia coli*) use one or multiple beating flagella,[1] respectively. In contrast, crawling cells and some intracellular pathogens advance by actin polymerization.

In this chapter we present recent experiments and concepts to understand the latter mechanism, the production of forces in the advancing edges of crawling cells or for the propulsion of intracellular organelles, which is also of fundamental interest for the medical and engineering sciences.

The foundation of the research of cell motility as a distinct discipline was laid in the 1970s by the group around Michael Abercrombie [1]. He was the first to divide the motion of fibroblasts[2] into three phases: extension, adhesion, and contraction, which form the dogma of cellular motion as it is recognized today. In this mechanism, a slow movement ($\sim$1 $\mu m\,min^{-1}$) is generated by the extension of flat membrane sheets, lamellipodia, in the direction of movement. The advancement of the membrane is accompanied by the formation of focal adhesions, contacts among the substratum, the cell membrane, and actin stress fibers. Finally the cell rear retracts, accompanied by a de-adhesion of the membrane from the substratum.

However, there exist variations of this dogma. Fish keratocytes[3] perform a rapid continuous motion ($\sim$10 $\mu m\,min^{-1}$) with a constant shape. They almost seem to glide over the surface and form only transient focal contacts with the

[1]Long cellular extensions.

[2]Most common cells of connective tissue in animals.

[3]Fish scale cells.

substratum with a much shorter lifetime than the focal adhesions formed in fibroblasts [3].

Another variant is the rapid motion ($\sim$10 μm min^{-1}) of the slime mold *Dictyostelium discoideum*, which moves in amoeboid fashion. During this amoeboid motion, only non-specific contacts with the substratum are formed and actin stress fibers are absent [18].

Despite the various mechanisms, cell motion requires first the self-organization of the cell into an advancing and receding edge. This manifests itself by different molecular concentrations or activation levels of enzymes at the two poles. The polarization can be guided by external signals (e.g., chemical gradients or a variation in the mechanical properties of the support), but it might also arise in a homogeneous environment, after a transient mechanical perturbation of a stationary symmetric cell [60].

The three processes—extension, adhesion, and contraction—are coupled by sophisticated and complex mechanisms. However, it seems as if the process of front extension relies on a completely different machinery than the mechanics of adhesion and contraction and can be studied separately.

It had been known for a long time that on a cellular level, forces can be generated on the basis of muscle-like proteins (i.e., actin and myosin), which are indeed responsible for the contraction of the cell rear [27]. However, more recently it was discovered that the protrusions at the leading edges as well as the motions of intracellular organelles [29] or pathogens, like the bacterium *Listeria monocytogenes* [53] or the *Vaccinia* virus [13], can be associated with the so-called actin polymerization machinery [52] as shown schematically in Figure 2.1. The molecular basis and minimal ingredients of the actin polymerization machinery are now very well understood based on biomimetic experiments by the group of Marie-France Carlier [33]. However, a theoretical

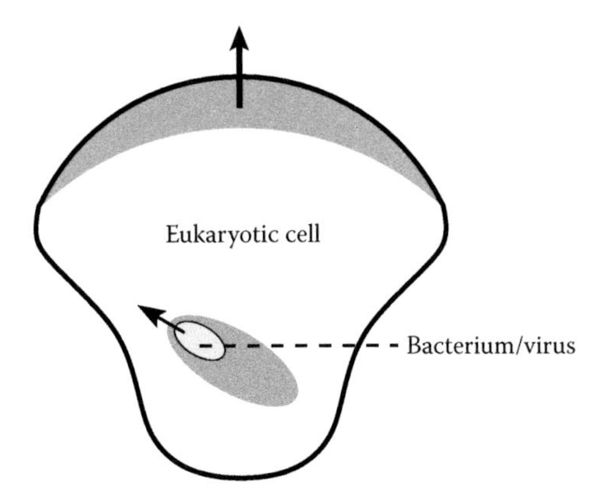

FIGURE 2.1 Actin polymerizes into an elastic filament network (shaded in dark gray) at the leading edges of motile eukaryotic cells or at the outer surfaces of pathogens and organelles, and induces cellular motion. The direction of motion is indicated by the black arrows.

understanding of the physical mechanisms of the polymerization process is still a matter of debate and is the main topic of this review.

In Section 2.2 we outline the biochemical basis of actin polymerization and present a selection of experimentally observed phenomena. We present the processes taking place at the leading edges of locomoting cells and then mainly rely on biomimetic experiments. In Section 2.3 we present some theoretical concepts to interpret and explain the existing experiments. The list of presented models is by no means complete but rather represents a selection of the leading ideas. We focus on the class of Brownian ratchet models and macroscopic models of symmetry breaking.

2.2 Experimental Observations

2.2.1 Biochemistry of Actin Polymerization and Organization of Leading Edges of Advancing Cells

Actin is a small globular protein of 42 kDa present in all eukaryotic cells [51]. Under physiological conditions, actin monomers (G-actin) polymerize into long helical filaments (F-actin). In a living organism these polymerization or depolymerization processes are tightly regulated. The literature on the dynamics of actin and the proteins that interact with actin is vast. As an introduction we refer the reader to the comprehensive biochemical reviews by Pollard et al. [46,47] and Rafelski et al. [49] and the books by Bray [7] and Howard [26]. Due to the limited scope of this chapter, we only present the basic phenomena and common terminology associated with the actin polymerization machinery, which will allow the reader to understand the pertinent questions and concepts.

2.2.1.1 Actin Polymerization *in vitro*

Under physiological conditions, that is, at an ionic strength of approximately 100 mM, monomeric actin polymerizes spontaneously into filaments. The filament growth typically starts with a nucleation process, since actin dimers and trimers are unstable. Shortening or elongation of existing filaments occurs predominantly via subunit addition or subtraction at the filament ends and not via filament breaking or annealing processes. Actin monomers at a concentration c may bind to a filament end with a rate $\sim k_+c$ and dissociate with a rate $\sim k_-$. In a stationary situation (i.e., zero net growth of the filament), one can write:

$$0 = k_+c - k_- \tag{2.1}$$

The concentration $c_c = k_-/k_+$ associated with this equilibrium is called the critical concentration.

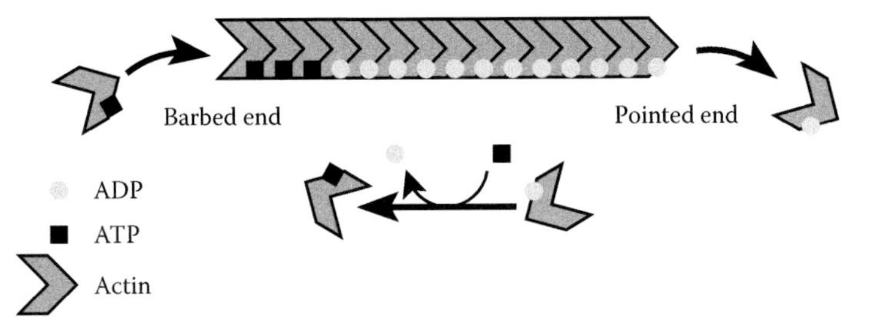

FIGURE 2.2 Actin subunits treadmill through a filament. ATP-actin polymerizes at the barbed end, ATP hydrolyzes, and ADP-actin depolymerizes at the pointed end. In the solution, ATP is exchanged for ADP at the monomers.

G-actin has a structural polarity [25]. The same polarity is also found in actin filaments and can be visualized as an arrowhead pattern by decoration of filaments with myosin.[4] From this arrowhead pattern the two filament ends are referred to as pointed and barbed, respectively, as shown in Figure 2.2.

In the presence of Mg-ATP the structural difference translates into a difference of the critical concentrations and rate constants between the barbed and pointed ends and causes a treadmilling of subunits through the filament.

Briefly, this phenomenon can be explained as follows (for details see Ref. [46]). Most actin monomers are bound to ATP (typically Mg-ATP). The critical concentration for this ATP-bound species is about six times lower for the barbed end than for the pointed end. For ADP-actin, the critical concentrations are about the same for both ends but about ten times higher than for ATP-actin at the barbed end. Therefore, in the steady state the ATP-actin concentration is above the critical concentration of the barbed end and below the critical concentration of the pointed end. Polymerized actin subunits are still bound to ATP but in the course of time ATP hydrolyzes irreversibly into ADP+P_i and, later on, the inorganic phosphate P_i dissociates from the filament with a half-life of several minutes. Consequently, ATP-actin polymerizes at the barbed end, travels along the filament whereby ATP is hydrolyzed and finally ADP-actin depolymerizes at the pointed end. Figure 2.2 shows schematically the key processes of the treadmilling cycle.

The process of irreversible ATP hydrolysis and P_i release (and subsequently the exchange ADP $\rightarrow$ ATP at the actin monomer in solution) keeps the system out of equilibrium and allows for a constant flux of monomers through the filament at constant filament length, which forms the basis of cellular motility.

[4]Myosin molecules bind to each actin subunit in the filament in an oriented fashion, leading to a typical pattern along the filament, which looks like a series of arrowheads in an electron micrograph. For details see Ref. [7].

2.2.1.2 Proteins that Regulate Actin Polymerization *in vivo*

In the living cell the actin polymerization machinery is tightly regulated by signaling processes. Many different factors interact with actin or participate in the regulation of this polymerization machinery. In the following paragraphs we only discuss the relevant factors that determine the actin architecture of the leading edges of advancing cells or the actin comets formed by intracellular pathogens and seem crucial to generate motion.

First of all, most actin monomers are bound to so-called monomer binding proteins, for example, thymosin-β4 and profilin, and are thus not able to nucleate new filaments. But profilin-ATP-actin complexes elongate existing filaments at the barbed end nearly as efficiently as ATP-actin.

Electron micrographs of the leading edge show a dense network of actin filaments linked to each other by Y junctions, where a new filament grows from an existing filament at a 70° angle. The distance between two crosslinks is of the order ~20 to 30 nm (in comparison, the persistence length of actin filaments is 15 μm [42]). This Y junction is initiated by the interaction of the Wasp/Scar protein and the Arp2/3 complex, whereby Wasp/Scar is a membrane-bound protein that incites the binding of the Arp2/3 complex to an existing filament, which in turn then serves as a nucleation point for a new filament.

Free barbed filament ends are quickly covered by so-called capping proteins, thus limiting filament elongation to a zone near the plasma membrane. At the pointed filament end, depolymerization takes place. This process can be accelerated by a protein complex called ADF/cofilin, which is able to sever old filaments containing ADP-actin subunits. These filament fragments are then depolymerizing rapidly. Besides Arp2/3 there exist other types of filament nucleators, called formins. In the presence of profilin and ADF/cofilin, they seem to favor the growth and bundling of several actin filaments into cables [38], present in spike-like membrane extensions called filopodia. In this chapter we limit discussion to actin networks produced by the Arp2/3 complex.

The above outlined process of polymerization at the membrane/actin network interface and depolymerization far away from the membrane provides the mechanism for pushing the membrane in the direction of movement. However, little is known about the mechanical properties of such filament networks, and recent experiments indicate that the loading history determines the growth velocity of the network [43].

Certain bacteria invade other living cells and hijack the actin polymerization machinery of their host cells to propel themselves forward. They carry a protein on their outer surface (e.g., ActA for *Listeria monocytogenes*), which adopts the same function as the Wasp/Scar protein in the membranes of eukaryotes: It triggers the polymerization of an Arp2/3 crosslinked network at the outer bacterial surface. This actin network typically develops asymmetrically only at one side of the bacterium, thus pushing the bacterium in the other direction.

A similar mechanism might also be responsible for the motion of endocytic vesicles in living cells (see [29] and references therein).

To summarize the growth processes and the resulting architecture of the actin system at the leading edge: near the cell membrane containing an activating enzyme (e.g., Wasp or ActA), actin polymerizes into a dense crosslinked filament network, which extends several micrometers into the cell and where fast polymerizing barbed ends are oriented toward the membrane. Polymerization is restricted to a narrow zone near the membrane because free barbed ends are rapidly blocked by capping proteins. Free pointed ends that are far away from the membrane are depolymerizing. These, out of equilibrium growth processes driven by the irreversible hydrolysis of ATP, lead to the extension of membrane protrusions or propel bacteria forward.

2.2.2 Biomimetic Experiments

2.2.2.1 General Observations

The minimal set of biochemical ingredients to induce actin-driven motion were identified about 10 years ago [33]. At the same time actin-driven motion has been successfully reconstituted *in vitro* by replacing the bacterium with mimetic objects: for example, beads [11,41,62], vesicles [23,35,56], or droplets [6].

In these experiments, objects (hard, soft, fluid) coated with either ActA or Wasp/Scar proteins are added to a solution containing ATP, actin, the Arp2/3 complex and a few well-defined regulatory proteins. With this design, actin polymerizes predominantly at the surface of the object, that is, the internal interface, and depolymerizes at the interface between the network and the solution, that is, the external interface.

After an initial phase (i), where polymerization occurs symmetrically around the object, the symmetry is broken and the actin cloud starts to grow asymmetrically (ii). In later stages an actin comet develops (iii), and the object starts to move with velocities up to $0.1\,\mu\mathrm{m\,s^{-1}}$. A schematic representation of the three phases of the actin cloud evolution is shown in Figure 2.3.

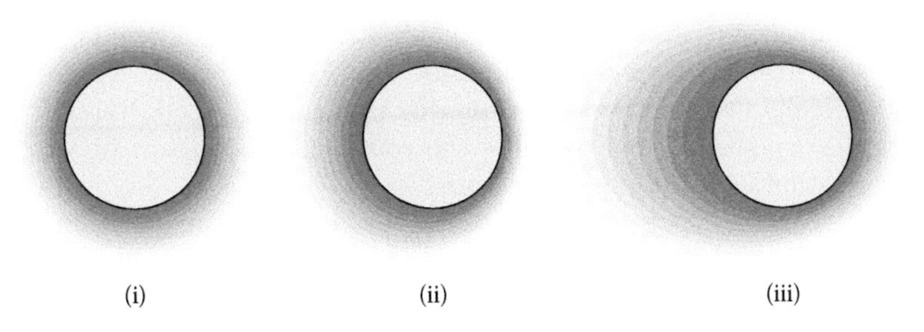

FIGURE 2.3 Schematic view of the evolution of an actin gel (external shades) around a bead (gray) as described in the text with (i) symmetric growth, (ii) symmetry breaking, and finally, (iii) comet formation.

Interestingly the mode of movement depends on the surface parameters of the object. Depending on the conditions, one can observe a continuous motion or a saltatory motion, where the object undergoes stop-and-go cycles [5,14,31,55], which is also reflected by variations in the actin density in the comet.

The network grown around these biomimetic objects has elastic properties with a Young's modulus of 10^3 to 10^4 Pa [20,36] and is often referred to as an actin gel.

The actin filaments interact, at least transiently, with the activating enzyme bound to or adsorbed on the object's surface: for example, the stress to detach an actin comet from a bead has been estimated to be about 100 pN μm^{-2} [36], whereby the adsorbed activating enzyme stays on the bead. Interestingly, in the same experimental set-up, actin comets under compression appeared to be hollow.

In the regime of continuous motion the bead velocity is not affected by the viscosity of the medium (over five orders of magnitude). This raises the question of the dissipative force, which is obviously not the Stokes force on the object (about 1 fN for a bead of 1 μm radius moving with a velocity of 1 $\mu m\, min^{-1}$ in a standard motility assay. In contrast, the stalling force for the growing actin tail is a few nN [nanonewton] [36,48]). Several studies indicate that friction between the actin gel and the propelled object is the major source of dissipation [19,61]. This would support the hypothesis that the saltatory bead or vesicle motion is the result of a stick–slip motion, where a certain critical force must be overcome to rupture bead–gel bonds and to displace the object with respect to the gel [4,19].

2.2.2.2 Symmetry Breaking

Biomimetic experiments are not only effective tools to study the generation of motion by polymerization but they have also revealed a spontaneous symmetry-breaking instability in the growing actin network (transition from (i) to (ii) in Figure 2.3), whose consequences *in vivo* are not clear and whose nature is still a subject of debate in the literature.

Even though, as often argued by biologists, symmetry breaking does not play a role in bacterial systems, because the ActA proteins are distributed asymmetrically around the bacterium due to a previous cell division, the instability is a powerful tool to study the coupling mechanisms between actin polymerization and mechanical stresses.

So far there seems to be a consensus in the literature that mechanical stresses build up in the actin network due to growth, because new monomers are inserted at the internal curved interface and push older network layers away from the object. In a symmetric situation one expects therefore high tangential stresses at the external actin/solution interface and high normal stresses at the internal object/actin interface. Then the symmetry breaking is driven by a release of elastic stresses in the actin gel, either by an asymmetric polymerization/depolymerization or by a fracture at the external gel interface, whereby the two mechanisms are difficult to distinguish.

There seem to exist subtle differences in the nature of the symmetry breaking for hard [11,41,62] and soft objects [23,35,56], as demonstrated in a more recent study [14]. In fluorescence labeling experiments it was shown that for soft objects (vesicles) the symmetry is broken at the internal gel interface, that is, polymerization is considerably slowed down on one side of the vesicle, such that an actin tail develops at the opposite side. In contrast, for hard beads the symmetry is broken at the external (depolymerizing) gel interface, whereby some authors suggest that the filament network is ruptured due to an accumulation of tangential stresses [58]; whereas theoretical models indicate that a local stress-dependent depolymerization is sufficient to induce symmetry breaking [28,50]. Note however that actin-driven motion alone does not require the polymerization at curved surfaces, as has been demonstrated theoretically [12].

Furthermore, the mechanical properties of the biomimetic object have an effect on the mode selection of the instability. Whereas for soft objects only instabilities, which produce one single actin tail, have been reported, there are observations of higher-order instabilities for hard spheres, depending on the experimental conditions [14]. Therefore, the boundary conditions at the internal gel interface seem crucial for the mode selection.

2.3 Theoretical Approaches

Two types of theoretical concepts to explain how polymerization processes transform biochemical into mechanical energy have been developing in parallel over the past 15 years.

The first concept, called "Brownian ratchet models," was introduced by Peskin et al. [44]. Its major postulate is that polymerization processes (e.g., actin polymerization) are able to rectify the Brownian motion and can thus induce motion. Brownian ratchet models describe microscopically the polymerization of actin filaments in the presence of an obstacle. Although very consuming of computational efforts, Brownian ratchet models allow for the incorporation of a very detailed description of the kinetics of the polymerization machinery. They provide ingenious tools to study the complex phenomena that have been observed in biomimetic experiments.

The second type we refer to as the macroscopic concept does not focus so much on the dynamics of the single filament, but rather considers the actin filament network as a continuous elastic body under growth, where the growth dynamics is driven by a thermodynamic force, the chemical potential. These coarse-grained models emphasize the global stress distribution in the filament network and the nonlocal aspect of elasticity, while neglecting the details of the polymerization process. Nevertheless, they are more suitable to describe the symmetry breaking in actin gels around spherical objects in terms of simple physical ingredients.

The ideal multiscale model would combine both concepts and use the global stress distribution as an input for the complex polymerization kinetics at the free interfaces.

In the following subsections we briefly review the two concepts. In the first subsection on Brownian ratchets we mainly discuss Refs. [16,21] as they provide, from our point of view, the most advanced description for the polymerization dynamics close to the obstacle. In the second subsection on macroscopic models we briefly outline and discuss the major drawbacks of Refs. [50] and [28] and then give a perspective on how these problems can be solved using homogenization techniques.

2.3.1 Brownian Ratchet Models

2.3.1.1 General Concept

As mentioned above, the idea of a Brownian ratchet model for polymerization forces in biological systems was introduced by Peskin and co-workers [44]. This work explored the rectification of Brownian motion of a particle by the intercalation of new monomers at the interspace between a filament tip and the particle, which gives an ideal ratchet velocity v of

$$v = \frac{2D}{\delta} \tag{2.2}$$

where D denotes the diffusion coefficient of the particle and δ the size of a monomer. However, later on it was shown that the actin-driven motion was relatively independent of the particle size [24] and independent of the viscosity of the medium over several orders of magnitude [61], that is, independent of D. Therefore, the concept of Brownian ratchets was generalized to "elastic Brownian ratchets," where the tips of polymerizing filaments undergo fluctuations, which induce a propulsive force on the obstacle [39,59] and to "tethered ratchets" [40], which involves also the transient attachment of fluctuating filament tips to the obstacle. Typically in this description the polymerization rate constant k_p is weighted by the load f using Kramers theory

$$k_p = k_p^{\max} e^{-f\delta/k_B T} \tag{2.3}$$

where $k_p^{\max}$ is the rate constant at zero load, δ represents the gap size to intercalate a monomer, and k_B and T denote the Boltzmann constant and the absolute temperature, respectively.

The general framework has been used to quantitatively model the steady motion of flat objects [12], lamellipodia, and bacterial motion [39,40]. However, stochastic effects in the number of polarizing filaments were necessary to break the symmetry in a spherical gel around a bead homogeneously covered by ActA or Wasp, whereas the global elastic stress distribution due to filament crosslinking was neglected [40,59].

Other attempts to model biomimetic motility quantitatively by explicitly modeling the filament dynamics in the actin tail succeeded in obtaining the crossover from a continuous to a hopping motion [2,8]. However, the velocity oscillations were on a time scale of milliseconds with step sizes of a few nanometers, as opposed to the experimental oscillations on the scale of several minutes with step sizes on the micrometer scale [5,14,31,55].

More recently, the tethered ratchet model of Mogilner et al. [40] was submitted to more rigorous treatment concerning the polymer physics of the filament brush close to the obstacle [16,21]. Further away from the obstacle the crosslinked gel is advancing with a so-called grafting speed (force-dependent and coupled to the brush length). In the following we highlight the basic features of this model because it quantitatively reproduced velocity oscillations.

2.3.1.2 Quantitative Model for Velocity Oscillations in Actin-Based Motility

As mentioned previously, the motion of a mutant form of the bacterium *Listeria* (with a mutation in the ActA protein) is oscillatory and shows remarkable temporal patterns [19,31]. The bacteria move very slowly for 30 to 100 s, jump forward during a few seconds, and then slow down again abruptly. Such periodic behavior, consisting of long intervals of a slowly changing dynamics that alternate with short periods of very fast transitions is found in several chemical and biological systems and is known as relaxation oscillation [30].

Gholami et al. [21] and Enculescu et al. [16] have developed a microscopic model based on the concept of tethered elastic ratchets for actin-based motility. Their model consists of a brush of growing actin filaments close to an object (the bacterium) and describes the generation of forces and consequently the propulsion of the object (Figure 2.4).

Briefly, the model considers the case of fluctuating filaments close to an obstacle. Filaments may attach to the obstacle with a rate constant k_a and detach from the obstacle with a force-dependent rate constant k_d, resulting in two distributions of populations of filaments, that is, of attached and detached filaments n_a and n_d, respectively. Opposite the obstacle, the filament ends are anchored in a crosslinked network, the actin gel. Detached filaments polymerize with a load-dependent velocity v_p. The distance between the grafting point (i.e., the interface between the network and the polymer brush) and the obstacle is denoted by ξ. The filaments are characterized by their free contour lengths l. One of the crucial ingredients of the model is that the grafting point is advancing in the direction of the obstacle with a so-called grafting velocity v_g, which depends on the free contour lengths of the polymers by

$$v_g(l) = v_g^{\max} \tanh(l/\bar{l}) \tag{2.4}$$

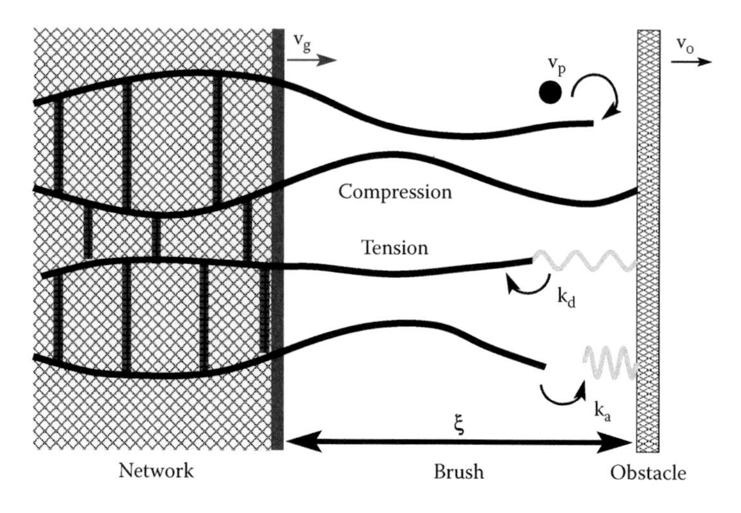

FIGURE 2.4 Schematic representation of the elastic ratchet model as explained in the text. Filaments are anchored with one end into a crosslinked network forming an actin gel and oriented with the other end against an obstacle. Adapted from M. Enculescu, A. Gholami, and M. Falcke (2008). Dynamic regimes and bifurcations in a model of actin-based motility. *Phys. Rev.* E 78:031915, and A. Gholami, M. Falcke, and E. Frey (2008). Velocity oscillations in actin-based motility. *New J. Phys.* 10:033022.

where $\bar{l}$ denotes a characteristic width of the interface between the crosslinked network and the filament brush.

Attached and detached filaments exert entropic forces [22] on the obstacle, $F_a(l, \xi)$ and $F_d(l, \xi)$, respectively, which lead to the propulsion of the object with an effective friction coefficient ζ.

The load dependence of the kinetic rate constants is again included using a Kramers-type expression [see Equation (2.3)]: that is,

$$k_d(F_a) = k_d^0 \exp(-\delta\, F_a/k_B T) \tag{2.5}$$

for the detachment and

$$v_p(l) = v_p^{\max} \exp(-\delta F_d/k_B T) \tag{2.6}$$

for the polymerization speed.

The full evolution of the length distributions of the two filament populations $n_a(l, t)$ and $n_d(l, t)$ is described by advection reaction equations. However, it is shown that the two distributions contract rapidly on the scale of 10^{-2} s into monodisperse distributions $N_a = n_a(t)\delta(l - l_a(t))$ and $N_d = n_d(t)\delta(l - l_d(t))$ localized at l_a and l_d for attached and detached filaments, respectively. Therefore, the dynamics of the system can be simplified to four ordinary differential equations for the evolution of the free contour lengths $l_a(t)$ and $l_d(t)$, respectively; the number of attached filaments $n_a(t)$ (the total number of filaments N is constant and therefore $n_d(t) = N - n_a(t)$); and the distance $\xi(t)$ between the obstacle and the grafting point.

The solution behavior of this system of equations has been analyzed numerically depending on the maximal grafting speed $v_g^{\max}$ and the rate constant for the attachment of filaments k_a. Fortunately, most other model parameters are known experimentally. The model displays two different dynamical regimes: steady and oscillatory motion, whereby the oscillatory regime is robust against changes in the parameters. The oscillations occur on the time scale of minutes and produce jumps of the obstacle displacement in the micrometer range and very much resemble relaxation oscillations.

A deeper analysis of the solutions suggests that the oscillations arise from a so-called push–pull mechanism, that is, a competition between pulling and pushing forces acting on the obstacle. In this mechanism, a long pull phase, where most of the filaments are attached to the obstacle and polymerization stalls, alternates with a short push phase, where most filaments are detached and polymerize rapidly. In the pull phase the grafting velocity v_g is higher than the polymerization speed v_p, and the magnitude of the forces on the obstacle increases due to their dependence on ξ, l_a, and l_d. At a certain point in the pushing phase, the pushing forces outweigh the pulling forces and cause an avalanche-like detachment of filaments and the obstacle "hops" forward, lowering the load on the detached filaments, which start to polymerize rapidly and thus $v_g < v_p$. Meanwhile free filaments start to attach to the obstacle and increase the pulling force, that is, the obstacle slows down. Free filaments start to buckle, the polymerization stalls, and the cycle reenters the pulling phase.

Obviously this complex cycle arises from the subtle interplay between pushing and pulling forces on the one hand, and the grafting and polymerization velocities on the other hand. It would be an interesting task to explore the parameter space further and identify the absolutely necessary ingredients to find oscillations in the push–pull mechanism.

Besides steady and oscillatory motion, the model also yields bistable and excitable behavior, which might lead to the reinterpretation of previous experiments and is reminiscent of the behavior caused by nonlinear friction in a variety of systems with complex surface chemistry [57]. Another aspect for future work is to couple the microscopic dynamics of the filament brush to the bulk mechanics of the crosslinked gel.

2.3.2 Macroscopic Models

While microscopic models give a detailed description of the polymerization and crosslinking dynamics, macroscopic models are more concerned about the global stress distribution in the gel, but adopt a more general formulation for the interface dynamics. In the following we briefly introduce and discuss the essence of three simple models—by Lee et al. [32], Sekimoto et al. [50], and John et al. [28]—describing the symmetry breaking in an actin gel around a solid bead. We conclude this section with the discussion of a more advanced mechanical model using homogenization techniques proposed by Caillerie et al. [10].

2.3.2.1 Phenomenological Model of Symmetry Breaking

As an introduction we present the problem of symmetry breaking from a purely phenomenological point of view as proposed by Lee et al. [32]. If a bead moves, it is natural to assume that a force is applied on the bead, probably due to deformed filaments that release stress on the bead. Let $g(\mathbf{r}, t)$ denote the force per unit area in the normal direction that is exerted on the bead, which could be, for example, a function of the local actin concentration. The total force on the bead is given by $\mathbf{F} = \int g(\mathbf{r}, t)\mathbf{n}dA$, with $\mathbf{n}$ being the unit normal vector on the bead. The velocity of the bead is related to the total force via a linear relation

$$\mathbf{v} = \xi\mathbf{F} = \xi \int g(\mathbf{r}, t)\mathbf{n}dA \tag{2.7}$$

where ξ is a dissipative coefficient, taken to be scalar for simplicity (it is necessarily so for a sphere in a Newtonian fluid).

It is then assumed that the rate of change of $g(\mathbf{r}, t)$ is a local function of the bead velocity $\mathbf{v}$, and that there is a feedback of the motion on the force g: faster motion is associated with a decrease in polymerization in the front and an increase of polymerization at the rear, and hence has an impact on g. Under the assumption of analyticity, the evolution of g can be written as

$$\partial_t g = -g - g^2 - cg^3 + a\mathbf{v} \cdot \mathbf{n} + bg\mathbf{v} \cdot \mathbf{n} \tag{2.8}$$

with $c > 0$ in order to ensure stability of the homogeneous stationary state $g = 0$. The signs of a and b are left arbitrary for the moment. Equations (2.8) and (2.7) constitute a complete set that can be solved numerically or analytically using a perturbation ansatz as outlined in the following.

It is a simple matter to see that the set [Equations (2.8) and (2.7)] admits the fixed point $g = \mathbf{v} = 0$. By superposing small perturbations on this solution, reporting into the above set, and expanding up to linear order in the perturbations, one finds that the fixed point is unstable for $a > 3/\xi \equiv a_c$ and stable otherwise. If motion takes place then this means that a symmetry breaking has occurred, and thus the local force g has lost the spherical symmetry. More precisely, by expanding g in spherical harmonics, $g = \sum_{\ell m} g_{\ell m} Y_{\ell m}(\theta, \phi)$, and reporting into Equation (2.8) by assuming a direction of motion, say along oz, one finds to leading order

$$\partial_t g_{\ell m} = \left[1 - \frac{a}{a_c}\delta_{\ell 1}\right] g_{\ell m} + \cdots \tag{2.9}$$

whereby the "$\cdots$" refers to nonlinear terms. Integration of harmonics other than the mode $\ell = 1$ in (2.7) vanishes exactly due to symmetry. Beyond the symmetry breaking bifurcation for $a > 3/\xi$ the perturbations grow exponentially in time, and nonlinear terms are needed. Because only the mode $\ell = 1$

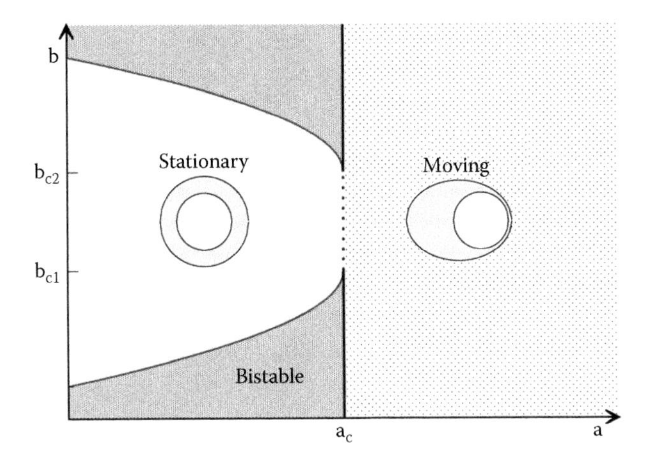

FIGURE 2.5 Phase diagram arising from the analysis of Equation (2.10). The transition between the stationary (white region) and moving (light shaded region) state occurs at $a = a_c$. For $b_{c1} < b < b_{c2}$ the bifurcation is supercritical. In the dark shaded region there coexist two solutions $\mathbf{v} = 0$ and $\mathbf{v} = \text{const} \neq 0$ that are linearly stable. Adapted from A. Lee, H.Y. Lee, and M. Kardar (2005). Symmetry breaking motility. *Phys. Rev. Lett.* 95:138101.

is excited, the other modes are treated as adiabatically enslaved to it (at least in the vicinity of the threshold). The following analysis consists of expanding the solution for higher-order terms and expressing the amplitudes of the higher harmonics in terms of the amplitude of the first-order mode. Once g has been replaced by $\mathbf{v}$ from Equation (2.7) and inserted into Equation (2.8), a closed equation for $\mathbf{v}$ (which is simpler to assess experimentally than g) is obtained [32]:

$$\partial_t \mathbf{v} = \epsilon \mathbf{v} + \frac{27}{5\xi^2}\left[\left(\frac{\xi b}{3} - 1\right) + \left(\frac{\xi b}{3} - 2\right) - c\right]\mathbf{v}^3 + u\mathbf{v}^5 \tag{2.10}$$

$\epsilon \equiv (1 - a/a_c)$ is a small parameter (expressing the fact that we focus on the instability threshold), and we have used the convention $\mathbf{v}^3 = v^2\mathbf{v}$, and so on. The expression of u in terms of the coefficients a, b, c, and ξ is not shown here. However, it is reported in [32] that u is negative in the considered parameter space. The amplitude equation (Equation (2.10)) serves to discuss the phenomenology of the motion, which we briefly summarize here and which is shown schematically for the $b - a -$ parameter plane in Figure 2.5.

If the coefficient of the cubic term is negative, then a continuous transition from the symmetric (motionless) to the asymmetric (moving) state occurs at $\epsilon = 0$, that is, the bifurcation is supercritical (an analog of a second-order transition). This happens for a certain range of the product ξb and by fixing c to unity for definiteness. For a certain range of the product ξb, the cubic coefficient changes sign, a signature of a subcritical bifurcation (an analog of a first-order transition). This is known to lead to multistability:

in a certain parameter range for $\epsilon < 0$ there coexist two solutions: $\mathbf{v} = 0$ and $\mathbf{v} = \text{const}$, both solutions being locally stable (i.e., stable with respect to small perturbations). One might think that the system may switch back and forth between the two solutions.

Several remarks are in order:

1. The above discussion is fully phenomenological, and actually Equation (2.10) could have been written directly. However, the derivation given in [32] has certain merit with regard to the description of the general laws behind the motion, and the feedback mechanism.
2. The coefficients are certainly complicated functions of experimental parameters, and a microscopic model is needed to relate phenomenological and experimental parameters.
3. The origin of the forces acting on the bead and the subsequent generation of motion in this model is not clear. We return to this vital question in the final conclusions.

2.3.2.2 Role of Tensile Stress during Symmetry Breaking in Actin Gels

A first physical macroscopic model including elasticity was put forward by Sekimoto et al. [50] to explain the birth of symmetry breaking of an initially symmetric gel layer growing on a spherical or cylindrical object. The crux of their analysis is that the gel that has been formed is continuously pushed outward due to the arrival of new monomers at the bead. This is supposed to lead to large lateral stresses on the outer gel interface. A schematic representation of the model is shown in Figure 2.6. In the following we outline the model in

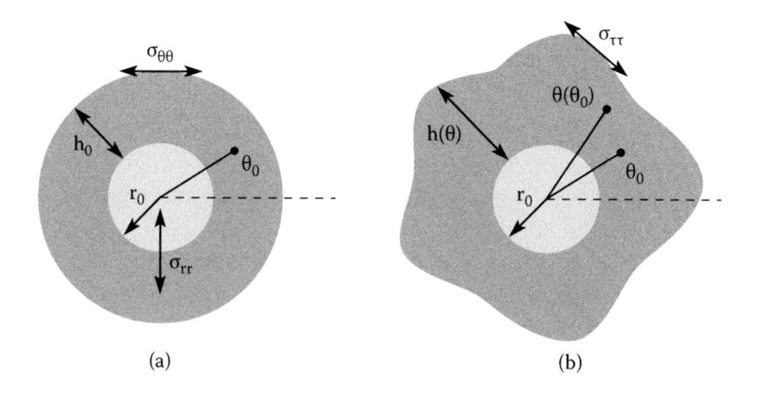

FIGURE 2.6 Schematic view of a bead surrounded by an elastic gel, showing several definitions as explained in the text. Shown is a symmetric gel (a) and an asymmetric gel (b). In the latter case the tensile stress is rather generally denoted by $\sigma_{\tau\tau}$ with τ designating the tangent direction.

a cylindrical geometry. If $\sigma_{\theta\theta}$ designates the tangential stress, and if the gel layer is axisymmetric, it is postulated that

$$\sigma_{\theta\theta} = E\frac{r - r_0}{r_0} \tag{2.11}$$

where E is Young's modulus and r denotes the distance to the center of the cylinder. At $r = r_0$ (the cylinder radius) the tensile stress vanishes, expressing the assumption that the gel layers are added unstretched at the bead surface. If h denotes the total gel thickness, one can write for the tensile stress at the external gel surface $\sigma_{\theta\theta}|_{r=r_0+h} = E\frac{h}{r_0}$. The external surface of the gel is not subject to a force, so that both the tangential and normal forces must vanish; that is, $\sigma_{r\theta}|_{r=r_0+h} = \sigma_{rr}|_{r=r_0+h} = 0$. The tangential force is also zero at the cylinder surface, $\sigma_{r\theta}|_{r=r_0}$. For a symmetric gel the shear stress vanishes everywhere in the gel. If a non-zero tensile stress exists (Equation (2.11)), the mechanical equilibrium in the bulk, that is,

$$div(\sigma) = 0 \tag{2.12}$$

provides the following relation between the tensile and radial stress distribution, $\sigma_{\theta\theta}$ and σ_{rr},

$$\sigma_{rr} = \frac{E}{2r_0 r}\left[(r - r_0)^2 - h^2\right] \tag{2.13}$$

The gel may grow because of a gain in polymerization energy at the cylinder surface. This growth takes place at a certain price: the higher the thickness of the gel, the larger the stored elastic energy. Therefore, one expects the growth to stop at a certain equilibrium thickness h_0. The following growth kinetic relation has been suggested [50]:

$$\partial_t h = k_p e^{c_p \sigma_{rr}|_{r=r_0}} - k_d e^{c_d \sigma_{\theta\theta}|_{r=r_0+h}} \tag{2.14}$$

where k_p, k_d, c_p, c_d are positive constants. The first term, which is positive, accounts for the polymerization at the cylinder, while the second one, which is negative, refers to depolymerization at the external surface. Note that on the one hand $\sigma_{rr}|_{r=r_0} = -Eh^2/(2r_0^2) < 0$, and thus stress penalizes polymerization and acts against gain in chemical bonds at the cylinder. On the other hand $\sigma_{\theta\theta}|_{r=r_0+h} > 0$ and this causes the depolymerization to increase with the gel thickness because $\sigma_{\theta\theta}|_{r=r_0+h}$ increases with h. Setting $\partial_t h = 0$ provides us with the steady-state thickness h_0 as a function of other parameters. One finds that:

$$\frac{c_p}{c_d}\left(\frac{h_0}{r_0}\right)^2 + \frac{h_0}{r_0} - \frac{2}{c_d E}\ln\left(\frac{k_p}{k_d}\right) = 0 \tag{2.15}$$

A steady solution exists as long as $k_p/k_d > 1$. It can easily be checked that the steady solution h_0 is stable with respect to a homogeneous (i.e., axisymmetric) increase or decrease in h_0. One further sees that h_0 is a linear function of r_0 because only the ratio h_0/r_0 enters into Equation (2.15). This seems to be in good agreement with experimental observations [41].

Let us now discuss the linear stability analysis of the steady solution with respect to perturbations that break the circular symmetry. In a cylindrical geometry the eigenmodes are $\sim e^{im\theta_0}$ (or $\cos(m\theta_0)$ in real variables), where θ_0 is the angular variable shown in Figure 2.6 and m is an integer. Because the equations are autonomous with respect to time, all the eigenmodes can be written as $\sim e^{\beta t}$, where β is the amplification or attenuation rate of the perturbation that must be determined from the model equations. We set the perturbed thickness as

$$h(\theta_0) = h_0[1 + \epsilon_m(t)\cos(m\theta_0)] \tag{2.16}$$

ϵ_m is a small, time-dependent quantity that justifies the linear stability analysis. The stress field inside the gel will thus be parametrized by the function $h(\theta_0)$. The linear analysis now consists of (1) solving the stress field in the gel with the appropriate boundary conditions, and (2) using the kinetic relation (Equation (2.14)) to obtain the dispersion relation $\beta = f(m, p)$, where p is an abbreviation for all the physical and geometrical parameters that enter into the equation.

To solve for the elastic field, one needs to specify a constitutive law. Unlike in the two models [10,28] we discuss in the following sections, where a constitutive law is used by evoking basic continuum mechanics concepts, the Sekimoto et al. model [50] is based on an extension of the postulate represented by Equation (2.11) to a modulated thickness $h(\theta_0)$. The basic ingredient of their analysis is to introduce an unknown function $\theta(\theta_0)$ such that a material point of the gel that is originally located at θ_0 in an undeformed reference state is moved to a new position $\theta(\theta_0)$ upon deformation. Then the elongation ratio $(r - r_0)/r_0$ is replaced by $(r\, d\theta(\theta_0) - r_0\, d\theta_0)/(r_0\, d\theta_0)$ so that the tensile stress takes the form

$$\sigma_{\theta\theta} = E\left[\frac{r}{r_0}\frac{d\theta(\theta_0)}{d\theta_0} - 1\right] \tag{2.17}$$

Using the equilibrium balance condition of Equation (2.12) and neglecting the shear stress (for a discussion of this assumption see Ref. [50]) one can again find a relation between σ_{rr} and $\sigma_{\theta\theta}$. By defining

$$T \equiv \int_{r_0}^{r_0+h} \sigma_{\theta\theta} dr \tag{2.18}$$

one deduces that

$$\sigma_{rr}|_{r=r_0} = -\frac{T}{r_0} \tag{2.19}$$

because only this quantity enters the kinetic relation (Equation (2.14)) that is needed for the derivation of the dispersion relation. Recall also that $\sigma_{rr}|_{r=r_0+h} = 0$ to linear order in the perturbation ϵ_m. Upon using Equation

(2.17) the integral of Equation (2.18) leads to

$$T = E\left[\left(h(\theta_0) + \frac{h(\theta_0)^2}{2r_0}\right)\frac{d\theta(\theta_0)}{d\theta_0} - h(\theta_0)\right] \tag{2.20}$$

Additionally, Equation (2.12) yields, under the assumption of zero shear stress, $\partial_\theta T = 0$; that is, T is independent of θ. This condition provides a relation between $\theta(\theta_0)$ and $h(\theta_0)$. Upon using $\int_0^{2\pi} \frac{d\theta(\theta_0)}{d\theta_0} d\theta_0 = 2\pi$, one can express T as a function of an integral $\int_0^{2\pi} F(h(\theta_0))d\theta_0$ where only $h(\theta_0)$ enters (F is given by Equation (D5) in Ref. [50]). Plugging this relation into Equation (2.20) provides a relation between $d\theta/d\theta_0$ and $h(\theta_0)$, and substituting h by Equation (2.16), $\sigma_{\theta\theta}|_{r=r_0+h}$ and $\sigma_{rr}|_{r=r_0}$ can be deduced to first order in ϵ_m. After some algebraic manipulations the dispersion relation is obtained:

$$\beta = \frac{\Omega_m k_d}{r_0} \tag{2.21}$$

where

$$\Omega_m = c_d E e^{c_d E \bar{h}_0} \frac{\bar{h}_0}{\bar{h}_0 + 2} \tag{2.22}$$

and where we have set $h_0/r_0 \equiv \bar{h}_0$. β is positive, meaning that the perturbation grows exponentially with time: the symmetric gel layer is thus unstable. Surprisingly, the dispersion relation does not depend on the wave number m (because Ω_m does not). Consequently, all wave numbers have the same growth rate. Thus, the linear stability analysis does not select a typical mode m (like the fastest growing mode) for the instability.

2.3.2.3 Nonlinear Study on Symmetry Breaking in Actin Gels

Unlike the previous model where a tensile stress distribution is *a priori* postulated, the idea of the model proposed by John et al. [28] is to treat the actin gel as an elastic continuum in the framework of a linear theory, and to formulate a simple kinetic relation expressing growth (or polymerization), different from Equation (2.14).

The model considers a bead (radius r_0) surrounded by a growing elastic actin gel (radius $r_0 + h$) as shown in Figure 2.7. The gel is stressed by a small molecular displacement L in normal direction at the bead/gel interface; that is,

$$u_r|_{r=r_0} = L \tag{2.23}$$

where u_r denotes the radial component of the displacement. This choice is motivated by the microscopic picture, that for the addition of monomers, enzymes facilitate a molecular displacement L at the bead/gel interface. This displacement is the source of stress. The bead/gel interface as well as the

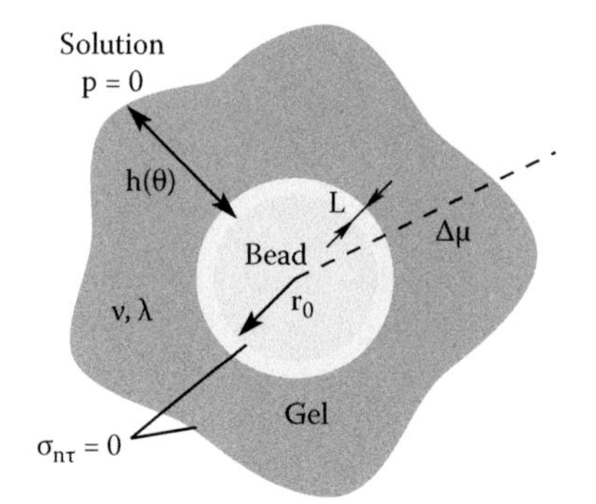

FIGURE 2.7 Schematic view of a bead surrounded by an elastic gel, showing also several definitions of Ref. [28] as explained in the text. From K. John, P. Peyla, K. Kassner, J. Prost, and C. Misbah (2008). A nonlinear study of symmetry breaking in actin gels—Implications for cellular motility. *Phys. Rev. Lett.* 100:068101. (With permission).

external gel surface is shear free and the normal stress at the external surface is set to zero.[5] No condition is imposed on the normal stress component at the bead because there a displacement is imposed instead. The boundary conditions on the stress are thus

$$\sigma_{n\tau} \equiv n_i \sigma_{ij} \tau_j |_{r=r_0, r=r_0+h} = 0, \quad \sigma_{nn} \equiv n_i \sigma_{ij} n_j |_{r=r_0+h} = 0 \tag{2.24}$$

where n_i and τ_i are the ith component of the unit normal and tangent vector of the surface under consideration (bead or external gel surface).

The stress distribution in the gel is obtained then by solving the Lamé equation for the displacement field

$$\nabla^2 \mathbf{u} + \frac{1}{1-2s} \nabla(\nabla \cdot \mathbf{u}) = 0 \tag{2.25}$$

where s is the Poisson ratio. The stress is related to the displacement $\mathbf{u}$ by Hooke's law

$$\sigma_{ij} = 2\nu\epsilon_{ij} + \lambda\epsilon_{kk}\delta_{ij} \tag{2.26}$$

where $\epsilon_{ij} = (\partial_i u_j + \partial_j u_i)/2$ is the strain tensor, and λ and ν are the Lamé coefficients which are related to Young's modulus E and s (for an isotropic material there are only two independent elastic parameters).

Equations (2.23) through (2.25) represent a complete set that allows us to determine the stress and displacement fields in the gel. Note that despite the fact that the bulk equations are linear, the problem acquires a nonlinear character via the geometry of the external gel boundary. Indeed, if we fix

[5]Actually it can be set to $-p$, where p denotes the liquid pressure, but this contribution is quite small.

a certain arbitrary geometry $h(\theta)$, then the stress and displacement will be a nonlinear function of h. The calculation can be handled analytically for a symmetric gel as well as in the linear stability analysis [28]. Beyond a linear analysis, a numerical study has been performed and is briefly discussed below.

Once the mechanical problem is solved, one needs to compute the cost in elastic energy per unit mass for inserting a monomer on the bead/gel interface, that is, the chemical potential difference. For the sake of simplicity we focus on the case of an inert external gel surface; that is, neither polymerization nor depolymerization takes place at the external interface. We assume that insertion of a monomer at the bead results in a displacement of the external gel surface in the radial direction proportional to the chemical potential change at the bead. We critically assess this assumption in the next section on homogenization models.

One can then write a kinetic relation of the shape evolution of the gel envelope

$$\partial_t h = -M\Delta\mu \tag{2.27}$$

where M denotes a mobility and $\Delta\mu$ the difference in the chemical potential between a volume element in the gel and in solution at the internal interface. Here we assume that the mobility is associated with the polymerization/depolymerization kinetics, which constitutes the prevailing dissipation mechanism. The chemical potential is composed of a contribution due to the gain in polymerization (denoted as $\Delta\mu_p < 0$) and an elastic part [28]:

$$\Delta\mu = \Delta\mu_p + \nu u_{ij}u_{ij} + \frac{\lambda}{2}u_{kk}^2 - \sigma_{nn}(1 + u_{kk}) \tag{2.28}$$

Note that Equation (2.27) differs from Equation (2.14) not only by the presence of the exponential function (which can be linearized because the stress energy is always small in comparison to the thermal excitation energy; hidden in the constant c_p), but most importantly by the stress combination. Equation (2.14) contains only a linear form and no quadratic forms as in Equation (2.28). Here, the quadratic form is essential for the mode selection leading to a comet formation, as is described below.

The stress problem in Equations (2.23) through (2.25) can easily be solved analytically for a spherical geometry (axisymmetric growth). Upon setting $\partial_t h = 0$ in Equation (2.27) one finds a steady solution with a gel thickness h_0 obeying [28]

$$h_0 = \left[\left(2\frac{E\alpha - (1-2s)\Delta\mu_p}{2E\alpha + (1+s)\Delta\mu_p}\right)^{1/3} - 1\right] r_0 \tag{2.29}$$

where $\alpha = L/r_0$. This solution exists for $2E\alpha/(1+s) \geq -\Delta\mu_p$: elasticity acts against monomer addition, so that the gel stops growing at that thickness. In the opposite limit, growth continues without bound. Both situations have

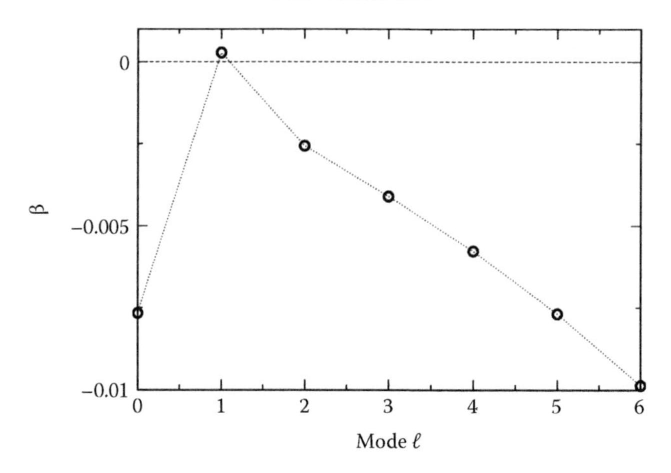

FIGURE 2.8 The dispersion relation as a function of the mode ℓ obtained from the linear stability analysis of the John et al. (2008) model. The fastest growing mode corresponds to $\ell = 1$. From K. John, P. Peyla, K. Kassner, J. Prost, and C. Misbah (2008). A nonlinear study of symmetry breaking in actin gels—Implications for cellular motility. *Phys. Rev. Lett.* 100:068101. (With permission).

been observed experimentally [45]; however in the latter case, growth stopped due to monomer depletion at the bead/gel interface.

The linear stability analysis of the symmetric case can be performed analytically (by decomposing the stress and the shape evolution into spherical harmonics $Y_{\ell m}$). The dispersion relation $\beta(\ell)$ is presented in Figure 2.8. The basic result is that a symmetric shape is unstable against symmetry-breaking [28]. Interestingly the mode that corresponds to a translation of the external surface with respect to the bead is the most unstable.[6] This translation motion is similar to that shown in Figure 2.9b. For this instability, the quadratic terms in Equation (2.28) play a crucial role because considering only the linear terms leads to a stable symmetric solution, with a zero growth rate for the translational mode.

To ascertain the subsequent evolution of the external boundary (i.e., in the fully nonlinear regime), a full numerical analysis has been performed [28]. The set of mechanical equations (Equations (2.23)–(2.25)) and the growth kinetics (Equation (2.27)) have been cast into a phase-field approach, which has now become a frequent method to treat free moving boundary problems. For the details of the phase-field formulation and their numerical implementation we refer the reader to the original paper [28]. Here we only report the basic results.

[6]We should not confuse the mode $\ell = 1$ with the usual global translation, which is a neutral mode. Here, only the external gel surface moves while the bead is fixed, so that the mode $\ell = 1$ is a physical one.

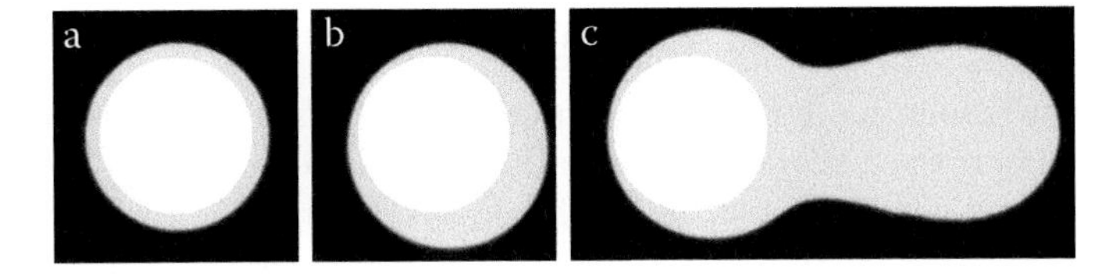

FIGURE 2.9 Symmetry breaking of a circular gel. Shown is the evolution of the gel thickness, starting from a homogeneous thin gel (a) with random small amplitude perturbations. (b) shows the initial symmetry breaking, while (c) shows the subsequent evolution of the shape into a comet in the far nonlinear regime. From K. John, P. Peyla, K. Kassner, J. Prost, and C. Misbah (2008). A nonlinear study of symmetry breaking in actin gels—Implications for cellular motility. *Phys. Rev. Lett.* 100:068101. From K. John, P. Peyla, K. Kassner, J. Prost, and C. Misbah (2008). A nonlinear study of symmetry breaking in actin gels—Implications for cellular motility. *Phys. Rev. Lett.* 100:068101. (With permission).

The numerical study in two dimensions with plane strain (using the finite element code Freefem++) shows that for axisymmetric initial conditions with small amplitude perturbations, the symmetry is broken for the mode $m = 1$ (identical to the mode $\ell = 1$ in three dimensions), which corresponds to a translation of the gel layer with respect to the bead. Where this symmetry breaking occurs depends only on the initial conditions. The instability then evolves further into an actin comet, reminiscent of the comet developed by *Listeria monocytogenes.* Figure 2.9 shows a typical result of a numerical simulation. The comet formation seems here to be the generic growth mode. This finding points to the fact that the comet formation is probably a quite robust feature; it results from simple physical prototypes.

The physical picture of the symmetry breaking may be understood as follows. Let us start with a symmetric layer as in Figure 2.9a. Suppose that due to some natural (inevitable) fluctuation, the gel layer becomes asymmetric, as in Figure 2.9b. The stress is due to addition of new monomers at the bead/gel interface. On the side where the gel thickness is small, the stress field is stronger than on the other side because the gel feels more the "outward pushing" of new monomers inserted at the bead. Because of the increase in the stress (and strain) in the thinner layer, polymerization becomes unfavorable there. New monomers will preferentially be inserted on the side where the thickness is larger. The appearance of modes larger than $\ell = 1$ would create several thin and thick regions, which are likely unfavorable. It seems thus that the mode $\ell = 1$ is optimal for the insertion of monomers at the bead/gel interface.

By considering also the stress-dependent depolymerization at the external gel interface, one finds another instability, whereby the location of the most unstable mode is determined by surface tension. This result is in agreement with the occasional experimental observation of higher-order modes [14], and is discussed next within a more elaborate homogenization model.

2.3.2.4 Homogenization Models

So far, continuous models have been suffering from the inadequate or insufficient description of the mechanical aspect of the actin gel. The greatest difficulty in the description of the mechanical equilibrium of the growing actin network arises from the fact that the growth history determines the network structure and therefore also the stress distribution. Consequently, a realistic model would have to include information on how the network evolved.

A second problem, which is most prominent in the description of the symmetry breaking around solid objects, is the coupling of the growth process between the internal and the external interfaces. As an example, consider the problem of actin growth around a bead functionalized with ActA or Wasp/Scar. Typically growth takes place at the barbed ends of the actin filaments, which point toward the bead/gel interface. This growth process pushes older gel layers further away from that interface. Experimentally, the growth process is observed by an increase in the gel thickness. However, it is not clear how the insertion of mass at a (fixed) solid/gel interface translates into the displacement of the (free) gel/liquid interface. A realistic elastic theory would have to account for this coupling problem in a rigorous way.

One observation that might help solve the above-mentioned problems, at least partly, is that the actin network forms more or less regular structures, which are not perfectly periodic but could be considered in a first approximation as "almost periodic." The actin gel can then be regarded abstractly as a network of elastic filaments connected by nodes with a certain periodicity. This network is completely defined by the positions of the nodes and their connectivities. In the network structure, the size of each elementary cell (e.g., the distance between two Arp2/3 crosslinks (several tens of nanometers)), is small compared to the total size of the structure ($\sim$1 μm), which introduces a small parameter η into the problem, which is the ratio of the length of the unit cell and the total network size. In the following we briefly outline the basic idea of an actin homogenization model in two dimensions [10].

We consider a planar network of stiff elastic bars around a solid cylinder (radius r_0) with the topology shown in Figure 2.10. The bars are connected to each other via nodes. The actin filaments are assumed to be linked to the cylinder at N_t sites evenly located on its surface at a distance $p = \eta r_0$ between two close sites, that is, at an angular distance $\eta = \frac{2\pi}{N_t}$. The actin gel is made of N_n layers of bars in the radial direction. As the growth of the gel is due to the polymerization of actin monomers at the surface of the cylinder, each layer is assumed to consist of the same number of nodes. So the nodes of the gel can be numbered by two integers (ν^1, ν^2) with ν^1 numbering the radial layers and ν^2 the position of the node in each layer, respectively.

It is assumed that the discrete net is made up of a large number of bars, meaning that N_t and N_n are very large and of the same order. To be more precise, the parameter η is assumed to be very small and the number N_n of layers is given by $N_n = \frac{\alpha}{\eta}$ with α being of order 1 with respect to η. Using

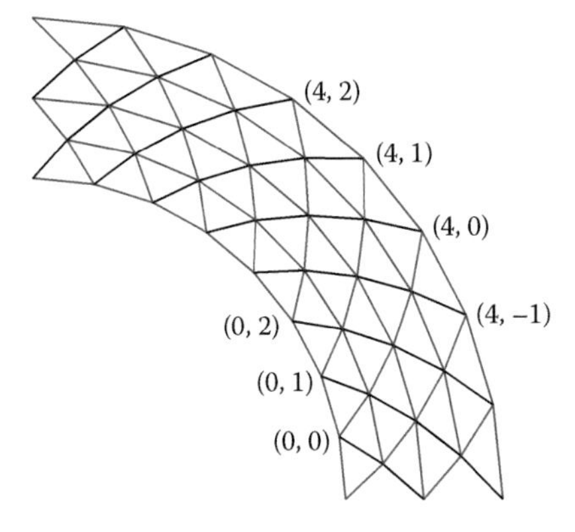

FIGURE 2.10 (See color insert following Page 398.) Sketch of a small part of the filament network with some examples for the node numbering (ν^1,ν^2). The "curves" along which $\nu^1 = \text{const.}$ and $\nu^2 = \text{const.}$ are shown in blue and red, respectively, in the color figure.

this notation, a node of the gel can be labeled by $(\mu^{1\eta}, \mu^{2\eta})$ with $\mu^{i\eta} = \eta\nu^i$. The coordinates $(\mu^{1\eta}, \mu^{2\eta})$ take values in $\omega =]0, \alpha[\times]0, 2\pi[$ and are meant to become the set of Lagrangian curvilinear coordinates of the equivalent continuous medium.

The upscaling of the net to a continuous medium consists of determining the equivalent stresses from the bar tensions, the equations of equilibrium (or motion) satisfied by these stresses and an equivalent constitutive equation ensuing from the properties of the bars. This can be carried out using an asymptotic expansion (for an introduction see [9,34,54]). Here, as the network structure is simple, a more heuristic presentation can be used. The basic idea of the homogenization process is that, for most of the motions of the network, the positions of its nodes (ν^1, ν^2) can be approximated by a continuous function $\psi(\nu^1, \nu^2)$ where $\mu^{i\eta} = \eta\nu^i$. The purpose then is to determine the equations governing this deformation function. The equivalent Cauchy stress tensor is given by the classical relation due to Cauchy [9,34]

$$\sigma = \frac{1}{g}\sum_{b=1}^{3} N^b \vec{e}^{\,b} \otimes \vec{B}^b \tag{2.30}$$

$\vec{B}^b$ is the "bar vector" linking the two extremities of the bar b ($b = 1, 2, 3$) of the elementary cell (ν^1, ν^2) of the network (shown in Figure 2.11) in a deformed state, $\vec{e}^{\,b} = \frac{\vec{B}^b}{\|\vec{B}^b\|}$ is the corresponding unit vector, N^b the tension in the bar and $g = \|\vec{B}^1 \wedge \vec{B}^3\|$ is the surface of the elementary cell. The constitutive equation of the equivalent continuous medium follows from the constitutive equations of the bars, which, for the sake of simplicity, are assumed to be

$$N^b = k^b \frac{l^b - l^b_m}{l^b_m} \tag{2.31}$$

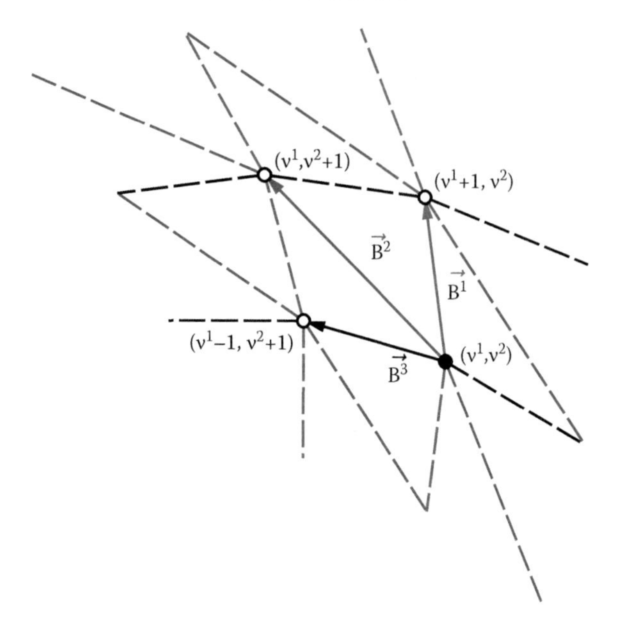

FIGURE 2.11 (See color insert following Page 398.) Sketch of an elementary cell, showing the node numbering (ν^1, ν^2) and the elementary bar vectors $\vec{B}^b$ (solid arrows). All other bars are shown as dashed lines. The "curves" along which $\nu^1 = \text{const.}$ and $\nu^2 = \text{const.}$ are shown in blue and red, respectively, in the color figure.

with $l^b = \|\vec{B}^b\|$. l_m^b is the length of the bar b at rest and serves as a parameter in the constitutive equation.

For a symmetrical equilibrium configuration of the gel to be possible, the constitutive equations of the bars 1 and 3 should be identical, that is, $k^1 = k^3$ and $l_m^1 = l_m^3$.

Following the homogenization assumption stating that the position of the node (ν^1, ν^2) is $\Psi(\nu^1, \nu^2)$ with $\mu^{i\eta} = \eta\nu^i$, a simple Taylor expansion yields:

$$\vec{B}^1 = \eta\partial_1^\mu\Psi,\ \vec{B}^2 = \eta\partial_2^\mu\Psi \quad \text{and} \quad \vec{B}^3 = \eta\left(-\partial_1^\mu\Psi + \partial_2^\mu\Psi\right) \tag{2.32}$$

where $\partial_i^\mu = \dfrac{\partial}{\partial\mu^i}$. Because they are associated with a quite simple numbering system for the nodes, the variables μ^1 and μ^2 arise naturally as Lagrangian variables of the equivalent continuous medium through the homogenization process. However, they are not the most convenient variables to study symmetrical equilibrium configurations. Therefore we have introduced the variables λ^1 and λ^2 defined by

$$\lambda^1 = \mu^1 \quad \text{and} \quad \lambda^2 = \frac{\mu^1}{2} + \mu^2 \tag{2.33}$$

Setting

$$\vec{\varphi}\left(\lambda^1,\lambda^2\right)=\vec{\psi}\left(\lambda^1,-\frac{\lambda^1}{2}+\lambda^2\right) \tag{2.34}$$

one finds

$$\vec{B}^1=\eta\left(\vec{g}_1+\frac{1}{2}\vec{g}_2\right),\ \vec{B}^2=\eta\vec{g}_2 \quad \text{and} \quad \vec{B}^3=\eta\left(-\vec{g}_1+\frac{1}{2}\vec{g}_2\right) \tag{2.35}$$

with $\vec{g}_1=\partial_1^\lambda\vec{\varphi}$ and $\vec{g}_2=\partial_2^\lambda\vec{\varphi}$.

Carrying these relations into Equation (2.30) yields

$$\sigma=\frac{1}{\|\vec{g}_1\wedge\vec{g}_2\|}\sum_{i=1}^{3}\vec{S}^i\otimes\vec{g}_i \tag{2.36}$$

with $\vec{S}^1=\eta(N^1\vec{e}^{\,1}-N^2\vec{e}^{\,2})$ and $\vec{S}^2=\eta(\frac{1}{2}N^1\vec{e}^{\,1}+N^2\vec{e}^{\,2}+\frac{1}{2}N^3\vec{e}^{\,3})$.

As the only forces acting on the gel are applied on its boundaries, the equilibrium of the continuous medium reads classically

$$\operatorname{div}\sigma=0 \tag{2.37}$$

Using the virtual power formulation of that equation and the change of variables

$$(\lambda^1,\lambda^2)\leftrightarrow\vec{x}=\vec{\varphi}(\lambda^1,\lambda^2)$$

it can be proven that the equilibrium equation reads

$$\sum_{i=1}^{2}\partial_i^\lambda\vec{S}^i=0 \tag{2.38}$$

2.3.2.4.1 Coupling between growth and mechanics To study the growth of such an homogenized network, one can stay within the picture of a mechanical equilibrium of the actin gel on the time scale of the growth process. From the homogenized elastic equations one can derive the elastic contribution $\Delta\mu_e$ to the chemical potential $\Delta\mu$ for the addition or subtraction of nodes at the gel interfaces starting from the free elastic energy F_e in the network; that is,

$$F_e=\frac{1}{\eta^2}\int_\Omega d\lambda^1 d\lambda^2\sum_{b=1,2,3}f^b \tag{2.39}$$

where f^b is the elastic energy associated with the extension or contraction of each of the filaments. The elastic chemical potential is then given by the variation of the elastic energy with respect to the size and shape of the network $\Delta\mu_e=\delta F_e/\delta\Omega$ by respecting the boundary conditions.

We assume now that the chemical potential contains also a contribution from the chemical process of polymerization $\Delta\mu_c$, where $\Delta\mu_c < 0$ at the internal interface and $\Delta\mu_c > 0$ at the external interface. This assumption accounts for the polar treadmilling behavior of the actin polymerization; that is, polymerization occurs at the internal interface and depolymerization at the external interface. Furthermore, the chemical potential contains a contribution from interfacial energy: that is, $\Delta\mu_s = -\gamma\kappa$, with γ being the surface tension coefficient and κ being the curvature of the interface. This leads to the following expression for the normal velocities of the two free interfaces in the Lagrangian coordinates (λ^1, λ^2)

$$v_n^i = -\eta M^i \left(\Delta\mu_e^i + \Delta\mu_c^i + \Delta\mu_s^i\right) = -\eta M^i \Delta\mu^i \tag{2.40}$$

with $i = 0$ for the internal and $i = 1$ for the external interface.

2.3.2.4.2 Homogeneous gel growth and linear stability analysis

First one can consider the symmetric problem, that is, the growth of a gel with homogeneous thickness $\alpha = \eta N_n$, which has an axisymmetric solution $\vec{\varphi} = \varphi_r(\lambda_1)\vec{e}_r(\lambda_2)$. The equilibrium equation in this case then reduces to

$$0 = 2\partial_1^\lambda\left(\tilde{N}^1\partial_1^\lambda\varphi_r\right) - \left(\frac{1}{2}\tilde{N}^1 + \tilde{N}^2\right)\varphi_r \tag{2.41}$$

where we have introduced

$$\tilde{N}^b = \eta\frac{N^b}{l^b} \tag{2.42}$$

In this situation the filaments with $b = 2$ are oriented in a tangential direction. Equation (2.41) can be solved numerically using continuation methods [15].

Figure 2.12 shows a solution of Equation (2.41) for a given network thickness α. Naturally, l^2 is extending as one moves away from the bead, whereas l^1 is first shortening and then extending to reach its equilibrium length at the outer gel surface. Consequently, the gel is under radial compression and under tangential extension far away from the bead surface. However, for regions close to the bead surface, the gel is under tangential compression.

Figure 2.13 shows the dependence of the chemical potential on the number of radial filament layers α. Assuming that new filaments are inserted in the same stressed state as the already-present material at the two interfaces, the elastic chemical potential is identical at the two interfaces for a homogeneous gel ($\Delta\mu_e^0 = \Delta\mu_e^1 = \Delta\mu_e > 0$). With increasing network size (i.e., increasing α), $\Delta\mu_e$ increases in a strongly nonlinear fashion. Note that for higher values of α, the homogenization approach breaks down and the gel is starting to "fold back." Beyond this point, at $\alpha \approx 1.25$ in Figure 2.13, no physical meaningful solutions exist.

We consider now a filament network that is allowed to grow symmetrically with identical mobilities ($M^0 = M^1 = M$) at the two interfaces following the

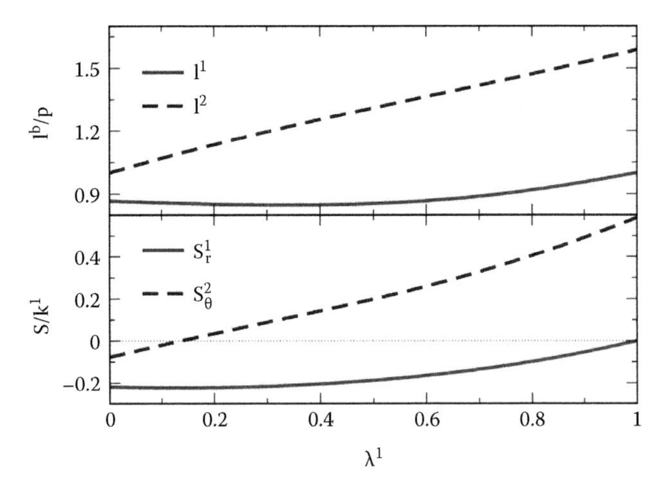

FIGURE 2.12 Filament lengths (top), radial and tangential tensions (bottom) depending on the positions in the network in mechanical equilibrium. Parameters are $k^2/k^1 = 1$ and $l_m^1 = l_m^2 = p$.

two dynamic equations:

$$\partial_t \alpha^0 = \eta M \left(\Delta\mu_c^0 + \Delta\mu_e\right) = -v_p + v_e \tag{2.43}$$

$$\partial_t \alpha^1 = -\eta M \left(\Delta\mu_c^1 + \Delta\mu_e\right) = -v_d - v_e \tag{2.44}$$

whereby the positions of the internal and external interface are denoted by α^0 and α^1, and where we have introduced the polymerization speed $v_p =$

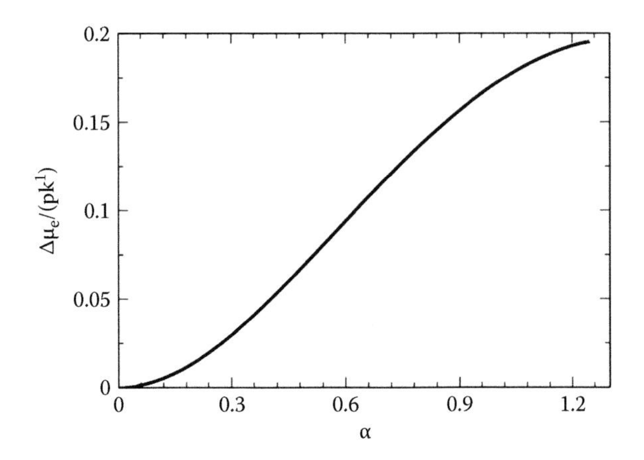

FIGURE 2.13 Dependence of the elastic part of the chemical potential $\Delta\mu_e$ on the number of radial filament layers α. Remaining parameters are $l_m^1 = l_m^2 = p$ and $k^2/k^1 = 1$.

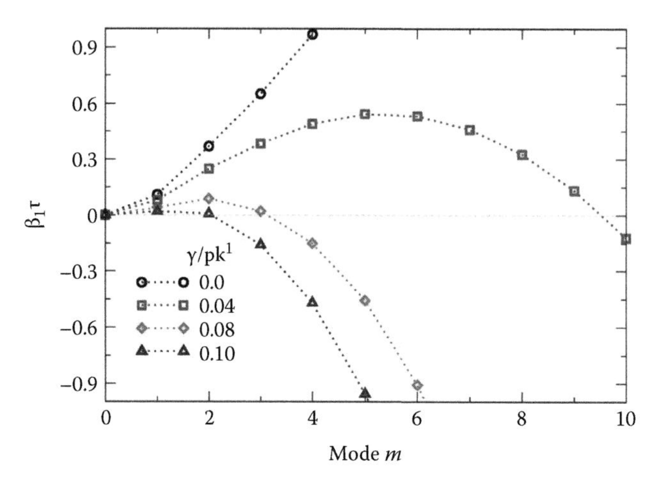

FIGURE 2.14 Dispersion relation with Dirichlet boundary conditions at the internal interface. Shown is only the larger of two eigenvalues β_1 (the second one is always negative) depending on the wave number m for various values of the surface tension as indicated in the legend. Parameters are $\alpha = 1$, $k^2/k^1 = 1$, $l_m^1 = l_m^2 = p$. The time scale is $\tau = r_0/(Mk^1 p^2)$.

$-\eta M \Delta\mu_c^0 > 0$, the depolymerization speed $v_d = \eta M \Delta\mu_c^1 > 0$, and an "elastic speed" $v_e = \eta M \Delta\mu_e$. The steady state for the gel thickness is given by $\partial_t \alpha = \partial_t \alpha^1 - \partial_t \alpha^0 = 0$. Defining now a mean velocity $\bar{v} = (v_p + v_d)/2$ and a velocity difference $\Delta v = (v_p - v_d)/2$, one obtains that in the steady state, $\Delta v = v_e$ and $\partial_t \alpha^0 = \partial_t \alpha^1 = -\bar{v}$. This means that although the gel thickness does not change, both interfaces are moving with the same velocity $-\bar{v}$ and therefore $\bar{v}$ has the physical meaning of the treadmilling speed.

If we now transform the dynamical equations into the comoving frame moving with velocity $-\bar{v}$ in the direction of λ^1 we can study the linear stability of the network thickness α with respect to perturbations of the type $\cos(m\lambda^2)$ at the internal and external interface, $\epsilon^0(\lambda^2)$ and $\epsilon^1(\lambda^2)$, respectively,

Figure 2.14 shows the dispersion relation for the largest growth rate depending on the wave number m. One of two eigenvalues is always negative (i.e., stable), whereas the other one can be positive (unstable), depending on surface tension. We did not find a threshold value for the filament layer number beyond which the gel becomes stable toward small perturbations independent of the surface tension. However, a higher surface tension can suppress instabilities for thin gels. Recent experiments have shown the occurrence of higher modes than one, that is, the formation of up to three actin comets around one bead [14], depending on the experimental conditions. The actual value for the surface tension of an actin network against water should be rather small because actin is a soluble protein. Furthermore, typically small beads with higher curvature break the symmetry faster than larger beads [58], which is in agreement with our model, where the time scale of symmetry breaking increases linearly with the radius of the bead r_0. Note also that assuming a constant polymerization potential but changing the radius r_0 by keeping all other parameters constant leads to a linear relation between the gel thickness (i.e., $\varphi_r(\alpha) - \varphi_r(0)$) and r_0 in steady state. These two results have also been obtained in simpler models based on scaling arguments [41,48,58] and hold for the case that monomer diffusion is fast enough to avoid depletion of monomers due to polymerization at the internal bead gel interface. Another interesting point is the type of instability one might observe, that is, an undulating versus peristaltic instability. For small modes $m \leq 4$ one finds an undulating instability; that is, the perturbations at the external and internal interface are in phase, whereas for higher-order modes one should observe a peristaltic instability where the two perturbations have the opposite phase (data not shown).

2.4 Conclusions and Perspectives

In this chapter we have tried to summarize the complex properties and out-of-equilibrium phenomena of actin gels linked by the Arp2/3 complex, which are at the origin of the motility of animal cells, intracellular organelles, and pathogens. Primarily we have focused on two subjects: the complex actin polymerization dynamics under load at the polymer brush, and the symmetry breaking of actin gels grown from the surfaces of small objects. While we have treated both subjects separately, it is obvious that a full understanding of the system will have to include both approaches: the macroscopic stress distribution in the actin gel couples to the polymerization kinetics in the polymer brush, which in turn changes the deformation state of the gel and the macroscopic stress distribution. We have shown in the previous paragraphs that homogenization models are at the moment the most appropriate models to capture the complex microscopic structures of biological materials on the one hand and take advantage of a continuous framework on the other hand. We believe that the future in the modeling of growing actin gels in complex geometries (e.g., the advancing cell edge) lies in the coupling of these homogenization models to complex dynamics in the polymer brush, as proposed, for example, in Refs. [16,21].

An important question that remains to be elucidated concerns how motion can be generated once the gel layer has become asymmetric. So far we have either limited our considerations to the case of a symmetry breaking around objects, and neglected the generation of motion, or, as in most microscopic models, we have only considered a stationary actin comet, which pushes an obstacle by polymerization. Both concepts are circumventing, by more or less hand waving arguments, one critical question. What is the origin of motion in the absence of external forces, provided that the actin comet and the object are only surrounded by a Newtonian viscous fluid (recall that we are in a regime with $Re \ll 1$) and not attached to some support? Recently, Prost et al. [48] have put forward a simple argument based on largely disparate friction coefficients for the obstacle and the actin tail and the property of treadmilling. In the following we outline this argument.

Suppose that an obstacle and its associated actin comet move with velocities $\mathbf{v}_o$ and $\mathbf{v}_c$, respectively, in the laboratory frame. In the viscous regime the force balance reads then

$$0 = \xi_o \mathbf{v}_o + \xi_c \mathbf{v}_c \tag{2.45}$$

where ξ_o and ξ_c denote the friction coefficients of the object and the comet with the surrounding fluid, respectively. The difference in the two velocities $\mathbf{v}_t = \mathbf{v}_o - \mathbf{v}_c$ is the treadmilling speed. Substituting $\mathbf{v}_c$ in Equation (2.45) and after some rearrangement, one finds for the object velocity

$$\mathbf{v}_o = \mathbf{v}_t \frac{\xi_c}{\xi_o + \xi_c} \tag{2.46}$$

This means that although the dissipation due to fluid friction is very small, it plays nevertheless a decisive role because the ratio of friction coefficients determines the object velocity. In the limit of $\xi_c \gg \xi_o$ this velocity approaches the treadmilling speed. Given the fact that the actin comet is much larger than the object (e.g., bead, droplet, or vesicle), which causes a much larger friction, the experimentally observed obstacle velocities are indeed close to the treadmilling speed.

Another biological aspect that might be of importance when considering more complex cellular systems is the fact that the actin polymerization system is not constitutively active as in *in vitro* assays but is regulated by signaling cascades, which constitute a nonlinear dynamical system. Typically these signaling cascades are modeled by reaction–diffusion systems that lead to pattern formation [37]; for example, polarization of the cell into leading and advancing edge. It remains to be shown how these two important mechanisms, elastic instabilities and instabilities due to reaction–diffusion processes, integrate to produce cellular motion.

Acknowledgments

C.M. and K.J. acknowledge financial support from the CNES and the Alexander von Humboldt Foundation, and with D.C., M.I., P.P., and A.R., we acknowledge financial support from ANR MOSICOB (Modélisation et Simulation de fluides Complexes Biomimétiques).

References

[1] M. Abercrombie (1980). The crawling movement of metazoan cells. *Proc. R. Soc. London B* 207:129–147.

[2] J.B. Alberts and G.M. Odell (2004). *In silico* reconstitution of *Listeria* propulsion exhibits nano-saltation. *PloS Biol.* 2:e412.

[3] K.I. Anderson and R. Cross (2000). Contact dynamics during keratocyte motility. *Curr. Biol.* 10:253–260.

[4] A. Bernheim-Groswasser, J. Prost, and C. Sykes (2005). Mechanism of actin-based motility: a dynamic state diagram. *Biophys. J.* 89:1411–1419.

[5] A. Bernheim-Groswasser, S. Wiesner, R.M. Golsteyn, M.-F. Carlier, and C. Sykes (2002). The dynamics of actin-based motility depend on surface parameters. *Nature* 417:308–311.

[6] H. Boukellal, O. Campás, J.-F. Joanny, J. Prost, and C. Sykes (2004). Soft *Listeria*: actin-based propulsion of liquid drops. *Phys. Rev. E* 69:061906.

[7] D. Bray (1992). *Cell Movements*, New York, Garland Publishing, Inc.

[8] N.J. Burroughs and D. Marenduzzo (2007). Nonequilibrium-driven motion in actin networks: comet tails and moving beads. *Phys. Rev. Lett.* 98:238302.

[9] D. Caillerie and B. Cambou (2001). Les techniques de changement d'échelles dans les matériaux granulaires, in *Micromécanique des Milieux Granulaires*, Paris, Hermès Sciences.

[10] D. Caillerie, K. John, N. Meunier, C. Misbah, P. Peyla, and A. Raoult (2009). A model for actin driven motility through discrete homogenization. *Manuscript in preparation.*

[11] L.A. Cameron, M.J. Footer, A. Van Oudenaarden, and J.A. Theriot (1999). Motility of ActA protein-coated microspheres driven by actin polymerization. *Proc. Natl. Acad. Sci. USA* 96:4908–4913.

[12] A.E. Carlsson (2001). Growth of branched actin networks against obstacles. *Biophys. J.* 81:1907–1923.

[13] S. Cudmore, P. Cossart, G. Griffiths, and M. Way (1995). Actin-based motility of *Vaccinia* virus. *Nature* 378:636–638.

[14] V. Delatour, S. Shekhar, A.-C. Reymann, D. Didry, K. Hô Diêp Lê, G. Romet-Lemonne, E. Helfer and M.-F. Carlier (2008). Actin-based propulsion of functionalized hard versus fluid spherical objects. *New J. Phys.* 10:025001.

[15] E.J. Doedel, A.R. Champneys, T.F. Fairgrieve, Y.A. Kuznetsov, B. Sandstede, and X.J. Wang (1997). *AUTO97: Continuation and Bifurcation Software for Ordinary Differential Equations*, Concordia University, Montreal.

[16] M. Enculescu, A. Gholami, and M. Falcke (2008). Dynamic regimes and bifurcations in a model of actin-based motility. *Phys. Rev. E* 78:031915.

[17] D.A. Fletcher and J.A. Theriot (2004). An introduction to cell motility for the physical scientist. *Phys. Biol.* 1:T1–T10.

[18] Y. Fukui and S. Inoué (1997). Amoeboid movement anchored by eupodia, new actin-rich knobby feet in *Dictyostelium*. *Cell Motil. Cytoskel.* 36:339–354.

[19] F. Gerbal, P. Chaikin, Y. Rabin, and J. Prost (2000). An elastic analysis of *Listeria monocytogenes* propulsion. *Biophys. J.* 79:2259–2275.

[20] F. Gerbal, V. Laurent, A. Ott, M.-F. Carlier, P. Chaikin, and J. Prost (2000). Measurement of the elasticity of the actin tail of *Listeria monocytogenes*. *Eur. Biophys. J.* 29:134–140.

[21] A. Gholami, M. Falcke, and E. Frey (2008). Velocity oscillations in actin-based motility. *New J. Phys.* 10:033022.

[22] A. Gholami, J. Wilhelm, and E. Frey (2006). Entropic forces generated by grafted semiflexible polymers. *Phys. Rev. E* 74:041803.

[23] P.A. Giardini, D.A. Fletcher, and J.A. Theriot (2003). Compression forces generated by actin comet tails on lipid vesicles. *Proc. Natl. Acad. Sci. USA* 100:6493–6498.

[24] M.B. Goldberg and J.A. Theriot (1995). *Shigella flexneri* surface protein IcsA is sufficient to direct actin-based motility. *Proc. Natl. Acad. Sci. USA* 92:6572–6576.

[25] K.C. Holmes, D. Popp, W. Gebhard, and W. Kabsch (1990). Atomic model of the actin filament. *Nature* 347:44–49.

[26] J. Howard (2001). *Mechanics of Motor Proteins and the Cytoskeleton*, Sinauer Associates, Inc., Sunderland, MA.

[27] G. Isenberg, P.C. Rathke, N. Hülsmann, W.W. Franke, and K.E. Wohlfahrt-Bottermann (1976). Cytoplasmic actomyosin fibrils in tissue culture cells. Direct proof of contractility by visualization of ATP-induced contraction in fibrils isolated by laser microbeam dissection. *Cell Tiss. Res.* 166:427–443.

[28] K. John, P. Peyla, K. Kassner, J. Prost, and C. Misbah (2008). A nonlinear study of symmetry breaking in actin gels—Implications for cellular motility. *Phys. Rev. Lett.* 100:068101.

[29] M. Kaksonen, C.P. Toret, and D.G. Drubin (2006). Harnessing actin dynamics for clathrin-mediated endocytosis. *Nature Rev. Mol. Cell. Biol.* 7:404–414.

[30] M. Krupa and P. Szmolyan (2001). Relaxation oscillation and canard explosion. *J. Differ. Equations* 174:312–368.

[31] I. Lasa, E. Gouin, M. Goethals, K. Vancompernolle, V. David, J. Vandekerckhove, and P. Cossart (1997). Identification of two regions in the N-terminal domain of ActA involved in the actin comet tail formation by *Listeria monocytogenes*. *EMBO J.* 16:1531–1540.

[32] A. Lee, H.Y. Lee, and M. Kardar (2005). Symmetry-breaking motility. *Phys. Rev. Lett.* 95:138101.

[33] T.P. Loisel, R. Boujemaa, D. Pantaloni, and M.-F. Carlier (1999). Reconstitution of actin-based motility of *Listeria* and *Shigella* using pure proteins. *Nature* 401:613–616.

[34] A.E.H. Love (1944). *A Treatise of the Mathematical Theory of Elasticity*, New York, Dover Publications.

[35] L. Ma, L.C. Cantley, P.A. Janmey, and M.W. Kirschner (1998). Corequirement of specific phosphoinositides and small GTP-binding protein Cdc42 in inducing actin assembly in *Xenopus* egg extracts. *J. Cell Biol.* 140:1125–1136.

[36] Y. Marcy, J. Prost, M.-F. Carlier, and C. Sykes (2004). Forces generated during actin-based propulsion: a direct measurement by micromanipulation. *Proc. Natl. Acad. Sci. USA* 101:5992–5997.

[37] A.F.M. Marée, A. Jilkine, A. Dawes, V.A. Grieneisen, and L. Edelstein-Keshet (2006). Polarization and movement of keratocytes: a multiscale modelling approach. *Bull. Math. Biol.* 68:1169–1211.

[38] A. Michelot, J. Berro, C. Guérin, R. Boujemaa-Paterski, C. J. Staiger, J.-L. Martiel, and L. Blanchoin (2007). Actin-filament stochastic dynamics mediated by ADF/cofilin. *Curr. Biol.* 17:825–833.

[39] A. Mogilner and G. Oster (1996). Cell motility driven by actin polymerization. *Biophys. J.* 71:3030–3045.

[40] A. Mogilner and G. Oster (2003). Force generation by actin polymerization II: the elastic ratchet and tethered filaments. *Biophys. J.* 84:1591–1605.

[41] V. Noireaux, R.M. Golsteyn, E. Friederich, J. Prost, C. Antony, D. Louvard, and C. Sykes (2000). Growing an actin gel on spherical surfaces. *Biophys. J.* 78:1643–1654.

[42] A. Ott, M. Magnasco, A. Simon, and A. Libchaber (1993). Measurement of the persistence length of polymerized actin using fluorescence microscopy. *Phys. Rev. E* 48:R1642–R1645.

[43] S.H. Parekh, O. Chaudhuri, J.A. Theriot, and D.A. Fletcher (2005). Loading history determines the velocity of actin-network growth. *Nature Cell Biol.* 7:1219–1223.

[44] C.S. Peskin, G.M. Odell, and G.F. Oster (1993). Cellular motions and thermal fluctuations: the Brownian ratchet. *Biophys. J.* 65:316–324.

[45] J. Plastino, I. Lelidis, J. Prost, and C. Sykes (2004). The effect of diffusion, depolymerization and nucleation promoting factors on actin gel growth. *Eur. Biophys. J.* 33:310–320.

[46] T.D. Pollard, L. Blanchoin, and R.D. Mullins (2000). Molecular mechanisms controlling actin filament dynamics in nonmuscle cells. *Annu. Rev. Biophys. Biomol. Struct.* 29:545–576.

[47] T.D. Pollard and G.G. Borisy (2003). Cellular motility driven by assembly and disassembly of actin filaments. *Cell* 112:453–465.

[48] J. Prost, J.-F. Joanny, P. Lenz, and C. Sykes (2008). The physics of *Listeria* propulsion, in *Cell Motility*, New York, Springer, pp. 1–30.

[49] S.M. Rafelski and J.A. Theriot (2004). Crawling toward a unified model of cell motility: spatial and temporal regulation of actin dynamics. *Annu. Rev. Biochem.* 73:209–239.

[50] K. Sekimoto, J. Prost, F. Jülicher, H. Boukellal, and A. Bernheim-Grosswasser (2004). Role of tensile stress in actin gels and a symmetry-breaking instability. *Eur. Phys. J. E* 13:247–259.

[51] L. Stryer (1995). *Biochemistry*, New York, W.H. Freeman & Company.

[52] J.A. Theriot and T.J. Mitchison (1991). Actin microfilament dynamics in locomoting cells. *Nature* 352:126–131.

[53] T.A. Theriot, T.J. Mitchison, L.G. Tilney, and D.A. Portnoy (1992). The rate of actin-based motility of intracellular *Listeria monocytogenes* equals the rate of actin polymerization. *Nature* 357:257–260.

[54] H. Tollenaere and D. Caillerie (1998). Continuous modeling of lattice structures by homogenization. *Adv. Eng. Software* 29:699–705.

[55] L. Trichet, O. Campàs, C. Sykes, and J. Plastino (2007). VASP governs actin dynamics by modulating filament anchoring. *Biophys. J.* 92:1081–1089.

[56] A. Upadhyaya, J.R. Chabot, A. Andreeva, A. Samadani, and A. van Oudenaarden (2003). Probing polymerization forces by using actin-propelled lipid vesicles. *Proc. Natl. Acad. Sci. USA* 100:4521–4526.

[57] M. Urbakh, J. Klafter, D. Gourdon, and J. Israelachvili (2004). The nonlinear nature of friction. *Nature* 430:525–528.

[58] J. van der Gucht, E. Paluch, J. Plastino, and C. Sykes (2005). Stress release drives symmetry breaking for actin-based movement. *Proc. Natl. Acad. Sci. USA* 102:7847–7852.

[59] A. van Oudenaarden and J.A. Theriot (1999). Cooperative symmetry-breaking by actin polymerization in a model for cell motility. *Nature Cell Biol.* 1:493–499.

[60] A.B. Verkhovsky, T.M. Svitkina, and G.G. Borisy (1999). Self-polarization and directional motility of cytoplasm. *Curr. Biol.* 9:11–20.

[61] S. Wiesner, E. Helfer, D. Didry, G. Ducouret, F. Lafuma, M.-F. Carlier, and D. Pantaloni (2003). A biomimetic motility assay provides insight into the mechanism of actin-based motility. *J. Cell Biol.* 160:387–398.

[62] D. Yarar, W. To, A. Abo, and M.D. Welch (1999). The Wiskott-Aldrich syndrome protein directs actin-based motility by stimulating actin nucleation with the Arp2/3 complex. *Curr. Biol.* 9:555–558.

Chapter 3

Cancer: Cell Motility and Tumor Suppressor Genes

Rémy Pedeux, Damien Ythier, and Alain Duperray

Contents

3.1 Cancer: A Disease of the Genome

Cancers are caused by abnormalities of the genome [18]. These abnormalities can be due to the effects of carcinogens such as chemical and viruses (environment), inherited traits, and randomly acquired errors during DNA replication and/or repair. Thus, cancer cells are characterized by genomic rearrangements, activation or repression of oncogenes and tumor suppressor genes, karyotypic and phenotypic instability, uncontrolled growth, and metastatic ability with the acquisition of new phenotypes (loss of adhesion, gain of new resistances). Several types of genes may be affected in cancer [17]. The *oncogenes* are genes that when mutated or expressed at high levels help turn a normal cell into a cancer cell by allowing the gain of new properties

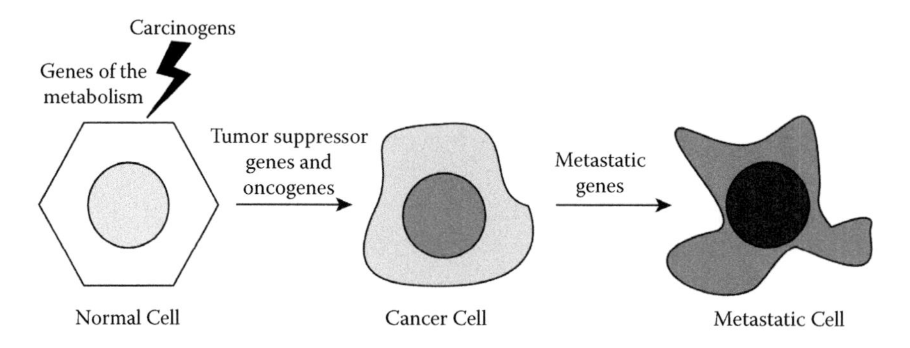

FIGURE 3.1 Genes involved in the tumorigenesis process. Each step of the process is characterized by the involvement of a set of genes: genes of the metabolism in response to carcinogen insults oncogenes and tumor suppressor genes in cell transformation, and metastatic genes in dissemination.

resulting in hyperactive growth and division, resistance to apoptosis, and bypass of senescence. These genes were originally identified because they have a high homology with sequences from RNA virus that can induce cell transformation (e.g., v-myc). The *tumor suppressor genes* are encoding proteins that negatively regulate cell proliferation and prevent the occurrence of tumors. Two types of tumor suppressor genes have been defined:

1. The "caretakers" of the genome (or type 1), which protect the genome from mutations; they usually are DNA repair genes such as Xeroderma pigmentosum genes (XP).
2. The "gatekeepers" of the genome (or type 2), which protect the genome by controlling cell growth, apoptosis, and senescence. p53, a type 2 tumor suppressor gene, is the most frequently inactivated gene in tumors.

Finally, genes encoding for proteins that metabolize carcinogens (e.g., P450) and genes encoding for proteins involved in the metastatic process are also critical in the cellular transformation process (Figure 3.1).

3.2 Tumorigenesis: A Multistep Process

Transformation of a normal cell into a tumor cell results from events (Figure 3.2) that will affect the genomes of cells (1). Cancer cells bear numerous chromosomal abnormalities such as mutations, translocations, deletions, duplications, and/or amplifications. The main characteristic of a tumor cell is proliferation without constraint from the homeostasis that controls cell proliferation in organs (2). Thus, a malignant cell is characterized by its ability to invade the neighbor tissues, migrate, and form metastases at a distant site

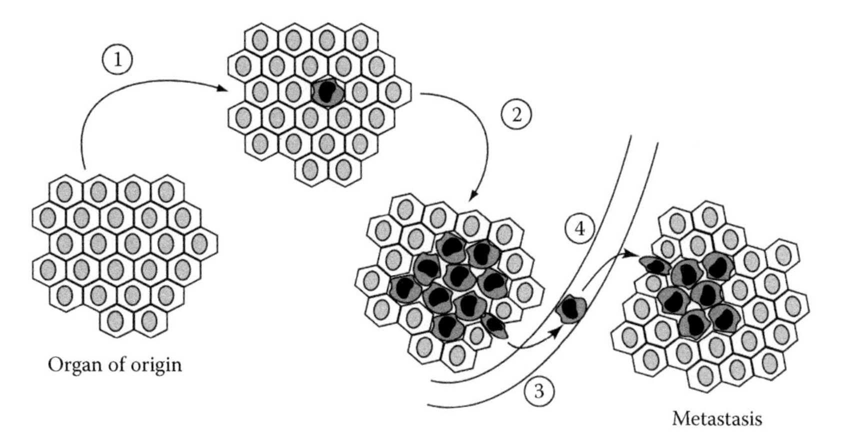

FIGURE 3.2 Steps of tumor and metastatic processes.

(3). Development of metastatic foci reflects tumor progression, indicating a change in the biology of tumor cells forming these foci (4). The expression of several adhesion molecules is modified. These cells synthesize enzymes responsible for the digestion of the intercellular space and the basal membranes surrounding the tissues. They can invade neighboring tissues and migrate along the lymphatic vessels or natural cavities. Tumor cells, and especially metastatic cells, have receptors called *integrins* that allow them to adhere to endothelial cells and extracellular matrix. Adhesion of these cells on a new substrate triggers signals through the cellular membrane that enhance or inhibit cell proliferation. Cells seeded in other organs are metastatic only when they proliferate. They can then give rise to a tumor mass and produce secondary metastasis. The metastatic process consists of a succession of events in which tumor cells develop characteristics of resistance to the immune system, toxic and mechanical insults, and express various receptors for growth factors or substrates. Such characteristics result from altered gene expression [3].

3.3 Molecular Mechanisms Involved in Cell Migration and Invasion

For many years, it has been established that the invasive process depends on adherence, motility, and migration of cells in the body. The loss of cell adhesion and the increased motility of tumor cells have been studied both in experimental models *in vitro* and *in vivo*. The cell motility induced by cytokines and growth factors is physiologically controlled during various events such as development or tissue repair, and it can be activated inappropriately during tumor progression. Here we emphasize the different proteins that are important for the control of cell migration and invasion, such as adhesion proteins expressed at the cell surface, cytoskeletal proteins, and molecules implicated in intracellular signaling cascades.

Cell adhesion is a fundamental process for cell recognition and the formation of tissues. Cell adhesion can take place at different levels, such as specific intercellular adherence between adjacent cells or with the extracellular matrix. Both types of interactions often coexist in the same tissue. The molecular interactions among cells, or between cells and the extracellular matrix, are highly regulated during development and cellular differentiation, and are altered in several pathologies including cancer progression. These interactions are mainly provided by adhesion molecules belonging to four main families:

1. Members of the immunoglobulin superfamily
2. The selectins, proteins involved in the interactions of leukocytes with endothelial cells in blood vessels
3. The cadherins, mainly implicated in cell–cell junctions
4. The integrins, which are strongly involved in cell adhesion and migration by interacting with proteins of the extracellular matrix such as collagen, fibronectin, etc.

3.3.1 Adhesion Proteins

In solid tumors, especially in carcinomas, the expression of cadherins and integrins by tumor cells has been shown to play an important role during tumor progression [11]. The attachment of cells to the extracellular matrix (ECM) is mediated by integrins, which are cell surface receptors for many proteins of this ECM, such as collagen, fibronectin, vitronectin, etc. In addition, integrins interact intracellularly with signaling proteins and components of the cell cytoskeleton, and these interactions can either modify their adhesive properties or participate in signal transmission. These properties are modified in tumoral cells [14].

Cadherins are adhesion proteins involved in cell–cell adhesion. E-cadherin is the prototypic type I cadherin expressed by epithelial cells involved in intercellular adhesion. These interactions, important for the cohesion of the epithelial layer, become deficient when epithelial cells acquire a migratory phenotype, after transformation. The first stages of this phenomenon are called epithelial–mesenchymal transition [14] (EMT), and the loss of E-cadherin has been associated with EMT. E-cadherin mediates homophilic interaction, and its extracellular domain of bladder tumor cells associates with other cadherins expressed by neighboring cells. Inside the cells, the cytoplasmic domain links this protein to the cell cytoskeleton and signaling cascades. This link is performed by intermediary proteins, such as beta-catenin and plakoglobin. E-cadherin functions and its association with intracellular proteins are modified in many epithelial carcinomas, and it has been shown that restoration of plakoglobin can suppress cell migration of bladder tumor cells [25].

3.3.2 Cytoskeleton

To adopt various forms, to apply forces on their substrate, and to make coordinated movements, eukaryotic cells use a complex network of protein filaments that extend throughout the cytoplasm. This network is called the cytoskeleton and is a very dynamic structure that reorganizes continuously when cells change shape, divide, or respond to the environment. These long filaments are all formed through polymerization of monomeric subunits. Three main types of protein filaments form the cytoskeleton:

1. Actin filaments: 5 to 9 nm in diameter, they are made from actin monomers and are essential for movement, through their interactions with the motor protein myosin. They associate through protein linkers with the cytoplasmic domains of integrins and cadherins.
2. Intermediate filaments with a diameter of 10 nm. They are made from monomers of lamin, vimentin, or keratin (in epithelial cells). They are responsible for the cell mechanical strength.
3. Microtubules: 25 nm in diameter, they are made from tubulin monomers. They are important for the transport of vesicles and the segregation of chromosomes during mitosis. They are also essential for the maintenance of the cell structure in general.

Actin is the most abundant protein in many cells (on average it accounts for 5% of total protein). The actin filament is polarized, with one end capable of rapid growth (+ or pointed end), while the other end tends to lose subunits if not stabilized (– end or barbed end). The monomeric actin molecules related to ATP are added to the + end with a protein complex ARP2/ARP3 (actin-related protein). The activity of these complexes is directed by the N-WASP (Wiskott-Aldrich syndrome protein). During the polymerization reaction, ATP is hydrolyzed, leaving ADP trapped in the polymer. Actin molecules linked to the ADP are removed at the – end. This process is catalyzed by cofilin, which binds to the actin-ADP and induces its detachment from the filament. Actin monomers must be reloaded with ATP before joining the + end of the filament. Profilin accelerates the exchange of ADP to ATP. In a resting state, the actin molecules are continually added to the + end and are continually lost at the – end, resulting in the movement of treadmilling. In a state of growth of the filament, the actin molecules are added faster than they are lost, either by inhibition of cofilin or by activating Arp2/3 [21].

With regard to its cytoskeleton, a migrating epithelial cell is characterized by a loss of stress fibers, an increase in the cortical actin network, and a diminution in the number and size of focal adhesion sites. Actin rearrangement is controlled by the action of small GTPases of the Rho family. Three roles have been identified for the three members of the Rho family [34]:

1. Cdc42 is responsible for the nucleation of actin filaments and the formation of filopodia. The sequence of events is as follows: Cdc42 activates

N-WASP, which in turn activates the Arp2/3 complex, causing an increase in the nucleation of actin filaments.

2. RAC1 is responsible for the formation of the actin network, leading to the extension of the lamellipodium.
3. RhoA is responsible for the formation of focal contacts followed by the formation of contractile bundles or stress fibers. The active form of RhoA (GTP-RhoA) induces the activation of the Rho kinase, which in turn phosphorylates and inactivates myosin phosphatase. It results in an increase in phosphorylation and activation of myosin II, which interacts with actin and forms stress fibers.

The general idea is that factors that induce migration have a positive effect on the activators Cdc42 and RAC and an inhibitory effect on RhoA, eliminating the stress fibers and promoting the formation of filopodia and lamellipodia.

3.4 *In vitro* Methods for Cell Migration Studies

Cell migration is essential both in physiological conditions, such as embryonic development, wound healing, and inflammation, and in pathological situations, including chronic inflammatory diseases, tumor cell dissemination, and metastasis. All living cells are motile, but only a few are able to migrate inside the organism. Examples of these specialized cells are fibroblasts, stem cells, leukocytes, and tumor cells. Leukocytes are highly motile cells, as they have to react promptly to pathogenic invaders, and can migrate with a speed of 15 $\mu m \times min^{-1}$ [37], while fibroblasts, whose migration is necessary during tissue repair, are much slower ($\sim$0.2 to 1 $\mu m \times min^{-1}$).

Many *in vitro* assays have been developed to study cell migration and to obtain quantitative data. In the majority of tissues, cell migration occurs inside a complex, three-dimensional extracellular matrix. Nevertheless, a majority of assays allow the study of cell migration in two dimensions, on planar substrates.

3.4.1 2-D Migration Assays

These methods are easy to perform and can provide a good indication of the migratory capacity of cells, in response to modifications of expressions of molecules putatively involved in the migratory process (adhesive proteins, components of intracellular signaling cascades, and of course components of the extracellular matrix). For these assays, cells are cultured on the surface of culture plates, coated either with a matrix (such as collagen, fibronectin),

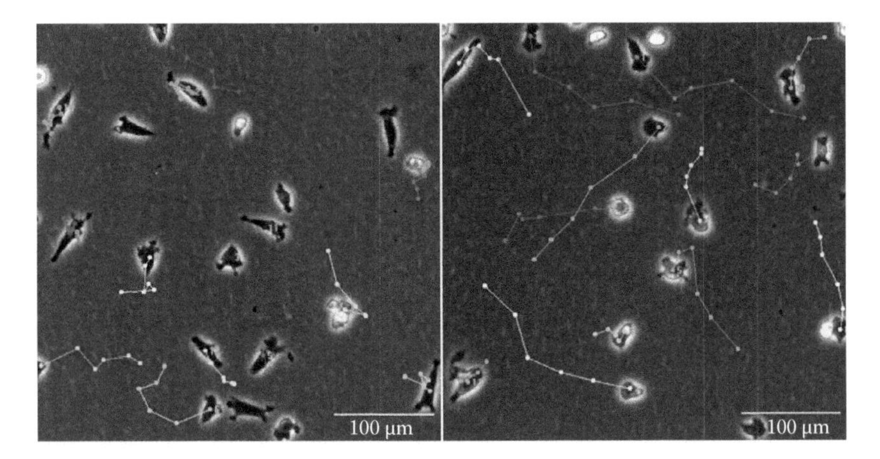

FIGURE 3.3 (See color insert following Page 398.) Visualization of cell tracking.

or left untreated. Then, cell migration is followed for 2 to 3 hours by taking pictures every 10 to 15 min.

3.4.1.1 Single Cell Tracking

Single cell tracking consists of culturing cells at low density and following their displacement by recording sequences of frames on automated video microscopes equipped to maintain the cells at a constant temperature (usually 37°C) in a humidified atmosphere. The location of each cell can be tracked over the entire sequence as a function of time to visualize the cell trajectory.

An example of such a treatment is given in Figure 3.3 for cells plated on a low concentration of fibronectin (1 $\mu g \times ml^{-1}$) on the right, or high concentration of fibronectin (50 $\mu g \times ml^{-1}$) on the left. Determination of the different trajectories of the cells is visualized by the lines. Several migration parameters are obtained from this method: first the velocity of each cell, from which can be deduced the mean velocity of the observed population, and also the cell direction and the persistence duration. In Figure 3.3, it is apparent that cells are faster on the right, on the order of 30 $\mu m \times hr^{-1}$ than on the left (8 $\mu m \times hr^{-1}$). The tracking of the cells can be automated by specialized software. One advantage of this method is that cell migration is easily distinguishable from cell proliferation.

3.4.1.2 Wound Healing Model

Wound healing assays are commonly used to assess the capacity of cell monolayers to migrate. Typically, a wound is made in a cell monolayer using a pipette tip, and the re-colonization of the lesion by the cells is monitored by time-lapse microscopy, over a time course of 24 to 48 hr. This technique has

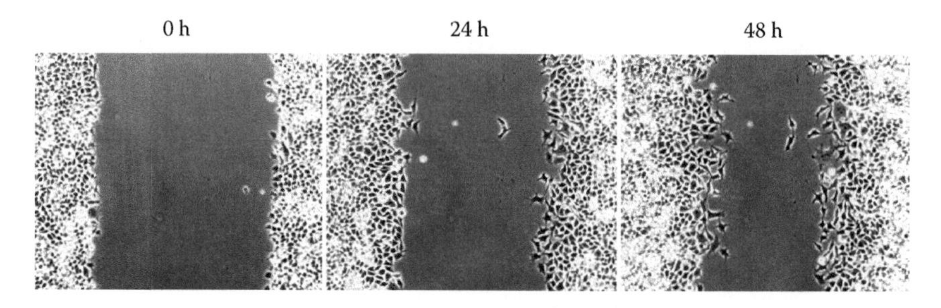

FIGURE 3.4 Wound healing assay. A scratch on the monolayer of RKO cells was performed (0 hr) and closure of the gap by cell migration was monitored by a phase contrast microscope coupled with a CCD camera. Still images were taken at 0, 24, and 48 hr.

been used for studying the involvement of many molecular processes in cell migration, such as the role of the Rho family GTPases [7] or the involvement of p53 [27]. The cell migration is measured as the rate of advance of the wound edge, or by quantification of the area re-colonized by the cells. An example is given in Figure 3.4. In fact, this assay is not a pure migration assay, as it analyzes the colonization ability of a cell monolayer in culture, a process involving both migration and proliferation [38]. Another putative problem with this assay comes from the fact that scratching a cell monolayer can induce cell reactions that are not related to migration.

3.4.1.3 Ring Assays

This assay is very similar to the wound healing assay, except that there is no scratching of the cell monolayer. Cells are first cultured inside a glass ring, which is removed after cells reach confluency, and cells are allowed to migrate from this circular zone for 24 to 48 hr. Cell migration is measured by the increase of the area covered by the cells. This area increase also derives from cell proliferation in addition to cell migration, but, contrary to the wound assay, there is no injury to the cell monolayer.

3.4.2 3-D Migration Assays

3.4.2.1 Filter Assays

These assays are carried out in modified Boyden chambers or Transwell cell culture chambers, where cells are allowed to migrate through filter membranes with pores of size 3 to 8 μm in diameter, as illustrated in Figure 3.5.

Cells are added to the upper compartment, and migrated cells are quantified in the lower chamber. This assay has been modified for the study of chemotaxis, with a chemoattractant placed in the lower compartment, or for the study of invasion, where the filter is coated with a layer of ECM proteins.

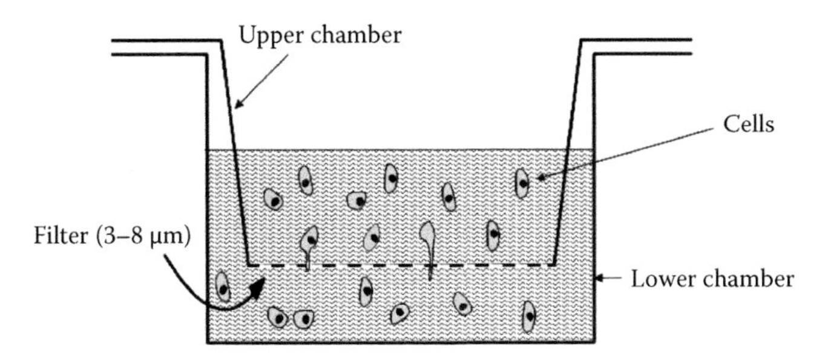

FIGURE 3.5 Migration assay with a modified Boyden chamber.

In this case invading cells must migrate through the matrix before reaching the filter. To mimic the passage of circulating cells such as leukocytes or tumor cells through the vessel wall (inflammation and metastasis), endothelial cells can be cultured on the surface of the filter to evaluate the ability of immune or cancer cells to cross the endothelial monolayer, a process called extravasation [1,26]. The quantification of cells in the lower chamber at the end of the assay is usually done by numbering the cells or by fluorescence techniques.

3.4.2.2 3-D Cell Tracking Methods

A putative problem with the filter assays is that the number of migrating cells is obtained at the end of the experiment, and the observation of the migration process itself is not possible. To overcome this limitation, migrating cells can be observed by time-lapse video microscopy, inside a 3-D matrix. Fluorescent cells can be observed for several hours, and automatic 3-D tracking software has been developed. Confocal laser-scanning microscopy is used for the observation of migrating cells inside 3-D gels. The advantage of this technique is the high resolution that can be obtained, along with the possibility of making 3-D reconstructions of entire cells inside the matrix [8].

3.5 New Family of Tumor Suppressor Genes

One of the most intriguing observations has shown that proteins involved in the migration process may be tumor suppressor proteins. Below we describe the identification of a new family of tumor suppressor proteins and the characterization of the role of one of its member, ING4, in the migration process.

The founding member of the ING family, ING1, was discovered during a screen designed to identify genes that promote neoplastic transformation when repressed: tumor suppressor genes (TSG). To perform this screen, the authors used subtractive hybridization of cDNAs from a nontransformed human mammary epithelial cell line and several human breast cancer cell lines.

The cDNAs specifically expressed only in normal cells were cloned in antisense and expressed in preneoplastic cell lines in order to repress the corresponding gene. The resulting cells were injected into nude mice. The emerging tumors were collected and the sequence of the expressed cDNA responsible for the transformation of the preneoplastic cells was determined. The cDNA sequence allowed the identification of the repressed gene: ING1 [9]. Four additional members of the family (ING2–5) were then identified by sequence homology search [23,24,31,32]. Phylogenetic analyses revealed that ING genes are well conserved along evolution (from yeast to human), suggesting their importance in cellular processes [13].

All the members of the human ING genes family were mapped on independent chromosomes. Since their identification, all ING genes but ING5 were shown to encode several variants, as a result of the usage of different promoters, exons, and splices (Figure 3.6).

ING genes code for proteins that share a strong homology in their C-terminal region. This C-terminal region contains a nuclear localization sequence (NLS) (responsible for their nuclear targeting) and a plant homeodomain (PHD) finger. Such domains are commonly found in proteins involved in chromatin remodeling. The N-terminal part of ING proteins is unique (Figure 3.7) and thus may be responsible for their specific functions. ING protein sequence analysis shows that ING1 and ING2 on one hand, and ING4 and ING5 on the other hand, share high homology. Therefore, they may have closely related or redundant functions.

Initially, when the ING1 gene was discovered, the authors identified the p33ING1b isoform [9] originally described to interact physically with p53 and to enhance its transcriptional activity [10]. Thus, p33ING1b is necessary for p53 to repress cell proliferation. Subsequently, in agreement with their TSG status, all the ING proteins have been involved in p53 pathways such as cell cycle arrest, apoptosis, and senescence. However, the results observed in cellular models must be taken carefully because studies on ING1 knockout mice revealed that under physiological conditions, ING1 functions appear mostly independent of p53 signaling pathways [4,15]. In addition to their involvement in p53 tumor suppressor pathways, ING proteins have been implicated in chromatin remodeling. Indeed, ING proteins are components of histone acetyltransferase (HAT) and histone deacetylase (HDAC) complex. HAT and HDAC are responsible for the level of chromatin compaction. The degree of chromatin compaction regulates its accessibility by transcription factor and thus regulates gene expression. In this regard, a recent study reported that the PHD of ING2 binds with high affinity to the histone 3 trimethylated on lysine 4. This interaction allows the recruitment of Sin3a/HDAC complex to the promoter of genes (coding for proteins involved in cell proliferation), and thus repressed them [30]. Furthermore, ING1 and ING2 proteins have been involved in DNA repair [2,36].

Since ING1 has been identified as a TSG and since ING proteins have been involved in several tumor suppressive pathways, many studies have been

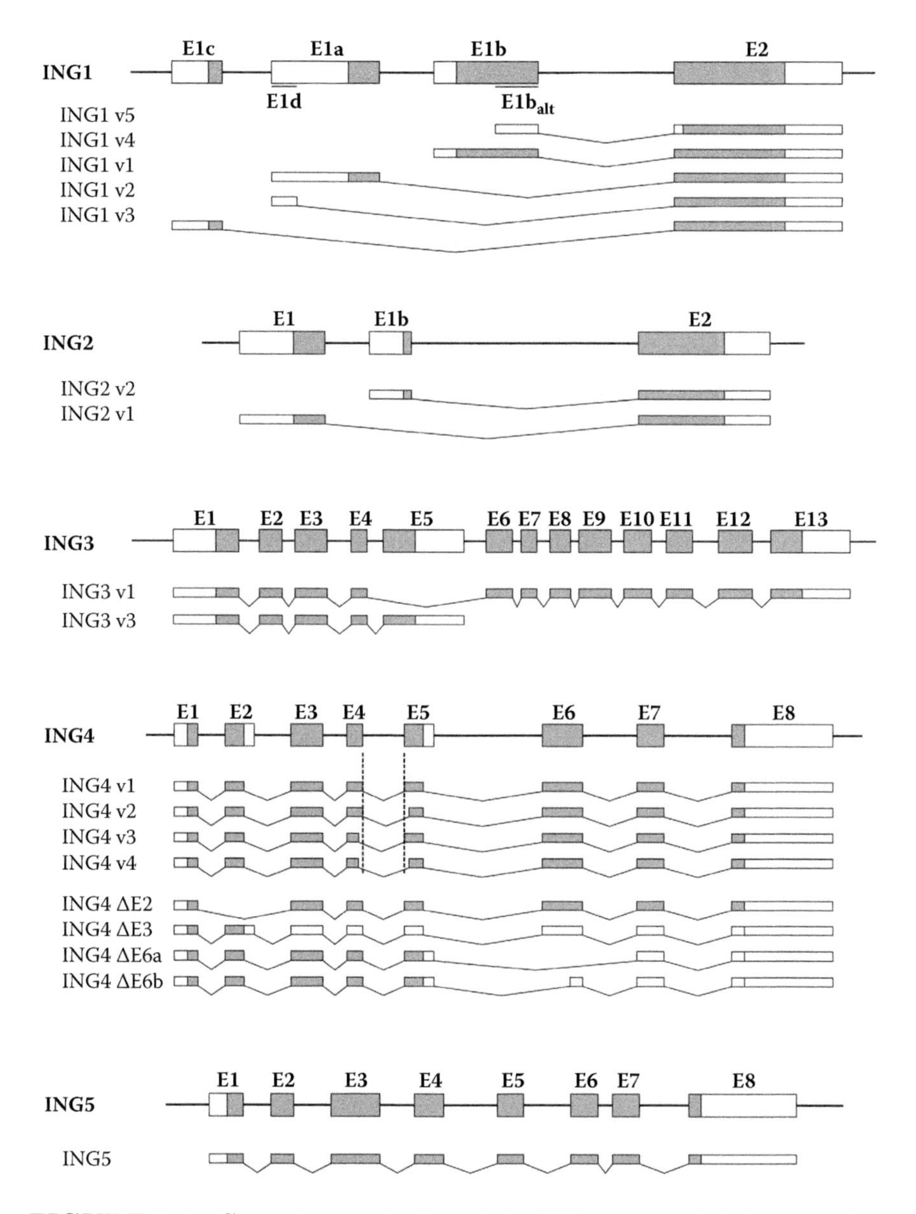

FIGURE 3.6 Genomic organization of the ING genes and their transcripts. ING genes are represented with their different exons (E) and transcript variants are below. Coding and non-coding regions are in gray and white squares, respectively. The names of ING genes and their corresponding variants are written on the left of the figure.

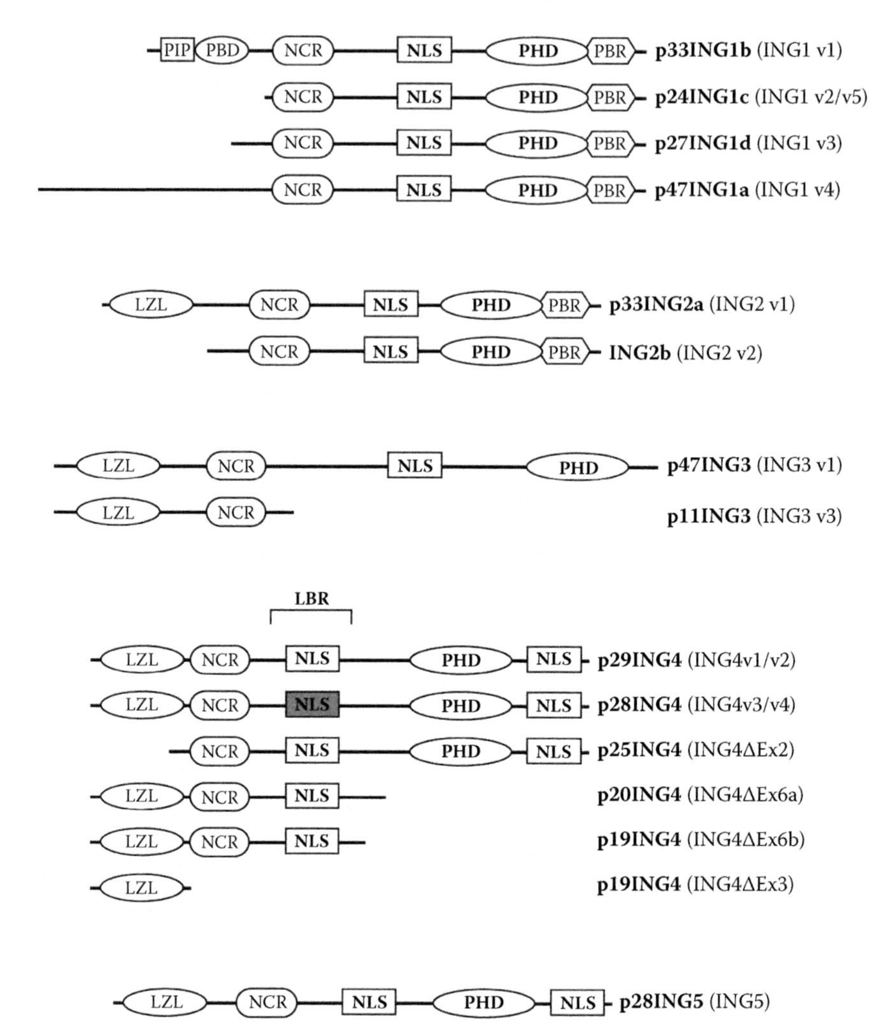

FIGURE 3.7 ING protein structures. Their names and corresponding variants appear on the right. The characteristic domains of the ING protein family are indicated. PIP, PCNA interacting-protein motif; PBD, partial bromo domain; NCR, novel conserved region; NLS, nuclear localization signal; PHD, plant homeodomain; PBR, poly basic region; LZL, leucine zipper-like region; LBR, Liprin α1 binding region. The nonfunctional NLS of ING4v3 and ING4v4 are represented by a gray box.

conducted to investigate ING protein status in tumors [39]. Consistent with its tumor suppressor function, the ING1 knockout mouse model developed B-cell lymphoma with earlier and higher incidence as compared to normal mice. In humans, ING protein inactivation has been reported in many cancer types: brain, breast, stomach, skin (melanoma), blood (lymphoma), etc. [39].

Mutational analysis conducted on ING genes in cancer revealed that mutations are rare events. However, when ING mutations were identified, they were found in the critical domains for their tumor suppressor functions: NLS or PHD [6]. A few studies have also reported that ING protein inactivation could occur by their exclusion from the nucleus [20,35]. However, the large majority of studies reveal that ING proteins are inactivated by the loss of their mRNA expression or stability [39]. Few studies have identified the mechanisms responsible for ING gene repression. For example, in human head and neck cancer, ING1 downregulation is due to the deletion of the gene [12]. In ovarian cancer, ING1 has been reported to be downregulated by hypermethylation of the gene [28]. In most studies, the mechanism(s) responsible for ING loss of expression remain unknown.

3.6 ING Gene Involvement in Cell Migration

As described above, mounting evidence reveals the involvement of the ING gene family in tumor suppressor pathways such as inhibition of cell growth, apoptosis, and senescence. These processes mostly implicate ING protein nuclear functions and their ability to regulate gene expression. Recently, one member of the ING family, ING4, has been shown to have a cytoplasmic function for regulating cell spreading and migration [29]. ING4 was shown to interact and co-localize with liprin α1 in the cytoplasm at the forefront of the leading edges of the protruding membranes. Liprin α1 was previously described as playing a major role in neuron axon guidance and dendrite extension [28]. Such a process allows the extension and migration of the extremities of axons and dendrites to their target and thus shares a common mechanism with those involved in cell migration. Using a cell spreading assay with cells overexpressing ING4, ING4 was reported to significantly delay the ability of cells to spread. Moreover, cells exhibited fewer filopodia and lamellipodia.

Overexpression of liprin α1 alone increased the number of filopodia and lamellipodia per cell, but overexpression of ING4 with liprin α1 abrogated this effect. On the contrary, ING4 knockdown (KD) increased cell spreading, filopodia and lamellipodia formation whereas liprin α1 KD had the opposite effect. Taken together, these results indicated that liprin α1 and ING4 cooperate to regulate cell spreading. Using a Boyden chamber and wound healing assay, ING4 was also shown to inhibit cell migration in a liprin-α1-dependent way. Overall, this study showed that ING4 and liprin α1 cooperate to regulate cell adhesion and migration. Subsequently, three new variants for ING4 have been reported (ING4v2–4) [33].

These variants result from alternative splice donor or acceptor sites between exons 4 and 5. It was shown that the ING4v4 variant exerts a dominant negative effect on ING4v1 cell cycle inhibitor function and also on its ability to suppress cell spreading and migration. Interestingly, a previous study had identified ING4 as a suppressor of contact inhibition loss elicited by MYCN [16]. Loss of contact inhibition is a characteristic of transformed cells. ING4 was identified after a screen in which cells that lost contact inhibition by MYCN overexpression were transfected with a cDNA library. When cells reached confluence, the cells were dividing, except those that expressed a cDNA that codes for a protein that restores contact inhibition. The use of AZT/5-FU allows killing of dividing cells. Thus, only cells that stop dividing at confluence remained alive. Sequencing of the expressed cDNA allows identifying which protein is able to inhibit the loss of contact inhibition.

The fact that ING4 was identified in that screen suggests that ING4 is a tumor suppressor gene. To further test this hypothesis, ING4 status in several cancer cell lines was tested. In seven of the nine cell lines tested, ING4 was harboring a deletion of four amino acids. At this time it has been interpreted as a mutation hotspot. However, this deletion was reinterpreted as the ING4v4 variant recently described [33]. Taken together, these results indicate that ING4 functions, especially the ability to suppress cell spreading and migration, may be regulated by the balance of expression of its different variants. These functional studies involving ING4 in cell spreading and migration conformed to the implication of ING4 in cell migration and invasion in melanoma. A reduced ING4 expression in more than half of the tumor samples and a correlation between ING4 downregulation and the development of metastasis was observed [19]. ING4 was shown to downregulate the metalloproteinases MMP-2 and MMP-9 that degrade components of the extracellular matrix and thus are strongly involved in the invasion and metastasis of malignant tumors.

At the present time, functional studies have only involved ING4 among the ING genes in metastatic processes such as cell spreading, invasion and migration. However, ING1 loss of expression has been reported in breast cancer, and this loss correlates with the metastatic status of these tumors. Further investigations are needed to establish a direct link between ING1 and the ability of tumors to develop metastasis.

References

[1] L. Celli et al. (2006). Evidence of a functional role for interaction between ICAM-1 and nonmuscle (alpha)-actinins in leukocyte diapedesis. *J. Immunol.* 177(6):4113–4121.

[2] K.J. Cheung Jr. et al. (2001). The tumor suppressor candidate p33(ING1) mediates repair of UV-damaged DNA. *Cancer Res.* 61(13):4974–4977.

[3] A.C. Chiang and J. Massague (2008). Molecular basis of metastasis. *New Engl. J. Med.* 359(26):2814–2823.

[4] A.H. Coles et al. (2007). Deletion of p37Ing1 in mice reveals a p53-independent role for ING1 in the suppression of cell proliferation, apoptosis, and tumorigenesis. *Cancer Res.* 67(5):2054–2061.

[5] A.W. Dunah et al. (2005). LAR receptor protein tyrosine phosphatases in the development and maintenance of excitatory synapses. *Nat. Neurosci.* 8(4):458–467.

[6] X. Feng, Y. Hara, and K. Riabowol (2002). Different HATS of the ING1 gene family. *Trends Cell Biol.* 12(11):532–538.

[7] G. Fenteany, P.A. Janmey, and T.P. Stossel (2000). Signaling pathways and cell mechanics involved in wound closure by epithelial cell sheets. *Curr. Biol.* 10(14):831–838.

[8] P. Friedl and E.B. Brocker (2000). The biology of cell locomotion within three-dimensional extracellular matrix. *Cell Mol. Life Sci.* 57(1):41–64.

[9] I. Garkavtsev et al. (1996). Suppression of the novel growth inhibitor p33ING1 promotes neoplastic transformation. *Nat. Genet.* 14(4):415–420.

[10] I. Garkavtsev et al. (1998). The candidate tumour suppressor p33ING1 cooperates with p53 in cell growth control. *Nature* 391(6664): 295–298.

[11] M.H. Ginsberg and J.E. Schwarzbauer (2008). Cell-to-cell contact and extracellular matrix. *Curr. Opin. Cell Biol.* 20(5):492–494.

[12] M. Gunduz et al. (2000). Genomic structure of the human ING1 gene and tumor-specific mutations detected in head and neck squamous cell carcinomas. *Cancer Res.* 60(12):3143–3146.

[13] G.H. He et al. (2005). Phylogenetic analysis of the ING family of PHD finger proteins. *Mol. Biol. Evol.* 22(1):104–116.

[14] J.D. Hood and D.A. Cheresh (2002). Role of integrins in cell invasion and migration. *Nat. Rev. Cancer* 2(2):91–100.

[15] J.V. Kichina et al. (2006). Targeted disruption of the mouse ING1 locus results in reduced body size, hypersensitivity to radiation and elevated incidence of lymphomas. *Oncogene* 25(6):857–866.

[16] S. Kim et al. (2004). A screen for genes that suppress loss of contact inhibition: identification of ING4 as a candidate tumor suppressor gene in human cancer. *Proc. Natl. Acad. Sci. USA* 101(46):16251–16256.

[17] K.W. Kinzler and B. Vogelstein (1998). Landscaping the cancer terrain. *Science* 280(5366):1036–1037.

[18] C. Lengauer, K.W. Kinzler, and B. Vogelstein (1998). Genetic instabilities in human cancers. *Nature* 396(6712):643–649.

[19] J. Li, M. Martinka, and G. Li (2008). Role of ING4 in human melanoma cell migration, invasion and patient survival. *Carcinogenesis* 29(7):1373–1379.

[20] F. Lu et al. (2006). Nuclear ING2 expression is reduced in human cutaneous melanomas. *Br. J. Cancer* 95(1):80–86.

[21] L.M. Machesky (2008). Lamellipodia and filopodia in metastasis and invasion. *FEBS Lett.* 582(14):2102–2111.

[22] C.D. Mathers and D. Loncar (2006). Projections of global mortality and burden of disease from 2002 to 2030. *PLoS Med.* 3(11):e442.

[23] M. Nagashima et al. (2001). DNA damage-inducible gene p33ING2 negatively regulates cell proliferation through acetylation of p53. *Proc. Natl Acad. Sci. USA* 98(17):9671–9676.

[24] M. Nagashima et al. (2003). A novel PHD-finger motif protein, p47ING3, modulates p53-mediated transcription, cell cycle control, and apoptosis. *Oncogene* 22(3):343–350.

[25] K.M. Rieger-Christ et al. (2005). Restoration of plakoglobin expression in bladder carcinoma cell lines suppresses cell migration and tumorigenic potential. *Br. J. Cancer* 92(12):2153–2159.

[26] Y. Roche et al. (2003). Fibrinogen mediates bladder cancer cell migration in an ICAM-1-dependent pathway. *Thromb. Haemost.* 89(6):1089–1097.

[27] A.A. Sablina, P.M. Chumakov, and B.P. Kopnin (2003). Tumor suppressor p53 and its homologue p73 alpha affect cell migration. *J. Biol. Chem.* 278(30):27362–27371.

[28] D.H. Shen et al. (2005). Epigenetic and genetic alterations of p33ING1b in ovarian cancer. *Carcinogenesis* 26(4):855–863.

[29] J.C. Shen et al. (2007). Inhibitor of growth 4 suppresses cell spreading and cell migration by interacting with a novel binding partner, liprin alpha 1. *Cancer Res.* 67(6):2552–2558.

[30] X. Shi et al. (2006). ING2 PHD domain links histone H3 lysine 4 methylation to active gene repression. *Nature* 442(7098):96–99.

[31] Y. Shimada et al. (1998). Cloning of a novel gene (ING1L) homologous to ING1, a candidate tumor suppressor. *Cytogenet. Cell Genet.* 83(3-4):232–235.

[32] M. Shiseki et al. (2003). p29ING4 and p28ING5 bind to p53 and p300, and enhance p53 activity. *Cancer Res.* 63(10):2373–2378.

[33] M. Unoki et al. (2006). Novel splice variants of ING4 and their possible roles in the regulation of cell growth and motility. *J. Biol. Chem.* 281(45):34677–34686.

[34] F.M. Vega and A.J. Ridley (2008). Rho GTPases in cancer cell biology. *FEBS Lett.* 582(14):2093–2101.

[35] D. Vieyra et al. (2003). Altered subcellular localization and low frequency of mutations of ING1 in human brain tumors. *Clin. Cancer Res.* 9(16):5952–5961.

[36] J. Wang, M.Y. Chin, and G. Li (2006). The novel tumor suppressor p33ING2 enhances nucleotide excision repair via inducement of histone H4 acetylation and chromatin relaxation. *Cancer Res.* 66(4):1906–1911.

[37] J. Werr et al. (1998). Beta 1 integrins are critically involved in neutrophil locomotion in extravascular tissue *in vivo*. *J. Exp. Med.* 187(12):2091–2096.

[38] J.C. Yarrow et al. (2004). A high-throughput cell migration assay using scratch wound healing, a comparison of image-based readout methods. *BMC Biotechnol.* 4:21.

[39] D. Ythier et al. (2008). The new tumor suppressor genes ING: genomic structure and status in cancer. *Int. J. Cancer* 123(7):1483–1490.

Part II

Single Cell Migration Modeling

Chapter 4. Coupling of Cytoplasm and Adhesion Dynamics Determines Cell Polarization and Locomotion

Observation of epidermal cells or cell fragments on flat adhesive substrates has revealed two distinct morphological and functional states: a nonmigrating symmetric "unpolarized state" and a migrating asymmetric "polarized state." They are characterized by different spatial distributions and dynamics of important molecular components as F-actin and myosin-II within the cytoplasm, and integrin receptors at the plasma membrane contacting the substratum, thereby inducing so-called focal adhesion complexes. So far, mathematical models have reduced this phenomenon to gradients in regulatory control and signaling molecules or to different mechanics of the polymerizing and contracting actin filament system in different regions of the cell edge.

Here we offer an alternative self-organizational model to reproduce and explain the emergence of both functional states for a certain range of dynamical and kinetic model parameters. We apply an extended version of a two-phase, highly viscous cytoplasmic flow model with variable force balance equations at the moving edge, coupled to a four-state reaction–diffusion–transport system for the bound and unbound integrin receptors beneath the spreading cell or cell fragment. In particular, we use simulations of a simplified 1-D model for a cell fragment of fixed extension to demonstrate characteristic changes in the concentration profiles for actin, myosin, and doubly bound integrin, as they occur during transition from the symmetric stationary state to the polarized migrating state. In the latter case the substratum experiences a low magnitude "pulling force" within the larger front region, and an opposing high-magnitude "disruptive force" at the shorter rear region. Moreover, simulations of the corresponding 2-D model with free boundary show characteristic undulating protrusions and retractions of the cell (fragment) edge, with local accumulation of doubly bound adhesion receptors behind it, combined with a modulated retrograde F-actin flow. Finally, for a stationary model cell (fragment) of symmetric round shape, larger fluctuations in the circumferential protrusion activity and adhesion kinetics can break the radial symmetry and induce a gradual polarization of shape and concentration profiles, leading to continuous migration in the direction of the leading front.

The aim of the chapter is to show how relatively simple laws for the small-scale mechanics and kinetics of participating molecules, responsible for the energy-consuming steps such as filament polymerization, pushing and sliding, binding and pulling on adhesion sites, can be combined into a nonlinearly coupled system of hyperbolic, parabolic, and elliptic differential equations that reproduce the emergent behavior of polarization and migration on the large-scale cell level.

Chapter 5. How Do Cells Move? Mathematical Modeling of Cytoskeleton Dynamics and Cell Migration

We present a novel approach to the modeling of the lamellipodial actin cytoskeleton meshwork. The model is derived from the microscopic description of mechanical properties of filaments and crosslinks and also of the life cycle of crosslinker molecules. The result is a multiphase evolution model for lamellipodia with arbitrary shape that allows us to relate computationally the structure and dynamics of the actin network to the traction forces and shape changes that constitute the amoeboid movement of cells.

Chapter 6. Computational Framework Integrating Cytoskeletal and Adhesion Dynamics for Modeling Cell Motility

Cell migration is a highly integrated process where actin turnover, actomyosin contractility, and adhesion dynamics are all closely interlinked. The computational framework presented here aims to investigate the coupling among these fundamental processes. Two different applications of the model are considered: first its relevance to describe cell migration and second its ability to predict the cell morphologies as observed on patterned substrata. In the model the cell membrane oscillations originating from the interaction between passive hydrostatic pressure and contractility are sufficient to lead to the formation of adhesion spots. Cell contractility then leads to the maturation of these adhesion spots into focal adhesions through integrin recruitment, which reciprocally stimulates reinforcement of the stress fibers. Due to active actin polymerization, which enhances protrusion at the leading edge, the traction force required for cell translocation can be generated. However, if the force is not strong enough, the maturation of the stress fibers allows for redistribution of the forces throughout the cytoskeleton, and the cell can thus recover a new stable shape.

Numerical simulations first performed in the context of unstimulated cell migration (i.e., for a homogeneous and isotropic substratum) show that the model hypotheses are satisfactory to reproduce the main features of fibroblast cell migration as well as the well-known biphasic evolution of the cell migration speed as a function of the adhesion strength. In the context of patterned substrata, the numerical simulations allow us to explain how the forces generated by the stress fibers of the virtual cells are regulated at the adhesion site through feedback mechanisms and how the competing stress fibers can generate an equilibrium state corresponding to a stable cell shape.

Chapter 4

Coupling of Cytoplasm and Adhesion Dynamics Determines Cell Polarization and Locomotion

Wolfgang Alt, Martin Bock, and Christoph Möhl

Contents

4.1 Biology of Cell Polarization and Migration

Cell polarization and migration play a central role in the development and maintenance of tissues in multicellular organisms. During ontogenesis, new tissues are formed by the coordinated division and locomotion of single cells. The polarization of a cell not only defines the direction of migration [30], but also the cell division axis [61] and thus the 3-D structures of tissues, organs, and finally the whole organism.

4.1.1 Asymmetry of Actin Polymerization and Substrate Adhesion

The coordinated development and release of focal adhesions (FAs) is a basic requirement for directed cell migration. Migrating cells feature pronounced adhesion dynamics and a structural polarity with a clearly distinguishable frontal and rear area. Actin polymerization predominates at the cell front, resulting in a protruding lamella in the direction of migration. New focal adhesions composed of clustered protein complexes develop at the lower membrane of the lamella near the leading edge and couple the F-actin network mechanically to extracellular matrix proteins. Simultaneously, the mature FAs residing at the opposed trailing edge are dissolved while myosin-driven contraction of F-actin moves the cell body forward [10,36].

Arising from the process described above, the polarity of the cell can be referred to two structural asymmetries, which are key requirements for effective cell migration: asymmetry of actin polymerization and asymmetry of adhesion strength. The growth of actin filaments has to predominate at the cell front for pushing the leading edge in the direction of migration [58,63]. To move the cell body forward, it must be released from the substrate during contraction by dissolving the FAs at the back (rear release) while the FAs at the cell front must remain stable to provide a mechanical attachment for the contractile machinery pulling the cell body. In the absence of rear release, traction forces could be dominated by adhesion forces and the cell would get stuck [54]. In this regard, the spatial distribution of adhesion strength and actin polymerization defines the direction of migration and could be specifically regulated due to directed cell movement.

4.1.2 Flow of Actin Filaments and Myosin Gradient

The mechanisms underlying the structural polarity of migrating cells are still under discussion, particularly those concerning an intrinsic directionality of the cytoplasm. For example, the finding that asymmetric adhesive structures define polarization of a touching cell [30] suggests that directional flow of the actin cytoskeleton is involved in the process of cell polarization. Moreover,

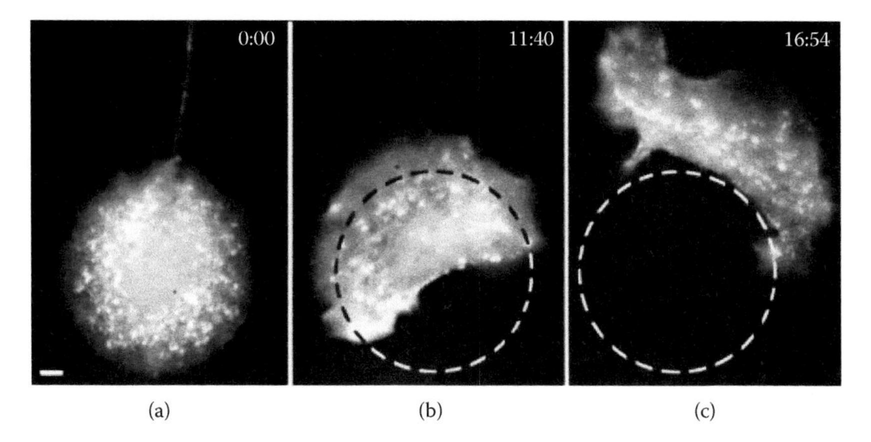

FIGURE 4.1 Cell fragment experiments in which mechanical stress induces a transition from a circular nonmigrating state (a) to a polarized migrating state (b) and (c). The cell fragment was labeled with fluorescent myosin II. Time scale is min:sec and length scale of bar is 2 μm (from [63]).

recent studies on fish keratocytes have revealed that polarization occurs spontaneously and is accompanied by a reorganization of the actin cytoskeleton, which finally leads to cell locomotion [64]. In these experiments, unpolarized solitary cells feature a circular shape with a radially symmetric actin distribution. Although the cells do not move, transient protrusions and retractions appear at the cell edge, and actin flows centripetally with a decreasing flow gradient from the cell edge to the center. In this apparently unstable state, spontaneous symmetry breaking results in a faster inward flow and in an increased concentration of actin at the rear region of the cell. On the opposed front edge, the reduced inward actin flow causes protrusion and the development of a lamellipodium. In this polarized state, the cell starts to migrate while attaining a more or less constant shape.

A similar behavior was observed with cell fragments extracted from the lamellae of fish keratocytes [63]; see Figure 4.1. Because these fragments were lacking most cell organelles and the microtubule system, they were limited mainly to the actin–myosin machinery. They appeared either in an unpolarized, nonmigrating state with homogeneous actin–myosin distribution or a polarized, migrating state with a rising myosin concentration from front to rear. Interestingly, polarization of fragments could be induced by mechanical stimulation, for example, by shear flow or stress release, leading to a transition from the stationary to the migrating state.

Cell polarization is thought to result from the local activation of GTPases of the Rho family [43,54,62]. These proteins are known to regulate actin polymerization, myosin activity, and FA assembly, whereas it remains unclear how the distribution of the GTPases is controlled. Recent experiments have revealed that by inhibition of myosin activating signaling pathways, the cell's ability to polarize is reduced [64]. However, the observations that cells or cell fragments polarize either spontaneously or due to mechanical perturbations raise doubts that biochemical signaling is the primary reason for inducing and maintaining polarity. Experiments focusing on cell mechanics show that direction and strength of locomotion forces are inherently connected with the retrograde F-actin flow [22,53]. This suggests that the mechanical action of myosin—namely to induce cytoplasmic contraction, flow and force transduction at adhesive sites—plays a key role in explaining the ubiquitously observed phenomena of cell polarization and locomotion.

4.2 Previous Models of Cytoplasm and Adhesion Dynamics

The first detailed mathematical model coupling cytoskeleton and adhesion dynamics was developed by Lauffenburger and co-workers almost two decades ago [17]. Here, the contractile actin–myosin network and its interaction with bound adhesion sites is represented as a mechanical system of connected viscoelastic units of generalized Kelvin-Voigt type, which constitute the (three) inner segments as well as a front segment (lamellipod) and a tail segment (uropod) for a rectangular model cell of fixed width and length. A coupled system of reaction–diffusion equations for free and substrate-bound integrins or adhesion receptors on the *dorsal* (lower) and *ventral* (upper) part of the model cell is solved under pseudo-steady-state assumptions. This dynamically provides the number of adhesion bonds in both end segments, lamellipod and uropod, whereby their rupture (dissociation) kinetics exponentially increase with force load onto a bound adhesion site. The resulting nonlinear dynamics for the local (forward or backward) displacement of each viscoelastic unit produces a persistent forward translocation of the whole model cell. However, this is achieved only if a front-tail asymmetry is presupposed, either by exposing more free adhesion receptors or by assuming higher adhesion bond affinity at the front compared to the tail. The authors present a series of results on how the simulated cell migration speed depends on various model parameters as, for example, cytoskeletal contractility or adhesion strength. A later variant of this model presents a more explicit study of adhesion bond disruption kinetics at the rear of the cell and already uses a four-state model for integrin binding to the cytoskeleton and to the substratum [51].

In recent years a series of more elaborate models has been developed, accounting for details of the meanwhile discovered molecular regulation mechanisms for the chemical and mechanical processes, particularly at the free boundary of a moving cell. One model type is based on spatially discrete algorithms (cellular Potts model) using the definition of local energies to determine the protrusion and retraction of boundary elements via the stochastic metropolis rule (e.g., [39,46,47]).

Another class of mechanical cell models takes into account the branching and anisotropy of cytoskeletal actin filaments at various times and in various regions of the cell (see, e.g., [23,28,65]). Besides explicit constructions of an elastically crosslinked network in the leading edge [32,48], biophysical models have been developed to capture different dynamics of Arp2/3-induced branching and myosin-induced contraction of F-actin networks by Mogilner and co-workers [25,33,44]. The last article is devoted to explaining the mentioned polarization experiments with cell fragments (see Figure 4.1) by assuming a different actin–myosin network organization at the rear in comparison to the front region (see Chapter 5).

Based on the "reactive flow" model by Dembo and co-workers [3,12,13], the most elaborate model extension has been presented by Oliver et al. [49]; they use full 3-D, two-phase flow equations with free ventral boundary and with two additional rapidly diffusing messenger concentrations that regulate actin network contractility and (de-)polymerization. Moreover, cell adhesion is modeled by a Navier-slip boundary condition at the substratum, in which only constant adhesive properties are taken into account. The analysis of this complex model, performed in the thin-film limit, is restricted to linear-stability arguments for ruffle generation and to local expansion analyses at the moving tip; there, phenomenological equations for boundary mass fluxes are considered without specifying the types of molecular mechanisms for tip protrusion. Finally, quantitative estimations for the pseudopod protrusion and cell translocation speed under various sublimit assumptions are given, which turn out to be consistent with observed values, particularly for osteoblasts, although no numerical simulations are given that would reinforce the analytic results. A similar thin-film approximation of the 2-D equations under incompressibility assumptions for a "viscous polar gel" was used in [34] to derive explicit expressions for the advancing speed of a cell lamella, again by predefining its polarity.

Except for the last one, the mentioned models do not explicitly quantify the varying force field, which is applied by the cytoskeleton onto the substrate covered by a migrating cell and which has been approximately reconstructed by different inverse methods from experimental assays of cells, for example, moving on flexible substrata [6,15,38,40,56]. A first model implementing force transduction to the substrate was proposed by Gracheva and Othmer [24] by specifying a spatially 1-D system of viscoelastic equations for cytoskeleton dynamics, whose polymerization, contractility, and adhesive binding are regulated by signaling molecules. However, they make a pseudo-steady-state assumption for the binding kinetics of myosin polymers to actin filaments, and of transmembrane integrin proteins to the substratum. Moreover, they impose artificially defined gradients from tail to front of certain regulatory proteins in order to stimulate polarized cell translocation.

Recently, adhesion kinetics have also been implemented into an extended cytoplasm flow model [4,5] describing the F-actin dynamics in an annular domain and its coupling to lamellipodial protrusions and retractions [59].

Force-dependent maturation of FA complexes and active polymerization of F-actin enable the simulation model to reproduce characteristics of fibroblast shape deformation and translocation (see Chapter 6).

In an earlier publication we have presented another extension of the basic two-phase flow model for the 2-D viscous cytoplasm dynamics [2] by coupling the constitutive hyperbolic–elliptic equations to a system of four reaction–diffusion–transport equations for the integrins beneath the cell or cell fragment [35]. Here we propose a generalized continuum model in which we couple cytoplasm and adhesion dynamics with mechanical tension and transport of the plasma membrane to reproduce spontaneous and induced cell polarization leading to migration, with assembly of adhesion sites at the cell front, and adhesion release at the cell's rear end. Moreover, the model exhibits typical features of migrating cells as protrusion–retraction cycles, rearward actin flow, pulling forces at the front, and a concentration of disruptive forces at the rear. In the model, the cytoplasm is described as a viscous and contractile fluid of polymers representing the actin cytoskeleton interpenetrated by an aqueous phase. This actin filament network is preferentially assembled at the cell edge, and can be contracted by crosslinking with diffusing myosin oligomers. The moving actin filaments then couple to transmembrane adhesion proteins that are freely diffusing in the membrane or bound to the substrate on the extracellular domain. Thus, the cytoskeleton mechanically connects to the substrate through dynamic binding processes and results in force transduction and finally cell locomotion.

Throughout our model presentation we rely on continuum descriptions in which macroscopic mass and momentum laws are combined with mesoscopic submodels for fast molecular kinetics due to adequate pseudo-steady-state assumptions.

4.3 Two-Phase Flow Model for Cytoplasm Coupled to Reaction, Diffusion, and Transport for Myosin-II and Integrin Proteins

For simplicity, we restrict our model derivation and analysis to a flat 2-D geometry, so that cells or cell fragments are assumed to be homogeneously spread on the substrate without considerable change in cell height. Thereby, 3-D effects around the cell nucleus (e.g., due to cell rolling) or along the ventral (upper) plasma membrane are neglected. To reproduce the main biophysical mechanisms and biochemical processes that enable a cell to polarize and translocate on an adhesive substratum, we nevertheless distinguish between the cytoplasm and the exterior plasma membrane. On the one hand,

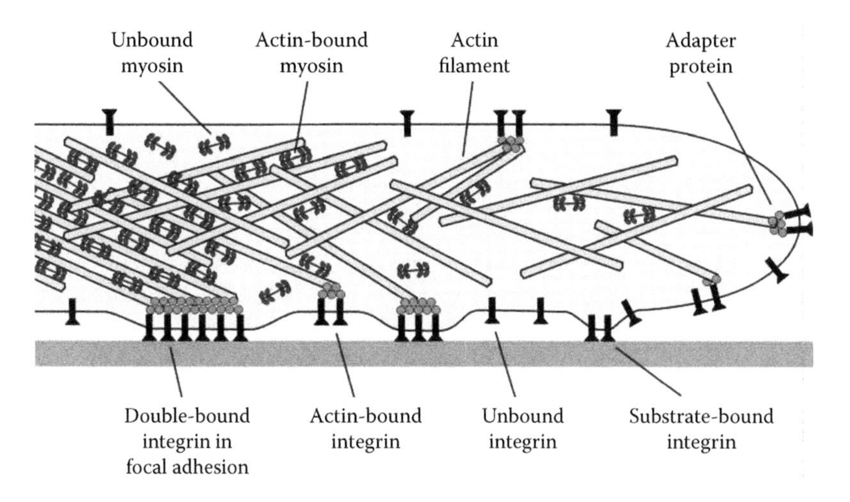

FIGURE 4.2 Schematic longitudinal section through a cell fragment (or a cell lamella) of constant height with involved proteins. For further explanation and model derivation see the following sections.

F-actin assembly, myosin kinetics, and viscous network flow take place within the cytoplasmic interior of the cell. On the other hand, integrin binding to substrate or cytoskeleton and its transport or diffusion are confined to the dorsal (lower) plasma membrane, where forces are transduced to or generated at the cell periphery, leading to tip protrusion or retraction. Thus, the moving cell (fragment) is simply represented by a time-dependent connected domain $\Omega(t) \subset \mathbb{R}^2$, over which the cytoplasmic volume extends with fixed constant height, and any ruffles or blebs on top of the cell are neglected. Rather, we assume that the cell dynamics are completely determined by a flat cytoskeleton sheet of F-actin network; see Figure 4.2. This crosslinked filament phase with volume fraction $\theta(t, \mathbf{x}), \mathbf{x} \in \Omega(t)$ is of constant thickness and connected to the upper and lower plasma membrane in such a way that the cytosol, i.e., the solvent phase with volume fraction $(1 - \theta(t, \mathbf{x}))$, is also confined to the same volume space.

By suitable model simplifications, we aim to capture the self-organizational power of the cytoplasm as a two-phase fluid coupled with the reaction, transport, and diffusion of a series of chemical ingredients. We explicitly model the kinetics and dynamics of only those F-actin associated proteins that induce the main biomechanical processes of force generation and transduction to the substratum, namely myosin oligomers and transmembrane integrin proteins. All other regulatory proteins, such as RAC, Rho, or the branching protein Arp2/3 or smaller substrate molecules (e.g., monomeric

G-actin) are assumed to rapidly diffuse within the cytosol under fast regeneration so that they attain constant reservoir concentrations serving as parameters in the model.

4.3.1 Mass Balance and Flow Equations

In the chosen 2-D continuum description, cell deformation and translocation with respect to the fixed substratum is represented by movement of the cell edge $\Gamma(t) = \partial\Omega(t)$, which, however, is induced by three distinguished mass flows with potentially different transport velocity fields on $\overline{\Omega}$, namely the mean velocities $\mathbf{v}$ of the F-actin cytoskeleton (θ) and $\mathbf{w}$ of the aqueous cytosol ($1-\theta$) plus the transport velocity $\mathbf{u}$ of the lower dorsal plasma membrane. Thereby both $\mathbf{v}$ and $\mathbf{w}$ are averaged over the constant cell height. For the membrane area flux $\mathbf{u}$ we make the simplifying assumption that the lipid bilayer between the flat substratum and an adhering cell is always stretched out to its maximal extension with constant area concentration of the lipid-protein mixture, thus constituting an incompressible 2-D Newtonian fluid. Then, freely diffusing transmembrane integrin proteins are additionally transported by the membrane velocity $\mathbf{u}$, myosin oligomers by the cytosol velocity $\mathbf{w}$, and F-actin bound myosin-II motor molecules by the cytoskeleton velocity $\mathbf{v}$. The detailed mass balance equations are discussed in the following paragraphs.

4.3.1.1 Mass Conservation for Cytoskeleton and Cytosol Phases

Cytoskeleton and cytosol in the flat 2-D geometry can experience counteracting flows due to local contraction, assembly, or relaxation of the F-actin network, while the bulk cytoplasmic fluid with constant volume fraction 1 can be assumed to be incompressible, at least in the range of occurring pressure differences $\lesssim$ kPa. Compare the analogous situation of a water-filled sponge that is internally condensing without changing its shape. Then, because of the fixed height assumption, the total 2-D volume flux $\mathbf{W} = \theta\mathbf{v} + (1-\theta)\mathbf{w}$ is divergence-free at any time t, yielding the first local mass conservation law

$$\nabla \cdot (\theta\mathbf{v} + (1-\theta)\mathbf{w}) = 0 \tag{4.1}$$

Therefore, the total cell volume (i.e., the 2-D cell area) is also conserved over time, so that the "bulk" fluid moves together with the free boundary $\Gamma(t)$, meaning that the total volume flux $\mathbf{W}$ has to fulfill the natural free boundary condition

$$\nu_\Gamma \cdot \mathbf{W} = \dot{\Gamma} \tag{4.2}$$

with ν_Γ denoting the exterior normal of Γ and $\dot{\Gamma}$ quantifying the normal speed of the cell edge.

In addition to possible convection, the F-actin network can locally be assembled by filament polymerization from the pool of monomeric G-actin within the cytosol, and it can be disassembled by the reverse process of severing or depolymerization. This mass exchange between the two phases is given by a second conservation law, which for the filament volume fraction θ can be written as

$$\partial_t \theta + \nabla \cdot (\theta \mathbf{v}) = R(\theta) \tag{4.3}$$

with a *net assembly rate* $R(\theta)$ to be modeled. According to our simplifying assumption, the dependence on G-actin, Arp2/3, and possible regulatory proteins enters only via constant parameters:

$$R(\theta) = (k_{on} a_g - k_{off}) B(\theta) - r\theta \tag{4.4}$$

with concentration of globular actin a_g, polymerization rate k_{on}, and depolymerization rate k_{off} at the filaments' plus (barbed) ends, as well as another lumped disassembly rate r (see, e.g., [42]). Due to relatively fast nucleation and capping of actin filaments, the *relative number B of barbed ends* is assumed to stay in pseudo-equilibrium with the local F-actin concentration $a = \theta \cdot a_{\max}$, where we suppose a maximal condensation of actin filaments in the order of $a_{\max} = 800\,\mu\text{M}$. Following [42] we write

$$B(\theta) = \frac{1}{\omega}\left(\varepsilon + \nu_0 \frac{\theta}{K_a / a_{\max} + \theta}\right) \tag{4.5}$$

with capping rate ω, a basic nucleation rate ε, and an induced branching rate $\nu_0 = \nu_a \text{Arp0}$, proportional to the concentration Arp0 [μM] of activated Arp2/3, together with a half-saturation concentration K_a for its primary actin binding site.

4.3.1.2 Reaction–Transport–Diffusion Equations for Myosin Oligomers

Myosin-II oligomers are the most important actin binding proteins that are responsible for the generation of contractile forces within crosslinked F-actin networks. Thus, their spatial distribution within a polarizing or moving cell plays a key role in the cytoplasm dynamics. In a most simple way we only distinguish between freely diffusing myosin oligomers (m_f) and those that are bound to cytoskeletal actin filaments (m_b) and, therefore, are convected with velocity $\mathbf{v}$. Because free myosin tetramers first have to attach to the actin network at one binding site, we consider this bimolecular reaction (m_f with a) as the rate-limiting step. Correspondingly we assume that the relatively faster processes of double binding (m_b with a) and power stroke formation are in a pseudo-steady state. Thereby the contractile stress $\psi = \psi_0 m_b \theta$ with $\theta = a/a_{\max}$ in the network is determined; see Equation (4.25) in the following section. Finally, we suppose a constant diffusivity D_m for free myosin oligomers, embedded into the cytosol flow $\mathbf{w}$. Under these assumptions the local mass balance equations for the corresponding myosin concentrations are:

$$\partial_t m_f = \nabla \cdot (D_m \nabla m_f - m_f \mathbf{w}) - \alpha_m \cdot a \cdot m_f + \delta_m(a) \cdot m_b \tag{4.6}$$

$$\partial_t m_b = -\nabla \cdot (m_b \mathbf{v}) + \alpha_m \cdot a \cdot m_f - \delta_m(a) \cdot m_b \tag{4.7}$$

For the association rate α_m we assume a constant parameter, whereas the dissociation rate is supposed to increase quadratically with increasing network concentration a because of steric inhibition or competition for binding sites:

$$\delta_m(a) = \delta_m^0 \left(1 + \frac{a^2}{a_{opt}^2}\right) = \delta_m^0 \left(1 + \frac{\theta^2}{\theta_{opt}^2}\right) \tag{4.8}$$

with an optimal actin network concentration $a_{opt} = \mathrm{a}_{\max}\theta_{opt}$. Moreover, in the case of fast diffusion and no transport, the constant equilibrium concentrations m_f^* and m_b^* satisfy the equality

$$m_b^* = \frac{\alpha_m a}{\delta_m(a)} m_f^* \tag{4.9}$$

so that the resulting contractile stress $\psi(\theta)$ becomes optimal for $\theta = \theta_{opt}$.

4.3.1.3 Mass Conserving Flow of Dorsal Plasma Membrane

As mentioned before, we will assume that the dorsal plasma membrane beneath an adhering cell (fragment) is stretched and that small fluctuations can be neglected. On the contrary top side, the ventral plasma membrane usually shows extensive wrinkles or folds, which most probably are induced by the contractile action of the cytoskeleton itself, indicating that the plasma membrane tip at the cell edge stays under a positive tension τ^Γ. Before discussing the corresponding dynamics of the dorsal membrane, we state the two extreme possibilities, namely, whether or not the membrane moves together with the cell:

1. *Membrane sticking to substrate:* During cell translocation over the substratum, the whole dorsal membrane stays fixed and no slip can occur due to relatively strong interaction forces with the substratum, for example, via the glycocalyx, so that $\mathbf{u} \equiv 0$.
2. *Membrane sticking to cell edges:* The dorsal membrane is pulled along the substratum due to cytoplasm protrusions at the leading front, but with no slip around the lamellar tips due to strong membrane curvature. In this way the membrane area flux satisfies the mass conservation law together with a boundary condition analogous to Equation (4.2), namely,

$$\nabla \cdot \mathbf{u} = 0 \text{ on } \Omega(t) \qquad \nu_\Gamma \cdot \mathbf{u} = \dot{\Gamma} \text{ on } \Gamma(t) \tag{4.10}$$

In the latter case, the membrane can slip over the substratum, thereby experiencing a finite frictional drag force, see Equation (4.40) below, whereas in the first case this force is infinitely large to suppress slipping—but then the membrane has to slip around the tips at moving cell edges. Clearly, a certain mixture of both possibilities is physically realizable but not considered here.

Moreover, frictional drag onto the dorsal membrane can also occur at its cytoplasmic side. Whereas the relative flow of cytosol will have negligible effects, the horizontal F-actin flow develops a vertical flow profile. This profile depends on the amount of "Navier slip" in a cortical layer on top of the dorsal membrane and emerges due to frictional forces between actin filaments and membrane proteins. In analogy with the more explicit thin-film approximation [49] we only consider the averaged profile velocity $\mathbf{v}$. In this way we can introduce a simplifying *cortical slip parameter*, $0 < \kappa < 1$, so that the effective relative velocity between cytoskeleton and dorsal membrane is reduced to $\kappa(\mathbf{v} - \mathbf{u})$. Thus, the resulting *effective velocity of cortical F-actin* in a thin layer above the dorsal membrane and substratum is given by

$$\mathbf{v}_c = \kappa\mathbf{v} + (1 - \kappa)\mathbf{u} \tag{4.11}$$

Here the factor κ, which clearly depends on the viscous shear properties of the cytoskeleton, should be larger than 0.5 to reflect observations of a generally slippy behavior (see, e.g., [22,29]). Moreover, we suppose that the vertical profile, thus $\mathbf{v}_c$, is not changed if some of the cortical actin filaments are (transiently) bound to integrins in the dorsal membrane (see Figure 4.3 and the following paragraph): either those bound integrins or integrin complexes are passively pulled through the lipid–protein bilayer with relative velocity $\mathbf{v}_c - \mathbf{u}$ or, in case of substrate-fixed adhesion bonds, with relative velocity $-\mathbf{u}$. The whole F-actin network above such a bond is slowed down by a certain local frictional force per adhesion site

$$\mathbf{F}_c = \Phi_0\theta\mathbf{v}_c \tag{4.12}$$

entering into the corresponding force balance law; see Equation (4.23). In any case, the frictional drag onto the dorsal membrane, induced by the relative motion of singular integrin complexes, will be neglected in our model.

4.3.1.4 Reaction–Transport–Diffusion Equations for Membrane Integrins

The mechanical connection between the actin cytoskeleton and extracellular matrix proteins is provided by transmembrane integrins appearing in four different states [35,51]; see Figure 4.3.

In the freely diffusing state f, integrins are neither coupled to the substrate nor to the actin cytoskeleton and move according to a simple diffusion-transport law within the dorsal membrane. These integrins can change their state by binding to the actin cytoskeleton (a) or to the substrate (s). In state a, the integrins move with the cortical F-actin velocity $\mathbf{v}_c$, whereas integrins in state s remain stationary with respect to the substrate. Actin- or substrate-bound integrins can switch back into the freely diffusing state f by unbinding,

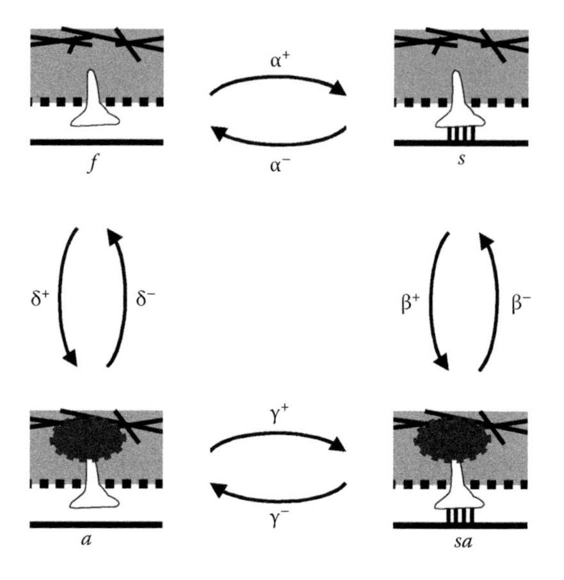

FIGURE 4.3 Scheme of four states of the transmembrane adhesion protein integrin. (f): unbound and freely diffusing within the dorsal membrane; (s): bound to the substrate; (a): bound to the actin network and thus moving with the actin flow; (sa): bound to actin and substrate (force-transducing state). Integrins can switch between these states due to reversible binding kinetics with binding/unbinding rates α, β, γ, and δ.

or into the state sa by coupling to the cytoskeleton and the extracellular matrix. Cell adhesion occurs only in this double bound state s representing a focal adhesion (FA), where the frictional force $\mathbf{F}_c$ (Equation (4.12)) of the moving actin network is transduced to the substrate.

The concentration of integrins $c_{\#}$ in the different states is described by the following coupled system of differential equations consisting of terms for spatial movement and binding kinetics.

$$\partial_t c_f = \nabla \cdot (D_f \nabla c_f - c_f \mathbf{u}) + \alpha^- c_s + \delta^- c_a - \left(\delta_0^+ a + \alpha^+\right) c_f \tag{4.13}$$

$$\partial_t c_a = -\nabla \cdot (c_a \mathbf{v}_c) + \delta_0^+ a c_f - (\delta^- + \gamma^+) c_a + \gamma_0^- \exp(\rho_\gamma |\mathbf{F}_c|) c_{sa} \tag{4.14}$$

$$\partial_t c_s = \alpha^+ c_f - \left(\alpha^- + \beta_0^+ a\right) c_s + \beta_0^- \exp(\rho_\beta |\mathbf{F}_c|) c_{sa} \tag{4.15}$$

$$\partial_t c_{sa} = \gamma^+ c_a + \beta_0^+ a c_s - \left(\beta_0^- \exp(\rho_\beta |\mathbf{F}_c|) + \gamma_0^- \exp(\rho_\gamma |\mathbf{F}_c|)\right) c_{sa} \tag{4.16}$$

The bonds of the force-transducing integrins (c_{sa}) can break when they experience a mechanical stress by the cytoskeleton, in our model given by the modulus of the force vector $\mathbf{F}_c$ defined in Equation (4.12). Then, according to the theory of Bell [7,57], the dissociation rates γ^- and β^- depend exponentially on the mechanical load $|\mathbf{F}_c|$; see Equations (4.14–4.16) above, with γ_0^- and β_0^- describing the basic dissociation rates without load and the exponential coefficients $\rho_{\#} = \zeta_{\#}/k_B T$ measuring potentially different rupture rates from the substrate or cytoskeleton binding site, respectively.

4.3.1.5 Mass Flux Conditions at Free Boundary

In addition to the already-mentioned bulk flux conditions of Equations (4.2) and (4.11) related to the normal speed of the cell edge $\Gamma(t)$, we have to impose compatible boundary conditions onto the concentrations of those molecular species that are not fixed to the substratum. For the two parabolic diffusion equations (Equations (4.6) and (4.13)) we impose, respectively, natural zero-flux boundary conditions onto freely diffusing myosin (m_f) and fixed Dirichlet conditions ($c_f = c_f^0$) onto freely diffusing integrin. Thereby we suppose a constant reservoir of fresh adhesion receptors expressed in the upper ventral membrane and diffusing (or eventually being transported) from there around the lamellar tip into the lower dorsal part of the membrane.

The F-actin flux $\theta \mathbf{v}$ cannot leave the cell, meaning that on the free boundary Γ the *relative inward normal F-actin velocity* always has to satisfy the inequality:

$$V = \dot{\Gamma} - \nu_\Gamma \cdot \mathbf{v} \geq 0 \tag{4.17}$$

If the strong inequality $V > 0$ holds at a certain boundary point, then two different modeling situations may arise for the cell edge:

1. *No-stick condition at the lamellar tip:* Local disruption of the actin network from the edge is allowed (e.g., under suitable load conditions [35]) so there is no new F-actin production directly at the plasma membrane edge. Therefore we have to impose zero Dirichlet condition for the F-actin concentration:

$$\theta = 0 \quad \text{if } V > 0 \text{ holds at a non-sticky point } \mathbf{x} \in \Gamma \tag{4.18}$$

2. *Network sticking to the lamellar tip:* As indicated in Figure 4.4, active polymerization of actin filaments directly at the plasma membrane is allowed either at fluctuating free filament ends touching the tip membrane [43] or at filaments that are bound to clamp-motor proteins anchored in the tip membrane [16]. Both cases could occur simultaneously in a local region, but whether active polymerization with inward mass flux $\theta V > 0$ can take place depends on local force balance conditions; see Section 4.3.2 and Equation (4.36).

Finally, also the two hyperbolic Equations (4.7) and (4.14) require zero influx conditions, so that for the transported concentrations m_b and c_a we have to impose zero Dirichlet conditions only in cases of $V > 0$ and $V_c > 0$, respectively. In the latter case the inward normal velocity of $\mathbf{v}_c$ is defined in analogy with Equation (4.17), noting that under the modeling hypothesis of Equation (4.10) we have $V_c = \kappa V$.

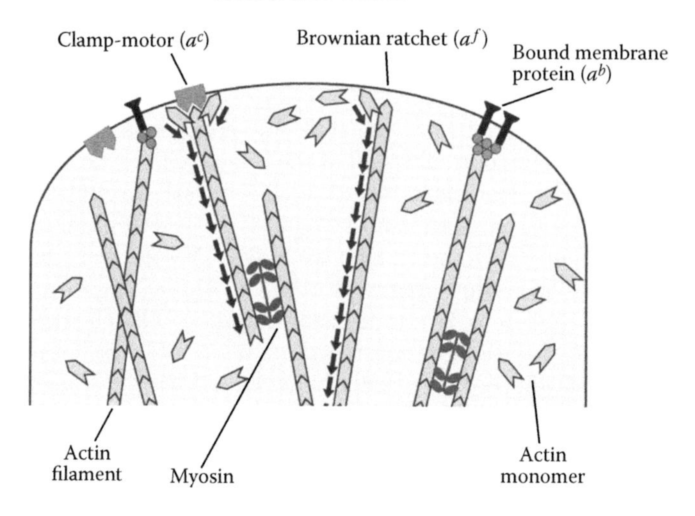

FIGURE 4.4 Schematic top view onto the lamellar tip of a cell (fragment) showing three possibilities for actin filament ends to interact with the free plasma membrane: anchoring at a membrane protein, as integrin; freely fluctuating with polymerization by a "Brownian ratchet" mechanism; binding to a "clamp-motor" protein, like the WASP-complex, with induced polymerization. The resulting local pseudo-equilibrium F-actin concentrations are a^b, a^f, and a^c respectively. See Equations (4.31) and (4.33) for a quantitative model.

4.3.2 Force Balance Equations

4.3.2.1 Two-Phase Flow Equations for Cytoskeleton and Cytosol

Because cell movement usually occurs on a time scale of minutes and cell sizes usually are on the order of tens of microns, already the cytosol, with its consistency similar to an aqueous viscous fluid, has a low Reynolds number. This is also true for the cytoskeleton with consistency similar to a dense viscoelastic gel. The viscoelastic properties of the cytoplasm have been experimentally studied [18,19] and also mathematically modeled [11,49], whereby the simple model of a Maxwell fluid seems to serve as a good description. This means that elastic properties are mainly effective on a shorter time scale of seconds and dominated by viscous effects on a medium scale of minutes. Moreover, in contrast to a passive elastic material, the cytoplasm primarily stays under active contractile stress as exerted by myosin-II oligomers (on the order of kPa), which is able to break weaker crosslinks of the F-actin network, for example, by α-actinin or filamin, so that the contraction forces are equilibrated only by viscous and frictional forces.

Under these assumptions, the highly viscous two-phase *creeping flow model* for cytoplasm, originally proposed 25 years ago by Dembo et al. [14] and extensively applied (see, e.g., [1,26]) can be condensed into a pseudo-stationary linear elliptic system for the *mean F-actin velocity* $\mathbf{v}$ and the *effective hydrostatic pressure* p on $\Omega(t)$:

$$\nabla \cdot \mu\theta\tilde{\nabla}\mathbf{v} + \nabla(S(\theta, m_b) - p) = \Phi(\theta, c_{sa})\mathbf{v}_c \tag{4.19}$$

$$\nabla p = \varphi\theta(\mathbf{v} - \mathbf{w}) \tag{4.20}$$

Here the generalized (elliptic) Stokes equation (4.19) involves the effective stress $S(\theta, m_b)$ as defined in Equation (4.24) and the symmetrized displacement rate $\tilde{\nabla}\mathbf{v}$ (see, e.g., [35]). Moreover, $\mathbf{v}_c$ is defined in Equation (4.11), and μ and Φ denote the coefficients of bulk viscosity and substratum friction; see Equation (4.23). Because of negligible cytosol viscosity, the simple Darcy law, Equation (4.20), suffices as a model description, where the coefficient φ measures the internal two-phase flow friction. From the last equation one explicitly solves for the cytosol velocity $\mathbf{w} = \mathbf{v} - (1/\varphi\theta)\nabla p$, so that the total volume flux is

$$\mathbf{W} = \mathbf{v} + (1 - \theta)(\mathbf{w} - \mathbf{v}) = \mathbf{v} - \frac{1 - \theta}{\varphi\theta}\nabla p \tag{4.21}$$

Insertion into the mass-balance Equation (4.1) then yields the generalized (elliptic) Laplace equation:

$$\nabla \cdot \frac{1 - \theta}{\varphi\theta}\nabla p = \nabla \cdot \mathbf{v} \tag{4.22}$$

Notice that here the ellipticity degenerates for marginal volume fractions $0 < \theta < 1$, which can be relevant in cases where $\theta \to 0$ at boundary points of cytoskeleton disruption; see Equation (4.18).

We remark that together with the hyperbolic mass balance Equation (4.3), the linear elliptic system (Equations (4.19) and (4.22)) constitutes generalized pseudo-stationary Navier-Stokes equations for the F-actin network as a compressible, highly viscous, and reactive fluid. It is mutually coupled to the mass concentrations in Equations (4.7) and (4.16) via a myosin-mediated contractile stress term appearing in $S(\theta, m_b)$ (see Equation (4.24)) and an adhesion-mediated friction coefficient

$$\Phi(\theta, c_{sa}) = \Phi_0 c_{sa}\theta \tag{4.23}$$

In this way the right-hand side of Equation (4.19) reads $\mathbf{F}_v = \Phi\mathbf{v}_c = c_{sa}\mathbf{F}_c$, with the frictional force $\mathbf{F}_c$ per doubly bound integrin defined in Equation (4.12).

Tracing these model equations back to their derivation [2,12,35] provides us not only with precise biophysical conditions on $\mathbf{v}$ and p at the moving boundary Γ (see next paragraph), but also with genuine nonlinear parameter functions based on thermodynamical reasoning at the molecular scale: The function S in Equation (4.19) represents the effective stress in the network phase, which is induced by molecular interactions between the F-actin filaments and by thermic interactions with solvent molecules in the cytosol. S is generally expressed as the weighted negative sum $S = -\theta P_a - (1-\theta) P_s$ of corresponding pressures P_a and P_s that are applied to any network volume element and any cytosol element, respectively. Passive elastic stresses can be neglected because these are already relaxed on the creeping flow time scale. When finding a free binding site on filaments, previously single bound myosin-II tetramers exert power stroke motor forces with their free myosin heads (cf. Figures 4.2 and 4.4). Thus, an attractive stress $-P_a = \psi_0 m_b$ is applied, where the simple coefficient ψ_0 comprises binding affinity, power stroke probability, and the mean applied force per stroke. On the other hand, standard Gibbs free energy arguments suggest a molecular *solvent pressure* $P_s = -\sigma_0 \ln(1-\theta)/(1-\theta)$; see [2]. Then we arrive at expressions for the

Effective stress: $$S = S(\theta, m_b) = \psi(\theta, m_b) - \sigma(\theta) \tag{4.24}$$

Contractile stress: $$\psi(\theta, m_b) = \psi_0 \theta m_b \tag{4.25}$$

Swelling pressure: $$\sigma(\theta) = \sigma_0 |\ln(1-\theta)| \tag{4.26}$$

See Figure 4.5 for corresponding plots with $m_b = m_b^*$ in Equation (4.9). We emphasize that the effective stress function in Equation (4.24) sums up the contributions from the cytoskeleton and the cytosol, so that both phases are

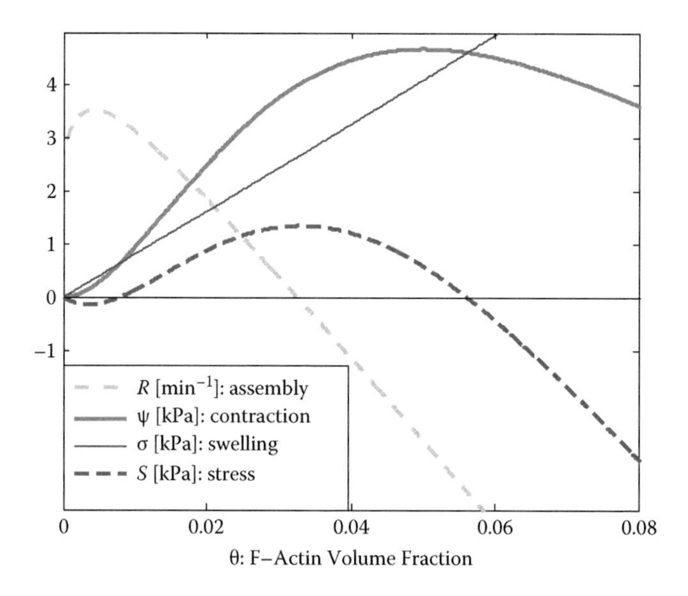

FIGURE 4.5 Plots of F-actin model functions. Dashed curves: net assembly rate $R(\theta)$, Equation (4.4) and effective stress $S(\theta) = \psi(\theta) - \sigma(\theta)$, Equation (4.24). Continuous curves: contractile stress ψ defined in Equation (4.41) and swelling pressure σ, Equation (4.26).

not separated anymore. However, when deriving the corresponding natural boundary conditions for the elliptic problem, the two phases must be considered separately again.

4.3.2.2 Stress and Pressure Balance Conditions at Free Boundary

Recall that according to Equation (4.2) the cell edge $\Gamma(t)$ always moves together with the normal component of the total cytoplasmic flux $\mathbf{W}$ as quantified in Equation (4.21). Then, from Equation (4.17) the relative inward normal F-actin flux also is determined as proportional to the inward normal pressure gradient on $\Gamma(t)$

$$\theta V = \theta(\dot{\Gamma} - \nu_\Gamma \cdot \mathbf{v}) = -\frac{1-\theta}{\varphi}\nu_\Gamma \cdot \nabla p \tag{4.27}$$

This explicit linear relation of free boundary speed, normal F-actin velocity, and pressure gradient serves as an extra free boundary condition in addition to the set of boundary conditions that are necessary for uniquely solving the linear elliptic system (Equations (4.19) and (4.22)) on the given domain $\Omega(t)$. These depend not only on the particular model choice of boundary pressure functions at the cell edge membrane, but also on the outcome of the F-actin flow requirement $V \geq 0$ in Equation (4.17).

Using the general derivations in [2], Equations (58) through (62), one obtains separate *pressure balance* conditions for each of the two phases at all free boundary points of $\Gamma(t)$ satisfying $V > 0$:

Cytoskeleton:
$$-\nu_\Gamma \cdot \mathbb{T}_a \cdot \nu_\Gamma + \theta p = \theta P_a^\Gamma \tag{4.28}$$

Cytosol:
$$(1-\theta)p + \sigma(\theta) = (1-\theta)P_s^\Gamma \tag{4.29}$$

with the intrinsic stress tensor $\mathbb{T}_a = \mu\theta\widetilde{\nabla}\mathbf{v} + \psi(\theta, m_b)\mathbb{I}$. These equations mean that at those parts of the tip plasma membrane that are exposed to the cytoskeleton network (θ) or to the cytosol ($1-\theta$), respectively, the sum of internal pressures is in balance with a certain *boundary cytoskeleton pressure* P_a^Γ or *boundary cytosol pressure* P_s^Γ at each volume element. Modeling expressions for these pressures are given in the following paragraph.

Summing up both pressure balance Equations (4.28) and (4.29) yields the general *Neumann-type condition* for $\mathbf{v}$ on $\Gamma(t)$

$$\nu_\Gamma \cdot \mathbb{T}_a \cdot \nu_\Gamma - (\sigma(\theta) + p) + \theta P_a^\Gamma + (1-\theta)P_s^\Gamma = 0 \tag{4.30}$$

which is now valid without restriction, see [2] Equation (58); that is, also at boundary points where the condition $V = 0$ holds. However, then the Dirichlet-type boundary condition, Equation (4.29), for p must be replaced by a corresponding inequality, and insertion of $V = 0$ into Equation (4.27) provides a *Neumann condition* for p instead. In particular, we have $\nu_\Gamma \cdot \nabla p = 0$, which also holds in case of disruption with $V > 0$ and $\theta = 0$.

Thus, in any constellation of the mentioned conditions on $\Gamma(t)$, for fixed time t and given profiles of all protein concentrations, we obtain a well-posed elliptic boundary value problem for $\mathbf{v}$ and p to be solved on $\Omega(t)$ (eventually by numeric iteration and/or conjugate gradient method; cf. [35]). Finally, the normal speed $\dot{\Gamma}$ of the free boundary can be calculated using Equation (4.27). This is also possible in the case of network disruption at some boundary point $\mathbf{x}_0 \in \Gamma(t)$ with $\theta(t, \mathbf{x}_0) = 0$, since then $V(t, \mathbf{x}_0) > 0$ is obtained as the L'Hospital-limit of $-\frac{1-\theta(t,\mathbf{x})}{\theta(t,\mathbf{x})} \nu_\Gamma \cdot \nabla p(t, \mathbf{x})$ as $\mathbf{x} \to \mathbf{x}_0$ from the interior.

4.3.2.3 Boundary Pressure Functions at Cell Edge

We consider creeping cell migration only in a medium that is not under external pressure or stress. Therefore, in a purely physical model, the boundary pressures P_a^Γ and P_s^Γ along the cell edge $\Gamma(t)$ should be set to zero. However, the edge is physically defined as the lamellar tip, where the ventral and dorsal parts of the surrounding plasma membrane meet. Thus, if τ_Γ denotes the scalar tension value of the dorsal plasma membrane at the cell edge (see the following paragraph), then this tension appears as a *boundary tip pressure* acting on both phases: the cytoskeleton and the cytosol. Moreover, in addition to the counteracting cytoskeleton and cytosol pressure on the left-hand sides of Equations (4.28) and (4.29), there could be extra pressures due to *active polymerization forces*, which usually are generated at barbed actin filament ends that can become exposed to the tip membrane in two variants; see Figure 4.4.

- *Brownian ratchet model:* Assume that at a point of the lamellar tip $\Gamma(t)$, a fraction a^B of filaments is bound to membrane proteins in a fast pseudo-steady-state equilibrium with the actual F-actin concentration $a = \theta a_{\max}$. Then from the remaining network with concentration $a^F(a) = a - a^B(a)$, Arp2/3-induced branching can occur. The barbed filament ends are more or less normally exposed to the tip membrane with concentration $a^f = a^f(a) = \alpha_0^f \cdot \mathrm{Arp0} \cdot \frac{a^F(a)}{K_a + a^F(a)}$ with suitably chosen coefficients; see also Equation (4.5). Then, due to insertion of G-actin in between fluctuating filament and membrane, there appears a *free polymerization pressure*:

$$p^f = \pi_0^f \cdot a^f(a) \tag{4.31}$$

  with a *ratchet coefficient* $\pi_0^f > 0$ depending on the free energy of one monomer addition; see relevant force estimations, for example, in [41].
- *Clamp motor model:* Alternatively or additionally, a certain fraction of tip bound actin filaments with concentration $a^c = a^c(a) < a^B(a)$ can be bound to WASP-like membrane proteins, which serve as filament end-tracking motors. By means of an energy-consuming polymer elongation process at the clamped end of the filament, these proteins push the bound filament outward, thus leading to a *clamp polymerization*

pressure:

$$p^c = \pi_0^c \cdot a^c(a) \tag{4.32}$$

Here again, the clamp motor coefficient $\pi_0^c > 0$ depends on the energy of one monomer translocation; see [16].

The sum of these active polymerization pressures would induce an averaged relative normal inward mass flux $\theta V \geq 0$ of the whole cytoskeleton, based on the particular inflows of a^f- and a^c-filaments satisfying $\theta V = a^f V_f = a^c V_c$. However, this flow experiences a viscous resistance from the remaining fixed filaments (concentration $a^B - a^c$), with a viscosity coefficient that could increase in the presence of bound myosin molecules. Thus we get the total *tip polymerization pressure*:

$$P_{poly} = p^f + p^c - \mu_\Gamma(m_b)(a^B - a^c)a_{\max} V \tag{4.33}$$

Clearly, here the relative inward velocity V is the one defined by the normal component of $\mathbf{v}$ in relation to $\dot{\Gamma}$; see Equation (4.17). For a similar resistance model based on elastic crosslinking see [23].

Finally, because this intrinsic boundary pressure P_{poly} acts on the cytoskeleton volume fraction, while simultaneously inducing a corresponding counterpressure on the cytosol volume fraction, we can state the following distribution for the boundary pressures generated at the membrane:

$$\theta P_a^\Gamma = \theta(1 - \kappa_\Gamma)\tau_\Gamma + P_{poly} \tag{4.34}$$

$$(1 - \theta)P_s^\Gamma = (1 - \theta + \theta\kappa_\Gamma)\tau_\Gamma - P_{poly} \tag{4.35}$$

where we introduce a weight factor, $0 \leq \kappa_\Gamma < 1$, measuring the relative effect of membrane tension τ_Γ on the cytosol phase and possibly depending on the not explicitly modeled tip geometry. Then, by substituting p from Equation (4.29) into Equation (4.30) we obtain the generalized *Neumann-type boundary condition* for $\mathbf{v}$ in all points of $\Gamma(t)$ where the network is attached, $a^B > 0$, and where $V > 0$ holds:

$$\begin{aligned} \nu_\Gamma \cdot \mathbb{T}_a \cdot \nu_\Gamma - \sigma(\theta) - \frac{a_{\max}}{1-\theta}\mu_\Gamma(m_b)(a^B - a^c)V & \\ + \frac{1}{1-\theta}\left(\sigma(\theta) + p^f + p^c - \theta\kappa_\Gamma\tau_\Gamma\right) = 0 & \end{aligned} \tag{4.36}$$

This means that the *mean inward F-actin polymerization speed* $V \geq 0$ on the moving cell edge $\Gamma(t)$ is implicitly determined by solving the linear elliptic system Equations (4.19) and (4.22) for $\mathbf{v}$ and p and satisfying all boundary conditions. Thereby the Neumann condition above contains all membrane-protruding pressure terms in series, namely the swelling pressure σ, the polymerization pressures p^f and p^c induced by a Brownian ratchet or a clamp motor mechanism, as well as a counteracting stress due to dorsal membrane tension τ_Γ.

4.3.2.4 Global Force Balance at Adhesive Substratum

There are two kinds of forces exerted by the migrating cell (fragment) onto the fixed flat substratum: the integrin adhesion-mediated active frictional force represented by the vector field $\mathbf{F}_v$ on the right-hand side of Stokes Equation (4.19), and an analogous passive frictional force $\mathbf{F}_u$ due to motion of the dorsal plasma membrane along the substratum:

$$\mathbf{F}_v = c_{sa}\mathbf{F}_c = \Phi(\theta, c_{sa}) \tag{4.37}$$

$$\mathbf{F}_u = \Phi_u \mathbf{u} \tag{4.38}$$

with Φ as in Equation (4.23) and an assumed friction constant $\Phi_u \geq 0$. Supposing that on the substratum no other forces are applied than these, then their integral sum has to vanish so that the zero force balance holds:

$$0 = \int_{\Omega(t)} (\mathbf{F}_v + \mathbf{F}_u) \tag{4.39}$$

Furthermore, if we assume that the dorsal membrane, viewed as a 2-D incompressible fluid satisfying the zero divergence condition in Equation (4.10), has relatively low viscosity, then the *membrane tension* τ_u induced by the frictional flow can be defined according to Darcy's law

$$\nabla \tau_u = \Phi_u \mathbf{u} \tag{4.40}$$

and thus determined as the solution of the Laplace equation $\Delta \tau_u = 0$ with Neumann boundary condition $\nu_\Gamma \cdot \nabla \tau_u = 0$; see Equation (4.10). Then the zero force balance of Equation (4.39), together with Equation (4.19), implies that the boundary tension values $\tau_\Gamma = \tau_u|_{\Gamma(t)}$, which are uniquely determined up to a constant, necessarily fulfill the *integrability condition* $\int_{\Gamma(t)} (\mu\theta\widetilde{\nabla}\mathbf{v} + S(\theta, \mu_b) - p + \tau_\Gamma)\nu_\Gamma = 0$, which by insertion of the Neumann boundary condition in Equation (4.30) and using the symmetry of $\widetilde{\nabla}\mathbf{v}$ reduces to the equivalent necessary condition $\int_{\Gamma(t)} (\theta P_a^\Gamma + (1-\theta) P_s^\Gamma - \tau_\Gamma)\nu_\Gamma = 0$. Indeed, this condition is fulfilled for the boundary pressure model functions that were chosen in the preceding paragraph, Equations (4.34) and (4.35), because then even the integrand in the previous condition vanishes.

4.4 Results of Model Simulations

4.4.1 Spontaneous Cell Polarization in 2-D Model

We have simulated the adhesive motion of a flat cell or cell fragment represented by a 2-D domain, $\Omega(t)$, with moving cell edge or lamellar tip, $\Gamma(t) = \partial\Omega(t)$, under certain simplifying assumptions:

1. The lower dorsal membrane sticks to the substratum ($\mathbf{u} = 0$) and there exists a small membrane tension $\tau_\Gamma > 0$, constant over the whole cell edge.
2. There occurs no active polymerization pressure at the cell edge ($p^f = p^c = 0$); only the similar swelling pressure $\sigma(\theta)$ and the hydrostatic pressure p can push the boundary.
3. Disruption of the F-actin network from the lamellar tip $\Gamma(t)$ can locally occur if the network tension exceeds a certain threshold that might depend on the fraction of membrane-bound actin filaments $a^B = A\frac{a}{K_B+a}$.
4. The free myosin-II concentration is at a fixed constant level $m_f^0 > 0$ and the amount of F-actin bound myosin-II oligomer is in a pseudo-steady equilibrium $m_b^*(a)$ according to Equation (4.9), so that the contractile stress is only a function of $\theta = a/a_{\max}$:

$$\psi(\theta) = \psi_0 \theta m_b^* = \frac{\psi_0 \alpha_m m_f^0}{a_{\max} \delta_m^0} \frac{\theta^2}{1 + \theta^2/\theta_{opt}^2} \tag{4.41}$$

More details and a list of chosen parameters can be found in [35], Section 1 and Table 2.1. Because unpolarized cells and cell fragments, as observed under various conditions [63,64], attain a regular circular shape, we chose as an initial condition a circle $\Omega(0)$ of radius 6 μm. Moreover, to mimic the radial spreading of cells after exposure onto a flat substratum, we start with constant integrin densities and radially symmetric initial configuration for the volume fraction θ, with slightly larger values closer to the center. Due to this initial perturbation, the F-actin concentration rapidly condenses into a central region of high θ, surrounded by a lamella-type region of low θ; see Figures 4.6(a) and (b). Later, also the actin and surface-bound integrin adhesion proteins c_{sa} concentrate around this center; see Figure 4.6(e). Both phenomena are supported by a strong retrograde F-actin flow, which collects actin filaments and actin-bound integrins c_a in radial direction from almost everywhere in the periphery. Moreover, the hydrodynamic pressure has its maximum within the center region (data not shown), so that its negative outward gradient represents the squeezed flow of cytosol from the contracting F-actin network.

At the free cell edge with relatively low F-actin concentration, there occurs a metastable equilibration between a positive swelling pressure pushing the lamellar tip outward and a resisting viscous network tension pulling the tip inward. See Figure 4.6(b). After 5 min, local regions of protrusion or retraction can be observed, which point in varying directions along the cell periphery. Notice that these spatio-temporal fluctuations are not due to the tiny stochastic perturbations imposed on the F-actin polymerization rate, but represent the emergent chaotic dynamics of the cytoplasm as a reactive and contractile two-phase fluid [3,11]. Furthermore, behind a cyclically protruding and retracting free edge we can observe a layer with slightly increased concentration of substrate and actin-bound integrin c_{sa}; see Figure 4.6(e). This is the onset of polarization: fresh free integrin proteins are appearing at the

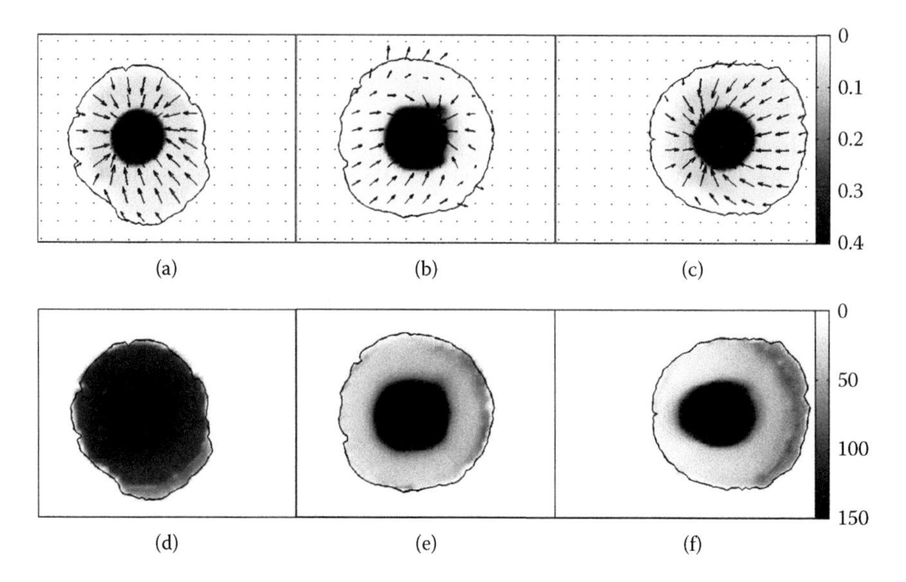

FIGURE 4.6 2-D simulation of a spontaneously polarizing cell (fragment) starting with radially symmetric initial conditions: Spatial distributions shown as pseudocolor plots of F-actin volume fraction θ in (a) through (c) and concentration of actin and surface-bound integrin proteins c_{sa} in (d) through (f) at three different time instants: (a) and (d) 1 min, (b) and (e) 5 min, and (c) and (f) 10 min after initialization. The width of the shown region is 22 μm.

protruding part of the edge from the upper membrane, thus also increasing c_{sa} in this region. As a consequence, a larger frictional force of the retrograde F-actin is generated in that direction, inducing a bias in the force vector field transduced to the substratum. Thus, the whole cell fragment starts to move in this direction, which later becomes the leading edge of the migrating cell fragment; see Figures 4.6(c) and (f). The emerging polarization of the cell is most clearly expressed in the c_{sa} distribution of Figure 4.6(f), where an increasingly dense band is formed behind the leading edge, and the central region of focal adhesions is slowly shifted rearward with increasing migration speed, while becoming deformed in shape similar to the whole cell fragment.

This simulation is an example for the autonomous formation of a metastable unpolarized, almost circular state and its spontaneous transition into a polarized, migrating state of a flat cell fragment. By changing some of the model parameters we can influence the degree of this symmetry breaking instability, but so far we are not able to reproduce the observed longer-time stability of circular cell fragments; see again [63]. One reason for this failure seems to be that in our model simplifications, we assumed a constant distribution of freely diffusing myosin-II oligomers, contradicting experimental results on clearly expressed gradients of myosin-II concentration decaying toward the cell edge; see, for example, [60]. Therefore we have started to investigate the full coupled model system by including active tip polymerization and the kinetics for myosin-II diffusion, binding, and transport, in addition to the already implemented analogous kinetics and dynamics of integrin adhesion molecules. The next section presents the achieved results in a simple but nevertheless instructive 1-D situation.

4.4.2 Induced Onset of Cell Polarization and Migration in 1-D Model

We have simulated adhesion, polarization, and migration of an idealized flat cell fragment having a fixed extension and only one degree of freedom to move, as can be observed experimentally, for example in Figure 4.1(c). Thus, in a 1-D cross-section along its moving direction, the fragment is represented by an interval $\Omega(t) = [x_b(t), x_b(t) + L]$ of fixed length L, moving with body speed $\mathrm{v}_b(t) = \dot{x}_b(t)$, so that all kinetics and dynamics within the cytoplasm and dorsal membrane can be described by the corresponding 1-D equations and conditions as in Section 4.3.2, but now written in cell-centric coordinates; for example, $\tilde{\theta}(t, y) = \theta(t, x_b(t) + y)$ or $\tilde{\mathrm{v}}(t, y) = \mathrm{v}(t, x_b(t) + y) - \mathrm{v}_b(t)$. Then, assuming the following particular restrictions, we have solved a simplified system of boundary value problems:

1. The actin network is always sticky at the lamellar tips (no disruption), so that $\tilde{\theta} > 0$ on the whole closed interval $[0, L]$.
2. We assume active tip polymerization with a simultaneous parallel effect of the Brownian ratchet ($\pi_0^f = 5\,\mathrm{Pa} \cdot \mu\mathrm{M}^{-1}$) and the clamp motor mechanism ($\pi_0^c = 5\,\mathrm{Pa} \cdot \mu\mathrm{M}^{-1}$). Moreover, we suppose that the membrane-bound cortex shear viscosity strongly increases if myosin-II oligomers are bound: $\mu_\Gamma(m_b) = 0.1(1 + 45 m_b)\,\mathrm{Pa} \cdot \mathrm{min} \cdot \mu\mathrm{m}^{-1}$. For the actin-binding membrane proteins at the tip with maximal concentration $A = 50\,\mu\mathrm{M}$ and self-enhanced binding with dissociation constant $K = 158.1\,\mu\mathrm{M}$, we use the pseudo-equilibrium $2a^B(a) = A + K^2/a + a - \sqrt{(A-a)^2 + 2(A+a)K^2/a + (K^2/a)^2}$ and $a^c = 0.1\, a^B$.
3. The dorsal membrane moves together with the cell (no slip at the tips), so that $\tilde{\mathrm{u}} \equiv 0$ and $\mathrm{v}_c = \mathrm{v} = \tilde{\mathrm{v}} + \mathrm{v}_b$. The tip membrane tension difference is $[\tau_\Gamma]_0^L = \Phi_u L \mathrm{v}_b$ with the minimum always equal to a fixed positive constant $\tau_0 = 25$ Pa. Finally, the substrate force balance in Equation (4.39) reads

$$\Phi_u L \mathrm{v}_b = \int_0^L \Phi_0 \tilde{c}_{sa}\, \tilde{\theta}(\tilde{\mathrm{v}} + \mathrm{v}_b) \tag{4.42}$$

   This is an implicit equation to be solved for the migration speed v_b, because the Neumann boundary conditions and the right-hand side of the elliptic Equation (4.19) for $\tilde{\mathrm{v}}$ (after eliminating the pressure $\tilde{p}$) contain expressions that depend (linearly) on v_b.

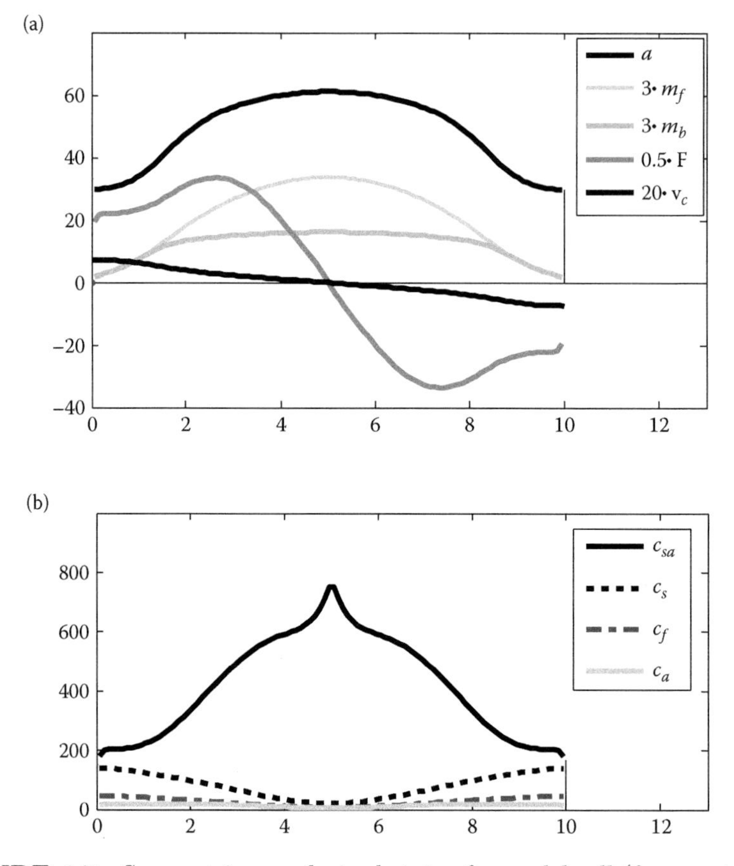

FIGURE 4.7 Symmetric unpolarized state of a model cell (fragment) with length 10 μm showing the concentration profiles of (a) F-actin a, free and bound myosin-II m_f and m_b, the force F onto the substrate and the effective cortical F-actin velocity v_c; (b) of integrin proteins in the four different states, namely substrate- and actin-bound c_{sa}, substrate-bound c_s, freely diffusing c_f, and actin-bound c_a.

Starting with the same unpolarized initial condition as in the previous Section 4.4.1, we obtain a similar 1-D symmetric configuration with no cell translocation and a central F-actin plateau, a central maximum of FA (focal adhesion) sites (i.e., c_{sa}-integrins), as well as a centripetal F-actin flow; see Figure 4.7. In addition, also the concentration of total myosin-II is enriched in the central region, with actin-bound myosin forming a central plateau, consistent with fluorescence pictures of unpolarized keratinocyte fragments; see [63]. Moreover, the F-actin flow is highest at the boundary (with values $V \sim 0.5\,\mu\text{m} \cdot \text{min}^{-1}$) due to the assumed active tip polymerization.

In contrast to the previous 2-D model simulations, here we find robust parameter constellations yielding stability of this symmetric unpolarized state:

even quite strong but still subthreshold perturbations of the F-actin concentration at one side induce only transient locomotion together with shifts in most concentration profiles. After some delayed overshooting, the cell fragment returns to its nonmoving nonpolarized stable state in Figure 4.7. However, if F-actin polymerization is continuously stimulated at one side (mimicking the effect of a chemo- or haptotactic gradient) then, not surprisingly, the cell fragment slowly polarizes and starts to persistently translocate in this direction (data not shown).

Similarly, to mimic the mechanical stimulation experiments with keratinocyte fragments as performed in [63], we locally increase the activation (and F-actin-binding) of myosin-II at the left-hand side for a certain time (up to 30 s), which is thought to be analogous to the experimental push by a micropipette flow pulse, since by local compression of the cytoskeleton, filament alignment and thus myosin action will be enhanced. If time span or amplitude of this myosin pulse stays subthreshold, the model cell responds only by a transient migration as described above and asymptotically returns to its stable resting state; see the speed curve in Figure 4.8(a). However, if the pulse strength exceeds a certain threshold, myosin-II and F-actin are condensed at the rear, whence active tip polymerization (V) is drastically reduced at this side, so that the cell rapidly starts to migrate in the other direction; see Figure 4.8(b). After cessation of the pulse, the migration speed is reduced but the cell maintains its polarized locomotion state. This induced polarization, as a nonlinear threshold behavior, is supported by a further positive feedback mechanism: due to cell translocation, the focal adhesion sites (c_{sa}) are successively shifted rearward relative to the cell, which stabilizes the asymmetric polarization and later increases the locomotion speed to a constant asymptotic value ($v_b \sim 0.12\,\mu\text{m} \cdot \text{min}^{-1}$).

In this migration state the cell (fragment) attains characteristic concentration profiles; see Figure 4.9. In addition to the already mentioned polar gradients of F-actin and myosin-II, the most impressive distribution is that of the FA sites: the c_{sa} profile shows a characteristic broad peak behind the leading edge (as in the 2-D case above), followed by a slight decrease and a second plateau of even more condensed adhesion sites in the back part of the cell (fragment) that, however, rapidly decays at the very rear; see Figure 4.9(b). This last phenomenon is the theoretically expected and experimentally observed *rear release* of adhesion sites or integrins, not induced by any directed regulatory protein, but only by the fact that the rear part experiences a steep increase in the force F transduced to the substratum; see the plot in Figure 4.9(a). By Equation (4.37) this is proportional to the F-actin mass flow θv_c with respect to the substratum: while in the major front part of the migrating cell (fragment) the F-actin flow is retrograde and the centripetally pulling negative force is modest in amplitude, near the trailing edge the direction of flow reverses and the positive force becomes very strong, now centripetally pulling off the focal adhesion sites. Thus, the reason for cell translocation is not "more adhesion" or "stronger force" at the front compared to the rear, as

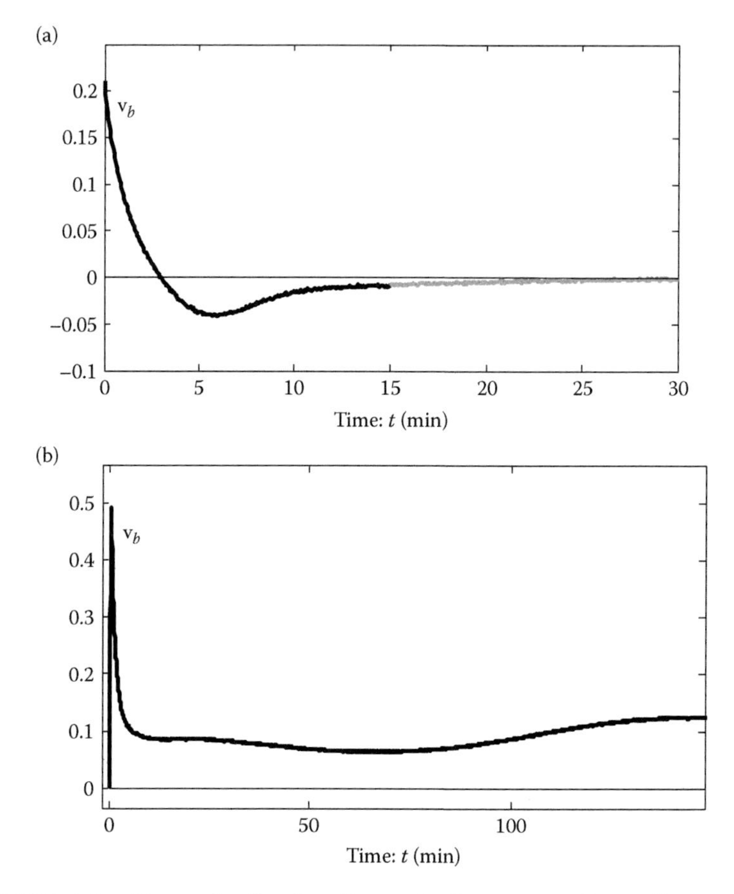

FIGURE 4.8 Plots of cell migration velocity v_b over time after a perturbation of the unpolarized state in Figure 4.7 by imposing an additional activation rate α_m^+ for actin-bound myosin-II oligomers over an induction period of 0.5 min locally at the left cell side: (a) with rate $\alpha_m^+ = 20\ \text{min}^{-1}$, showing a stable convergence of v_b toward the unpolarized speed zero, and (b) with rate $\alpha_m^+ = 50\ \text{min}^{-1}$, showing the convergence toward a positive polarized speed $\sim 0.12\,\mu\text{m} \cdot \text{min}^{-1}$.

is asserted in some models, cf. [17]. Indeed, for vanishing passive friction Φ_u due to Equation (4.42), the total force integral even vanishes. The true physical reason for cell migration is the aforementioned asymmetry in the polarized cell state, expressed by a wide front region with modest rearward force and short rear region with strong forward force.

We finally investigate the dependence of migration speed on two cell physiologically important parameters, namely the *adhesiveness of the substratum* quantified by the relative number of available adhesion sites Adh0, and the *responsiveness of the F-actin network* measured, for instance, by the

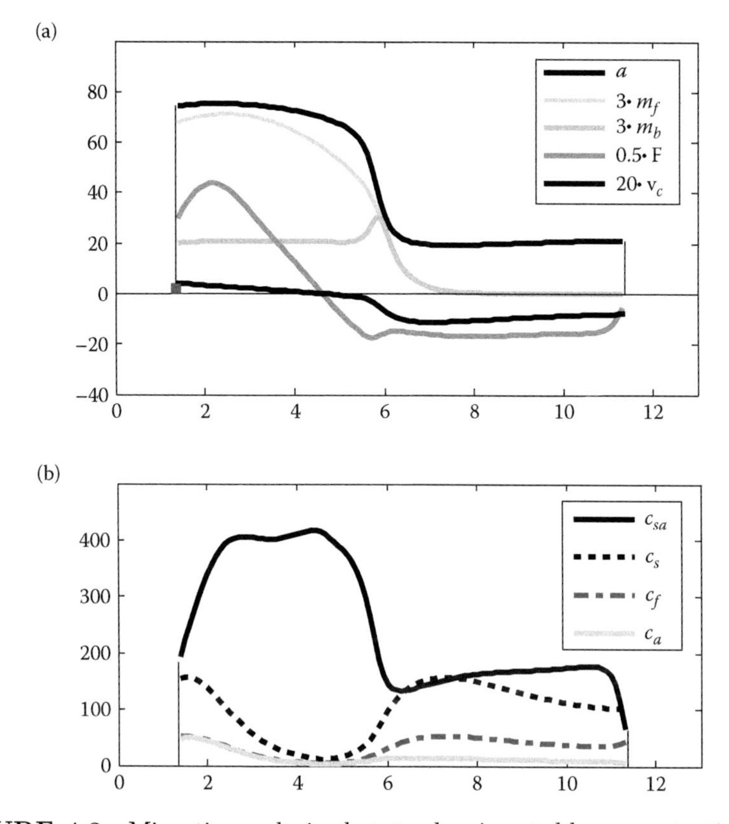

FIGURE 4.9 Migrating polarized state showing stable concentration profiles of (a) F-actin, a, free and bound myosin-II, m_f and m_b, the transduced locomotion force F, and the effective cortical F-actin velocity v_c; (b) integrin-concentrations as in Figure 4.7. Only FA integrins in the *sa* state are able to transduce force from the contractile actin network to the substrate. The migration speed is $v_b = 0.125\,\mu m \cdot min^{-1}$.

concentration Arp0 of activated Arp2/3 proteins. These are mainly responsible for controlling F-actin polymerization by available free filament ends as they appear in the net assembly rate of Equation (4.4) and the ratchet polymerization pressure of Equation (4.31). The results depicted in Figure 4.10 reveal the existence of optimal ranges for both parameters, consistent with earlier modeling results (see [17,24]) and with experimental observations (see, for example, [27,52]).

In particular, the migration response curve of Figure 4.10(a) has been performed for fixed parameter Arp0 $= 10\,\mu M$ and for adhesion values $0 \leq$ Adh0 ≤ 20, where according to Table 4.2 the adhesiveness proportionally influences not only the two adhesion rates $\alpha^+ = \gamma^+ = \text{Adh0} \cdot \alpha$, but also the passive membrane friction coefficient $\Phi_u = \text{Adh0} \cdot 6\,\text{Pa} \cdot \text{min} \cdot \mu m^{-2}$. This is based on

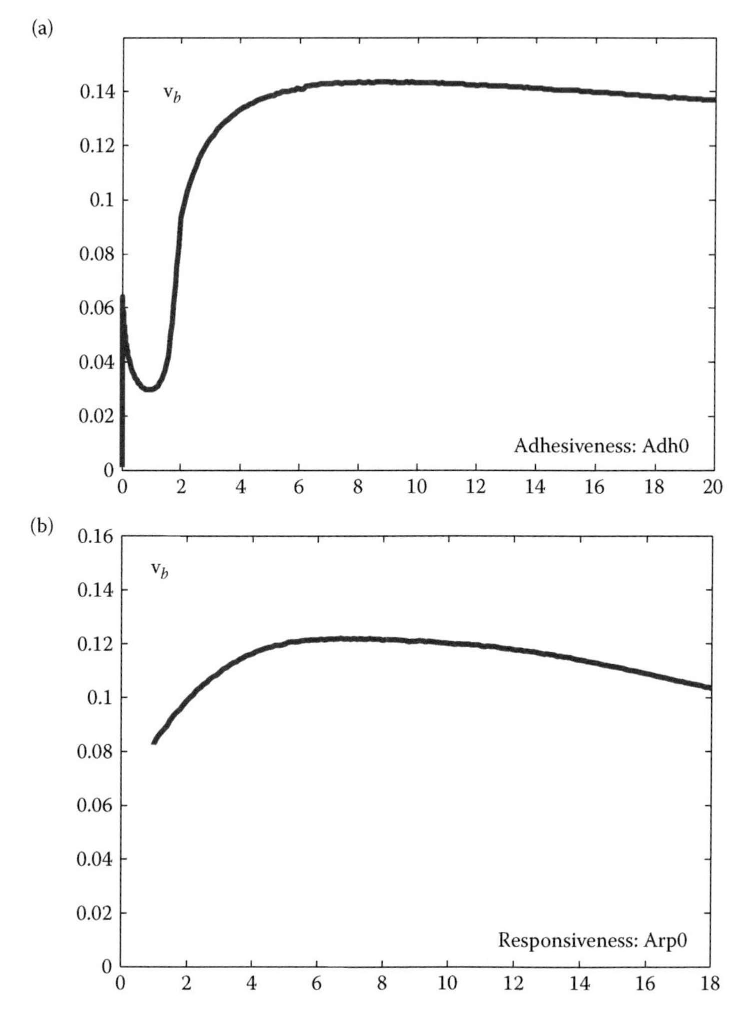

FIGURE 4.10 Migration speed v_b of a model cell in its stable polarized state, plotted over varied parameters of (a) *adhesiveness* Adh0, relative number of adhesion binding sites on the substratum (e.g., fibronectin coating); (b) *F-actin responsiveness* expressed by Arp0 [μM], the cytoplasmic concentration of activated Arp2/3 complexes.

the idealizing assumption that there are no other relevant exterior forces that would resist locomotion relative to the substratum than those due to friction of the dorsal membrane with respect to an adhesive coating of, for example, fibronectin or collagen.

Under these hypotheses and parameter choices, our model simulations predict a minimal migration speed of $v_b \sim 0.035\,\mu\text{m} \cdot \text{min}^{-1}$ for Adh0 ~ 1, which increases toward the doubled speed when lowering the adhesiveness to

TABLE 4.1 Model Variables and Functions

Symbol	Meaning	Unit
$\theta(t, \mathbf{x})$	Volume fraction of F-actin	Dimensionless
$a(t, \mathbf{x})$	Concentration of F-actin	μM
$m_f(t, \mathbf{x})$	Concentration of free myosin-II	μM
$m_b(t, \mathbf{x})$	Concentration of F-actin bound myosin-II	μM
$c_f(t, \mathbf{x})$	Concentration of free integrin	μm^{-2}
$c_s(t, \mathbf{x})$	Concentration of substrate bound integrin	μm^{-2}
$c_a(t, \mathbf{x})$	Concentration of actin bound integrin	μm^{-2}
$c_{sa}(t, \mathbf{x})$	Concentration of substrate-and-actin bound integrin	μm^{-2}
$R(\theta)$	F-actin net polymerization rate	min^{-1}
$\psi(\theta, m_b)$	Contractile stress	Pa
$\sigma(\theta)$	Swelling pressure	Pa
$S(\theta, m_b)$	Effective cytoplasmic stress	Pa
$p(t, \mathbf{x})$	Effective two-phase flow pressure	Pa
$\tau_u(t, \mathbf{x})$	Dorsal membrane tension	Pa
$\mathbf{v}(t, \mathbf{x})$	Mean F-actin velocity	$\mu m \cdot min^{-1}$
$\mathbf{v}_c(t, \mathbf{x})$	Cortical F-actin velocity	$\mu m \cdot min^{-1}$
$\mathbf{u}(t, \mathbf{x})$	Dorsal membrane velocity	$\mu m/min^{-1}$
$v_b(t)$	Migration velocity of cell (fragment) body (1-D)	$\mu m \cdot min^{-1}$
$\Phi(\theta, c_{sa})$	Adhesional friction	$Pa \cdot min \cdot \mu m^2$
$\mathbf{F}_v$	Active frictional force	$Pa \cdot \mu m$
$\mathbf{F}_c$	Local frictional force per adhesion site	$Pa \cdot \mu m$
$\mathbf{F}_u$	Passive frictional force at dorsal membrane	$Pa \cdot \mu m$
$\dot{\Gamma}$	Normal speed of free boundary $\Gamma(t)$	$\mu m \cdot min^{-1}$
V	Relative inward F-actin velocity at boundary	$\mu m \cdot min^{-1}$
P_a^{Γ}	Boundary cytoskeleton pressure	Pa
P_s^{Γ}	Boundary cytosol pressure	Pa
τ_{Γ}	Membrane tension at boundary (lamellar tips)	Pa
$a^B(a)$	Concentration of tip-bound F-actin	μM
$a^c(a)$	Concentration of F-actin bound to clamp motor proteins	μM
$a^f(a)$	Concentration of free filaments exposed to tip	μM
$\mu_{\Gamma}(m_b)$	F-actin shear viscosity relative to a^B filaments	$Pa \cdot min \cdot \mu m^{-1}$
p^f	Free polymerization pressure at tip membrane	Pa
p^c	Clamp motor polymerization pressure	Pa

almost zero. This surprising phenomenon is consistent with our experimental measurements of human keratinocyte polarization and migration [37] revealing slightly increased motility of polarized cells on glass compared to those on a low-density fibronectin coat. The reason for this can be seen from the corresponding profiles of FA concentration (c_{sa}) and substrate force distribution (F) for the two adhesiveness values Adh0 $= 0.01, 0.2$; see Figures 4.11(a) and (b). Due to ongoing retrograde flow of F-actin, working against internal

TABLE 4.2 Parameters for 1-D Simulations

Symbol	Meaning	Value	Unit
a_g	Concentration of G-actin	30	μM
$a_{\max}$	Maximal concentration of F-actin	800	μM
k_{on}	F-actin polymerization rate at plus ends	696	$(\text{min} \cdot \mu\text{M})^{-1}$
k_{off}	F-actin depolymerization rate at plus ends	258	min^{-1}
ω	F-actin capping rate	1250	min^{-1}
r	F-actin disassembly rate	0.2	min^{-1}
ε	Basal F-actin nucleation rate	0.75	min^{-1}
Arp0	Concentration of activated Arp2/3 complexes	10	μM
ν_a	Arp2/3-induced nucleation rate	60	$(\text{min} \cdot \mu\text{M})^{-1}$
K_a	Half-saturation concentration for nucleation	3	μM
m_f^0	Equilibrium reservoir concentration of free myosin-II	10	μM
D_m	Diffusion coefficient for free myosin-II	0.5	$\mu\text{m}^2 \cdot \text{min}^{-1}$
α_m	F-actin binding rate of myosin-II	0.5	$(\text{min} \cdot \mu\text{M})^{-1}$
δ_m^0	Dissociation rate of bound myosin-II	3	min^{-1}
a_{opt}	Optimal F-actin concentration for myosin-II binding	40	μM
c_f^0	Reservoir concentration of free integrin at lamellar tips	50	μm^{-2}
D_f	Diffusion coefficient for free integrin	0.5	$\mu\text{m}^2 \cdot \text{min}^{-1}$
Adh0	Number of available substrate sites per integrin	3	Dimensionless
Talin	F-actin association factor	0.0125	Dimensionless
Fifactor	Focal complex factor (intracellular)	50	Dimensionless
Fefactor	Focal adhesion factor (extracellular)	50	Dimensionless
α^-	Dissociation rate for substrate-bound integrin	$\alpha = 5$	min^{-1}
α^+	Free integrin binding rate to substrate	$\text{Adh0} \cdot \alpha$	min^{-1}
β_0^+	F-actin binding rate for substrate-bound integrin	$\alpha \cdot \text{Talin}$	$(\text{min} \cdot \mu\text{M})^{-1}$
β_0^-	FA dissociation of the F-actin link	$\alpha/\text{Fifactor}$	min^{-1}
δ_0^+	Free integrin binding rate to F-actin	$\alpha \cdot \text{Talin}$	$(\text{min} \cdot \mu\text{M})^{-1}$
δ^-	Dissociation rate for F-actin-bound integrin	α	min^{-1}
γ^+	Substrate binding rate for F-actin-bound integrin	$\text{Adh0} \cdot \alpha$	min^{-1}
γ_0^-	FA dissociation rate of substrate link	$\alpha/\text{Fefactor}$	min^{-1}

TABLE 4.2 Parameters for 1-D Simulations (Continued)

Symbol	Meaning	Value	Unit
ρ_β	Exponential FA-rupture coefficient for F-actin link	11.7	$(\text{Pa} \cdot \mu\text{m})^{-1}$
ρ_γ	Exponential FA-rupture coefficient for substrate link	11.7	$(\text{Pa} \cdot \mu\text{m})^{-1}$
A	Total free F-actin binding sites on boundary $\Gamma(t)$	50	μM
π_0^f	Ratchet polymerization pressure coefficient	0.0125	$\text{Pa} \cdot \mu\text{M}^{-1}$
π_0^c	Clamp motor polymerization pressure coefficient	0.0125	$\text{Pa} \cdot \mu\text{M}^{-1}$
κ_Γ	Tip curvature weight factor for membrane tension	0.5	Dimensionless
α_0^f	Relative amount of exposed barbed ends per Arp2/3	2	Dimensionless
ψ_0	Contractile stress per bound myosin-II	0.0163	$\text{Pa} \cdot \mu\text{M}^{-1}$
$\tilde{\psi}_0$	Strength of contractile stress (simplified model)	0.625	Pa
σ_0	Strength of swelling pressure	0.125	Pa
μ	Viscosity of F-actin network phase	0.625	$\text{Pa} \cdot \text{min}$
κ	Cortical slip parameter for F-actin flow	0.5	Dimensionless
φ	Drag coefficient between network and solvent	2	$\text{Pa} \cdot \text{min} \cdot \mu\text{m}^{-2}$
Φ_0	Friction per actin substrate-bound integrin	0.02	$\text{Pa} \cdot \text{min}$
Φ_u	Additional friction associated to cell body	Adh0 · 6	$\text{Pa} \cdot \text{min} \cdot \mu\text{m}^{-2}$
L	Length of the cell (fragment)	10	μm

viscosity and drag, the polarized cell (fragment) gathers the few FA sites on the back side in a way that the local FA and force distributions are almost symmetric (for almost vanishing adhesiveness Adh0 $= 0.01$) but still with an additional negative (pulling) force plateau on the front side (though of tiny absolute value $|\text{F}| \sim 0.05\,\text{Pa}$). For increased but still small adhesiveness (Adh0 $= 0.2$), the much more frequent FA sites start to accumulate at the rear, thus the asymmetry of forces is enhanced and the increased friction reduces the migration speed.

On the other hand, for further increasing adhesiveness, Adh0 > 1, the front plateau of pulling forces at the enriched FA "carpet" is proportionally increased; see Figure 4.11(c) and (d). However, the dominant reason for the nonlinear speeding-up response is the prominent increase in disruptive forces $|\text{F}_c(\text{rear})|$ up to values of Adh0 ~ 2, leading to a drastic reduction in

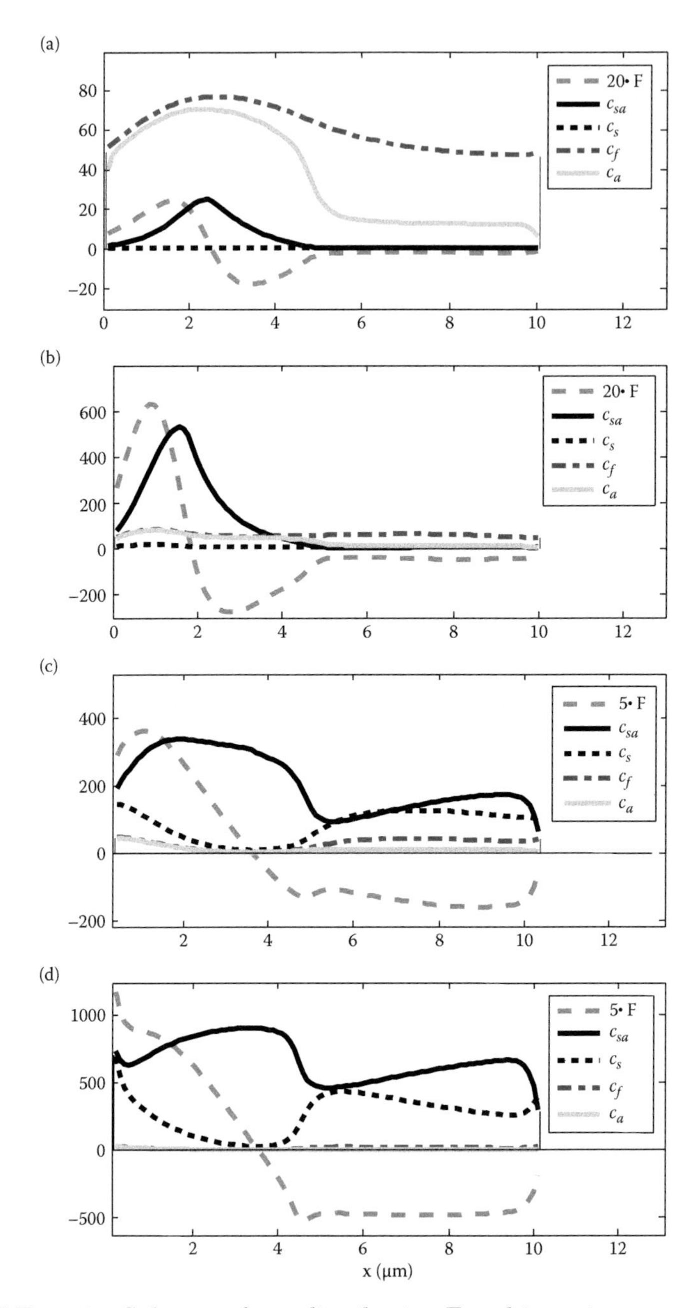

FIGURE 4.11 Substrate force distribution F and integrin concentrations in the four different states, plotted for varying adhesiveness parameter Adh0 with corresponding migration speed v_b as in Figure 4.10(a): (a) 0.01, speed 0.06 $\mu m \cdot min^{-1}$; (b) 0.2, speed 0.04 $\mu m \cdot min^{-1}$; (c) 3.0, speed 0.125 $\mu m \cdot min^{-1}$; and (d) 20.0, speed 0.138 $\mu m \cdot min^{-1}$.

resistive FA sites at the very rear; see the corresponding indicator curves in Figure 4.12(a). For very large adhesiveness, Adh0 $\sim$ 20, the FA sites again start to accumulate at the very rear despite a strong disruptive force there, as can be seen in Figure 4.11(d), causing a slight decrease in migration speed. Thus, while an adhesiveness between 5 and 12 leads to a saturated optimality in migration speed, within the range $1 < \text{Adh0} < 4$ the cell (fragment) has a high sensitivity for responding to an increase in adhesiveness with an up to four-fold increase in speed. Carried over to the mean forward speed of competing leading lamellae in a whole cell model (as for the 2-D simulations in Figure 4.6), this could be used to explain polarization and haptotaxis of cells in spatial adhesion gradients; cf. [20].

To understand the analogous optimal migration performance for responsiveness parameters in the range between $4\,\mu\text{M}$ and $10\,\mu\text{M}$ of Arp2/3 concentration Arp0, see Figure 4.10(b), we plot the same indicator curves for FA concentrations and substrate forces in Figure 4.12(b). Again, with increasing Arp2/3-induced actin polymerization there is an almost linear increase in pulling force at the tip (due to increased polymerization velocity V there) while the concave migration speed curve directly corresponds to the proportional curve for the disruptive forces at the rear and the resulting convex curve for the accumulated FA sites there. For larger Arp0 values, the growing accumulation of FA sites again leads to a slight speed reduction. Finally, the about 50% gain in speed for increasing responsiveness in the range $1\,\mu\text{M} < \text{Arp0} < 4\,\mu\text{M}$ could again be carried over to a competing lamellar protrusion response in spatial chemotactic gradients, which are known to stimulate F-actin polymerization at the leading front.

4.4.3 Migration Speed in Simplified 1-D Model

To explore which properties of our coupled cytoplasm adhesion model are essential for obtaining the 1-D simulation results in the previous section, we simplified the model by freezing the following variables and parameters; cf. [45]:

1. Assuming fast diffusion and F-actin binding of free myosin-II oligomers, we take the pseudo-steady-state condition of Equation (4.9) for bound myosin-II, but with the equivalent dissociation rate $\delta_m(a) = \delta_m^0 \exp(-2a/a_{opt})$ instead of Equation (4.8), and obtain as contractility $\psi(\theta) = \tilde{\psi}_0 \cdot \theta^2 \exp(-2\theta/\theta_{opt})$ with a coefficient $\tilde{\psi}_0$ analogous to Equation (4.41).
2. Except for the FA friction function Φ we set all other friction coefficients (φ, Φ_u) to zero. Thus, the cell migration speed is determined by the zero integral in Equation (4.42).
3. All "exterior" pressures or tensions at the free membrane tip $(\pi_0^f, \pi_0^c, \tau_\Gamma)$ are assumed to vanish; in particular, there occurs no active polymerization ($V \equiv 0$).

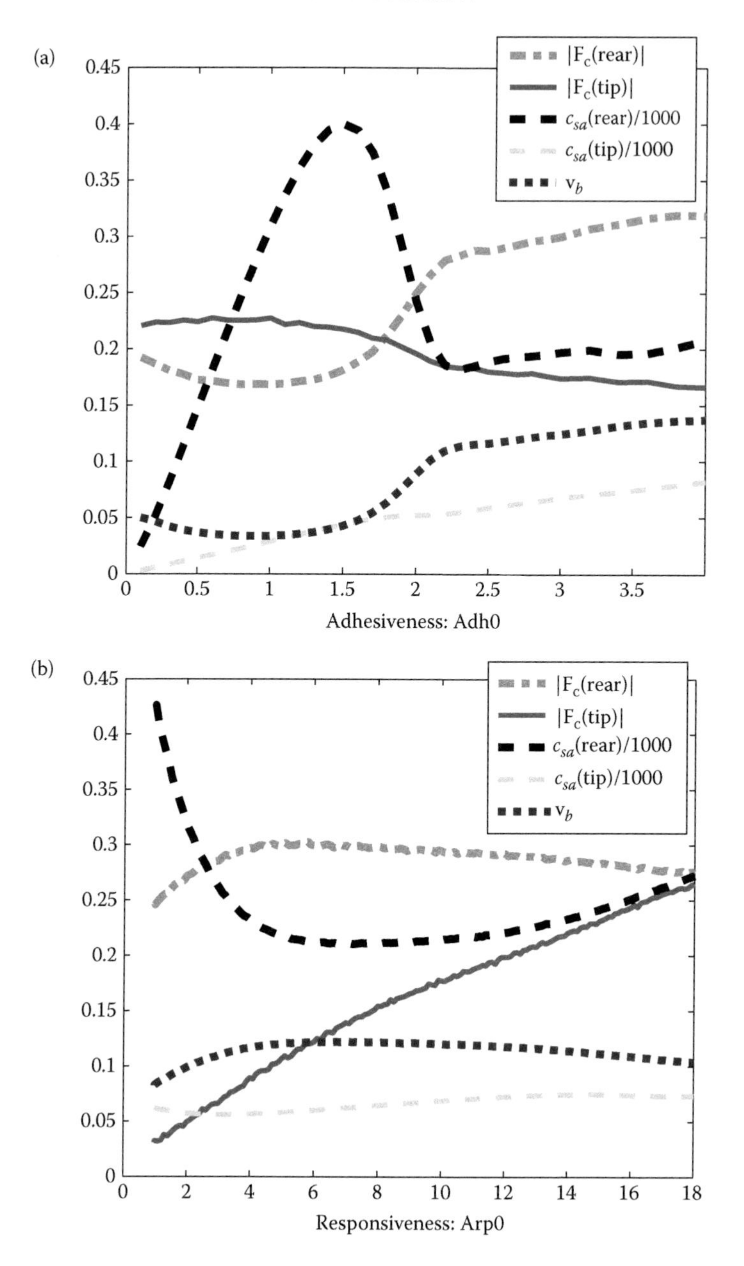

FIGURE 4.12 Variation of migration speed v_b and certain indicator functions at the cell edge as a function of (a) adhesiveness Adh0 and (b) F-actin responsiveness Arp0 [μM]; plotted are the load $|F_c|$ per FA site at the cell rear (bold) and tip (light), and the FA concentration c_{sa} at the rear (bold, dashed) and tip (light, dashed). Ripples in the force curves reflect the slight stochastic noise that was added in the force Equation (4.19).

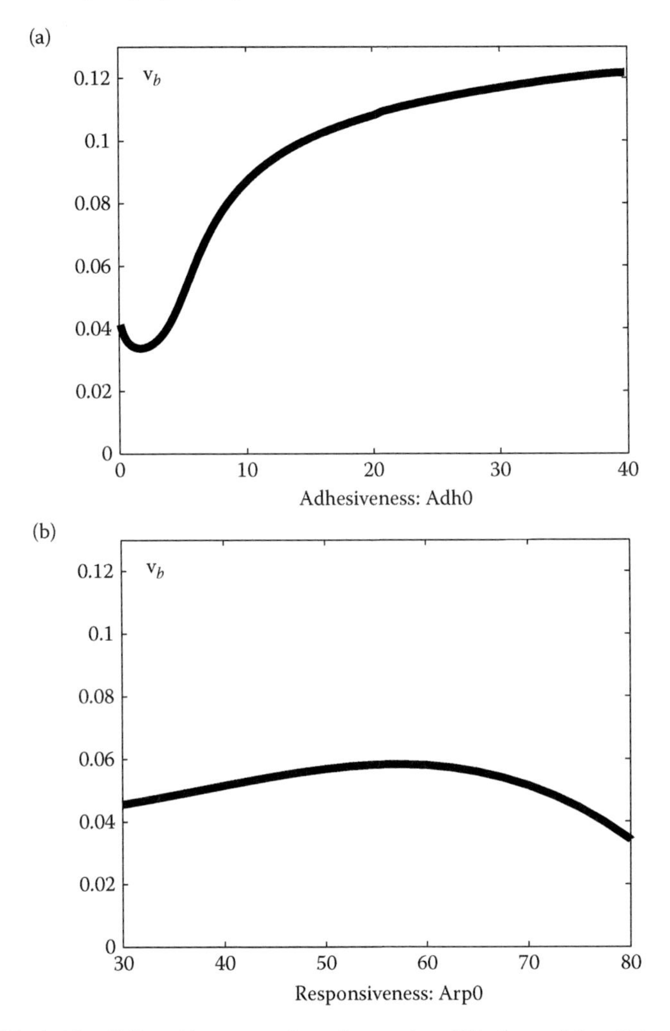

FIGURE 4.13 Migration speed v_b for a simplified model cell in its stable polarized state as a function of (*a*) adhesiveness and (*b*) F-actin responsiveness as described in Figure 4.10.

Starting with uniform FA concentration and a minor asymmetry in the distribution of actin, the system reaches a polarized and migrating state and the distributions of the various concentrations exhibit the same characteristics as in the full model (see Figure 4.9). We again analyze the effect of substratum adhesiveness and F-actin responsiveness on the migration speed; see Figure 4.13. In contrast to the observations for the full model (Figure 4.10), a saturation behavior of the migration velocity emerges with increasing adhesiveness instead of an optimal range. However, the responsiveness of the F-actin-network has an optimal value as in the full model.

Further simulations after successive variation of the frozen parameters (see above) reveal that an optimal range of adhesiveness could be gained only if a passive frictional load $\Phi_u > 0$ was chosen to increase monotonically with adhesion strength Adh0, as it was assumed in the full model; see Table 4.2 and Figure 4.10(a). Although such a friction appears to be physically realistic, the question remains whether such an optimal velocity response is of general biological relevance. It could be that the response of living systems is typically guided by adaptation: migrating cells show the capability to dissolve their focal adhesions by proteolytic cleavage of integrins [55], which could be an adaptive strategy to optimize cellular adhesiveness on various substrates. On the other hand, the adhesiveness could be raised by active segregation of extracellular matrix proteins.

Nevertheless, the universal optimality curve as a function of F-actin responsiveness, seen in both Figures 4.10(b) and 4.13(b), seems to hold for any mechanical model that implements a saturating F-actin polymerization by association of regulating proteins as Arp2/3. This indicates that an optimal control of migration velocity by tuning certain chemical processes such as F-actin polymerization is more easily realized than tuning adhesion, and thus may have evolved as a generally effective strategy to regulate mechanical processes such as cell adhesion and locomotion.

4.5 Discussion and Outlook

Based on a larger set of interwoven mass and force balance equations for mechano-chemical processes within the cytoplasm and the surrounding plasma membrane, the presented continuum model is a simplified although already sufficiently complex version of a more comprehensive 3-D whole-cell model. From the 1-D and 2-D simulation results we conclude that even without organizing centers (as the cell nucleus or microtubuli; cf. [31]) or regulating systems (as the Rho/RAC control cascades; cf. [54]), the F-actin cytoskeleton and its associated proteins having mechanical functions (myosin and integrin) constitute a self-organizing biophysical system with the ability of autonomous polarization and locomotion. The transition from a symmetric, unpolarized stationary state to a polarized migrating state can be mechanically induced or spontaneous due to stochastic or chaotic fluctuations, depending on the (meta-)stability of the stationary state. Clearly, for reproducing the mentioned experimental data, the 2-D system offers a wider spectrum of possibilities, one of which has been elaborated by Kozlov and Mogilner [33]. They prove a defined bi-stability between the radially symmetric cell shape and a polarized, circularly indented shape by allowing for different anisotropic organization of the actin–myosin cytoskeleton.

In comparison to this, further numerical "experiments" using our coupled viscous flow transport–diffusion–reaction system with free boundary should be performed to explore the capability of isotropic continuum models. Although local orientation or alignment of actin filaments will have an enforcing effect on the dynamics, interactions, and feedback among F-actin, crosslinkers, motor proteins, and adhesion complexes (for a 1-D model, see [21]), already the presented simulation results show that internal concentration gradients and directional flow, leading to persistent polarization, can emerge from local fluctuations. Thereby super-threshold pattern formation arises, not as in common excitable media by a purely reaction–diffusion feedback, but rather by a spatially distributed coupling between chemically induced force generation/relaxation and mechanically induced bond formation/dissociation.

More than a decade ago, Bereiter-Hahn and Lüers [8] claimed that, based on their experiments and measurements with migrating keratocytes, polarization and directionality of cells are determined by local variations in actin network stiffness and hydrostatic pressure. Together with the more recent high-resolution measurements of dynamic vector field (as F-actin velocity, forces onto pliable substrates, or directionality of FA shape deformations), it may soon be possible to probe and evaluate the various assumptions and hypotheses on mass and force balance conditions, which we postulate in Sections 4.3.1 and 4.3.2. For example, the biphasic correlations between retrograde actin flow speed and traction stress detected by Gardel et al. [22] could well be resolved by the measured and simulated FA gradient away from an advancing lamellar tip, and by our central modeling assumption that the local traction force $\mathbf{F}_v$ is proportional to the product $c_{sa} \cdot \mathbf{v}_c$ of FA concentration and cortical F-actin velocity.

Moreover, the distribution of flow and force, extensively discussed for the 1-D case (in Section 4.4.2), has to be quantitatively characterized for the 2-D free boundary situation (Section 4.4.1), where the overall picture is the same. Wider distribution of traction forces occurs in the leading front region (with maximum flow speed at the tip, depending on the active polymerization pressure) and strong disruptive counter-forces in a confined region of "rear release". The actin flow patterns of our 2-D simulations show, in principle, the same characteristics as experimental data by Yam et al. [64] for fish keratocytes. During the process of polarization, the simulated cell loses its radial symmetry of a uniform inward flow pattern, leading to a stronger inward flow in the rear region, whereby the most pronounced forces come from the two sides relative to the establishing locomotion direction; see Figure 4.6(c). Indeed, traction force experiments on polarized fish keratocytes have revealed the same pattern, with major traction forces from the two flanks at the rear and weaker retrograde forces at the front [50].

Further evaluation of the presented model should compare the simulated integrin concentration profiles with more precise quantitative data on the temporal turnover and spatial distribution of FAs in migrating cells, as already mentioned above. For this reason, the model must be extended to include stochastic clustering dynamics of integrins, which probably is an essential feature of FAs determining the detailed patterns of F-actin flow and traction forces. We emphasize that our model results do not depend on the specific mechanisms of myosin-II contraction. Essential for the emergence of coupled flow and mass gradients is the bimodal (cubic-like) stress function $S = S(\theta)$ in Figure 4.5, which similarly appears in a modified mechanical situation as, for example, the nematode sperm motility system; see [9].

The experimental response curves for the migration speed of polarized cells depending on adhesiveness [52] show a broad optimum range with a logarithmic slow decay for larger Adh0 values and a steep decay for lower values, similar to the curves found in our 1-D model (Figure 4.10) and in earlier analogous simulation results using alternative visco-elastic mechanics [17, 24]. The deviating behavior for very low adhesiveness values remains to be investigated with a more precise quantification of other (weaker) substrate-mediated forces that may come into account.

An optimum curve for the protrusion speed as a function of the barbed end density B, thus also of the concentration Arp0, has been obtained by Mogilner and Edelstein-Keshet [42] with the aid of 1-D diffusion-reaction-transport systems, including membrane resistance and the Brownian ratchet mechanism. However, they do not model retrograde actin flow or force transduction to the substratum, so that the shape of their curve ([42], Figure 6) essentially differs from our results in Figures 4.10(b) and 4.12(b).

Finally, in generalization of the model simulations by Stéphanou et al. [59] and in comparison with other modeling approaches, our full 2-D model should be used for reproducing the experimentally observed translocation paths and deformation structures of single blood or tissue cells in culture, which show well-expressed phases of persistent polarization and locomotion, interrupted by events of speed reduction, contractile rounding, successive re-polarization, and migration. Statistical analysis, in the same spirit as having been performed for the original annular lamella model [4], could help in revealing the microscopic mechanisms that are responsible for cell motility behavior at the macroscopic level.

Acknowledgments

This joint work has been generously supported by a grant on Simulation models for cell motility—Coupling substrate adhesion and cytoplasm dynamics (I/80543, Volkswagen-Stiftung). Moreover, the hospitality of the Soft Matter Physics group at Leipzig University during the final preparation of this chapter is greatly appreciated. Finally we thank our colleagues at Bonn University and the Research Center Jülich for helpful discussions and critical remarks.

References

[1] W. Alt (1996). Biomechanics of actomyosin-dependent mobility of keratinocytes. *Biofizika* 41:169–177.

[2] W. Alt (2003). Nonlinear hyperbolic systems of generalized Navier-Stokes type for interactive motion in biology. In *Geometric Analysis and Nonlinear Partial Differential Equations*, Eds. S. Hildebrandt and H. Karcher, Heidelberg, Springer, pp. 431–461.

[3] W. Alt and M. Dembo (1999). Cytoplasm dynamics and cell motion: two-phase flow models. *Math. Biosci.* 156:207–228.

[4] W. Alt and R. Tranquillo (1995). Basic morphogenetic system modeling shape changes of migrating cells: how to explain fluctuating lamellipodial dynamics. *J. Biol. Syst.* 3:905–916.

[5] W. Alt and R. Tranquillo (1997). Protrusion-retraction dynamics of an annular lamellipodial seam. In *Dynamics of Cell and Tissue Motion*, Eds. W. Alt, A. Deutsch, and G. Dunn, Basel, Birkhäuser, pp. 73–81.

[6] D. Ambrosi, A. Duperray, V. Peschetola and C. Verdier (2009). Traction patterns of tumor cells. *J. Math. Biol.* 58:163–181.

[7] G.I. Bell (1978). Models for the specific adhesion of cells to cells. *Science* 200:618–627.

[8] J. Bereiter-Hahn and H. Lüers (1998). Subcellular tension fields and mechanical resistance of the lamella front related to the direction of locomotion. *Cell Biochem. Biophys.* 29:243–262.

[9] D. Bottino, A. Mogilner, T. Roberts, M. Stewart, and G. Oster (2002). How nematode sperm crawl. *J. Cell. Sci.* 115:367–384.

[10] K. Burridge, M. Chrzanowska-Wodnicka, and C. Zhong (1997). Focal adhesion assembly. *Trends Cell Biol.* 7:342–347.

[11] M. Dembo (1986). The mechanics of motility in dissociated cytoplasm. *Biophys. J.* 50:1165–1183.

[12] M. Dembo (1989). Field theories of the cytoplasm. *Comments Theor. Biol.* 1:159–177.

[13] M. Dembo and F.W. Harlow (1986). Cell motion, contractile networks, and the physics of penetrating reactive flow. *Biophys. J.* 50:109–121.

[14] M. Dembo, F.W. Harlow, and W. Alt (1984). The biophysics of cell surface mobility. In *Cell Surface Dynamics, Concepts and Methods*, Eds. A.D. Perelson, Ch. DeLisi, and F.W. Wiegel, New York, Marcel Dekker, pp. 495–542.

[15] M. Dembo and Y.L. Wang (1999). Stresses at the cell-to-substrate interface during locomotion of fibroblasts. *Biophys. J.* 76:2307–2316.

[16] R.B. Dickinson (2009). Models for actin polymerization motors. *J. Math. Biol.* 58:81–103.

[17] P.A. DiMilla, K. Barbee, and D.A. Lauffenburger (1991). Mathematical model for the effects of adhesion and mechanics on cell migration speed. *Biophys. J.* 60:15–37.

[18] E. Evans and A. Yeung (1989). Apparent viscosity and cortical tension of blood granulocytes determined by micropipet aspiration. *Biophys. J.* 56:151–160.

[19] W. Feneberg, M. Westphal, and E. Sackmann (2001). Dictyostelium cells' cytoplasm as an active viscoplastic body. *Eur. Biophys. J.* 30:284–294.

[20] D.E. Frank and W.G. Carter (2004). Laminin 5 deposition regulates keratinocyte polarization and persistent migration. *J. Cell. Sci.* 117:1351–1363.

[21] J. Fuhrmann, J. Käs and A. Stevens (2007). Initiation of cytoskeletal asymmetry for cell polarization and movement. *J. Theor. Biol.* 249:278–288.

[22] M.L. Gardel, B. Sabass, L. Ji, G. Danuser, U.S. Schwarz, and C.M. Waterman (2008). Traction stress in focal adhesions correlates biphasically with actin retrograde flow speed. *J. Cell Biol.* 183:999–1005.

[23] A. Gholami, M. Falcke, and E. Frey (2008). Velocity oscillations in actin-based motility. *New J. Phys.* 10:033022.

[24] M.E. Gracheva and H.G. Othmer (2004). A continuum model of motility in amoeboid cells. *Bull. Math. Biol.* 66:167–193.

[25] H.P. Grimm, A.B. Verkhovsky, A. Mogilner, and J.-J. Meister (2003). Analysis of actin dynamics at the leading edge of crawling cells: implications for the shape of keratocyte lamellipodia. *Eur. Biophys. J.* 32:563–577.

[26] M. Herant, W.A. Marganski, and M. Dembo (2003). The mechanics of neutrophils: synthetic modeling of three experiments. *Biophys. J.* 84:3389–3413.

[27] B. Hinz, W. Alt, C. Johnen, V. Herzog, and H.W. Kaiser (1999). Quantifying lamella dynamics of cultured cells by SACED, a new computer-assisted motion analysis. *Exp. Cell Res.* 251:234–243.

[28] F. Huber, J. Käs, and B. Stuhrmann (2008). Growing actin networks form lamellipodium and lamellum by self-assembly. *Biophys. J.* 95:5508–5523.

[29] G. Jiang, G. Giannone, D.R. Critchley, E. Fukumoto, and M.P. Sheetz (2003). Two-piconewton slip bond between fibronectin and the cytoskeleton depends on talin. *Nature* 424:334–337.

[30] X. Jiang, D.A. Bruzewicz, A.P. Wong, M. Piel, and G.M. Whitesides (2005) Directing cell migration with asymmetric micropatterns. *Proc. Natl. Acad. Sci. USA* 102:975–978.

[31] I. Kaverina, O. Krylyshkina, and J.V. Small (2002). Regulation of substrate adhesion dynamics during cell motility. *Int. J. Biochem. Cell Biol.* 34:746–761.

[32] S.A. Koestler, S. Auinger, M. Vinzenz, K. Rottner, and J.V. Small (2008). Differentially oriented populations of actin filaments generated in lamellipodia collaborate in pushing and pausing at the cell front. *Nat. Cell Biol.* 10:306–313.

[33] M.M. Kozlov and A. Mogilner (2007). Model of polarization and bistability of cell fragments. *Biophys. J.* 93:3811–3819.

[34] K. Kruse, J.-F. Joanny, F. Jülicher, and J. Prost (2006). Contractility and retrograde flow in lammellipodium motion. *Phys. Biol.* 3:130–137.

[35] E. Kuusela and W. Alt (2008). Continuum model of cell adhesion and migration. *J. Math. Biol.* 58:135–161.

[36] D.A. Lauffenburger and A.F. Horwitz (1996). Cell migration: a physically integrated molecular process. *Cell* 84:359–369.

[37] T. Libotte, H.W. Kaiser, W. Alt, and T. Bretschneider (2001). Polarity, protrusion-retraction dynamics and their interplay during keratinocyte cell migration. *Exp. Cell Res.* 270:129–137.

[38] C.M. Lo, H.B. Wang, M. Dembo, and Y.L. Wang (2000). Cell movement is guided by the rigidity of the substrate. *Biophys. J.* 79:144–152.

[39] A.F. Marée, A. Jilkine, A. Dawes, V.A. Grieneisen, and L. Edelstein-Keshet (2006). Polarization and movement of keratocytes: a multiscale modelling approach. *Bull. Math. Biol.* 68:1169–1211.

[40] R. Merkel, N. Kirchgessner, C.M. Cesa, and B. Hoffmann (2007). Cell force microscopy on elastic layers of finite thickness. *Biophys. J.* 93:3314–3323.

[41] A. Mogilner (2009). Mathematics of cell motility: have we got its number? *J. Math. Biol.* 58:105–134.

[42] A. Mogilner and L. Edelstein-Keshet (2002). Regulation of actin dynamics in rapidly moving cells: a quantitative analysis. *Biophys. J.* 83:1237–1258.

[43] A. Mogilner and G. Oster (1996). Cell motility driven by actin polymerization. *Biophys. J.* 71:3030–3045.

[44] A. Mogilner and G. Oster (2003). Polymer motors: pushing out the front and pulling up the back. *Curr. Biol.* 13:R721–R733.

[45] C. Möhl (2005). Modellierung von Adhäsions- und Cytoskelett-Dynamik in Lamellipodien migratorischer Zellen. Diploma thesis, Universität Bonn, Germany.

[46] S.I. Nishimura and M. Sasai (2007). Modulation of the reaction rate of regulating protein induces large morphological and motional change of amoebic cell. *J. Theor. Biol.* 245:230–237.

[47] S.I. Nishimura, M. Ueda, and M. Sasai (2009). Cortical factor feedback model of cellular locomotion and cytofission. *PLoS Comp. Biol.* 5(3):e1000310.

[48] D. Oelz, C. Schmeiser, and A. Soreff (2005). Multistep navigation of leukocytes: a stochastic model with memory effects. *Math. Med. Biol.* 22:291–303.

[49] J.M. Oliver, J.R. King, K.J. McKinlay, P.D. Brown, D.M. Grant, C.A. Scotchford, and J.V. Wood (2005). Thin-film theories for two-phase reactive flow models of active cell motion. *Math. Med. Biol.* 22:53–98.

[50] T. Oliver, M. Dembo, and K. Jacobson (1999). Separation of propulsive and adhesive traction stresses in locomoting keratocytes. *J. Cell Biol.* 145:589–604.

[51] S.P. Palecek, A.F. Horwitz, and D.A. Lauffenburger (1999). Kinetic model for integrin-mediated adhesion release during cell migration. *Ann. Biomed. Eng.* 27:219–235.

[52] S.P. Palecek, J.C. Loftus, M.H. Ginsberg, D.A. Lauffenburger, and A.F. Horwitz (1997). Integrin-ligand binding properties govern cell migration speed through cell-substratum adhesiveness. *Nature* 385:537–540.

[53] A. Ponti, M. Machacek, S.L. Gupton, C.M. Waterman-Storer, and G. Danuser (2004). Two distinct actin networks drive the protrusion of migrating cells. *Science* 305:1782–1786.

[54] A.J. Ridley, M.A. Schwartz, K. Burridge, R.A. Firtel, M.H. Ginsberg, G. Borisy, J.T. Parsons, and A.R. Horwitz (2003). Cell migration: integrating signals from front to back. *Science* 302:1704–1709.

[55] L. Satish, H.C. Blair, A. Glading, and A. Wells (2005). Interferon-inducible protein 9 (CXCL11)-induced cell motility in keratinocytes requires calcium flux-dependent activation of μ-calpain. *Mol. Cell. Biol.* 25:1922–1941.

[56] U.S. Schwarz, N.Q. Balaban, D. Riveline, A. Bershadsky, B. Geiger, and S.A. Safran (2002). Calculation of forces at focal adhesions from elastic substrate data: the effect of localized force and the need for regularization. *Biophys. J.* 83:1380–1394.

[57] U. Seifert (2000). Rupture of multiple parallel molecular bonds under dynamic loading. *Phys. Rev. Lett.* 84:2750–2753.

[58] J.V. Small, T. Stradal, E. Vignal, and K. Rottner (2002). The lamellipodium: where motility begins. *Trends Cell Biol.* 12:112–120.

[59] A. Stéphanou, E. Mylona, M. Chaplain, and P. Tracqui (2008). A computational model of cell migration coupling the growth of focal adhesions with oscillatory cell protrusions. *J. Theor. Biol.* 253:701–716.

[60] T.M. Svitkina, A.B. Verkhovsky, K.M. McQuade, and G.G. Borisy (1997). Analysis of the actinmyosin II system in fish epidermal keratocytes: mechanism of cell body translocation. *J. Cell Biol.* 139:397–415.

[61] M. Théry, V. Racine, A. Pépin, M. Piel, Y. Chen, J.B. Sibarita, and M. Bornens (2005). The extracellular matrix guides the orientation of the cell division axis. *Nat. Cell Biol.* 7:947–953.

[62] C.E. Turner (2000). Paxillin and focal adhesion signalling. *Nat. Cell Biol.* 2:E231–E236.

[63] A.B. Verkhovsky, T.M. Svitkina, and G.G. Borisy (1999). Self-polarization and directional motility of cytoplasm. *Curr. Biol.* 9:11–20.

[64] P.T. Yam, C.A. Wilson, L. Ji, B. Hebert, E.L. Barnhart, N.A. Dye, P.W. Wiseman, G. Danuser, and J.A. Theriot (2007). Actin-myosin network reorganization breaks symmetry at the cell rear to spontaneously initiate polarized cell motility. *J. Cell Biol.* 178:1207–1221.

[65] C. Zhu and R. Skalak (1988). A continuum model of protrusion of pseudopod in leukocytes. *Biophys. J.* 54:1115–1137.

Chapter 5

How Do Cells Move? Mathematical Modeling of Cytoskeleton Dynamics and Cell Migration

Dietmar Ölz and Christian Schmeiser

Contents

5.1 Introduction

The description of cell migration by the action of a lamellipodium given in [15] is reproduced here for completeness. Cells migrate by protruding at the front and retracting at the rear. Protrusion occurs in thin membrane-bound cytoplasmic sheets, 0.2 to 0.3 μm thick and several microns long, termed lamellipodia [22]. The major structural components of lamellipodia are actin filaments, which are organized in a more or less 2-D diagonal array with the fast-growing plus ends of the actin filaments directed forward, abutting the membrane [21]. Protrusion is effected by actin polymerization, whereby actin monomers are inserted at the plus ends of the filaments at the membrane interface and removed at the minus ends, throughout and at the base of the lamellipodium, in a treadmilling regime [16]. Stabilization of the actin meshwork is achieved by the crosslinking of the filaments by actin-associated proteins, such as filamin [12], as well as protein complexes such as the Arp2/3 [18], although the density and location of such crosslinks remain to be established. Because actin polymerization is involved in diverse motile processes aside from cell motility, including endocytosis and the propulsion of pathogens that invade cytoplasm [2], the question of how actin filaments are able to push against a membrane has spawned the development of various models [9].

Modeling efforts go back to 1996 and fall into two groups. The first group includes continuum models for the mechanical behavior of cytoplasm: a two-phase formulation for cytosol and the actin network [1], a 1-D viscoelastic model [4], a 1-D model for the actin distribution [10]; and a 2-D elastic continuum model [19]. The second group makes presumptions about the microscopic organization of the actin network. The Brownian ratchet model for the polymerization process introduced by Mogilner and Oster [11] considers actin crosslinking proteins as stabilizers of the lamellipodium meshwork, allowing enough flexibility for actin filaments to bend away from the membrane to accept actin monomers. Other models are based on the current idea [18] that the actin filaments in lamellipodia form a branched network with the Arp2/3 complex at the branch points [6,7,20]. A related model considers the lamellipodia as constructed from short filaments that take one of six orientations [8].

Recent studies have indicated that filaments in lamellipodia are not organized in branched arrays [5]. Rather, the pseudo-2-D actin network contains unbranched filaments whereby the filament density decreases from the front to the rear of the lamellipodium, indicative of a graded distribution of filament lengths. According to this structural information, we present a quasi-stationary modeling approach for the simulation of the turnover of the lamellipodium surrounding a cell on a flat substrate. This corresponds to *in vitro* situations such as cytoplasmic fragments of keratocytes [23]. Our approach differs from previous ones in that we describe the lamellipodium in terms of a continuous distribution of filaments of graded length and their linkages.

In this work the models from [15] and [13] are generalized from rotationally symmetric to arbitrary geometries. Otherwise, the assumptions on the mechanics of the network are the same: there is an elastic resistance against bending of actin filaments, against stretching and twisting of crosslinks between the filaments, against polymerization of the barbed ends by the membrane, and against the stretching of trans-membrane linkages (called adhesions) between filaments and the substrate. Our model of the cell–substrate interaction assumes a homogeneous isotropic substrate. However, inclusion of substrate inhomogeneity and anisotropy (cf. [3]) does not represent a principal obstacle.

Section 5.3 derives the basic model with a detailed probabilistic account of the life cycles of crosslinks and adhesions. The model is simplified by carrying out the limit of instantaneous crosslink and adhesion turnover. A picture from a time-dependent simulation is reproduced from [14], where numerical methods for simulations with the limiting model are presented. Finally, the connection to the results from [15] and [13] is established in Section 5.4 by showing that the general model possesses rotationally symmetric solutions. An analysis of the stability of a trivial steady state illustrates simulation results presented in [15].

5.2 Modeling

A central feature of the model is an age-structured production/decay of crosslinks and adhesions, consistent with dynamic association/dissociation of linkage molecules with the actin network. To obtain a feasible mathematical description we adopt a homogenization limit, based on the assumption that the density of filaments within the lamellipodium is very high. We let the number of filaments tend to infinity to obtain a model based on continuous quantities instead of discrete ones.

The assumptions made are as follows:

A1: The lamellipodium is 2-D and has the topology of a ring; that is, it lies between two closed curves.

A2: All actin filaments belong to one of two families, called clockwise and counterclockwise. Filaments of the same family do not cross each other. Crossings of clockwise with counterclockwise filaments are transversal. All barbed ends touch the leading edge of the lamellipodium, that is, the outer curve of the previous assumption. Filaments are inextensible.

A3: Filaments polymerize at the barbed ends with given polymerization speed. Depolymerization at the pointed ends is a stochastic process with prescribed distribution.

As a consequence of A1 and A2, the lamellipodium has the organization depicted in Figure 5.1.

There are two families of locally parallel filaments. Looking from the center of the lamellipodium ring, the filaments in the first group bear to the right and the second group to the left (relative to each other); referred to as clockwise and counterclockwise filaments, respectively.

We assume the presence of n^+ clockwise filament with indices $i = 0, \dots, n^+-1$ and n^- counterclockwise filament with indices $j = 0, \dots, n^- -1$. An arc length parametrization of the clockwise filaments at time t is given by $\{F_i^+(t,s) : \ -L_i^+(t) \le s \le 0\} \subset \mathbb{R}^2$, where $s = -L_i^+(t)$ corresponds to the pointed end and $s = 0$ to the barbed end. In a similar manner we represent the counterclockwise filaments at time t by $\{F_j^-(t,s) : \ -L_j^-(t) \le s \le 0\} \subset \mathbb{R}^2$ such that $|\partial_s F_i^+| \equiv |\partial_s F_j^-| \equiv 1$.

The lengths $L_i^+(t)$ and $L_j^-(t)$ of the filaments are random variables whose distributions are considered as given by

$$P(-L_i^+(t) \le s) = \eta_i^+(t,s), \quad P(-L_j^-(t) \le s) = \eta_j^-(t,s) \tag{5.1}$$

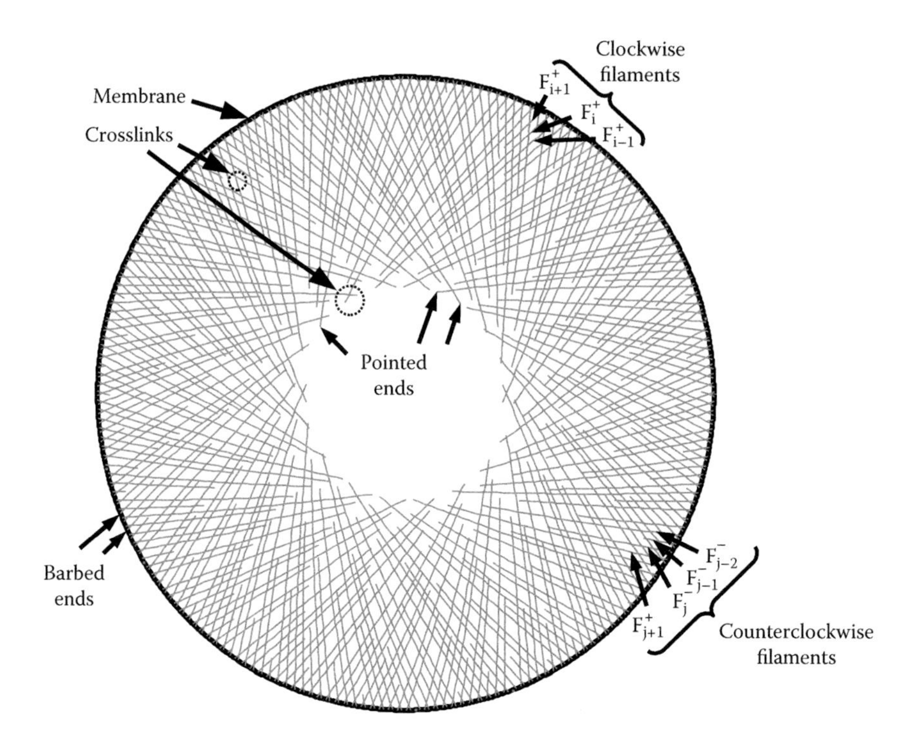

FIGURE 5.1 Constituent elements of model.

The arc length s is a geometric parameter. Because of the polymerization at the barbed ends, polymerized actin molecules travel along the filament toward the pointed ends with the polymerization speed denoted by $v_i^+(t)$ and $v_j^-(t)$, respectively. Because of this and because of the inextensibility assumption in A2, Lagrange variables along the filaments are given by $\sigma_i^+ = s + \int_0^t v_i^+(\tilde{t})d\tilde{t}$ and $\sigma_j^- = s + \int_0^t v_j^-(\tilde{t})d\tilde{t}$. In other words, the path of the actin molecule with label σ_i^+ on the clockwise filament F_i^+ is given by $F_i^+(t, \sigma_i^+ - \int_0^t v_i^+(\tilde{t})d\tilde{t})$. The fact that only depolymerization happens at the pointed ends is reflected by the assumption that

$$-L_i^+(t) + \int_0^t v_i^+(\tilde{t})d\tilde{t} \quad \text{and} \quad -L_j^-(t) + \int_0^t v_j^-(\tilde{t})d\tilde{t} \tag{5.2}$$

are increasing functions of time. As a consequence, the distribution functions η_i^+ and η_j^- are decreasing with respect to time when written in terms of the Lagrangian variables. In other words, $\partial_t \eta_i^+ - v_i^+ \partial_s \eta_i^+ \leq 0$ and $\partial_t \eta_j^- - v_j^- \partial_s \eta_j^- \leq 0$.

The next step is to describe the kinematics of crosslinks, and we start with the modeling assumptions:

A4: A crosslink is a connection between a material point on a clockwise and a material point on a counterclockwise filament. Crosslinks can be created spontaneously at the crossing between two filaments and they can also break. Creation and breaking are stochastic processes. There exists at most one crosslink for any pair of filaments at any time.

For the creation of crosslinks, we need the information about filament crossings. We assume that each clockwise filament crosses each counterclockwise filament at most once. At time t, the crossings are described by a set of index pairs:

$$\overline{\mathcal{C}}(t) := \left\{(i, j) : \exists\, s^+_{i,j}(t),\, s^-_{i,j}(t) \text{ such that } F^+_i(t, s^+_{i,j}(t)) = F^-_j(t, s^-_{i,j}(t))\right\} \quad (5.3)$$

If a crosslink between the filaments with indices i and j is created at time t^*, then this happens at the crossing point. Once established, however, the two binding sites will travel with the material along the two filaments. Thus, at a later time $t = t^* + a$, the binding sites will be located at

$$F^+_i(t, s^+_{a,i,j}(t))\,, \quad s^+_{a,i,j}(t) := s^+_{i,j}(t-a) - \int_{t-a}^{t} v^+(\tilde{t})d\tilde{t} \quad (5.4)$$

$$F^-_j(t, s^-_{a,i,j}(t))\,, \quad s^-_{a,i,j}(t) := s^-_{i,j}(t-a) - \int_{t-a}^{t} v^-(\tilde{t})d\tilde{t} \quad (5.5)$$

until the crosslink eventually breaks. We call a the age of the crosslink. Below we assume a resistance of crosslinks against stretching and twisting. This means there are elastic forces related to the stretching

$$S_{i,j}(t, a) := F^+_i(t, s^+_{a,i,j}(t)) - F^-_j(t, s^-_{a,i,j}(t)) \quad (5.6)$$

and to the twisting

$$T_{i,j}(t, a) := \varphi_{i,j}(t, a) - \varphi_0 \quad (5.7)$$

where

$$\varphi_{i,j}(t, a) := \arccos[\partial_s F^+_i(t, s^+_{a,i,j}(t)) \cdot \partial_s F^-_j(t, s^-_{a,i,j}(t))] \quad (5.8)$$

is the angle between the directions of the filaments at the binding sites, and φ_0 is an equilibrium angle determined by the properties of the crosslinking molecule. We allow also obtuse angles $0 \leq \varphi,\ \varphi_0 \leq \pi$ allowing for crosslinkers sensitive to the orientation of actin filaments.

The probability distribution for the existence of a crosslink with respect to age will be denoted by $r_{i,j}(t, a)$, where

$$\int_0^\infty r_{i,j}(t, a)\, da \leq 1 \quad (5.9)$$

is the probability that a crosslink between the ith clockwise filament and the jth counterclockwise filament exists at time t. We postulate the following model for the evolution of the distribution:

$$\begin{cases} \partial_t r_{i,j} + \partial_a r_{i,j} = -\zeta(S_{i,j}, T_{i,j}) r_{i,j}, \\ r_{i,j}(t,0) = \beta(T_{i,j}(t,0)) \left(1 - \int_0^\infty r_{i,j}(t,a)\, da\right) \end{cases} \tag{5.10}$$

This model has the standard form of age-structured population models (see, for example, [17]). The differential equation describes aging and breaking of crosslinks, the boundary condition at $a = 0$ describes their creation. The dependence of the breaking rate on stretching and twisting reflects that a crosslink might be broken by being loaded too much. The twisting dependence of the creation rate β could eliminate the possibility for a crosslink to be established, if the angle between the filaments is too far from the equilibrium angle. Integration of the differential equation with respect to a shows that the second factor in the creation rate guarantees Equation (5.9), that is, the fact that there is at most one crosslink. Just as for the pointed-end (de)polymerization process, all we need to know about the processes of creation and breaking of crosslinks is the distribution $r_{i,j}$.

The domain of the differential equation in model (5.10) is determined by the requirement that both binding sites (on the ith clockwise filament and on the jth counterclockwise filament) have not been depolymerized yet: $s^+_{a,i,j}(t) > -L^+_i(t)$, $s^-_{a,i,j}(t) > -L^-_j(t)$.

The next modeling step is the passage to a continuum description by letting the total numbers of filaments n^+ and n^- tend to infinity. In the limit, the discrete indices $\alpha^+_i = \pi(2i/n^+ - 1)$, $i = 0, \dots, n^+ - 1$ and $\alpha^-_j = \pi(2j/n^- - 1)$, $j = 0, \dots, n^- - 1$ are replaced by a continuous parameter $\alpha \in [-\pi, \pi)$. Then we interpret the discrete filament positions $F^+_i(t,s)$ and $F^-_j(t,s)$ as approximations for the values $F^+(t, \alpha_i, s)$ and, respectively, $F^-(t, \alpha_j, s)$ of continuous functions

$$F^\pm : [0, \infty) \times B \to \mathbb{R}^2, \quad \text{with } B := [-\pi, \pi) \times [-L, 0] \tag{5.11}$$

where L is a maximal length of filaments. Note that we assume periodicity of B in the sense that all functions of α are 2π-periodic. The fact that filaments of the same family do not cross implies that $F^\pm(t, \cdot)$ has to be one-to-one. The shape of the lamellipodium at time t is given by $\Omega(t) = F^+(t, B) \cup F^-(t, B)$. According to assumption A1, its boundary consists of an inner and an outer curve: $\partial\Omega(t) = \partial\Omega_{in}(t) \cup \partial\Omega_{out}(t)$. The fact that, by assumption A2, all barbed ends touch the leading edge, takes the mathematical form $\partial\Omega_{out}(t) = \{F^\pm(t, \alpha, 0) : \ -\pi \le \alpha < \pi\}$.

Continuous versions of the length distributions η^+_i and η^-_j are given by

$$\eta^\pm : \ [0, \infty) \times B \to [0, 1] \tag{5.12}$$

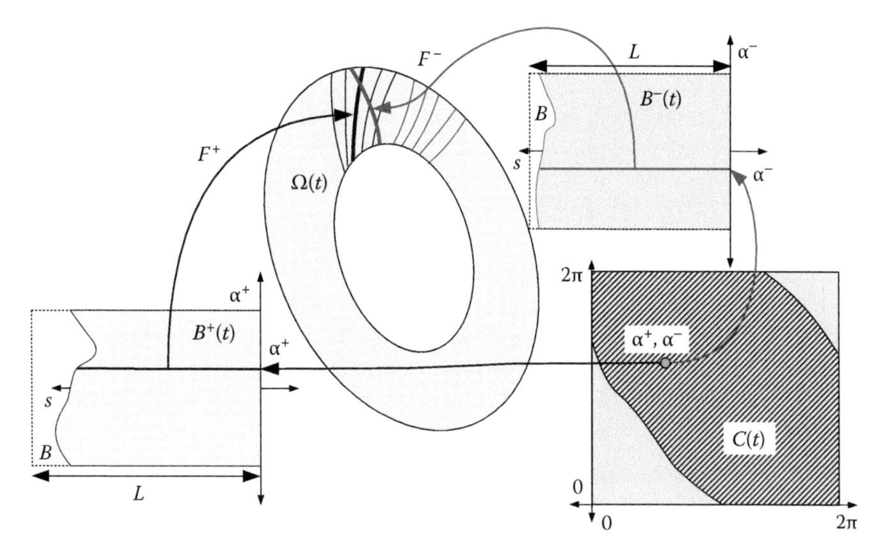

FIGURE 5.2 Functional framework of model.

now getting the deterministic interpretation as the expected fraction of filaments in each index element $d\alpha$, whose length at time t is bigger than $-s$. They are nondecreasing functions of s satisfying $\eta^{\pm}(t, \alpha, 0) = 1$.

Crossings of filaments can only occur in $\Omega_c(t) = F^+(t, B) \cap F^-(t, B) \subset \Omega(t)$. Similar to $\overline{\mathcal{C}}(t)$, we construct a set of index pairs by

$$\begin{aligned} \mathcal{C}(t) =& \big\{(\alpha^+, \alpha^-) \in [-\pi, \pi)^2 : \ \exists s^{\pm}(t, \alpha^+, \alpha^-) \text{ such that} \\ & F^+(t, \alpha^+, s^+(t, \alpha^+, \alpha^-)) = F^-(t, \alpha^-, s^-(t, \alpha^+, \alpha^-))\big\} \end{aligned} \tag{5.13}$$

which corresponds to the discrete set $\overline{\mathcal{C}}(t)$ in the sense that $(\alpha_i, \alpha_j) \in \mathcal{C}(t)$ if $(i, j) \in \overline{\mathcal{C}}(t)$. Consistent with the assumption that two given filaments cross at most once, we assume that for each pair $(\alpha_+, \alpha_-) \in \mathcal{C}(t)$ there is only one $s^{\pm}(t, \alpha^+, \alpha^-)$. Defining

$$B^{\pm}(t) := \big\{(\alpha^{\pm}, s^{\pm}(t, \alpha^+, \alpha^-)) : \ (\alpha_+, \alpha_-) \in \mathcal{C}(t)\big\} \subset B \tag{5.14}$$

the maps $(\alpha_+, \alpha_-) \mapsto (\alpha^{\pm}, s^{\pm}(t, \alpha^+, \alpha^-))$ from $\mathcal{C}(t)$ to $B^{\pm}(t)$ are invertible. Combining one of them with the other's inverse gives an invertible map $(\alpha^+, s^+) \mapsto (\alpha^-(t, \alpha^+, s^+), s^-(t, \alpha^+, s^+))$ from $B^+(t)$ to $B^-(t)$. We complete the description of the geometry of crossings by defining the angle between crossing filaments:

$$\varphi(t, \alpha^+, \alpha^-) = \arccos[\partial_s F^+(t, \alpha^+, s^+(t, \alpha^+, \alpha^-)) \cdot \partial_s F^-(t, \alpha^-, s^-(t, \alpha^+, \alpha^-))] \tag{5.15}$$

We introduce the polymerization rates $v^{\pm}(t, \alpha)$ such that v_i^+ approximates $v^+(t, \alpha_i)$ and v_j^- approximates $v^-(t, \alpha_j)$ and abbreviate the s-values at time t

of crosslinks with age a by

$$s_a^+(t,\alpha^+,\alpha^-) := s^+(t-a,\alpha^+,\alpha^-) - \int_{t-a}^{t} v^+(\tilde{t},\alpha^+)d\tilde{t} \tag{5.16}$$

$$s_a^-(t,\alpha^+,\alpha^-) := s^-(t-a,\alpha^+,\alpha^-) - \int_{t-a}^{t} v^-(\tilde{t},\alpha^-)d\tilde{t} \tag{5.17}$$

The age-dependent probability distribution of crosslinks can be understood as an expected crosslink density $\rho(t,\alpha^+,\alpha^-,a)$ with $(\alpha^+,\alpha_-) \in \mathcal{C}(t-a)$. This condition means that for a crosslink of age a connecting the filaments with labels α^+ and α^- to exist at time t, the filaments must have crossed at time $t-a$. From Equation (5.10) the transport equation for the crosslink density becomes

$$\partial_t\rho + \partial_a\rho = -\zeta(S,T)\rho \tag{5.18}$$

Boundary conditions are required at $a = 0$, describing the creation of new crosslinks, and for $(\alpha^+,\alpha_-) \in \partial\mathcal{C}(t-a)_+$, with $\partial\mathcal{C}(t-a)_+$ denoting the part of the boundary of $\mathcal{C}(t-a)$, where the filaments with labels α^+ and α^- start to have a crossing. In other words, the time direction points into the domain of ρ at these points. The boundary data there have to be zero because there are no preexisting crosslinks. So the boundary conditions are

$$\rho(a=0) = \beta(T_0)\left(1 - \int_0^\infty \rho\, da\right), \quad \rho(t,\alpha^+,\alpha^-,a) = 0 \quad \text{for } (\alpha^+,\alpha_-) \in \partial\mathcal{C}(t-a)_+ \tag{5.19}$$

with $T_0 = T(a=0)$. Note that the upper bound in the integration should actually be $t - t_0(\alpha^+,\alpha^-)$ with $(\alpha^+,\alpha_-) \in \partial\mathcal{C}(t_0)_+$, but for simplicity we consider the definition of ρ as continued by zero to arbitrary values of $a > 0$. The stretching and twisting terms are now given by

$$S(t,\alpha^+,\alpha^-,a) = F^+(t,\alpha^+,s_a^+(t,\alpha^+,\alpha^-)) - F^-(t,\alpha^-,s_a^-(t,\alpha^+,\alpha^-)) \tag{5.20}$$

$$T(t,\alpha^+,\alpha^-,a) = \varphi_a(t,\alpha^+,\alpha^-,a) - \varphi_0 \tag{5.21}$$

with

$$\varphi_a(t,\alpha^+,\alpha^-) = \arccos\left[\partial_s F^+(t,\alpha^+,s_a^+(t,\alpha^+,\alpha^-)) \cdot \partial_s F^-(t,\alpha^-,s_a^-(t,\alpha^+,\alpha^-))\right] \tag{5.22}$$

The boundedness property (Inequality (5.9)) of the microscopic crosslink density determined by Equation (5.10) carries over to the modified model (Equations (5.18) and (5.19)). The accumulated distribution

$$\bar{\rho}(t,\alpha^+,\alpha^-) = \int_0^\infty \rho(t,\alpha^+,\alpha^-,a)\, da \tag{5.23}$$

satisfies the equation

$$\partial_t \bar{\rho} = -\int_0^\infty \zeta\rho\, da + \beta(1-\bar{\rho}) \tag{5.24}$$

preserving the property $0 \leq \bar{\rho}(t, \alpha^+, \alpha^-) \leq 1$.

Taking into account the length distribution of the filaments, we arrive at the effective crosslink density

$$\begin{aligned}\rho_{eff}(t, \alpha^+, \alpha^-, a) = \rho(t, \alpha^+, \alpha^-, a)\eta^+(t, \alpha^+, s_a^+(t, \alpha^+, \alpha^-))\\ \times \eta^-(t, \alpha^-, s_a^-(t, \alpha^+, \alpha^-))\end{aligned} \tag{5.25}$$

where each of the two filaments involved in a crosslink contributes a factor $\eta^\pm$. Note that ρ_{eff} satisfies

$$\partial_t\rho_{eff} + \partial_a\rho_{eff} = -\rho_{eff}\left(\zeta(S,T) - \frac{\partial_t\eta^+ - v^+\partial_s\eta^+}{\eta^+} - \frac{\partial_t\eta^- - v^-\partial_s\eta^-}{\eta^-}\right) \tag{5.26}$$

hence the same type of transport equation as ρ but with a modified decay rate, which takes into account the loss of crosslinks due to depolymerization of the pointed ends. Note that just as in the microscopic model, the fact that only depolymerization happens at the pointed ends is described by the inequality $\partial_t\eta^\pm - v^\pm\partial_s\eta^\pm \leq 0$.

Now we turn to the dynamics of adhesion molecules:

A5: An adhesion is a connection between a material point on a filament and a point on the substrate via a transmembrane linkage. Adhesions can be created spontaneously and they can also break. Creation and breaking are stochastic processes. The number of adhesions per filament length is restricted.

The densities $\rho_{adh}^\pm(t, \alpha, s, a)$ of adhesions on clockwise and counterclockwise filaments, respectively, satisfy the differential equations

$$\partial_t\rho_{adh}^\pm + \partial_a\rho_{adh}^\pm - v^\pm\partial_s\rho_{adh}^\pm = -\zeta^{adh}(S_{adh}^\pm)\rho_{adh}^\pm \tag{5.27}$$

with the boundary conditions

$$\rho_{adh}^\pm(a=0) = \beta^{adh}\left(\bar{\rho}_{\max}^{adh} - \int_0^\infty \rho_{adh}^\pm\, da\right), \quad \rho_{adh}^\pm(s=0) = 0 \tag{5.28}$$

where $\bar{\rho}_{\max}^{adh}$ is the maximal density of crosslinks along the filament and the breaking rate ζ^{adh} depends on the stretching of the adhesions:

$$S_{adh}^\pm(t, \alpha, s, a) = F^\pm(t, \alpha, s) - F^\pm\left(t-a, \alpha, s + \int_{t-a}^t v^\pm(\tilde{t}, \alpha)d\tilde{t}\right) \tag{5.29}$$

As for the crosslink density, the second boundary condition means that there are no pre-existing adhesions on newly polymerized parts of the filaments.

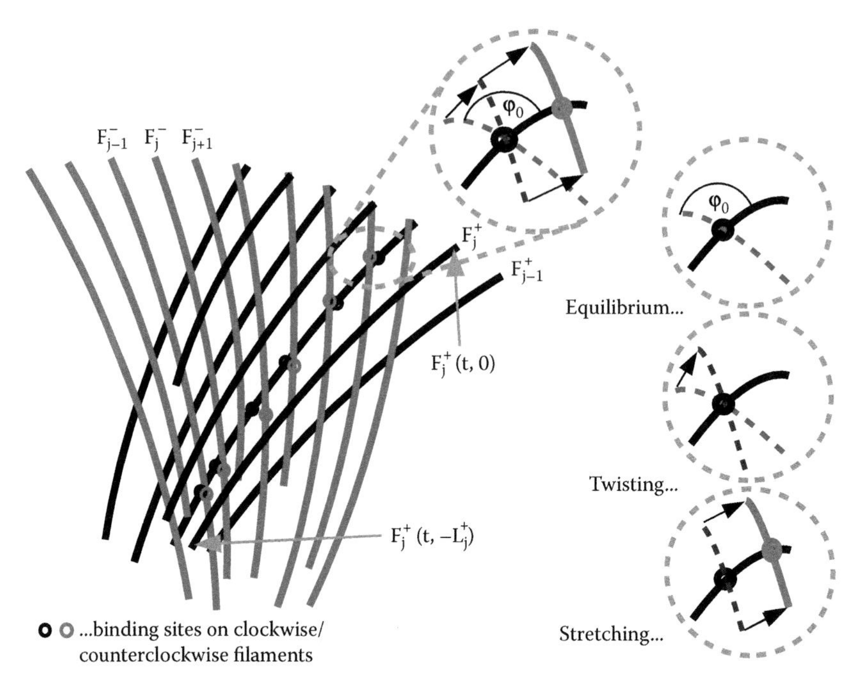

FIGURE 5.3 Crosslinks: microscopic description and possible transformations.

It remains to formulate assumptions determining the position of the filaments:

> **A6:** The position of the filaments is determined by a quasi-stationary balance of elastic forces resulting from bending the filaments, stretching and twisting the crosslinks, stretching the adhesions, and stretching the cell membrane around the leading edge.

The quasi-stationarity assumption means that elastic oscillations are neglected because the filament network is damped by viscous forces in the cytosol. Thus, the dynamics of the network result only from (de)polymerization and from the creation and breaking of crosslinks and adhesions.

Mathematically, the assumption A6 will be formulated by assuming that the filament positions minimize a potential energy functional containing contributions from the above-mentioned elastic effects. We require the functions $F^+(t, \cdot, \cdot)$ and $F^-(t, \cdot, \cdot)$ to minimize the functional

$$U(t)[G^+, G^-] := U^+_{bending}(t)[G^+] + U^-_{bending}(t)[G^-] + U_{scl+tcl}(t)[G^+, G^-] \\ + U_{membrane}[G^\pm] + U^+_{adh}(t)[G^+] + U^-_{adh}(t)[G^-] \qquad (5.30)$$

The energy contribution from bending the filaments is the Kirchhoff bending energy from standard linearized beam theory:

$$U^{\pm}_{bending}(t)[G^{\pm}] := \frac{\kappa^B}{2}\int_B \left|\partial_s^2 G^{\pm}\right|^2 \eta^{\pm}\, d(\alpha, s) \tag{5.31}$$

where κ^B is the bending stiffness of one filament and the notation $\int_B \cdots d(\alpha, s)$ denotes the double integral with respect to α and s.

The stretching and twisting energy of the crosslinks is modeled by

$$U_{scl+tcl}(t)[G^+, G^-] := \int_0^\infty \int_{C(t-a)} \left(\frac{\kappa^S}{2}|S|^2 + \frac{\kappa^T}{2}T^2\right)\rho_{eff}\, d(\alpha^+, \alpha^-)\, da \tag{5.32}$$

with the quantities S and T from Equations (5.20) and (5.21) applied to G^+ and G^-, respectively. The constants κ^S and κ^T are Hooke constants describing the stretching and, respectively, torsional stiffnesses of the crosslink molecules. Here, too, the notation $\int c(t-a) \cdots d(\alpha^+, \alpha^-)$ denotes the double integral with respect to α^+ and α^-.

Furthermore the energies associated with the stretching of integrins on clockwise and counterclockwise filaments, respectively, read

$$U^{\pm}_{adh}(t)[G^{\pm}] := \frac{\kappa^F}{2}\int_B \int_0^\infty \left|G^{\pm} - F^{\pm}\left(t-a, \alpha^{\pm}, s + \int_{t-a}^{t} v^{\pm}(\tilde{t}, \alpha)\, d\tilde{t}\right)\right|^2 \times \rho^{\pm}_{adh}\eta^{\pm}\, da\, d(\alpha, s) \tag{5.33}$$

Note that the evaluation of the adhesion energy and of the crosslink stretching and twisting energy at time t requires information on previous filament positions. Because of the treadmilling of filaments, the lifetime of monomers in a filament and, thus, of binding sites of crosslinks and adhesion molecules is finite. The densities ρ_{eff} and $\rho^{\pm}_{adh}$ have compact supports in the age direction and the above integrals with respect to a can be restricted to these supports. It is obvious that past filament positions enter into the computation of the adhesion energy. However, this is also the case for the energy in crosslinks, where past filament positions enter into the computation of $s^{\pm}_a$.

It remains to model the action of the cell membrane on the leading edge of the network:

A7: The cell membrane simulates a rubber band stretched around the barbed ends of the filaments.

This leads to a model of the form

$$U_{membrane}[G^{\pm}] := \kappa^M \left(\frac{C^+[G^+] + C^-[G^-]}{2} - C_0\right)_+^2 \tag{5.34}$$

with the circumference of the lamellipodium $C^+[G^+] = C^-[G^-]$, given by either one of the two equivalent formulations

$$C^{\pm}[G^{\pm}] := \int_{-\pi}^{\pi} |\partial_\alpha G^{\pm}(t, \alpha, 0)|\, d\alpha \tag{5.35}$$

The arithmetic mean is used in the energy for symmetry reasons. This models resistance against stretching the membrane as soon as its circumference exceeds the equilibrium value C_0.

The positions of the filaments at time t is now determined by minimizing the energy:

$$U(t)[F^+(t, \cdot, \cdot), F^-(t, \cdot, \cdot)] = \min U(t)[G^+, G^-] \tag{5.36}$$

under the constraint of inextensibility

$$|\partial_s G^+| = |\partial_s G^-| = 1 \tag{5.37}$$

and under the constraint that all barbed ends touch the leading edge:

$$\{G^+(t, \alpha, 0) : \; -\pi \le \alpha < \pi\} = \{G^-(t, \alpha, 0) : \; -\pi \le \alpha < \pi\} \tag{5.38}$$

The formulation of a well-posed problem still requires a start-up procedure. It first involves a decision about the maximal age A of all crosslinks and adhesions that are present at time $t = 0$. In other words, initial data $\rho(0, \alpha^+, \alpha^-, a)$ and $\rho^{\pm}_{adh}(0, \alpha, s, a)$ have to be prescribed, which vanish for $a > A$. To be able to compute the binding sites of all the initial crosslinks and adhesions, the positions $F^{\pm}(t, \alpha, s)$ of the filaments and the polymerization rates $v^{\pm}(t, \alpha)$ for $-A \le t \le 0$ have to be given.

5.3 Limit of Instantaneous Crosslink and Adhesion Turnover

For two reasons, numerical simulations with the model presented in the previous section are very costly. On the one hand, the densities of crosslinks and adhesions are functions of three variables ((α^+, α^-, a) and (α, s, a), respectively). On the other hand, the problem for the deformation of the filaments is a delay problem, such that the history of the filament dynamics must be stored up to the maximal age of crosslinks and adhesions.

Therefore a simplification will be carried out below that can also be motivated by the fact that in the model, macroscopic and microscopic scales are still mixed. The typical length and bending radius of a filament will in general be large compared to the size of a crosslinking or adhesion molecule, even when the latter is stretched, as assumed in our model. This also makes it plausible that the lifetime of such a connection is short compared to the typical time scale for the dynamics of the network. We make this assumption although it is not true for all applications. It means, for example, that we exclude the build-up of large adhesion complexes from our considerations.

Asymptotic methods will be used to derive an approximative problem, where the crosslink and adhesion densities can be computed explicitly. Also the delays disappear, and the problem becomes local in time at the expense of time derivatives appearing in the equations for the filament displacements. A physical interpretation of the approximation is that the rapid turnover of crosslinks and adhesions can be described as effective friction.

A nondimensionalization is carried out, where the typical filament length L is used as reference length for scaling s, $F^{\pm}$, $S_{\pm}$, S_{adh}, and C_0. With v_0 being the reference speed for the polimerization speed $v_{\pm}$, the reference time L/v_0 for scaling t is the typical time an actin molecule spends in a filament between polymerization at the barbed end and depolymerization at the pointed end. The birth and death rates of crosslinks and adhesions β, ζ, β^{adh}, and ζ^{adh} will be assumed to be of the same order of magnitude with a typical value $1/\bar{a}$ used for nondimensionalization, where $\bar{a}$ can be interpreted as typical lifetime of crosslinks and adhesions. Our main scaling assumption is that the dimensionless parameter

$$\varepsilon := \frac{\bar{a}v_0}{L} \tag{5.39}$$

is small. The reference values for the densities of cross-links and of adhesions are $1/(\bar{a}L)$ and, respectively, $\bar{\rho}^{adh}_{\max}/\bar{a}$. Finally, energy ($U$ and all its contributions) is scaled with the reference value $\kappa^B L$, and the reference values for κ^S, κ^T, κ^A, and κ^M, are $\kappa^B/(\varepsilon L)$, $\kappa^B L$, $\kappa^B/(\varepsilon\bar{\rho}^{adh}_{\max} L)$, and κ^B/L, respectively.

For notational simplicity, the same symbols will be used for the dimensionless quantities as for their dimensional counterparts. The filament displacement $F^{\pm}(t,\cdot,\cdot)$ is determined as a minimizer (with the side conditions (5.37) and (5.38)) of the sum of the scaled energy contributions

$$\begin{aligned}
U_{membrane}[G^{\pm}] &= \kappa^M \left(\frac{C^+[G^+] + C^-[G^-]}{2} - C_0 \right)_+^2 \\
U^{\pm}_{bending}(t)[G^{\pm}] &= \frac{1}{2} \int_B \left|\partial_s^2 G^{\pm}\right|^2 \eta^{\pm} \, d(\alpha, s) \\
U_{scl}(t)[G^+, G^-] &= \frac{\kappa^S}{2\varepsilon} \int_0^{\infty} \int_{\mathcal{C}(t-\varepsilon a)} |S|^2 \rho_{eff} \, d(\alpha^+, \alpha^-) \, da \\
U_{tcl}(t)[G^+, G^-] &= \frac{\kappa^T}{2} \int_0^{\infty} \int_{\mathcal{C}(t-\varepsilon a)} T^2 \rho_{eff} \, d(\alpha^+, \alpha^-) \, da \\
U^{\pm}_{adh}(t)[G^{\pm}] &= \frac{\kappa^F}{2\varepsilon} \int_0^{\infty} \int_B |G^{\pm} - F^{\pm *}|^2 \rho^{\pm}_{adh} \eta^{\pm} \, d(\alpha, s) \, da
\end{aligned} \tag{5.40}$$

with

$$
\begin{aligned}
S &= G^+(t, \alpha^+, s^+_{\varepsilon a}(t, \alpha^+, \alpha^-)) - G^-(t, \alpha^-, s^-_{\varepsilon a}(t, \alpha^+, \alpha^-)) \\
T &= \varphi_{\varepsilon a}(t, \alpha^+, \alpha^-) - \varphi_0 ,
\end{aligned} \tag{5.41}
$$

and

$$
F^{\pm *} := F^{\pm}\left(t - \varepsilon a, \alpha^{\pm}, s + \int_{t-\varepsilon a}^{t} v^{\pm}(\tilde{t}, \alpha)\, d\tilde{t}\right) \tag{5.42}
$$

The problems for the crosslink density and for the adhesion density become

$$
\begin{cases}
\varepsilon \partial_t \rho + \partial_a \rho = -\zeta(S, T)\rho \\
\rho(a = 0) = \beta(T_0)\left(1 - \int_0^\infty \rho\, da\right) \\
\rho(t, \alpha^+, \alpha^-, a) = 0 \quad \text{for} \quad (\alpha^+, \alpha^-) \in \partial\mathcal{C}(t - \varepsilon a)_+ \\
\rho(t = 0) = \rho_I
\end{cases} \tag{5.43}
$$

with $T_0 = \arccos[\partial_s F^+(t, \alpha^+, s^+) \cdot \partial_s F^-(t, \alpha^-, s^-)] - \varphi_0$ and, respectively,

$$
\begin{cases}
\varepsilon D_t^{\pm} \rho^{\pm}_{adh} + \partial_a \rho^{\pm}_{adh} = -\zeta^{adh}(S^{\pm}_{adh})\rho^{\pm}_{adh} \\
\rho^{\pm}_{adh}(a = 0) = \beta^{adh}\left(1 - \int_0^\infty \rho^{\pm}_{adh}\, da\right) \\
\rho^{\pm}_{adh}(s = 0) = 0 \\
\rho^{\pm}_{adh}(t = 0) = \rho^{\pm}_{adh,I}
\end{cases} \tag{5.44}
$$

with $S^{\pm}_{adh}(t, \alpha^{\pm}, s, a) = F^{\pm}(t, \alpha^{\pm}, s) - F^{\pm}(t - \varepsilon a, \alpha^{\pm}, s + \varepsilon \int_{t-a}^{t} v^{\pm}(\tilde{t})\, d\tilde{t})$. The initial data $\rho_I = \rho_I(\alpha^+, \alpha^-, a)$ and $\rho^{\pm}_{adh,I} = \rho^{\pm}_{adh,I}(s, a)$ have to be prescribed according to the discussion on Page 144. For notational convenience, the material derivative

$$
D_t^{\pm} := \partial_t - v^{\pm}\partial_s \tag{5.45}
$$

is used here and in the following.

The reason for the scaling assumption that the stiffnesses of the adhesions and of the crosslinks against stretching are $O(\varepsilon^{-1})$ relative to the other stiffnesses is *a priori* not obvious. Actually, when replacing $G^{\pm}(\alpha^{\pm}, s)$ by the minimizer $F^{\pm}(t, \alpha^{\pm}, s)$, the energies $U_{scl}(t)[F^+, F^-]$ and $U_{adh}(t)[F^{\pm}]$ are formally $O(\varepsilon)$. It will be shown below that in the limit $\varepsilon \to 0$ they still contribute to the minimization conditions.

The above-mentioned approximation is derived by passing to the limit $\varepsilon \to 0$. We start with Problem (5.44). The formal limiting equations

$$
\partial_a \rho^{\pm}_{adh} = -\zeta^{adh}(0)\rho^{\pm}_{adh} , \quad \rho^{\pm}_{adh}(a = 0) = \beta^{adh}\left(1 - \int_0^\infty \rho^{\pm}_{adh}\, da\right) \tag{5.46}
$$

have the solution

$$\rho_{adh}^{\pm}(t, \alpha^{\pm}, s, a) = \frac{\beta^{adh}\zeta^{adh}(0)}{\beta^{adh} + \zeta^{adh}(0)} e^{-\zeta^{adh}(0)a} \tag{5.47}$$

This is a singular limit because the small parameter ε multiplies the material derivative. As a consequence, the conditions at $s = 0$ and at $t = 0$ cannot be satisfied by the limiting (outer, in the language of singular perturbation theory) solution. We ignore eventual boundary and initial layers (i.e., thin regions close $s = 0$ and $t = 0$ with strong variation of the solution). In a similar manner, we deal with Problem (5.43). The main difference is that the limiting equations

$$\partial_a \rho = -\zeta(0, T_0)\rho\,, \quad \rho(a = 0) = \beta(T_0)\left(1 - \int_0^{\infty} \rho\, da\right) \tag{5.48}$$

and therefore also the limiting solution

$$\rho = \frac{\beta(T_0)\zeta(0, T_0)}{\beta(T_0) + \zeta(0, T_0)} e^{-\zeta(0, T_0)a} \tag{5.49}$$

depend on the displacement of the filaments via T_0. As announced above, the dependence is local in time.

If the limit $\varepsilon \to 0$ is carried out formally in Expressions (5.40), the contributions from the adhesions and from stretching the crosslinks disappear. To reveal their influence, the solution of the variational problem needs to be discussed.

The displacement $F^{\pm}(t, \cdot, \cdot)$ at time t has to satisfy the variational equation

$$\delta U[F^+, F^-](\delta F^+, \delta F^-) = 0 \tag{5.50}$$

for all admissible variations $(\delta F^+, \delta F^-)$, where δU is the variation of the total energy, that is, the sum of all terms in (5.40). Admissibility conditions for the variations are a consequence of the constraints $|\partial_s F^{\pm}| \equiv 1$ and $\{F^+(t, \alpha, 0) : \; -\pi \le \alpha < \pi\} = \{F^-(t, \alpha, 0) : \; -\pi \le \alpha < \pi\}$. The derivation of a strong formulation of the variational equations will be facilitated by a Lagrange multiplier approach, where we employ the Lagrange functions $\lambda^{\pm} = \lambda^{\pm}(\alpha, s)$, $\lambda_{edge} = \lambda_{edge}(\alpha^{\pm})$ and the additional functionals

$$U_{ext}^{\pm}[G^{\pm}] = \frac{1}{2}\int_B \lambda^{\pm}(\alpha, s)\left(|\partial_s G^{\pm}(\alpha, s)|^2 - 1\right)\eta^{\pm}\, d(\alpha, s) \tag{5.51}$$

describing extension of filaments, and

$$U_{edge}[G^+, G^-] = \int_{-\pi}^{\pi} \lambda_{edge}(\alpha^+)(G^+(t, \alpha^+, 0) - G^-(t, \hat{\alpha}(t, \alpha^+), 0)) \cdot \nu(t, \alpha^+) d\alpha^+ \tag{5.52}$$

describing the deviation between the outer edges of both filament families. Here $\hat{\alpha}(t, \alpha^+)$ is chosen such that $G^+(t, \alpha^+, 0) - G^-(t, \hat{\alpha}(t, \alpha^+), 0)$ is parallel to $\nu(t, \alpha^+)$, the outward unit normal vector along the barbed ends of the clockwise filaments, which can be computed from $\partial_\alpha G^+(t, \alpha^+, 0)$.

In the following the variations of the energy contributions and their limits as $\varepsilon \to 0$ are computed individually.

1. The variation of the stretching energy of the membrane reads

$$\delta U_{membrane}[F^\pm]\delta F^\pm = \kappa^M \left(C^\pm - C_0\right)_+ \times \int_{-\pi}^{\pi} \frac{\partial_\alpha F^\pm(s=0)}{|\partial_\alpha F^\pm(s=0)|} \cdot \partial_\alpha \delta F^\pm(s=0)\, d\alpha \quad (5.53)$$

with the same expression in the limit $\varepsilon \to 0$.

2. For the variation of the bending energy of the filaments we obtain

$$\delta U^\pm_{bending}[F^\pm]\delta F^\pm = \int_B \partial_s^2 F^\pm \cdot \partial_s^2 \delta F^\pm \, \eta^\pm \, d(\alpha, s) \quad (5.54)$$

again with the same expression for $\varepsilon \to 0$.

3. The variation of the energy contribution by stretching the cross links can now be written as

$$\delta U_{scl}(t)[F^+, F^-]\delta F^\pm = \pm \frac{\kappa^S}{\varepsilon} \int_0^\infty \int_{C(t-\varepsilon a)} S\delta F^\pm(t, \alpha^\pm, s^\pm_{\varepsilon a}) \times \rho_{eff}\, d(\alpha^+, \alpha^-)\, da \quad (5.55)$$

We can write Expressions (5.41) as

$$\begin{aligned} S = {} & F^+(t, \alpha^+, s^+_{\varepsilon a}) - F^+(t-\varepsilon a, \alpha^+, s^+(t-\varepsilon a, \alpha^+, \alpha^-)) \\ & - \left(F^-(t, \alpha^-, s^-_{\varepsilon a}) - F^-(t-\varepsilon a, \alpha^+, s^-(t-\varepsilon a, \alpha^+, \alpha^-))\right) \end{aligned} \quad (5.56)$$

This implies $S = \varepsilon a(D_t F^+ - D_t F^-) + O(\varepsilon^2)$, and therefore passing to the limit $\varepsilon \to 0$ gives

$$\delta U_{scl}(t)[F^+, F^-]\delta F^\pm = \pm \int_{C(t)} \mu^S(T_0)(D_t F^+ - D_t F^-)\delta F^\pm \, \eta^+\eta^-\, d(\alpha^+, \alpha^-) \quad (5.57)$$

with

$$\mu^S(T_0) = \kappa^S \int_0^\infty a\rho\, da = \frac{\kappa^S \beta(T_0)}{\zeta(0, T_0)(\beta(T_0) + \zeta(0, T_0))} \quad (5.58)$$

4. Before we compute the variation of the twisting energy, we observe that our formula for the angle between filaments is only valid if the constraint $|\partial_s F^\pm| = 1$ holds. Because in the Lagrange multiplier approach,

variations also are allowed that violate this condition, we reformulate the definition of the angle as

$$\cos \varphi_{\varepsilon a} = \frac{\partial_s F^+}{|\partial_s F^+|}(s = s_{\varepsilon a}^+) \cdot \frac{\partial_s F^-}{|\partial_s F^-|}(s = s_{\varepsilon a}^-) \tag{5.59}$$

With $T = \varphi_{\varepsilon a} - \varphi_0$ and $(\delta \frac{x}{|x|})_{|x|=1} = (x^\perp \cdot \delta x) x^\perp$, (here and from now on we write orthogonal vectors as $(x_1, x_2)^\perp = (-x_2, x_1)$), we obtain

$$\delta T[F^+, F^-]\delta F^\pm = -\frac{(\partial_s F^{\pm\perp} \cdot \partial_s F^\mp)(\partial_s F^{\pm\perp} \cdot \partial_s \delta F^\pm)}{\sin \varphi_{\varepsilon a}} = \mp \partial_s F^{\pm\perp} \cdot \partial_s \delta F^\pm \tag{5.60}$$

where $\partial_s F^\pm$ and $\partial_s \delta F^\pm$ are evalutated at $(t, \alpha^\pm, s_{\varepsilon a}^\pm)$. The sign in the last term is due to the fact that the superscript + indicates the family of clockwise filaments. It then holds that

$$\delta U_{tcl}(t)[F^+, F^-]\delta F^\pm = \mp \kappa^T \int_0^\infty \int_{C(t-\varepsilon a)} T(\partial_s F^{\pm\perp} \cdot \partial_s \delta F^\pm) \rho_{eff}\, d(\alpha^+, \alpha^-)\, da \tag{5.61}$$

and therefore, as $\varepsilon \to 0$, we conclude

$$\delta U_{tcl}(t)[F^+, F^-]\delta F^\pm = \mp \int_{C(t)} \mu^T(T_0) T_0 (\partial_s F^{\pm\perp} \cdot \partial_s \delta F^\pm) \eta^+ \eta^-\, d(\alpha^+, \alpha^-) \tag{5.62}$$

where now $\partial_s F^\pm$ and $\partial_s \delta F^\pm$ are evalutated at $(t, \alpha^\pm, s^\pm(t, \alpha^+, \alpha^-))$ and

$$\mu^T(T_0) = \kappa^T \int_0^\infty \rho\, da = \frac{\kappa^T \beta(T_0)}{\beta(T_0) + \zeta(0, T_0)} \tag{5.63}$$

5. The variation of the stretching energy of the adhesions is straightforward and reads

$$\delta U_{adh}^\pm[F^\pm]\delta F^\pm = \frac{\kappa^A}{\varepsilon} \int_0^\infty \int_B (F^\pm - F^{\pm *}) \cdot \delta F^\pm\, \rho_{adh}^\pm \eta^\pm\, d(\alpha, s)\, da \tag{5.64}$$

In the limit $\varepsilon \to 0$, a material derivative occurs similarly to the stretching of the crosslinks:

$$\delta U_{adh}^\pm[F^\pm]\delta F^\pm = \mu^A \int_B D_t^\pm F^\pm \cdot \delta F^\pm\, \eta^\pm\, d(\alpha, s) \tag{5.65}$$

with

$$\mu^A = \kappa^A \int_0^\infty a \rho^{adh}\, da = \frac{\kappa^A \beta^{adh}}{\zeta^{adh}(0)(\beta^{adh} + \zeta^{adh}(0))} \tag{5.66}$$

6. In the two Lagrangian terms, we do not include the contributions from the variation of the Lagrange multipliers. For the inextensibility term we obtain

$$\delta U^{\pm}_{ext}[F^{\pm}]\delta F^{\pm} = \int_B \lambda^{\pm}\partial_s F^{\pm} \cdot \partial_s \delta F^{\pm}\, \eta^{\pm}\, d(\alpha, s) \tag{5.67}$$

7. The term that guarantees that all pointed ends touch the leading edge gives

$$\delta U_{edge}[F^+, F^-]\delta F^{\pm} = \pm \int_{-\pi}^{\pi} \lambda^{\pm}_{edge} \nu \cdot \delta F^{\pm}(s=0)\, d\alpha \tag{5.68}$$

where $\lambda^{+}_{edge} = \lambda_{edge}(\alpha)$ and $\lambda^{-}_{edge} = \lambda_{edge}(\alpha^+(t, \alpha, 0))$.

Collecting the terms computed under (1) through (7) leads to the variational equation

$$\begin{aligned} &\int_{-\pi}^{\pi} \left[\kappa^M (C^{\pm} - C_0)_+ \frac{\partial_\alpha F^{\pm}}{|\partial_\alpha F^{\pm}|} \cdot \partial_\alpha \delta F^{\pm} \pm \lambda^{\pm}_{edge} \nu \cdot \delta F^{\pm} \right]_{s=0} d\alpha \\ &\pm \int_{C(t)} \left(\mu^S(T_0)(D_t F^+ - D_t F^-)\delta F^{\pm} - \mu^T(T_0) T_0 \partial_s F^{\pm\perp} \cdot \partial_s \delta F^{\pm} \right) \eta^+ \eta^-\, d(\alpha^+, \alpha^-) \\ &+ \int_B \left(\partial_s^2 F^{\pm} \cdot \partial_s^2 \delta F^{\pm} + \mu^A D_t^{\pm} F^{\pm} \cdot \delta F^{\pm} + \lambda^{\pm} \partial_s F^{\pm} \cdot \partial_s \delta F^{\pm} \right) \eta^{\pm}\, d(\alpha, s) = 0 \end{aligned} \tag{5.69}$$

where now there is no restriction on the variations δF^+ and δF^-. The first integral corresponds to the leading edge and will contribute boundary conditions to a strong formulation of the problem. From the second and third integrals, the Euler-Lagrange equations will be derived. For that purpose the integration domains have to be mapped to each other. Noting that in the second integral $F^{\pm}$ and $\delta F^{\pm}$ and their derivatives are evaluated at $(t, \alpha^{\pm}, s^{\pm})$, we employ the transformations $(\alpha^+, \alpha^-) \mapsto (\alpha, s) = (\alpha^{\pm}, s^{\pm}(t, \alpha^+, \alpha^-))$. We incorporate the corresponding Jacobians and the fact that these terms only contribute in $B^{\pm}(t)$ into the macroscopic stiffness parameters for the crosslinks:

$$\mu^S_{\pm} = \begin{cases} \mu^S \left| \dfrac{\partial \alpha^{\mp}}{\partial s^{\pm}} \right| & \text{in } B^{\pm}(t), \\ 0 & \text{in } B \setminus B^{\pm}(t), \end{cases}$$

$$\mu^T_{\pm} = \begin{cases} \mu^T \left| \dfrac{\partial \alpha^{\mp}}{\partial s^{\pm}} \right| & \text{in } B^{\pm}(t) \\ 0 & \text{in } B \setminus B^{\pm}(t) \end{cases} \tag{5.70}$$

where the interpretation of the additional factor is the number of crossings per unit length. The Euler-Lagrange equations are given by

$$\begin{aligned}
&\partial_s^2\left(\eta^{\pm}\partial_s^2 F^{\pm}\right) - \partial_s\left(\eta^{\pm}\lambda^{\pm}\partial_s F^{\pm}\right) + \eta^{\pm}\,\mu^A\,D_t^{\pm}F^{\pm}\\
&\pm\,\partial_s\left(\eta^{+}\eta^{-}\,(\varphi-\varphi_0)\,\mu_{\pm}^T(\varphi-\varphi_0)\,\partial_s F^{\pm\perp}\right)\\
&\pm\,\eta^{+}\eta^{-}\mu_{\pm}^S(\varphi-\varphi_0)\left(D_t^{+}F^{+} - D_t^{-}F^{-}\right) = 0
\end{aligned} \tag{5.71}$$

The terms in the first row correspond to standard linear models for the deformation of beams. The first term corresponds to bending, the second to stretching (just the right amount such that $|\partial_s F^{\pm}| = 1$ holds), and the third to friction caused by adhesion to the substrate. All these terms are evaluated at (t, α, s), and obviously none of them generates any coupling in α, that is, between different filaments. The terms in the second and third lines describe the effects of crosslinking. Note that in the equation for F^{+}, the derivatives of F^{-} have to be evaluated at $(t, \alpha^{-}(t, \alpha, s), s^{-}(t, \alpha, s))$ and vice versa, employing the mapping between $B^{+}(t)$ and $B^{-}(t)$. The last term shows that the macroscopic effect of the resistance against stretching of crosslinks is friction caused by the relative motion of the two filament families.

The solutions of the Euler-Lagrange Equations (5.71) have to satisfy the boundary conditions

$$\begin{cases}
-\partial_s\left(\eta^{\pm}\partial_s^2 F^{\pm}\right) + \eta^{\pm}\lambda^{\pm}\partial_s F^{\pm} \mp \eta^{+}\eta^{-}\mu_{\pm}^T(\varphi-\varphi_0)\partial_s F^{\pm\perp} = 0\,, \quad \text{for } s = -L\\
\partial_s\left(\eta^{\pm}\partial_s^2 F^{\pm}\right) - \lambda^{\pm}\partial_s F^{\pm} \pm \mu_{\pm}^T(\varphi-\varphi_0)\partial_s F^{\pm\perp}\\
\qquad\qquad = \pm\lambda_{edge}^{\pm}\nu - \kappa^M\left(C^{\pm} - C_0\right)_+\partial_\alpha\left(\dfrac{\partial_\alpha F^{\pm}}{|\partial_\alpha F^{\pm}|}\right), \quad \text{for } s = 0\\
\eta^{\pm}\partial_s^2 F^{\pm} = 0\,, \quad \text{for } s = -L,\, 0
\end{cases} \tag{5.72}$$

The Lagrange parameters $\lambda^{\pm}$ and $\lambda_{edge}^{\pm}$ have to be determined such that the constraints are satisfied:

$$\begin{aligned}
&|\partial_s F^{+}| = |\partial_s F^{-}| = 1\\
&\{F^{+}(t,\alpha,0):\ -\pi \le \alpha < \pi\} = \{F^{-}(t,\alpha,0):\ -\pi \le \alpha < \pi\}
\end{aligned} \tag{5.73}$$

The problem in (5.71)–(5.73) is the formal limit as $\varepsilon \to 0$ of Equations (5.40)–(5.44). Figure 5.4 shows one frame of a time-dependent simulation based on this model carried out in [14]. It describes a situation where an originally circular cell is pushed from the left side and returns to its circular shape after the pushing force has been turned off. The pushing forces induce deformation and a steady movement to the right. We remark that the observed deformation is not of elastic nature, although the shape becomes round again after the applied force ceases to be active. The stability of the round shape seems to be the result of the dissipative nature of the model.

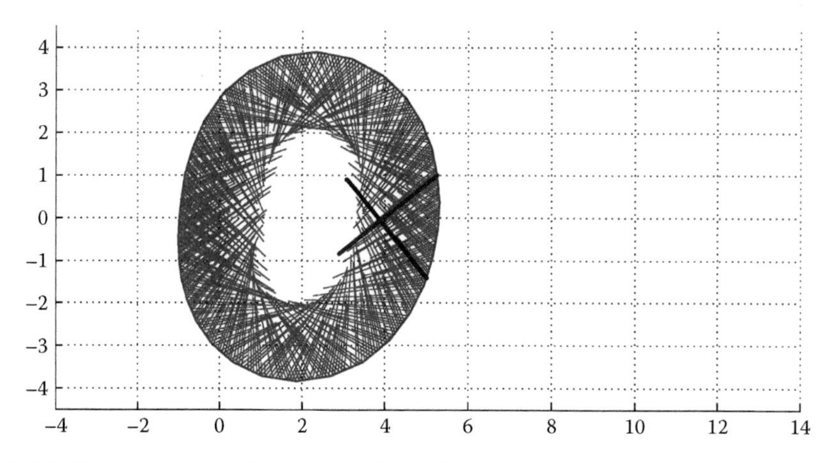

FIGURE 5.4 Pushed lamellipodium. Simulation result at time $t = 1.4$.

5.4 Rotationally Symmetric Solutions

In [15] a model has been derived where in addition to the assumptions of Section 5.2, the lamellipodium was assumed to be rotationally symmetric, that is, to have the shape of a circular ring. The limit of instantaneous crosslink and adhesion turnover in this symmetric model has been carried out in [13]. In this section we demonstrate that the model of [13] can be recovered as a special case of Equations (5.71) through (5.73).

A symmetric solution is only possible with symmetric data. This means that the prescribed polymerization speeds and the length distributions must be the same all around the lamellipodium:

$$v^{\pm}(t,\alpha) = v(t), \qquad \eta^{\pm}(t,\alpha,s) = \eta(t,s) \tag{5.74}$$

The first equation gives $D_t := D_t^+ = D_t^- = \partial_t - v\partial_s$. We search for solutions where all filament positions can be computed from the positions of one reference filament, such that all clockwise filaments are constructed by rotation of the reference filament, whereas a reflection followed by rotations is used for the counterclockwise filaments. Using the matrices of rotation and reflection/rotation

$$R(\alpha) := \begin{pmatrix} \cos\alpha & -\sin\alpha \\ \sin\alpha & \cos\alpha \end{pmatrix}, \qquad D(\alpha) := R(\alpha)\begin{pmatrix} 1 & 0 \\ 0 & -1 \end{pmatrix} \tag{5.75}$$

we make the ansatz

$$F^+(t,\alpha^+,s) = R(\alpha^+)z(t,s), \qquad F^-(t,\alpha^-,s) = D(-\alpha^-)z(t,s) \tag{5.76}$$

where $z(t, s)$, $-L \le s \le 0$, denotes the arc-length parametrization of the reference filament, in the following sometimes written in terms of polar coordinates: $z = |z|(\cos\phi, \sin\phi)$, $-\pi < \phi \le \pi$.

A straightforward computation shows that at crossings of filaments, that is, $F^+(t, \alpha^+, s) = F^-(t, \alpha^-, s)$,

$$\alpha^+ + \alpha^- = -2\phi \tag{5.77}$$

holds. With $D_t z = zD_t|z|/|z| + z^\perp D_t\phi$, this implies

$$\begin{aligned} D_t^+ F^+ - D_t^- F^- &= (R(\alpha^+) - D(-\alpha^-))D_t z = R(\alpha^+)(\mathbb{I} - D(-\alpha^+ - \alpha^-))D_t z \\ &= R(\alpha^+)(\mathbb{I} - D(2\phi))z^\perp D_t\phi = 2R(\alpha^+)z^\perp D_t \arg z \end{aligned} \tag{5.78}$$

where the symbol $\mathbb{I}$ denotes the identity matrix, and

$$\begin{aligned} \cos\varphi &= \partial_s F^+ \cdot \partial_s F^- = \partial_s z \cdot D(2\phi)\partial_s z = (\partial_s|z|)^2 - (|z|\partial_s\phi)^2 \\ &= 2(\partial_s|z|)^2 - 1 \end{aligned} \tag{5.79}$$

where the last equality is due to $|\partial_s z|^2 = (\partial_s|z|)^2 + (|z|\partial_s\phi)^2 = 1$. Because $\partial_s\alpha^\pm = -2\partial_s\phi$, the symmetry also gives

$$\mu_+^S = \mu_-^S = \mu^S 2|\partial_s\phi|\,, \qquad \mu_+^T = \mu_-^T = \mu^T 2|\partial_s\phi| \tag{5.80}$$

With these preparations, the Euler-Lagrange equation for F^+ can be written as (after multiplication with $R(-\alpha^+)$):

$$\begin{aligned} &\partial_s^2(\eta\partial_s^2 z) - \partial_s(\eta\lambda\partial_s z) + \eta\mu^A D_t z + \partial_s\left(\eta^2\mu^T(\varphi - \varphi_0)\partial_s z^\perp\right) \\ &+ 4\eta^2\mu^S|\partial_s\phi|z^\perp D_t\phi = 0 \end{aligned} \tag{5.81}$$

with $\varphi = \arccos(2(\partial_s|z|)^2 - 1)$. In the same way, the equation for F^- also turns out to be equivalent to Equation (5.81), using $(D(\alpha)z)^\perp = -D(\alpha)z^\perp$.

The boundary conditions now take the form

$$\begin{cases} -\partial_s\left(\eta\partial_s^2 z\right) + \eta\lambda\partial_s z - \eta^2\mu^T(\varphi - \varphi_0)\partial_s z^\perp = 0\,, & \text{for } s = -L \\ \partial_s\left(\eta\partial_s^2 z\right) - \lambda\partial_s z + \mu^T(\varphi - \varphi_0)\partial_s z^\perp = \kappa^M (C - C_0)_+ \dfrac{z}{|z|}\,, & \text{for } s = 0 \\ \eta\partial_s^2 z = 0\,, \quad \text{for } s = -L,\, 0 & \end{cases} \tag{5.82}$$

The Lagrange multiplier λ must be determined such that the constraint $|\partial_s z| = 1$ is satisfied. The other constraint is now satisfied automatically, and λ_{edge} is not needed anymore. Thus the model in Equations (5.81) and (5.82) represents the rotationally symmetric version of Equations (5.71) through (5.73).

Finally, we are looking for a stationary situation where the geometry of the lamellopodium does not change. This requires the assumption on the data that the polymerization speed and the length distribution are independent of time, that is, $v = \text{const}$ and $\eta = \eta(s)$. We then expect a time dependence of the reference filament of the form

$$z(t,s) = R(-\omega t)y(s)\,, \qquad y = |y|(\cos\psi, \sin\psi) \tag{5.83}$$

so, $|z|(t,s) = |y|(s)$ and $\phi(t,s) = \psi(s) - \omega t$. This means that the reference filament does not change its shape but rotates with constant angular velocity ω. This corresponds to lateral flow of barbed ends along the leading edge, as observed in experiments.

The ansatz implies $D_t\phi = \partial_t\phi - v\partial_s\phi = -\omega - v\partial_s\psi$ and $D_t z = -\omega R(-\omega t)y^\perp - vR(-\omega t)\partial_s y$, and we obtain

$$\begin{aligned}\partial_s^2\left(\eta\partial_s^2 y\right) - \partial_s(\eta\lambda\partial_s y) - \eta\,\mu^A\, v\partial_s y + \partial_s\left(\eta^2\mu^T(\varphi - \varphi_0)\partial_s y^\perp\right)\\ - 4\eta^2\mu^S|\partial_s\psi|y^\perp v\partial_s\psi = \omega\left(\eta\mu^A + 4\eta^2\mu^S|\partial_s\psi|\right)y^\perp\end{aligned} \tag{5.84}$$

where we have multiplied by $R(\omega t)$ already. The boundary conditions for y are the same as for z, that is, Equation (5.82).

In numerical experiments in [15], stable steady states have been observed. However, it seems that the resistance against twisting the crosslinks is necessary. With $\mu^T = 0$, the network typically degenerated in one of two ways. Either the filaments became more and more radial, or more and more aligned to the leading edge. We try to understand the former situation by setting $\mu^T = 0$ and observing that in this case there is a stationary solution of Equations (5.81), (5.82) of the form

$$z_0(s) = (r+s)\begin{pmatrix}1\\0\end{pmatrix} \tag{5.85}$$

where the equilibrium radius r and the Lagrange multiplier $\lambda_0(s)$ are given by

$$\kappa^M(2\pi r - C_0) = \mu^A v\int_{-L}^0 \eta(s)ds\,, \quad \eta(s)\lambda_0(s) = -\mu^A v\int_{-L}^s \eta(\hat{s})d\hat{s} \tag{5.86}$$

We try to analyze the stability of this state, where all filaments are in the radial direction, by linearization. In the ansatz

$$z(t,s) = z_0(s) + e^{\gamma t}\bar{z}(s), \quad \lambda(t,s) = \lambda_0(s) + e^{\gamma t}\bar{\lambda}(s) \tag{5.87}$$

the perturbation of the filament position has to be of the form $\bar{z}(s) = (b, a(s))$, to satisfy the linearized constraint $\partial_s z_0 \cdot \partial_s \bar{z} = 0$. The components of the linearization of Equations (5.81), (5.82) give two decoupled eigenvalue problems. The second component leads to

$$-\partial_s(\eta\bar{\lambda}) + \eta\mu^A\gamma b = 0\,, \qquad \eta(-L)\bar{\lambda}(-L) = 0\,, \quad \bar{\lambda}(0) = -\kappa^M 2\pi b \tag{5.88}$$

An easy computation produces the only eigenvalue

$$\gamma = -2\pi\kappa^M \left(\mu^A \int_{-L}^{0} \eta(s)ds \right)^{-1} \tag{5.89}$$

So the steady state is stable under perturbations in the radial direction. The other eigenvalue problem can be written as

$$\begin{cases} \partial_s^2 (\eta \partial_s^2 a) - \partial_s (\eta \lambda_0 \partial_s a) + \eta \mu^A \gamma a - \eta \mu^A v \partial_s a = 0 \\ -\partial_s (\eta \partial_s^2 a) + \eta \lambda_0 \partial_s a = 0 \,, \quad \text{for } s = -L \\ \partial_s (\eta \partial_s^2 a) - \lambda_0 \partial_s a = \mu^A v \dfrac{a}{r} \displaystyle\int_{-L}^{0} \eta(s)ds \,, \quad \text{for } s = 0 \\ \eta \partial_s^2 a = 0 \,, \quad \text{for } s = -L,\, 0 \end{cases} \tag{5.90}$$

No explicit solution is available here. Some information can be drawn from multiplication of the differential equation by the complex conjugate of a and subsequent integration, using integrations by parts and the boundary conditions:

$$\begin{aligned} \frac{\gamma}{v} \int_{-L}^{0} \eta |a|^2 ds = & -\frac{1}{\mu^A v} \int_{-L}^{0} \eta |\partial_s^2 a|^2 ds + \int_{-L}^{0} \left(\int_{-L}^{s} \eta(\hat{s}) d\hat{s} \right) |\partial_s a|^2 ds \\ & - \frac{1}{2} \int_{-L}^{0} \partial_s \eta |a|^2 ds - \frac{|a(0)|^2}{r} \int_{-L}^{0} \eta \, ds - \frac{1}{2} \eta(-L) |a(-L)|^2 + \frac{1}{2} |a(0)|^2 \end{aligned} \tag{5.91}$$

The first term on the right-hand side has the highest differential order. The fact that it is negative reflects the well-posedness of the problem. Among the remaining terms, the second term dominates the terms in the second row for high-frequency perturbations. This indicates that instability of the steady state is likely for large values of the product $\mu^A v$, that is, for strong adhesion and/or fast polymerization at the leading edge. These observations are compatible with the simulation results in [15].

References

[1] W. Alt and M. Dembo (1999). Cytoplasm dynamics and cell motion: two-phase flow models. *Math. Biosci.* 156(1-2):207–228.

[2] M. Carlier and D. Pantaloni (2007). Control of actin assembly dynamics in cell motility. *J. Biol. Chem.* 282:23005–23009.

[3] A. Chauvière, L. Preziosi, and T. Hillen (2007). Modeling the motion of a cell population in the extracellular matrix. *Discrete Contin. Dyn. Syst.* (Dynamical Systems and Differential Equations. Proceedings of the 6th AIMS International Conference, suppl.):250–259.

[4] M.E. Gracheva and H.G. Othmer (2004). A continuum model of motility in amoeboid cells. *Bull. Math. Biol.* 66(1):167–193.

[5] S. Koestler, S. Auinger, M. Vinzenz, K. Rottner, and J. Small (2008). Differentially oriented populations of actin filaments generated in lamellipodia collaborate in pushing and pausing at the cell front. *Nat. Cell Biol.* 10:306–313.

[6] C. Lacayo, Z. Pincus, M. VanDuijn, C. Wilson, D. Fletcher, F. Gertler, A. Mogilner, and J. Theriot (2007). Emergence of large-scale cell morphology and movement from local actin filament growth dynamics. *PLoS Biol.* 5(9):e233.

[7] I. Maly and G. Borisy (2001). Self-organization of a propulsive actin network as an evolutionary process. *Proc. Natl. Acad. Sci.* 98:11324–11329.

[8] A.F.M. Marée, A. Jilkine, A. Dawes, V.A. Grieneisen and L. Edelstein-Keshet (2006). Polarization and movement of keratocytes: a multiscale modelling approach. *Bull. Math. Biol.* 68(5):1169–1211.

[9] A. Mogilner (2006). On the edge: modeling protrusion. *Curr. Opin. Cell Biol.* 18:32–39.

[10] A. Mogilner, E. Marland, and D. Bottino (2001). A minimal model of locomotion applied to the steady gliding movement of fish keratocyte cells, in *Pattern Formation and Morphogenesis: Basic Processes.* H. Othmer and P. Maini (Eds.), New York, Springer.

[11] A. Mogilner and G. Oster (1996). Cell motility driven by actin polymerization. *Biophys. J.* 71:3030–3045.

[12] F. Nakamura, T. Osborn, C. Hartemink, J. Hartwig, and T. Stossel (2007). Structural basis of filamin A functions. *J. Cell Biol.* 179:1011–1025.

[13] D. Oelz and C. Schmeiser (2009). Derivation of a model for symmetric lamellipodia with instantaneous crosslink turnover. *In preparation.*

[14] D. Oelz and C. Schmeiser (2009). A steepest descent approximation of a model for the lamellipodial cytoskeleton. *In preparation.*

[15] D. Oelz, C. Schmeiser, and J.V. Small (2008). Modelling of the actin-cytoskeleton in symmetric lamellipodial fragments. *Cell Adhesion and Migration* 2(2):117–126.

[16] D. Pantaloni, C. Le Clainche, and M. Carlier (2001). Mechanism of actin-based motility. *Science* 292:1502–1506.

[17] B. Perthame (2007). *Transport Equations in Biology.* Frontiers in Mathematics. Birkhäuser Verlag, Basel.

[18] T. Pollard (2007). Regulation of actin filament assembly by Arp2/3 complex and formins. *Annu. Rev. Biophys. Biomol. Struct.* 36:451–477.

[19] B. Rubinstein, K. Jacobson, and A. Mogilner (2005). Multiscale two-dimensional modeling of a motile simple-shaped cell. *Multiscale Model. Simul.* 3(2):413–439.

[20] T. Schaus, E.W. Taylor, and G.G. Borisy (2007). Self-organization of actin filament orientation in the dendritic-nucleation/array-treadmilling model. *Proc. Natl. Acad. Sci. USA* 104(17):7086–7091.

[21] J. Small, G. Isenberg, and J. Celis (1978). Polarity of actin at the leading edge of cultured cells. *Nature* 272:638–639.

[22] J. Small, T. Stradal, E. Vignal, and K. Rottner (2002). The lamellipodium: where motility begins. *Trends Cell Biol.* 12:112–120.

[23] A.B. Verkhovsky, T.M. Svitkina, and G.G. Borisy (1998). Self-polarization and directional motility of cytoplasm. *Curr. Biol.* 9(1): 11–20.

Chapter 6

Computational Framework Integrating Cytoskeletal and Adhesion Dynamics for Modeling Cell Motility

Angélique Stéphanou

Contents

6.1 Introduction

Motility is a biological term that refers to a cell's ability to move. This ability is crucial because it determines the good or bad functioning of the cell. Physiologically, cell migration is paramount in embryogenesis or wound healing. Pathologically, inflammation and cancer cell invasion are examples of cell misbehavior related to the impairment of its normal motile properties.

The cell is a very complex object but its structure can be greatly simplified when dealing with motility events. These events are essentially due to the actin cytoskeleton, where actin turnover influences, in several ways, the protrusion and retraction forces acting on the membrane. Polymerizing actin filaments can push on the membrane [24]. Reciprocally, actin fibers pull back the membrane on which they are tethered, through contractility mechanisms involving myosin [8]. Consequently, the description of cell motility essentially consists of describing the interactions between the cytoskeleton and the cell membrane and the regulation processes involved, in connection with the constraints imposed by the extracellular environment.

Interaction with the extracellular environment, that is, the extracellular matrix (ECM), occurs through the formation of adhesions. The cell transmembrane proteins, the integrins, interact with the matrix proteins, such as the fibronectin, to link the cytoskeleton to the cell environment. The cytoskeleton dynamic is thus directly influenced by the mechanical properties of the matrix. The rigidity and topography are known to play a major role in the motile behavior of the cell [27].

From these brief biological considerations, it is clear that modeling cell motility implies the integration of elements of cytoskeletal and adhesion dynamics as well as information regarding the cell environment. Although the key elements are identified, the description of the processes involved remains complicated because they occur at many different time and space scales. From fast microsecond molecular interactions at the nanometer scale (adhesion clustering or actin filament polymerization) up to slow cell movements and migration at the minute and micrometer scales. Events occurring at each scale are very complex on their own. That is why they are often considered independently from the events occurring at other levels. Several models have been proposed over the past ten years, associated with a diversity of hypotheses and consequently theoretical approaches—either discrete [6,7] and often related to events occurring at smaller scales—or continuous (coupled partial differential equations) for higher scales where densities rather than entities matter [9,11]. The many different models are therefore often very specific and support one major hypothesis at a time [13]. Up to now, there have been very few attempts at unification of the various existing models, which simply represent many different facets of the same problem.

A significant challenge in modern computational cell biology is therefore to merge existing and successfull subcellular models. This requires the construction of some frameworks that allow us to integrate the different types of algorithms in order to handle the complexity emerging from the interactions among numerous processes of varying nature (biochemical, biomechanical, etc.), that is, to generate higher-order models. The use of a modular framework allows us, for example, to isolate the various and key processes. These can be considered either in isolation or synergistically. Such modular frameworks allow some flexibility because each module can be developed independently, with varying levels of refinement from one module to another. Some modules can also be considered black boxes until some new knowledge from experiments can be used to test and validate various hypotheses, thus illuminating the box.

This chapter aims to present such a framework and its potential to investigate various aspects of cell motility. Two applications are considered: first the relevance of the computational framework to describe spontaneous cell migration and second its ability to describe cell morphologies as observed on patterned substrates.

6.2 Computational Framework

Cell motility is a complex process in which cytoskeletal assembly, contractility, and adhesion dynamics are tightly interdependent and regulated by the properties of the extracellular environment (Figure 6.1). These properties include the proteic nature of the matrix and its topography or rigidity, to name but a few [5,27]. We therefore propose to construct a modeling framework that integrates these three major components: the cytoskeleton, adhesion, and extracellular environment [20].

6.2.1 Cell Motility

Cell membrane deformations basically result from the competition of protrusion (outward forces, F_{out}) and retraction (inward forces, F_{in}). Whereas retraction is clearly known to result from the pulling of the actin filaments anchored to the membrane, through the actomyosin contractility, protrusion has not yet been fully elucidated. Two mechanisms for protrusions have been identified, but it remains unclear how they are interrelated. In the first mechanism, pressure forces from the cytoplasm, or the cytoplasmic flows expelled via the cell contractions, are pushing the membrane outward in a blebbing form [1,17,18]. In the second mechanism, the polymerizing actin cytoskeleton is directly pushing on the membrane [4,24]. In that case a Brownian ratchet hypothesis has been proposed to explain how actin monomers could intercalate

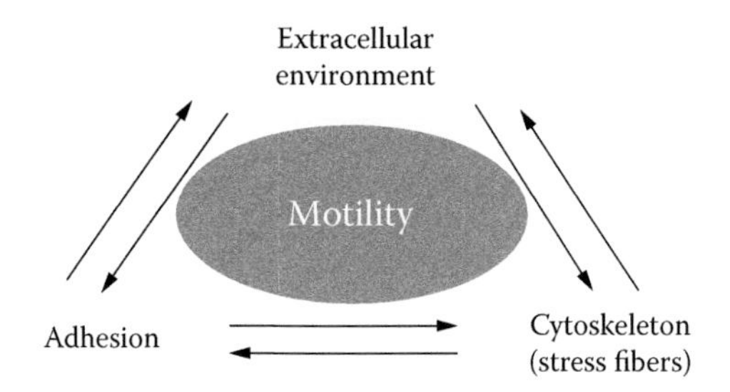

FIGURE 6.1 Three interacting components, adhesion, cytoskeleton, and extracellular environment, are required to describe cell motility.

among the growing tips of the filaments and the cell membrane [14]. Evidence tends to show that these two mechanisms coexist in the cell; however, one dominates the other, depending on the dynamical state of the cell: resting fluctuating state or actively migrating state. Actively migrating involves induced migration by external factors (chemical or haptotactic gradient), as opposed to noninduced random migration. In that latter case, the two protrusive mechanisms tend to be more entangled while the cell alternates between fast and oriented migrating phases with slow and erratic displacements.

The nonelicited (spontaneous) migration of the cell has been studied using a computational model based on the description of the intracellular actin dynamics coupled with the movement of the cell membrane [21,22]. In this model, actin polymerization and depolymerization dynamics around an equilibrium density for F-actin is considered; the contractile properties of the actomyosin network and the viscoelastic properties of the cell cortex of actin are also considered. The cell deformations, that is, the movements of the membrane, result from the nonequilibrium state between the competing outward and inward forces on the membrane; then intracellular space variations influence the turnover for actin, which in turn modifies the pushing and pulling forces.

The model considers the 2-D annular domain bounded by the cell body and the cortex–membrane complex at the lamellar tip. Specifically, the cell body is represented by a fixed circular shape to withstand the pressure in the annular cytoplasm ring, whereas the cortex–membrane complex at the lamellar tip defines a free radially moving boundary. We denote by $L(\theta, t)$ the width of the annular domain along any radial direction located by the angle θ ($0 \leq \theta \leq 2\pi$) (Figure 6.2), $a(\theta, t)$ represents the F-actin concentration and is assumed to be radially constant, and $v(\theta, t)$ stands for the F-actin tangential velocity. The normalized system of equations is given by:

$$\frac{\partial}{\partial t}(La) = -\frac{\partial}{\partial \theta}(Lav) + L(1-a) \tag{6.1}$$

$$av = \frac{\partial}{\partial \theta}\left[\mu a \frac{\partial v}{\partial \theta} + \sigma(a) - \kappa_n\right] \tag{6.2}$$

$$\delta(a)\frac{\partial L}{\partial t} = F_{out} + F_{in} + \kappa \tag{6.3}$$

where the first two equations describe the actin turnover and displacement in the visco-contractile cortex, respectively, with μ the viscous coefficient and $\sigma(a)$ the contractility of the actomyosin network, which locally depends on the actin density. The movements of the cell membrane are given by Equation (6.3) describing the force equilibrium on the membrane, where $\delta(a)$, F_{out}, F_{in}, and κ represents the adhesive condition, the protrusion force, the retraction force and the tension induced by the membrane curvature, respectively. κ_n in Equation (6.2) is a curvature-induced stress. A fully detailed description of the model can be found in Stéphanou et al. (2008) [22].

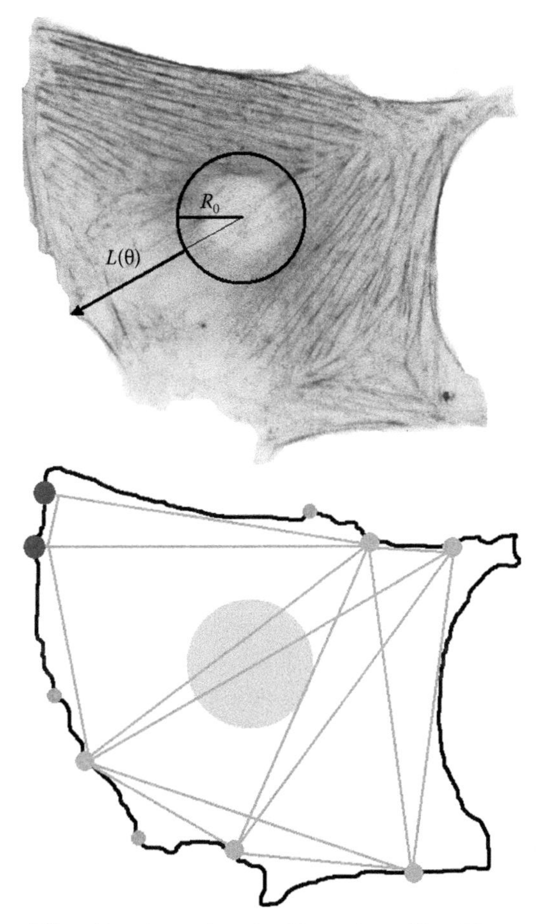

FIGURE 6.2 Visualization of actin fibers in a fibroblast cell (top). Actin is labeled with the GFP protein. The representation of the cell, as used in the theoretical model, is superimposed to the cell picture (top). The cell body is enclosed in a circular area with radius R_0. The membrane extension $L(\theta)$ is measured from the surface of the cell body for each angular direction θ. The bottom drawing shows how the different types of adhesion points and the related network of actin fibers are represented in the model simulations. Dark gray points represent the strongest adhesions; light gray points and small points represent adhesion types with decreasing resistance to traction and shorter lifetimes.

6.2.2 Adhesion

Adhesions connect the cell to the environment. The cell behavior therefore strongly depends on its ability to make and develop adhesions. Several types of adhesions can be identified. They vary in size and molecular composition.

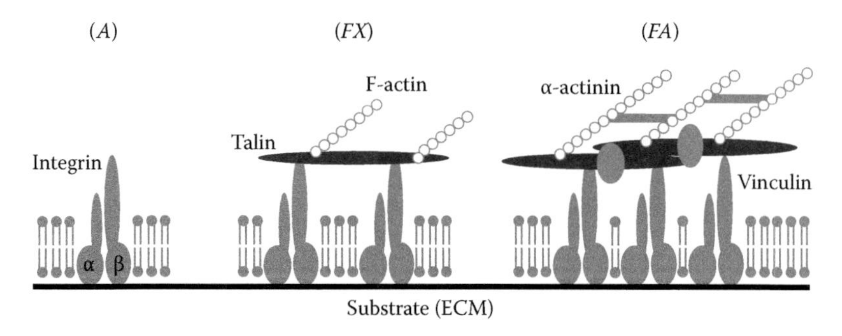

FIGURE 6.3 Three different types of adhesion that depend on the level of maturation of the adhesion can be identified. The first type is the adhesion point (A). Clustering of integrins then leads to a focal complex (FX) with the recruitment of talin, which contributes in connecting the adhesion to the cell cytoskeleton. α-Actinin and vinculin, among other proteins, are further recruited to produce a focal adhesion (FA).

These two characteristics influence the lifetime and adhesion strength. The adhesion progressively evolves from one type to another, in a hierarchical way, by the recruitment of new proteins that strengthen and stabilize its structure [28]. This maturation process of the adhesion is known to strongly depend upon mechanical stimulations from the actin fibers of the cytoskeleton [2,15]. Reciprocally, the growth of the adhesion site promotes the fibers' association into bundles able to sustain the cell translocation and the resulting migration.

Three types of adhesions can be identified (Figure 6.3). The most primitive type is the adhesion point (A), which simply results from the contacts among transmembrane proteins (integrins) and matrix proteins (such as fibronectin). Thermal fluctuations can account for the energy required for protein binding, of the $\alpha_v\beta_3$-integrin. Once such adhesion is formed, integrins rapidly gather to form a cluster, called a focal complex (FX). This is concomitant with the recruitment at the growing adhesion sites of other proteins such as talin, which permits connection of the integrins to the actin filaments. The contractility of the actin fibers creates a stress upon the adhesion, which further induces the recruitment of α-actinin, vinculin, FAK, VASP, and Arp2/3. This ultimately leads to the focal adhesion (FA), which is the most stable type [12].

6.2.3 Cytoskeleton

The actin cytoskeleton is a very dynamic structure. The fibers polymerize and depolymerize, associate with the myosin to generate the cell contraction, and aggregate to form bundles [10] able to generate a bigger stress on the adhesions.

The process of maturation of the adhesions strongly depends on mechanical solicitation from the actin fibers. To decide whether an adhesion should grow or not, we define a force-related criterion by calculating the resulting force $R_F(\theta_i, t)$ existing at time t for each adhesion i. If this resulting force is above a given threshold, then the adhesion is allowed to mature to the next level. If this condition is not fulfilled during the lifetime τ_X of the type X–adhesion, then the adhesion breaks.

Reciprocally, the maturation process of the adhesion with the recruitement of proteins such as zyxin and tensin allows for increasing the bundling of the filaments and consequently the magnitude of the force that can be developed. In the model it is thus assumed that stress fibers connect the cell body to FA; hence, only this adhesion type is considered able to transmit and sustain adequate traction force for effective cell translocation.

6.2.4 Extracellular Environment

Integration of the maturation process of the adhesions into the model of cell deformations allows us to investigate a range of scenarios for cell motility. Two applications have been considered, which depend on the extracellular environment.

In the first application, the computational framework has been used to investigate the basic principles that allow us to describe unstimulated cell migration. In that case, a homogeneous substrate has been considered, where adhesion can occur anywhere. The model has been validated with a range of well-known experimental results and gave us confidence to further investigate the effects of the temporal parameters ruling the adhesion lifetimes and recycling time on the motility process.

In the second application, the framework was further developed and adjusted to study the motility of the cell on a network of adhesive patches. The aim of this adhesively patterned substrate was to assess the effects of the discretization and localization of the adhesion sites on the cell morphodynamics.

This framework thus helps us better understand how the cytoskeletal and adhesion dynamics are influenced by the extracellular adhesive constraint.

6.3 Application of Computational Framework

6.3.1 Cell Migration on Homogeneous Substrate

Cell migration is a multistep process. The number of steps often varies, depending on the level of details given for the description of the processes involved. We describe it using three steps, which are membrane protrusion, adhesion to the substrate, and translocation of the cell via application of traction forces from the cytoskeleton. The computational framework addresses the coordination of these three steps to allow the cell to migrate and explore its environment, which we assume is homogeneous in this case. This implies that the substrate properties do not need to be described explicitly.

In the context of random cell migration on a homogeneous substrate, the model for cell membrane deformations is based on the following assumptions:

- The formation of an adhesion is materialized by an increased friction between the cell membrane and the substrate; that is, $\delta(a) = a + \alpha\delta_{adh}$, with α the friction coefficient.
- The protrusion of the membrane is primarily due to the cell hydrostatic pressure β reinforced by a polymerization-induced protrusion due to the increase of actin at the focal adhesion site. The resulting protrusive term is given by $F_{out} = \beta + \beta(a)\delta_{FA}$.
- The retraction is due to the actin network–membrane attachment, the intensity of which linearly depends on the actin density; that is, $F_{in} = -\gamma La$, with γ the elasticity coefficient for the actin network,
- The membrane curvature modulates the cell surface tension and is given by $\kappa_m = \chi_m \frac{\partial}{\partial\theta}\left(a\frac{\partial L}{\partial\theta}\right)$, where χ_m characterizes the membrane stiffness.

With $\delta_{adh} = \delta_A + \delta_{FX} + \delta_{FA}$ and $\delta_X = 1$ or 0, whether there is an adhesion of type X or not, the movements of the cell membrane are then ruled by the following equation:

$$(a + \alpha\delta_{adh})\frac{\partial L}{\partial t} = \beta + \beta(a)\delta_{FA} - \gamma La + \chi_m \frac{\partial}{\partial\theta}\left(a\frac{\partial L}{\partial\theta}\right) \tag{6.4}$$

It is assumed that the probability to form an adhesion is higher in a protrusive zone (such as a lamellipod) because the cell surface in contact with the substrate is bigger. In the computational model, this means that the membrane extension is maximum ($L_{\max}$), and this maximum should also correspond to a sufficient density of actin (a_{thr}) as detailed in the protrusion box of the flowchart (Figure 6.4).

Each adhesion type corresponds to a particular level of maturation characterized by its lifetime τ and resistance to traction. If the resulting tension force (R_F) exerted between the adhesion site and the cell body is positive (i.e., a traction force), then A matures into FX. If this condition is not fulfilled during the lifetime τ_A of A, then the adhesion breaks (Figure 6.4). Similarly, the maturation of FX into FA occurs if R_F applied on the adhesion reaches a threshold tension (R_{thr}) during the lifetime τ_{FX} of FX. Once FA is formed, recruitment of actin occurs at the adhesion site to promote the formation of stress fibers. The fibers contract and contribute in pulling the cell body forward. Translocation is assumed to occur when a threshold traction magnitude T_{thr} is reached. During the translocation event, all the adhesions are assumed to break. Before a new cycle can start, the adhesion proteins need to be recycled. During this time, the cell is considered unable to form new adhesions. This "refractory period" is noted τ_R in the flowchart (Figure 6.4).

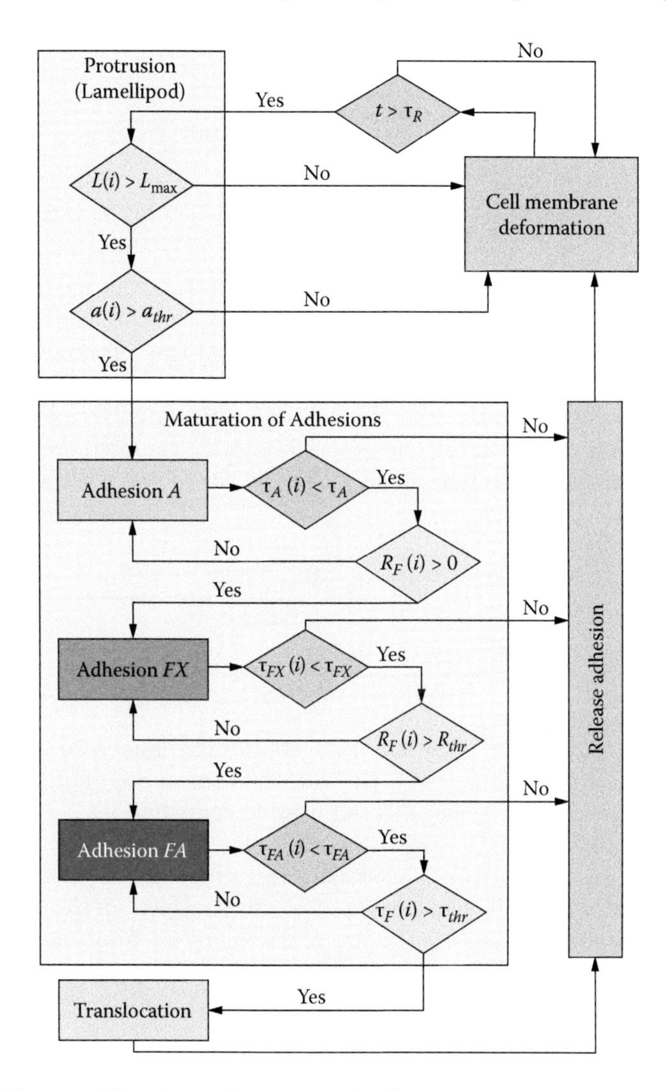

FIGURE 6.4 Flowchart describing the formation and maturation process of an adhesion i, up to the cell translocation, that is, effective migration. In a first step, the cell membrane deformations lead to a protrusion (lamellipod) and then to an adhesion that needs to mature (step 2) to be able to sustain the cell translocation (step 3).

To decide whether or not an adhesion should grow, a force-related criterion is defined by calculating the resulting force $R_F(\theta_i, t)$ existing at time t for each adhesion i. This resulting force corresponds to the sum of the contributions of all the individual forces $F(\theta_j, t)$ balancing the movements of the cortex–membrane complex for each direction θ_j. These forces are then projected on the θ_i-direction supporting the adhesion i ($\theta_i = 2\pi i/m$ with m the number of points defining the membrane boundary); namely,

$$R_F(\theta_i, t) = \sum_{j=0}^{m} F(\theta_j, t)\cos(\theta_j - \theta_i) \tag{6.5}$$

$$\text{with} \quad F(\theta_j, t) \quad \text{such that} \quad a\frac{\partial L}{\partial t} + F(\theta_j, t) = 0 \tag{6.6}$$

$F(\theta_j, t)$ is derived from Equation (6.3), neglecting the curvature-related term since its contribution is small. It thus comprises: (1) the adhesion-related term responsible for an increased tension between the cell body and the adhesion site, if an adhesion has formed; (2) the passive tension from the actin filaments existing everywhere in the cell and modulated by the local membrane extension and local density of actin; and (3) the pressure term, which tends to repel the cell body from the adhesion site. $F(\theta_j, t)$ for site j at time t is given by:

$$\begin{aligned} F(\theta_j, t) = \delta_{adh}(\theta_j, t) & \underbrace{\gamma_2[L(\theta_j, t + \Delta t) - L(\theta_j, t)]}_{\text{adhesion-related tension}} \\ & + \underbrace{\gamma L(\theta_j, t)a(\theta_j, t)}_{\text{filaments passive tension}} - \underbrace{[\beta + \beta(a)\delta_{FA}(\theta_j, t)]}_{\text{pressure force}} \end{aligned} \tag{6.7}$$

with $\gamma_2 = \alpha/\Delta t$, and Δt being the time step of the numerical scheme. Note that $F(\theta_j, t)$ strongly depends on the nature of the site, j. For example, if j is not an adhesion site, then only the passive contributions remain, that is, $F(\theta_j, t) = \gamma L(\theta_j, t)a(\theta_j, t) - \beta$.

The model assumes at this stage that stress fibers radially connect the FA to the cell body. The cell translocation (i.e., displacement of the cell centroid) and direction of migration θ_M (Figure 6.5) thus result from the competition among the traction forces exerted in each stress fiber supported by an FA.

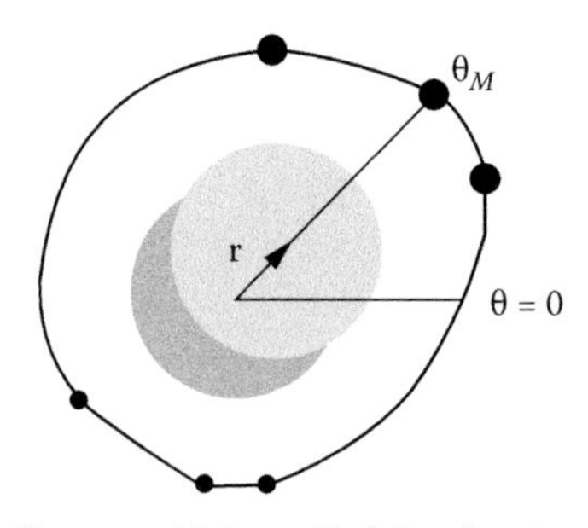

FIGURE 6.5 The cell centroid is pulled in the θ_M-direction corresponding to the greatest traction force from the filaments anchored to the FA (bigger spots). The vector **r** represents the cell centroid displacement from its initial position (dark gray) to its new position after translocation (light gray).

The resulting traction force T_F existing in each stress fiber i is calculated as:

$$T_F(\theta_i, t) = \sum_{j=0}^{m} \left\{ \underbrace{\gamma_2[L(\theta_j, t + \Delta t) - L(\theta_j, t)]}_{\text{adhesion-related}} + \underbrace{\gamma L(\theta_j, t) a(\theta_j, t)}_{\text{filament-related}} \right\} \delta_{FA}(\theta_j, t) \cos(\theta_j - \theta_i) \quad (6.8)$$

The displacement of the cell centroid r is simply given by $r = T_F(\theta_M, t)/k$. This corresponds to an "elastic release," with the elastic coefficient k, whereby the stress fibers suddenly break once the threshold force T_{thr} for translocating the cell is reached (parameters T_{thr} and k are given in Table 6.2 in the Appendix to this chapter; (Section 6.5).

6.3.1.1 Simulation Results

Simulations are performed using periodic boundary conditions for the variable θ. A small pertubation ε of the homogeneous steady state, given by $(L_0, a_0, v_0) = (1, 1 \pm \varepsilon, 0)$, is taken as the initial condition. The round shape and homogeneous actin density biologically characterize the cell state right after mitosis. The dimensionless parameters used for the simulations are given in Table 6.1 in the Appendix to this chapter (Section 6.5). Details on the choice of parameters are given in [22].

Integration of the model hypotheses described above, into the computational framework, allows us to reproduce some important experimental features of the random migration of fibroblast cells (those are detailed in [22]) and concern:

- Typical features of the cell migratory tracks
- Relationship between cell surface and speed
- Relationship between cell speed and adhesion strength

Figure 6.6 shows three simulated cell trajectories, recorded over 6 hours (biological cell time). The three different migratory tracks are obtained by changing the seed of the random generator in our computer program. They exhibit alternating phases where the cell can either explore a short perimeter (slow migrating phase), move bi-directionally, or transiently assume a persistent directional migration (fast migrating phase). This migrating behavior is representative of fibroblast cells and is mainly ruled by adhesion kinetics, which include adhesion lifetimes, maturation, and recycling time as shown in [22]. This latter parameter has been shown to strongly influence the directional persistence of the migratory tracks.

The principle of fibroblast cell migration is to develop sufficiently strong adhesions with the substrate to be able to translocate. One means to achieve this goal is to maximize its spreading surface. However, when the spreading

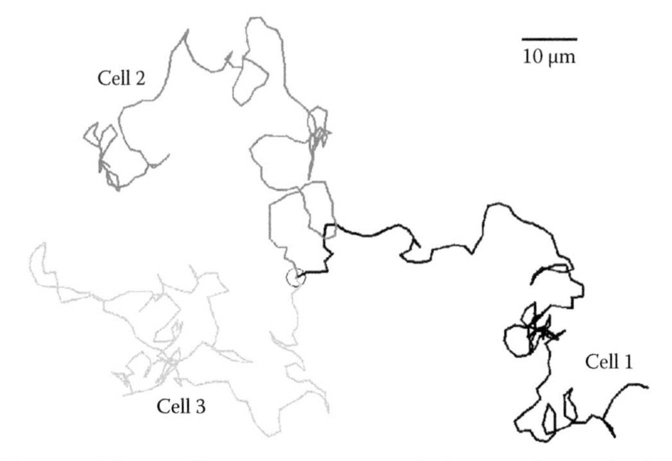

FIGURE 6.6 Three cell trajectories recorded over 6 hours (cell time) and reunited at the same ending point. The trajectories exhibit the same characteristics of alternating slow and erratic migrating phases with fast and more persistent ones. The average velocity of migration is 0.50 μm/min.

surface is too big, the cell can find itself strongly anchored to the substrate and breaking the adhesion becomes a longer process. Experimentally, this phenomenon is observed by a sudden decrease in the migrating speed down to zero, while the cell surface area is peaking to a maximum. The simulations performed are true to this observation, as shown in Figure 6.7, where the cell speed and area corresponding to cell 1 are plotted on a single graph.

Along the same line, Palecek et al. [16] have shown, in a famous experiment, that increasing the fibronectin density of the substrate led to an increased

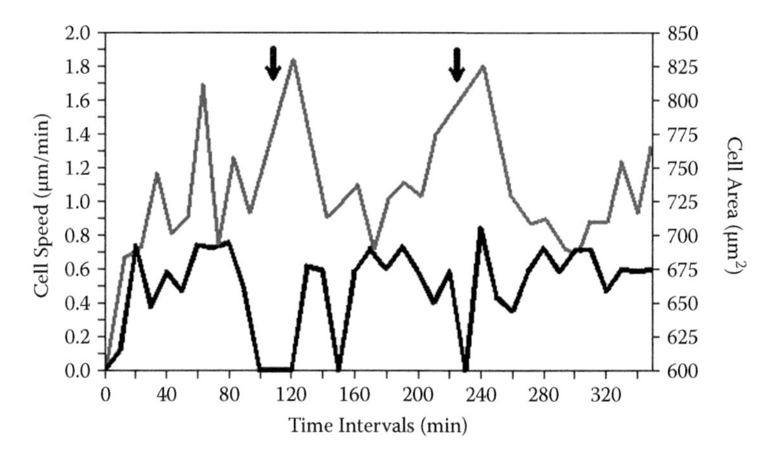

FIGURE 6.7 Cell speed (black curve) and corresponding cell spreading area (gray curve). The arrows indicate the correlation between slow speeds and big spreading areas.

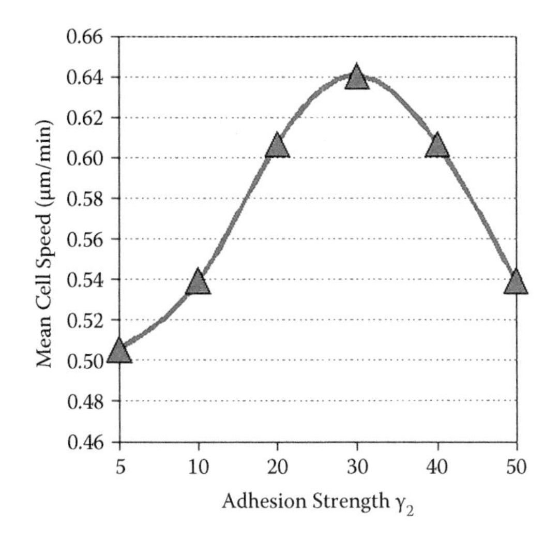

FIGURE 6.8 Biphasic evolution of cell speed as a function of cell adhesion strength controlled by parameter γ_2 in computational model.

cell speed up to a saturation point above which the cell speed decreased. Fibronectin allows for control of the attachment of the cell to its substrate. Consequently, increasing its density increases the attachment strength of the cell. Initially the cell can migrate more rapidly because the translocation threshold becomes easy to reach, until the attachment becomes too strong and prevents migration. In the model, the attachment strength of the cell to the substrate is controlled by the friction coefficient γ_2. Simulations show that increasing this parameter reveals this important feature of cell migration as shown in Figure 6.8.

6.3.2 Cell Motility on Network of Adhesive Patches

In this new context, we aim to set the basis to understand how the maturation of cell adhesions depends on the maturation of stress fibers. For that we consider a patterned substrate of adhesive patches, regularly arranged, in order to control the positioning and size of the adhesion zones of the cell. The modeling framework that allows the isolation of the model components, such as the maturation process of the adhesion and the protrusive and translocation conditions, can be easily adapted to integrate and test the hypotheses associated with the new experimental context. The new conditions are the following:

- The cell can only adhere on adhesive patches. This means that the formation of an adhesion is prohibited over the nonadhesive area.
- The formation of an adhesion point is modeled by the additional friction term $\delta(a) = a + \alpha\delta_A$. However, if the adhesion matures into a focal complex (FX) or focal adhesion (FA), then the movements of the membrane are prevented. This signifies that $\partial L/\partial t = 0$.

- The protrusion is only due to the internal hydrostatic pressure, $F_{push} = \beta$.
- The retraction is unchanged, $F_{pull} = \gamma La$.
- An additional tension force F_{fiber} is considered and concerns the tension exerted by the actin fibers tethered to the adhesive site. (Details of this tension force are given below.)

The movements of the cell membrane are then given by the following equation:

$$(a + \alpha\delta_A)\frac{\partial L}{\partial t} = \beta - \gamma La + \delta_{FX,FA}F_{fiber} \quad (6.9)$$

with $\delta_{FX,FA} = \delta_{FX} + \delta_{FA}$.

On the adhesive pattern, the cell does not need to protrude to be able to form an adhesion. This is assumed to occur spontaneously when the cell membrane is in contact with an adhesive patch. The protrusion box is therefore replaced by a box that rules the direct interaction of the cell with the extracellular matrix (Figure 6.9). The maturation process of the adhesions is improved to correspond more closely to reality. The primary adhesion A creates an increased friction between the membrane and the matrix. The membrane displacement $d(i)$ for the adhesion i is locally slower and leads to the recruitment of adhesion proteins and integrin clustering, which transform the primary adhesion into a focal complex FX. This is assumed to occur as long as the membrane displacement is smaller than a maximum admissible displacement $d_{\max}$. The adhesion i grows until a critical size S_{crit} is reached; that is, $S(i) > S_{crit}$. The focal complex thus becomes a focal adhesion FA. Each type X adhesion has a limited lifetime τ_X that increases with the maturation level of the adhesion.

The cytoskeleton is added as a new component of the framework (Figure 6.9). More specifically, this component rules the formation of the transverse actin bundles and stress fibers. At this stage, those are assumed to form spontaneously between two adhesion sites that reach the FX maturation level (Figure 6.2). The actin fibers are assumed to possess elastic properties modeled as spring forces $F_{i\rightarrow j}$ on the considered adhesion site i that are proportional to the distance l_{ij} between the adhesion sites i and j supporting the fiber i–j. The radial contributions of all fibers i–j are considered and expressed by:

$$F_{fiber}(\theta_i, t) = \sum_{j=1}^{n} F_{i\rightarrow j}\cos\alpha_{ij} \quad \text{with} \quad F_{i\rightarrow j} = -k[l_{ij} - l_0] \quad (6.10)$$

where n represents the number of fibers i–j, α_{ij} is the angle between the fibers i–j and the radial direction connecting the adhesion site i to the cell body. l_0 is the length of the unstrained filaments, evaluated for a circular cell shape that corresponds to the steady state.

The translocation rule for cell migration remains unchanged.

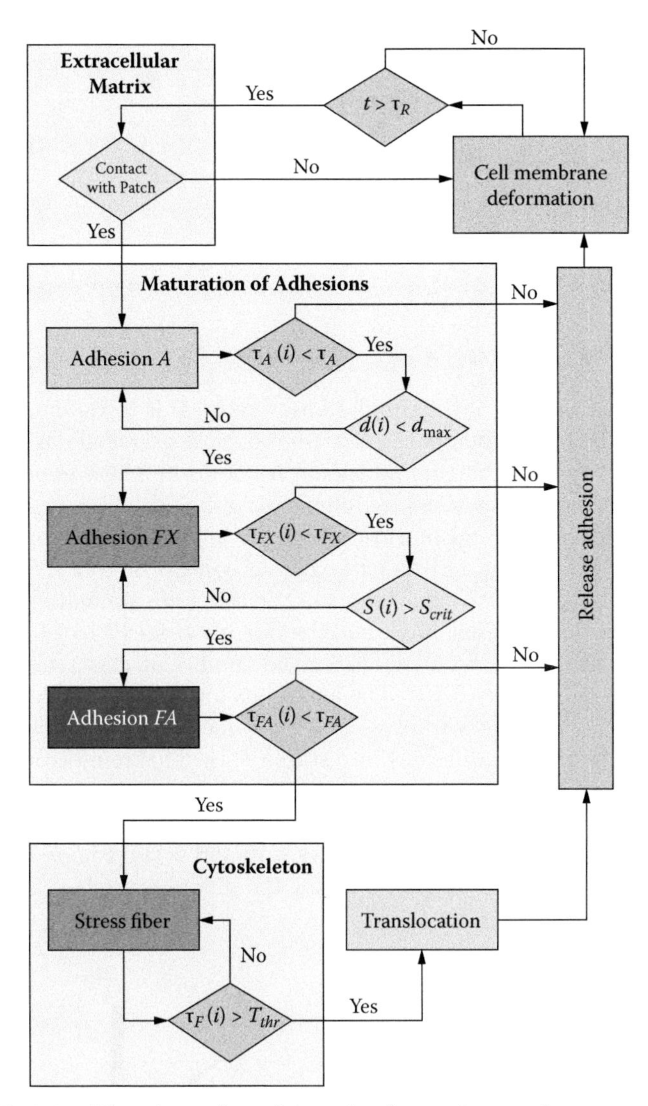

FIGURE 6.9 Flowchart describing the formation and maturation of the adhesion i involving four main steps. In a first step the cell membrane deforms and leads to a protrusion (lamellipod; step 2) and then to an adhesion that needs to mature (step 3) to be able to sustain the cell translocation (step 4).

6.3.2.1 Simulation Results

Simulations are performed on hexagonal patterns of adhesive patches with a constant 4-μm wide square shape, but with varying pitch lengths. The pitch of the pattern is the addition of the patch size and of the distance between two consecutive patches. Initial and boundary conditions are unchanged.

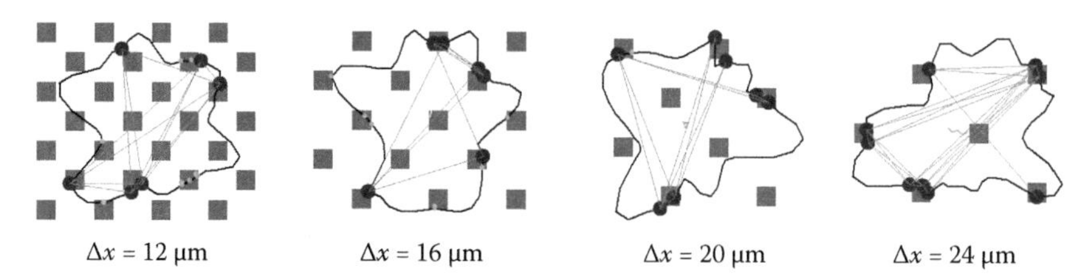

FIGURE 6.10 Visualization of the cell shapes, adhesions, and actin fibers for increasing pitch lengths Δx. The cell nucleus is not represented to increase readability.

At this stage the aim of the simulations performed is to validate, on a qualitative basis, the integration of the new components' extracellular environment and cytoskeleton into the computational framework. These simulations thus allow us to monitor and visualize simultaneously the adhesion dynamics of the cell, the formation and maturation of the actin network, and the resulting cell morphologies for a given extracellular context of adhesiveness. Figure 6.10 presents representative snapshots of the cell state simulated for adhesive patterns with four different pitch lengths ranging from 12 to 24 μm.

Two observations can be made. First, the number of FAs tends to increase with pitch length (Figure 6.10). The adhesions form clusters that concentrate the actin fibers. The tension forces generated locally thus increase and favor the maturation of new adhesions into stable FAs. This reinforcement process leads to more stable cell shapes. Second, the displacement of the cell centroid is very small compared to the homogeneous substrate case, with bigger amplitudes for higher pitch lengths (Figures 6.10 and 6.11). Figure 6.11 presents the trajectories recorded over 2 hours for the different pitches. The diameter

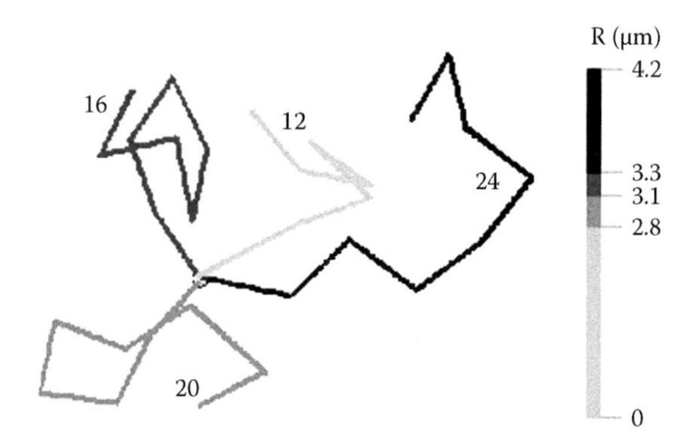

FIGURE 6.11 Displacements of the cell centroid for adhesive patterns with pitches of 12, 16, 20, and 24 μm (as indicated next to each corresponding trajectory) over a period of 2 hours. The vertical axis represents the diameters of the circles in which each trajectory is contained.

of the circular domain in which each trajectory is contained tends to increase with the pitch length.

On a qualitative basis, the increased displacements of the cell centroid with pitch length indicate that the forces generated in the cell are bigger because the amplitude of the displacements is proportional to the traction force responsible for the cell translocation.

6.4 Conclusion

A computational framework constructed from an existing mathematical model describing fibroblast cell deformations [21] has been presented. The initial continuous model of partial differential equation, coupling the movements of the cell membrane to the actin turnover, has been integrated into the framework as a main module interacting with three new components: the adhesion, cytoskeleton, and extracellular components.

Integration of the adhesion component was first considered to deal with spontaneous, that is, unstimulated, cell migration. Adhesions occur at a smaller scale compared to the scale used to describe the movements of the cell membrane. Moreover, adhesions are discrete entities that evolve individually from one another and depend on the local context of constraints. Consequently, a continous description is not adapted; however, the formation and maturation events can be easily transposed into a cellular automaton. The coupling of the automaton with the continuous cell model was then realized through the addition of a new constraint into the model equations.

Simulations of unstimulated cell migration, for a homogeneous and isotropic substrate, could be performed in a realistic way. A range of well-known cell migrating features could indeed be reproduced—more specifically, the relationships existing between the cell speed and the cell surface on the one hand and the cell speed and adhesion strength on the other hand.

Further refinements of the model have been proposed with the integration of the extracellular and cytoskeletal components. The recent literature has demonstrated the importance of adhesively patterned substrates as a tool to control the extracellular environment and investigate cell responses in terms of mitosis [25], morphology [3,26], and migration [27]. Different models have been proposed to explain the cell morphologies imposed by the extracellular adhesive constraint [19]; however, up to now, none is able to describe the evolution of the cell morphologies in a dynamical way because the models essentially describe static equilibrium states for the cell without any explicit description for the coupling of actin turnover, cytoskeletal remodeling, and adhesion dynamics.

In the improved computational framework, these couplings are considered and numerical simulations allow us to explain how the forces generated by the stress fibers of the virtual cells are regulated at the adhesion site through feedback mechanisms and how the competing stress fibers can generate transient equilibrium states corresponding to stabilized cell shapes.

A major limitation of the model at this stage is the crude hypothesis used to describe the contribution of the stress fibers. For example, the formation of the fibers is not explicitly described. Instead we assumed that the fibers formed spontaneously among mature adhesion sites. Another limitation is the use of constant parameters to describe the mechanical properties of the fiber without differentiation of filaments, fibers, and bundles. Improvements in the framework components will soon be considered. Indeed, one major advantage of such modular computational framework is the possibility of individually developing its components. New knowledge can be integrated or new hypotheses can be tested while keeping the other components unchanged.

Efforts toward the integration and interactions of existing models into new computational frameworks could well be one key for a better understanding of the cell complexity.

Acknowledgment

The author would like to thank Dr. E. Mylona and Dr. T. Tzvetkova-Chevolleau for their contributions and support in the development of this computational framework.

6.5 Appendix: Simulation Parameters

The parameters used in the simulations are given in Tables 6.1, 6.2, and 6.3. Table 6.1 presents the parameters used to describe the spontaneous membrane deformations observed on fibroblast cells [23].

Table 6.2 gathers the force-related parameters and Table 6.3 the time-related parameters that rule the spontaneous (random) migration of the fibroblast cells. More specific details on the choice of parameters can be found in [22].

TABLE 6.1 Dimensionless Parameters Defining Cell Mechanical and Chemical Properties

Parameter	Notation	Value
Protrusive hydrostatic pressure	β	0.5
Actin network elasticity	γ	0.5
Cytoplasm viscosity	μ	2.0
Membrane elasticity	χ_m	1.5

Source: From A. Stephanou, E. Mylona, M. Chaplain, and P. Tracqui (2008). A computational model of cell migration coupling the growth of focal adhesions with oscillatory cell protrusions. *J. Theor. Biol.* 253:701–716. (With permission).

TABLE 6.2 Simulation Parameters Defining Conditions for Formation and Maturation of Cell Adhesions and Translocation

Parameter	Notation	Value
Membrane extension threshold	L_{thr}	1.0
Actin density threshold	a_{thr}	1.1
Tension threshold	R_{thr}	1.0
Friction coefficient for adhesion	γ_2	5.0
Translocation threshold	T_{thr}	2.3
Stress fiber elasticity	k	14

TABLE 6.3 Temporal Simulation Parameters Governing Adhesion Protein Recycling Time and Lifetimes of Different Adhesion Types

Parameter	Notation	Iterations	Time (s)
Adhesion protein recycling time	τ_R	400	58
Adhesion point lifetime	τ_A	100	14
Focal complex lifetime	τ_{FX}	500	72
Focal adhesion lifetime	τ_{FA}	500	72

References

[1] J. Bereiter-Hahn and H. Luers (1998). Subcellular tension fields and mechanical resistance of the lamella front related to the direction of locomotion. *Cell Biochem. Biophys.* 29:243–262.

[2] A. Besser and S. Safran (2006). Force-induced adsorption and anisotropic growth of focal adhesions. *Biophys. J.* 90:3469–3484.

[3] I. Bischofs, F. Klein, D. Lehnert, M. Bastmeyer, and U. Schwarz (2008). Filamentous network mechanics and active contractility determine cell and tissue shape. *Biophys. J.* 95:3488–3496.

[4] M. Carlier and D. Pantaloni (2007). Control of actin assembly dynamics in cell motility. *J. Biol. Chem.* 282(32):23005–23009.

[5] O. Collin, P. Tracqui, A. Stéphanou, Y. Usson, J. Clément-Lacroix, and E. Planus (2006). Saptiotemporal dynamics of actin-rich adhesion microdomains: influence of substrate flexibility. *J. Cell Sci.* 119:1914–1925.

[6] H. Coskun, Y. Li, and M. Mackey (2006). Amoeboid cell motility: a model and inverse problem, with an application to live cell imaging data. *J. Theor. Biol.* 244:169–179.

[7] V. Deshpande, M. Mrksich, R. McMeeking, and A. Evans (2008). A bio-mechanical model for coupling cell contractility with focal adhesion formation. *J. Mech. Phys. Solids* 56:1484–1510.

[8] G. Giannone, B. Dubin-Thaler, O. Rossier, Y. Cai, O. Chaga, G. Jiang, W. Beaver, H. Dobereiner, Y. Freund, G. Borisy, and M. Sheetz (2007). Lamellipodial actin mechanically links myosin activity with adhesion-site formation. *Cell* 128:561–575.

[9] M. Gracheva and H. Othmer (2004). A continuum model of motility in amoeboid cells. *Bull. Math. Biol.* 66:167–193.

[10] P. Hotulainen and P. Lappalainen (2006). Stress fibers are generated by two distinct actin assembly mechanisms in motile cells. *J. Cell Biol.* 173:383–394.

[11] E. Kuusela and W. Alt (2009). Continuum model of cell adhesion and migration. *J. Math. Biol.* 58:135–161.

[12] S. Lo (2006). Focal adhesions: what's new inside. *Dev. Biol.* 294:280–291.

[13] A. Mogilner (2009). Mathematics of cell motility: have we got its number? *J. Math. Biol.* 58:105–134.

[14] A. Mogilner and G. Oster (1996). The physics of lamellipodial protrusion. *Eur. Biophys. J.* 25:47–53.

[15] A. Nicolas, A. Besser, and S. Safran (2008). Dynamics of cellular focal adhesions on deformable substrates: consequences for cell force microscopy. *Biophys. J.* 95:527–539.

[16] S. Palecek, J. Loftus, M. Ginsberg, D. Lauffenburger, and A. Horwitz (1997). Integrin-ligand binding properties govern cell migration speed through cell-substratum adhesiveness. *Nature* 385:537–540.

[17] E. Paluch, M. Piel, J. Prost, M. Bornens, and C. Sykes (2005). Cortical actomyosin breakage triggers shape oscillations in cells and cell fragments. *Biophys. J.* 89:724–733.

[18] E. Paluch, C. Sykes, J. Prost, and M. Bornens (2006). Dynamic modes of the cortical actomyosin gel during cell locomotion and division. *Trends Cell Biol.* 16 (1):5–10.

[19] A. Pathak, V. Deshpande, R. McMeeking, and A. Evans (2008). The simulation of stress fiber and focal adhesion development in cells on patterned substrates. *J. R. Soc. Interface* 5:507–524.

[20] M. Schwartz and A. Horwitz (2006). Integrating adhesion, protrusion and contraction during cell migration. *Cell* 125:1223–1225.

[21] A. Stéphanou, M. Chaplain, and P. Tracqui (2004). A mathematical model for the dynamics of large membrane deformations of isolated fibroblasts. *Bull. Math. Biol.* 66:1119–1154.

[22] A. Stéphanou, E. Mylona, M. Chaplain, and P. Tracqui (2008). A computational model of cell migration coupling the growth of focal adhesions with oscillatory cell protrusions. *J. Theor. Biol.* 253:701–716.

[23] A. Stéphanou and P. Tracqui (2002). Cytomechanics of cell deformations and migration: form models to experiments. *C.R. Biologies* 325:295–308.

[24] J. Theriot and T. Mitchison (1991). Actin microfilaments dynamics in locomoting cells. *Nature* 352:126–131.

[25] M. Théry, A. Jiménez-Dalmaroni, V. Racine, M. Bornens and F. Julicher (2007). Experimental and theoretical study of mitotic spindle orientation. *Nature* 447:493–496.

[26] M. Théry, A. Pépin, E. Dressaire, Y. Chen, and M. Bornens (2006). Cell distribution of stress fibres in response to the geometry of the adhesive environment. *Cell Motil. Cytoskeleton* 63:341–355.

[27] T. Tzvetkova-Chevolleau, A. Stéphanou, D. Fuard, J. Ohayon, P. Schiavone, and P. Tracqui (2008). The motility of normal and cancer cells in response to the combined influence of the substrate rigidity and anisotropic microstructure. *Biomaterials* 29:1541–1551.

[28] R. Zaidel-Bar, M. Cohen, L. Addadi, and B. Geiger (2004). Hierarchical assembly of cell-matrix adhesion complexes. *Biochem. Soc. Trans.* 32:416–420.

Part III

Mechanical Effects of Environment on Cell Behavior

Chapter 7. History Dependence of Microbead Adhesion under Varying Shear Rate

As a simple theoretical model of a cell adhering to a biological interface, we consider a rigid sphere moving in a viscous shear flow near a wall. Adhesion forces arise through intermolecular bonds between receptors on the cell and their ligands on the wall, which form flexible tethers that can stretch and tilt as the base of the cell moves past the wall; binding kinetics is assumed to follow a standard model for slip bonds. Typically, under physiological conditions, the time scale for the advection of bonds from the front to the back of a rolling sphere (as viewed in the frame of reference of the sphere) is comparable to the bonds' characteristic lifetime. This advection mechanism may lead to an accumulation of consummated bonds at the trailing edge of the sphere.

Microscale calculations detailing bond formation, advection, and breakage explain the nonlinear relation of the motion of the cell and the net force and torque resulting from adhesion. Three distinct types of macroscale cell motion are then predicted: either bonds accumulate at the back of the cell and the latter is nearly arrested; or bonds adhere strongly but are short-lived and the cell rolls over the wall without slipping; or the cell moves near its free-stream speed with bonds providing only weak frictional resistance to sliding. The model predicts bistability between these states, implying that at critical shear rates the system can switch abruptly among firm arrest, no-slip rolling, and free sliding, and also suggesting that sliding friction arising through bond tilting may play a significant dynamical role in some cell adhesion applications.

Chapter 8. Understanding Adhesion Sites as Mechano-Sensitive Cellular Elements

Cell sensitivity to substrate stiffness is fundamental in the control of many biological functions and pathological processes. Although adhesion sites are recognized to play a key role in the control of cell functions, their specific contribution in terms of cell sensitivity to substrate stiffness is not fully understood. We present a simplified theoretical approach to explain how dynamic adhesion sites behave as cell-sensitive elements while stationary adhesion sites do not. The biomechanical factors governing cell sensitivity can also be deduced and discussed from this theory. Main theoretical concepts are then illustrated by various experimental results issued from various cellular models (tissue cells, inflammatory cells) able to express either stationary or dynamic adhesion sites, depending on intracellular and extracellular conditions.

Chapter 9. Cancer Cell Migration on 2-D Deformable Substrates

Tumor cell migration is a very important phenomenon occurring during the formation of metastases, and requires a correlation between adhesion anchoring and cytoskeleton reorganization, as the cell moves forward. To understand such processes, different methods have been used to measure the displacement of fluorescent beads embedded within a gel, as a way to determine indirectly the traction stresses exerted by the cells. Here, a method for obtaining this traction is based on a minimization algorithm under force penalization. The method is applied to the case of migrating T24 cancer cells on polyacrylamide substrates. Results obtained on substrates with different rigidities are discussed. It is found that such cancer cells exert less traction than other cell types.

Chapter 10. Single-Cell Imaging of Calcium in Response to Mechanical Stimulation

How the mechanical stimuli or physical forces can be perceived by cells and transduced into biochemical responses (i.e., mechano-transduction) has been extensively investigated at single-cell levels. Along with the introduction of a wide variety of technologies to provide mechanical stimulation, the development of genetically encoded and fluorescence resonance energy transfer (FRET)-based biosensors for single-cell imaging has allowed the monitoring and quantification of the signaling cascades in live cells with high spatiotemporal resolution. Calcium ion (Ca^{2+}) is one of the most universal and important elements for many biological processes. It serves as a second messenger not only in signaling transduction in response to chemical stimuli, but also in mechano-transduction. This chapter provides the design strategies for approaching the single-cell imaging of calcium in response to mechanical stimulation. The focus is on the integration of genetically encoded calcium FRET biosensors, an engineered extracellular environment with controllable substrate rigidity, and optical laser tweezers. The dynamic and subcellular visualization of calcium in live cells upon mechanical stimulation can shed new light on the molecular mechanism by which cells perceive external mechanical cues and coordinate signaling pathways to regulate physiological functions.

Chapter 7

History Dependence of Microbead Adhesion under Varying Shear Rate

Sylvain Reboux, Giles Richardson, and Oliver E. Jensen

Contents

7.1 Introduction

The recruitment of blood-borne leukocytes to the vascular endothelium is a crucial step in the immune response. It is mediated by specific receptor–ligand interactions [2,32] that allow circulating leukocytes to form bonds with the endothelium under flow conditions. This results in the so-called adhesive rolling of leukocytes along the blood vessel walls prior to targeting sites of inflammation [20,29,31,34].

Similar adhesion mechanisms are found also in cancer cell metastasis [19], bacterial colonization under flow [16], and targeted drug delivery by functionalized particles [21,26]. This wide range of applications has made cell adhesion an active field of research, resulting in the identification of key adhesion molecules (e.g., E-, L- and P-selectin and their ligands) and the biomechanical characterization of the resulting intermolecular bonds.

However, the connections between physiological observations (e.g., the minimum shear threshold for leukocyte rolling [1,9]) and mechanochemical effects operating within individual intermolecular bonds [13] are yet to be fully understood. Much of the complexity arises from the multiscale nature of the nonlinear interactions of hydrodynamics, adhesion forces, and cell deformation. This has motivated the development of theoretical models of cell adhesion and cell rolling that have now reached considerable levels of sophistication [5,17].

Existing models of cell adhesion fall essentially into two classes, depending on whether bonds are represented within a continuum [6,7,15] or discrete framework [14]. In the first case, the bonds are generally modeled as vertical springs that resist sideways displacement, preventing a cell membrane that is bound to a wall from sliding along it. This ensures that an adherent cell in a shear flow exhibits genuine tank-treading motion, with a peeling process taking place at the trailing edge of the contact region. In the second case, binding and unbinding occur stochastically between individual points on the cell and substrate. The bonds are allowed to tilt freely (they have no preferred spatial orientation), enabling (in principle) some degree of sliding of the cell over the substrate.

In [24] we proposed a continuum deterministic model for binding kinetics in which bonds are allowed to tilt. To pass smoothly from the vertical-bond limit to the case in which bonds can tilt freely, we assumed that the bonds resist tilting via a biomechanical hinge of prescribed stiffness, while being subject to rotational diffusion. A microscale calculation (for two parallel sliding plates) revealed a nonlinear force–speed relation arising from bond formation, tilting, and breakage.

This nonlinear sliding friction law was used in a multiscale model describing the 2-D motion of a cylinder coated with receptors moving over a rigid flat wall in a shear flow [24]. Two distinct types of macroscale cell motion are predicted: either bonds adhere strongly and the cell rolls (or tank-treads) over the wall without slipping, or the cell moves near its free-stream speed with bonds providing weak frictional resistance to sliding. The model predicts bistability between these two states, implying that at critical shear rates the system can switch abruptly between no-slip rolling and free sliding, and suggesting that sliding friction arising through bond tilting may play a significant dynamical role in some cell-adhesion applications.

To our knowledge, bond resistance to tilting has yet to be characterized experimentally in the context of cell adhesion, although it is relevant in other biomimetic adhesives involving fields of oriented deformable binders [33] and has motivated prior modeling of the adhesive properties of rotatable elastic nanofibers [8] or micropillars [27].

We extend here the results obtained in [24] to the 3-D motion of a sphere. In addition we incorporate the effects of nonequilibrium binding kinetics (although we consider a steady problem in the reference frame of the center of the sphere). For the sake of simplicity we assume that the sphere is rigid. In the context of cell rolling adhesion, this assumption is commonly made on the grounds that many (but not all) features of leukocyte rolling have been demonstrated in flow-chamber experiments using ligand-coated microbeads [12].

Net vertical adhesion forces (as formulated by Dembo et al. [6]) tend to bring a sphere in direct contact with the wall. In practice, however, vertical adhesion forces will always be opposed by other forces like electrostatic or steric repulsion or, depending on the system studied, volume exclusion effects caused by local microstructure (e.g., glycocalyx). Also, bonds with a high resistance to tilting can resist compression by exerting a vertical force that diverges as the distance between the particle and the wall tends to zero [24]. To keep the model general and the analysis tractable, we choose to neglect the vertical force balance on the sphere and assume that the separation distance Δ^* between the sphere and the wall is a fixed parameter (we assume that Δ^* is comparable to the average unstressed length of the bonds λ^*, namely $\Delta^*/\lambda^* \equiv d = \mathcal{O}(1)$). To apply lubrication theory in the interstitial region, we assume that the radius R^* of the sphere is large compared to Δ^*. The frame of reference used is the sphere's center, with $(O, \mathbf{e}_x, \mathbf{e}_y, \mathbf{e}_z)$ directed such that $\mathbf{e}_x$ is the streamwise direction, $\mathbf{e}_y$ is the transverse direction, and $\mathbf{e}_z$ is vertical (Figure 7.1). The flow has a uniform shear rate G^* at infinity and is assumed to be purely viscous.

This chapter is organized as follows. In Section 7.2 we recall some known results about the hydrodynamics of a rigid sphere in a shear flow near a wall in the absence of adhesion, or with ad hoc friction forces that prevent sliding

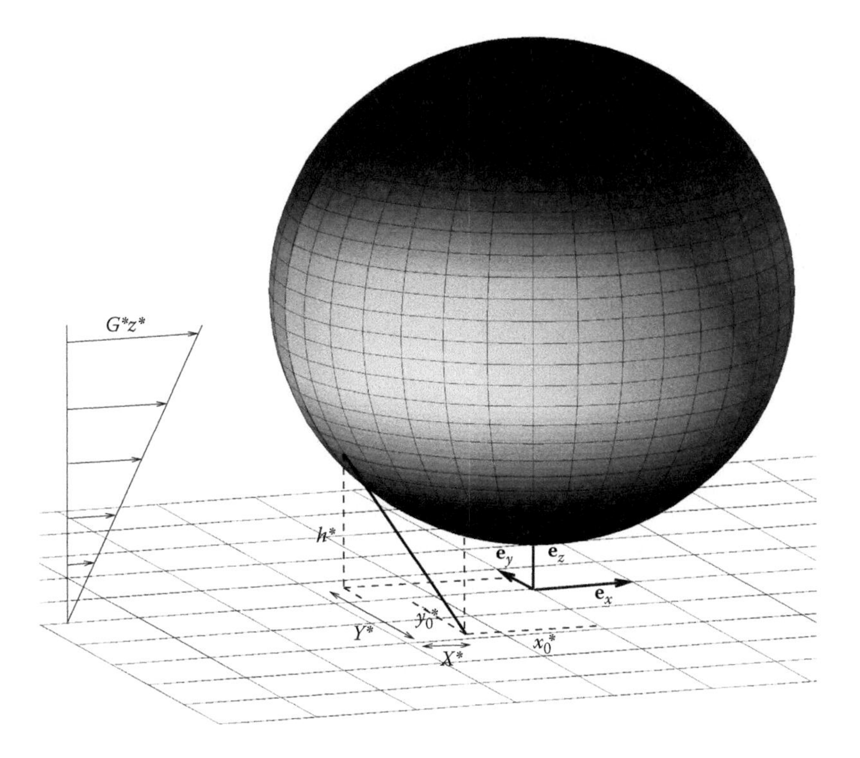

FIGURE 7.1 Binding between a sphere and a plane in a shear flow.

entirely. In Section 7.3 we derive in some detail a model for binding kinetics in 3-D that accounts for bond tilting and nonequilibrium effects. The model relies on assumptions similar to the 2-D model presented in [24], albeit with some qualitative differences that are outlined below. Numerical results are given in Section 7.4 for both (1) the force-velocity relations and (2) the steady motion of the sphere resulting from a balance of adhesive and hydrodynamic forces and torques. The implications of these results are discussed in Section 7.5.

7.2 Hydrodynamics of Sphere near Wall

To understand the effects of adhesive forces on the motion of a cell, we first solve the force and torque balance on a sphere near a wall (1) when it is subject only to hydrodynamic forces (free transport, Section 7.2.1) and (2) when the motion is constrained by strong adhesive friction forces (no-slip rolling, Section 7.2.2). Later we investigate the intermediate case in which the friction forces are coupled (nonlinearly) with the motion of the sphere.

Let $\epsilon = \lambda^*/R^*$ be the ratio of bond length to sphere radius (so that $\Delta^*/R^* = d\epsilon$). Typically, we expect $\epsilon \ll 1$. For a sphere translating parallel to a wall or rotating with its axis of rotation parallel to the wall, the drag from the fluid is singular as ϵ goes to zero, diverging as $\log \epsilon$ [11]. We therefore expect the velocities to scale as $R^*G^*/|\log \epsilon|$. In comparison, for a cylinder in a shear flow, the velocities scale as $\epsilon^{1/2}R^*G^*$.

7.2.1 Lack of Adhesion

Let us consider a sphere moving near a wall, at a fixed distance Δ^*, with horizontal velocity $V_h^*\mathbf{e}_x$ relative to the wall and rotational velocity $\Omega_h^*\mathbf{e}_y$ in a Newtonian fluid of dynamic viscosity μ^*. Following [11], the horizontal force and torque balance (in dimensional form) are, respectively,

$$6\pi\mu^* R^* \left(R^*G^*F_s + V_h^*|\ln d\epsilon|F_t + R^*\Omega_h^*|\ln d\epsilon|F_r\right) = 0 \tag{7.1a}$$

$$-4\pi\mu^* R^{*2} \left(R^*G^*T_s + 2V_h^*|\ln d\epsilon|T_t + 2R^*\Omega_h^*|\ln d\epsilon|T_r\right) = 0 \tag{7.1b}$$

where the dimensionless coefficients can be approximated in the small-ϵ limit [11] by

$$F_s \approx 1.7005 + \mathcal{O}(\epsilon), \qquad T_s \approx 0.9440 + \mathcal{O}(\epsilon) \tag{7.2a}$$

$$F_r \approx \frac{2}{15} - \frac{0.2526}{|\ln d\epsilon|} + \mathcal{O}\left(\frac{\epsilon}{|\ln \epsilon|}\right), \qquad T_r \approx -\frac{2}{5} - \frac{0.3817}{|\ln d\epsilon|} + \mathcal{O}\left(\frac{\epsilon}{|\ln \epsilon|}\right) \tag{7.2b}$$

$$F_t \approx -\frac{8}{15} - \frac{0.9588}{|\ln d\epsilon|} + \mathcal{O}\left(\frac{\epsilon}{|\ln \epsilon|}\right), \quad T_t \approx \frac{1}{10} - \frac{0.1895}{|\ln d\epsilon|} + \mathcal{O}\left(\frac{\epsilon}{|\ln \epsilon|}\right) \tag{7.2c}$$

The subscripts s, r, and t denote coefficients related to the effects of the shear flow, the rotational motion, and the translational motion of the sphere, respectively.

Introducing the dimensionless parameters V_h and Ω_h defined by $V_h^* = V_h R^* G^*$ and $\Omega_h^* = \Omega_h G^*$, Equation (7.1) becomes

$$F_s + V_h|\ln d\epsilon|F_t + \Omega_h|\ln d\epsilon|F_r = 0 \text{ and } T_s + 2V_h|\ln d\epsilon|T_t + 2\Omega_h|\ln d\epsilon|T_r = 0$$

which yields the following expressions for the horizontal and rotational velocities:

$$V_h = \frac{\frac{1}{2}T_s F_r - F_s T_r}{(F_t T_r - F_r T_t)|\ln d\epsilon|} = \left[\frac{3.716}{|\ln d\epsilon|} - \frac{9.197}{|\ln d\epsilon|^2} + \frac{23.41}{|\ln d\epsilon|^3} + \cdots\right] + \mathcal{O}(\epsilon) \tag{7.4a}$$

$$\Omega_h = \frac{-\frac{1}{2}T_s F_t + F_s T_t}{(F_t T_r - F_r T_t)|\ln d\epsilon|} = \left[\frac{2.109}{|\ln d\epsilon|} - \frac{6.072}{|\ln d\epsilon|^2} + \frac{16.00}{|\ln d\epsilon|^3} + \cdots\right] + \mathcal{O}(\epsilon) \tag{7.4b}$$

$$U_h = \left[\frac{1.607}{|\ln d\epsilon|} - \frac{3.124}{|\ln d\epsilon|^2} + \frac{7.406}{|\ln d\epsilon|^3} + \cdots\right] + \mathcal{O}(\epsilon) \tag{7.4c}$$

where $U_h \equiv V_h - \Omega_h$ is the sliding speed (scaled on $R^* G^*$) of the base of the sphere relative to the wall.

7.2.2 Rolling without Sliding

We now consider a sphere moving near a wall, equipped with a device imposing a no-slip condition between the sphere and the wall (e.g., one can imagine "ideal" adhesion molecules that provide infinite resistance against any slippage between the base of the sphere and the wall, as in Dembo et al.'s model [6]). As a result, the sphere is forced to roll without sliding and we have $V_{ns}^* - \Omega_{ns}^* R^* = 0$, where V_{ns}^* and Ω_{ns}^* denote the sphere's horizontal and rotational velocities respectively. Let $\mathcal{F}_{ns}^*$ denote the horizontal friction force exerted on the sphere. The force balance Equation (7.1) is modified as follows:

$$6\pi\mu^* R^*(R^* G^* F_s + V_{ns}^*|\ln d\epsilon|F_t - R^*\Omega_{ns}^*|\ln d\epsilon|F_r) + \mathcal{F}_{ns}^* = 0 \tag{7.5a}$$

$$-4\pi\mu^* R^{*2}(R^* G^* T_s + 2V_{ns}^*|\ln d\epsilon|T_t - 2R^*\Omega_{ns}^*|\ln d\epsilon|T_r) + R^*\mathcal{F}_{ns}^* = 0 \tag{7.5b}$$

Writing Equation (7.5) in terms of dimensionless (unstarred) variables, defined by $\mathcal{F}_{ns}^* = \mathcal{F}_{ns}\mu^* R^{*2} G^*$, $V_{ns}^* = V_{ns} R^* G^*$, and $\Omega_{ns}^* = \Omega_{ns} G^*$, gives

$$6\pi F_s + V_{ns}6\pi|\ln d\epsilon|F_t - \Omega_{ns}6\pi|\ln d\epsilon|F_r + \mathcal{F}_{ns} = 0 \tag{7.6a}$$

$$4\pi T_s + V_{ns}8\pi|\ln d\epsilon|T_t - \Omega_{ns}8\pi|\ln d\epsilon|T_r - \mathcal{F}_{ns} = 0 \tag{7.6b}$$

with the constraint $V_{ns} - \Omega_{ns} = 0$. This yields

$$V_{ns} = -\frac{1}{|\ln d\epsilon|} \frac{2T_s + 3F_s}{3(F_t + F_r) + 4(T_r + T_t)} = \left[\frac{2.912}{|\ln d\epsilon|} - \frac{7.182}{|\ln d\epsilon|^2} + \cdots\right] + \mathcal{O}(\epsilon) \tag{7.7a}$$

$$\Omega_{ns} = V_{ns}, \tag{7.7b}$$

$$\mathcal{F}_{ns} = \frac{12\pi T_s(F_t + F_r) - 24\pi F_s(T_r + T_t)}{3(F_t + F_r) + 4(T_r + T_t)} = \left[-10.10 + \frac{12.35}{|\ln d\epsilon|} + \cdots\right] + \mathcal{O}(\epsilon) \tag{7.7c}$$

We can now quantify the effects of these adhesion forces on the motion of the sphere by comparing the velocities obtained in Equation (7.7a,b) with those for a sphere moving at its free hydrodynamic velocity Equation (7.4a,b):

$$\frac{V_{ns}}{V_h} = 0.784 + \frac{0.0069}{|\ln d\epsilon|} + \cdots \quad \text{and} \quad \frac{\Omega_{ns}}{\Omega_h} = 1.381 + \frac{0.510}{|\ln d\epsilon|} + \cdots \tag{7.8}$$

Equation (7.8) shows that horizontal adhesion forces tend to make the sphere translate slower and rotate faster. In both cases, the change is on the order of 20 to 40%.

In the next section, we include nonequilibrium binding kinetics effects that allow for the build-up of an additional torque on the sphere. Under certain conditions, this torque can dominate the hydrodynamic drag and slow the sphere, reducing the translation speed and rotation rate by several orders of magnitude, much more dramatically than in Equation (7.8). Combined with the nonlinear relationship between adhesion forces and the sphere's motion, it also leads to interesting hysteretic behavior under slowly varying shear rates.

7.3 Adhesive Sphere in Shear Flow

We now focus on the steady motion of a sphere in a shear flow when the sphere and the wall are coated with adhesion molecules that can interact with each other to form bonds (i.e., mechano-resistant complexes). We write a model for the nonlinear forces exerted by adhesion molecules on the sphere and investigate, from force and torque balances, the different scalings of the translation and rotation speed of the sphere in different regions of parameter space. The steady states, defined by the translation and rotation speeds of the sphere, result from a balance of forces between the shear flow, the hydrodynamic drag and adhesion forces. The latter originate from the formation of bonds between the sphere and the wall, which itself depends on the velocity of the sphere. With some assumptions regarding receptor and ligand spatial distributions (such that the bonds can be considered a continuum with homogeneous physical properties), and assuming deterministic binding kinetics (as described by, for example, Dembo et al.'s model [6], see below), the net force and torque exerted collectively by all the bonds on the sphere do not depend on time. Within this framework, the sphere moves steadily even though its motion occurs through the continuous formation and breakage of adhesive bonds, which are naturally time-dependent processes.

Some care is needed in defining an evolution equation for the consummated bond density, determining the resulting adhesive force on the sphere and coupling this with the hydrodynamic forces to find the motion of the sphere as a function of imposed shear rate. We derive the model in detail below. The full model is stated in dimensionless variables in Section 7.3.4.

7.3.1 Geometry and Kinematics

Because we aim to stress the effects of nonequilibrium binding kinetics on the steady motion of the sphere, we make use of two frames of reference, one translating relative to the other. We first describe the dynamical formation and breakage of adhesive bonds with a time-dependent evolution equation in a frame of reference attached to the receptors (anchored to the wall). Translating to a frame of reference attached to the center of the sphere, we then derive the adhesive force densities exerted on the sphere as steady quantities that depend on spatial variables only. The relationship between the reference frame of the receptors on the wall and that of the center of the sphere is determined by the translational motion of the sphere. We therefore expect a strong coupling of the motion of the sphere and the adhesive forces that it is subject to.

$\mathcal{R}_s$ denotes the frame of reference of the center of the sphere. It is associated with a system of coordinates (x_s^*, y_s^*, z^*) with origin O_s on the horizontal wall vertically beneath the base of the sphere (see Figure 7.2).

$\mathcal{R}_w$ denotes the frame of reference of the wall. It is associated with a system of coordinates (x^*, y^*, z^*) and an origin O_w chosen, with no loss of generality, so that it coincides with O_s at time $t^* = 0$. Assuming that the motion of the sphere is steady and directed along $\mathbf{e}_x$, we then have $O_s = (V^*t^*, 0, 0)$ in $\mathcal{R}_w$ (see Figure 7.2).

Let $h_w^*(x^*, y^*, t^*)$ be the vertical distance, at a given time t^*, between the point $(x^*, y^*, 0)$ on the wall (in $\mathcal{R}_w$) and the lower surface of the sphere:

$$h_w^*(x^*, y^*, t^*) = \Delta^* + R^* - \sqrt{R^{*2} - y^{*2} - (x^* - V^*t^*)^2} \tag{7.9}$$

Similarly, $h_s^*(x_s^*, y_s^*)$ denotes the vertical distance between the point $(x_s^*, y_s^*, 0)$ on the wall (in $\mathcal{R}_s$) and the lower surface of the sphere:

$$h_s^*(x_s^*, y_s^*) = \Delta^* + R^* - \sqrt{R^{*2} - y_s^{*2} - x_s^{*2}} \tag{7.10}$$

We drop the subscripts to eliminate confusion. Note that h_s^* does not depend on time.

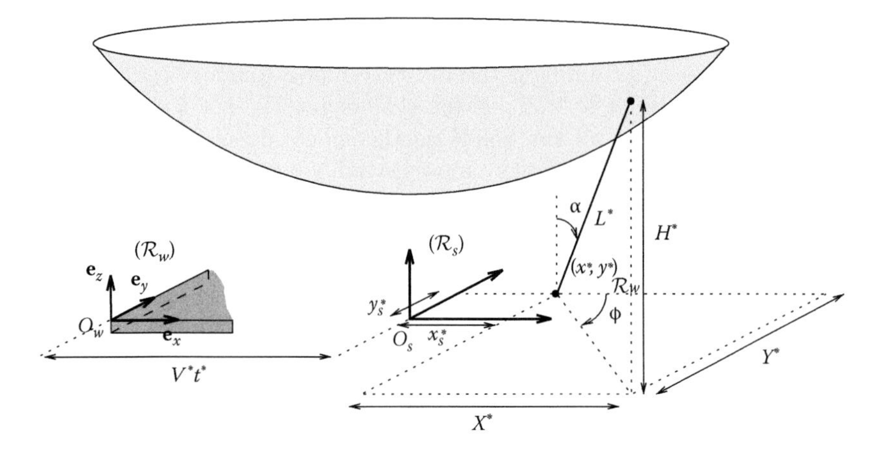

FIGURE 7.2 Schematic description of notation. The frame of reference $\mathcal{R}_w$ is attached to the wall; $\mathcal{R}_s$ moves with the center of the sphere. Both origins O_w and O_s are on the wall. A given receptor on the wall can be identified by its coordinates $(x^*, y^*, 0)$ in $\mathcal{R}_w$ or $(x_s^*, y_s^*, 0)$ in $\mathcal{R}_s$. A receptor–ligand bond can be characterized by Cartesian or spherical coordinates (X^*, Y^*, H^*) and (L^*, α, ϕ), respectively, measured with respect to $(x^*, y^*, 0)$ in $\mathcal{R}_w$.

We assume that stretching of an individual bond occurs over length scales comparable to the unstretched bond length λ^*. The adhesion region, where we expect most of the adhesive phenomena to take place, is then defined as the area of the wall beneath the sphere where the separation distance is of the same order of magnitude as the characteristic bond length λ^*. Because $\Delta^* = \mathcal{O}(\lambda^*)$, this corresponds to a circular area of diameter $\mathcal{O}(\sqrt{\lambda^* R^*})$ (or equivalently, $\mathcal{O}(\sqrt{\Delta^* R^*})$) beneath the sphere.

In the adhesion region, bonds form between two surfaces that, to leading order, are locally flat and parallel with error $\mathcal{O}(\epsilon^{1/2})$. In what follows we retain the terms of $\mathcal{O}(\epsilon^{1/2})$ but neglect higher-order corrections. In general, a sphere that moves near a wall will slide relative to it with a horizontal velocity $U^*\mathbf{e}_x$ at its base. Let us consider a point attached to the wall (e.g., a receptor) within the adhesion region and let $(x_s^*, y_s^*, 0)$ be its coordinates in $\mathcal{R}_s$. Then the horizontal velocity of the sphere relative to that point is given by $V^* - \Omega^* R^*$ to leading order in ϵ (since $x_s^*/R^* = \mathcal{O}(\epsilon^{1/2})$ within the adhesion region). The sliding velocity between the sphere and the wall can therefore be assumed uniform and equal to $U^* = V^* - \Omega^* R^*$ within the adhesion region, to leading order in ϵ.

7.3.2 Model of Binding Kinetics between Moving Surfaces

The binding between receptor-coated and ligand-coated surfaces, with surface densities m_r and m_l, respectively, is commonly referred to as 2-D binding, in contrast to 3-D binding where the molecules are in solution. The apparent rate of binding $K^*_{on,eq}$ between one receptor and the ligand-coated surface (in s^{-1}) is generally defined from the intrinsic binding rate $K^*_{on,int}$ [m^2s^{-1}] between one receptor and one ligand as $K^*_{on,eq} = m_l K^*_{on,int}$. The binding affinity K_{eq} is then the dimensionless ratio of the apparent binding rate and the off-rate $K^*_{off,eq}$ [s^{-1}] of a formed bond. These reaction rates are defined when no load is exerted on the bonds (e.g., for bonds that form vertically between two plates separated by a distance λ^*) and their dependence on force has to be modeled.

Following Dembo et al. [6], the forward and reverse reaction rates for receptor–ligand binding are written as Boltzmann distributions, allowing highly stretched "slip" bonds (for example) to be readily broken by thermal energy fluctuations. However, unlike Dembo et al. [6], we assume that the bonds are allowed to subtend an angle α with the vertical direction as well as an angle ϕ with the streamwise direction (Figure 7.2).

A given bond between a receptor at $(x^*, y^*, 0)$ on the wall in $\mathcal{R}_w$ and the sphere can be characterized in spherical coordinates (with origin at $(x^*, y^*, 0)$) by the two angles, α and ϕ, and its length L^*. Equivalently, a bond is characterized by the two components (X^*, Y^*) of its projection onto the horizontal plane, and the vertical component H^* (see Figure 7.2). The relationship between the two systems of coordinates is:

$$L^* = \sqrt{H^{*2} + X^{*2} + Y^{*2}}, \quad \alpha = \arctan\left(\frac{\sqrt{X^{*2} + Y^{*2}}}{H^*}\right), \quad \phi = \arctan\frac{Y^*}{X^*} \tag{7.11}$$

At a given time t^*, the vertical component H^* of a given bond can be written in terms of the height function h^*_w as:

$$H^* = h^*_w(x^* + X^*, y^* + Y^*, t^*) \tag{7.12}$$

To account for the extra degrees of freedom from Dembo's model (where all bonds are vertical), the forward rate is expressed as the probability density that a bond may form for a given value of (L^*, α, ϕ) times the probability density that this geometrical configuration is realized in the unbound state. The probability densities of forming or breaking bonds between the wall at $(x^*, y^*, 0)$ and the sphere take the form

$$K^*_{off,sph}(L^*, \alpha, \phi) = K^*_{off,eq} \exp\left[(\kappa^* - \kappa^*_{ts})\frac{(L^* - \lambda^*)^2}{2k^*_B T^*}\right] \tag{7.13a}$$

$$K^*_{on,sph}(L^*, \alpha, \phi) = K^*_{on,eq} \exp\left[-\kappa^*_{ts}\frac{(L^* - \lambda^*)^2}{2k^*_B T^*}\right] \mathcal{P}^*_{sph}(L^*, \alpha, \phi) \tag{7.13b}$$

respectively. Here k^*_B is Boltzmann's constant, T^* is the absolute temperature, κ^* [Nm^{-1}] is the spring constant of one molecular bond, and κ^*_{ts} [Nm^{-1}] is the spring constant of the transition state (see [6]) used to distinguish catch ($\kappa^* < \kappa^*_{ts}$) from slip ($\kappa^* > \kappa^*_{ts}$) bonds.

$\mathcal{P}^*_{sph}(L^*, \alpha, \phi)$ is defined as the probability density that a free bond (i.e., one anchored to the wall only) lies within the region defined by the point $(x^*, y^*, 0)$ (on the wall) and the two angles α and ϕ. Note that $K^*_{on,sph}$ has dimension [s^{-1}].

We assume that the energy associated with tilting a bond from its vertical position is independent of ϕ and of the form $\frac{1}{2}\kappa^*_\theta\alpha^2$ for some $\kappa^*_\theta \geq 0$ (with α and ϕ defined in Equation (7.11)). Hence we assume a Boltzmann distribution for $\mathcal{P}_{sph}$ of the form

$$\mathcal{P}_{sph}(L^*, \alpha, \phi) = \frac{\exp\left[-\kappa_\theta\alpha^2\right]}{\mathcal{N}}, \qquad \kappa_\theta \equiv \frac{\kappa^*_\theta}{2k^*_B\mathcal{T}^*} \tag{7.14}$$

with the normalization factor

$$\mathcal{N} = \int_{-\pi}^{\pi}\int_{0}^{\pi/2} \exp\left[-\kappa_\theta\xi^2\right] d\xi d\varphi = \pi^{3/2}\kappa_\theta^{-1/2}\mathrm{erf}\left(\pi\kappa_\theta^{1/2}/2\right) \tag{7.15}$$

In Equation (7.14), κ^*_θ [N $\cdot$ m] is the torsional spring constant, that is, the moment that has to be exerted about the bond's anchorage point on the wall in order to tilt the bond from the vertical by one radian. Its dimensionless counterpart κ_θ compares κ^*_θ with thermal fluctuation energy. The limit $\kappa_\theta \to 0$ therefore represents the limit in which the bonds are allowed to explore freely all possible angles under thermal fluctuations. For $\kappa_\theta \to \infty$, all the bonds are restricted to the vertical, $\mathcal{P}_{sph}(L^*, \alpha, \phi) \to \delta_{Dirac}(\alpha)$, and no sliding can occur between the bound cylinder and the wall, as was assumed in the models of [6] and others. To our knowledge, no experimental data are presently available to determine the actual value of κ_θ for the bonds that mediate cell adhesion. However, some adhesion molecules (e.g., P-selectins) have been reported to have a persistence length of 0.35 nm [10], that is, an order of magnitude less than their length. This suggests that $\kappa_\theta \ll 1$, at least during the initial stage of cell rolling, which is principally mediated by P-selectin/PSGL-1 interactions.

For the sake of generality, however, we make no assumption on the magnitude of κ_θ in the derivation of the present model. This is motivated by various applications in which tiltable microstructures may represent a resistive force to a sliding motion: for example, cell adhesion on synthetic substrates made of micropillars of well-characterized bending stiffnesses (see, for example, [27]) or the mechanical effects of microvilli in neutrophil rolling [4].

To make the forthcoming analysis easier, we define binding rates $K^*_{off,cart}$ [s^{-1}] and $K^*_{on,cart}$ [$\mathrm{m}^{-2}\mathrm{s}^{-1}$] for a given bond (L^*, α, ϕ) at $(x^*, y^*, 0)$ in terms of the bond's Cartesian coordinates (X^*, Y^*, H^*) (Figure 7.2). Equating the binding rates within the same infinitesimal volume (see Figure 7.3) in both sets of coordinates yields:

$$K^*_{on,cart}(X^*, Y^*, H^*)dX^*dY^* = K^*_{on,sph}(L^*, \alpha, \phi)d\alpha d\phi \tag{7.16}$$

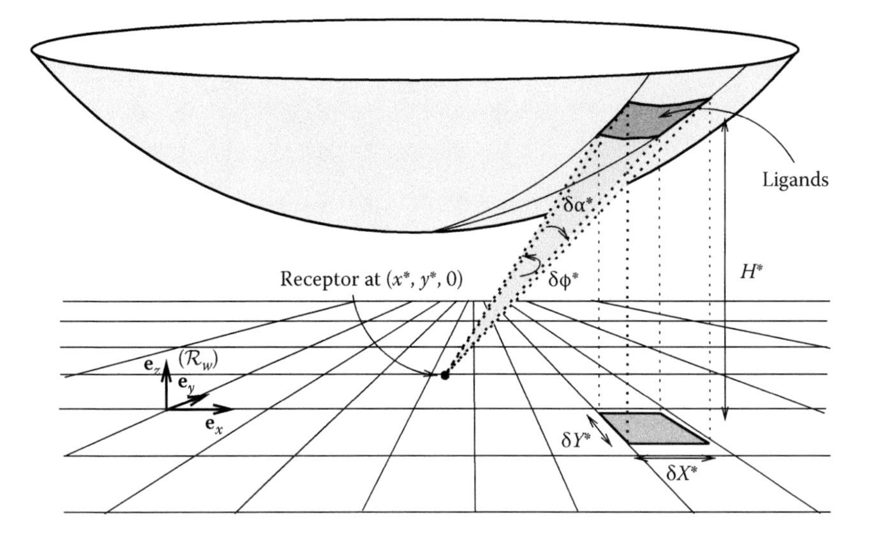

FIGURE 7.3 Rate of binding between a receptor and an infinitesimal surface on the sphere.

with the relationship between (L^*, α, ϕ) and (X^*, Y^*, H^*) given by Equation (7.11). Because the dissociation rate is defined on a per-bond basis, we have $K^*_{off,cart}(X^*, Y^*, H^*) = K^*_{off,sph}(L^*, \alpha, \phi)$. Substituting in Equation (7.13),

$$K^*_{off,cart}(H^*, X^*, Y^*) = K^*_{off,eq} \exp\left[(\kappa^* - \kappa^*_{ts})\frac{(L^* - \lambda^*)^2}{2k^*_B\mathcal{T}^*}\right] \tag{7.17a}$$

$$K^*_{on,cart}(H^*, X^*, Y^*) = K^*_{on,eq} \exp\left[-\kappa^*_{ts}\frac{(L^* - \lambda^*)^2}{2k^*_B\mathcal{T}^*}\right]\mathcal{P}^*_{cart}(X^*, Y^*, H^*) \tag{7.17b}$$

where $\mathcal{P}^*_{cart}$ [m^{-2}] is defined such that, for any cone Ω (with vertex $(x^*, y^*, 0)$),

$$\iint_\Omega \mathcal{P}^*_{cart}(X^*, Y^*, H^*)dX^*dY^* = \iint_\Omega \mathcal{P}_{sph}(L^*, \alpha, \phi)d\alpha d\phi \tag{7.18}$$

A change of variables on the RHS of Equation (7.18) leads from Equation (7.14) to

$$\mathcal{P}^*_{cart}(X^*, Y^*, H^*) = \frac{\exp\left[-\kappa_\theta\alpha^2\right]}{\mathcal{N}}\frac{H^*}{\sqrt{X^{*2} + Y^{*2}}L^{*2}} \tag{7.19}$$

where the second term on the RHS is the determinant of the Jacobian matrix of the transformation from (L^*, α, ϕ) to (H^*, X^*, Y^*). In what follows we use Cartesian coordinates.

We now evaluate the consummated bond density in $\mathcal{R}_w$. For a given point $(x^*, y^*, 0)$ on the wall (in $\mathcal{R}_w$), let $A^*_{tot}\, g^*_w(x^*, y^*, X^*, Y^*, t^*)\, \delta x^*\delta y^*\delta X^*\delta Y^*$ be

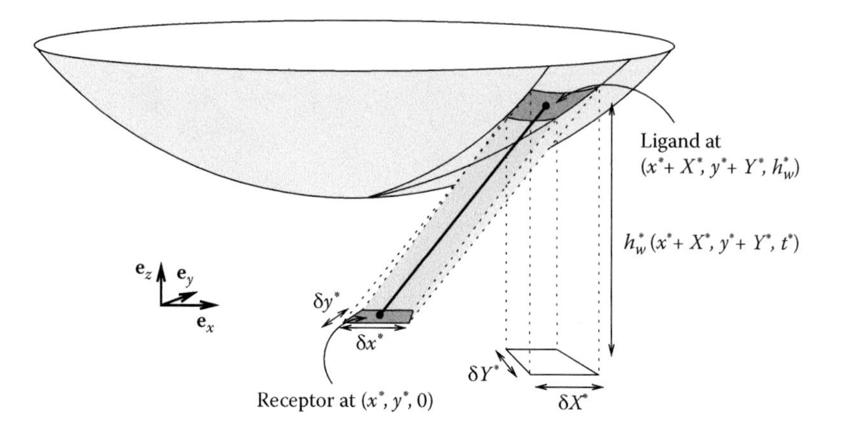

FIGURE 7.4 The bond density $g^*_w(x^*, y^*, X^*, Y^*, t^*)$ is defined as the density of bonds in the grayed volume at some time t^*.

the number of bonds that are attached between the infinitesimal patch of area $\delta x^* \delta y^*$ at $(x^*, y^*, 0)$ on the wall and the infinitesimal patch of area[1]

$$\frac{\delta X^* \delta Y^*}{\sqrt{1+\left(\dfrac{\partial h^*_w}{\partial x^*}\right)^2}} \quad \text{at} \quad (x^*+X^*, y^*+Y^*, h^*_w(x^*+X^*, y^*+Y^*, t^*))$$

on the sphere at time t^* (see Figure 7.4).

The evolution equation for the bond density g^*_w between the fixed wall and the moving sphere is obtained (in $\mathcal{R}_w$) by equating the rate of change in g^*_w with spontaneous bond formation, spontaneous bond breakage, and bond advection by horizontal sliding:

$$\frac{\partial g^*_w}{\partial t^*} + \frac{\partial}{\partial X^*}[U^* g^*_w] = \frac{K^*_{on}(h^*_w(x^*+X^*, y^*+Y^*, t^*), X^*, Y^*)}{\sqrt{1+\left(\dfrac{\partial h^*_w}{\partial x^*}\right)^2}} - \frac{K^*_{off}\left(h^*_w(x^*+X^*, y^*+Y^*, t^*), X^*, Y^*\right) g^*_w}{\sqrt{1+\left(\dfrac{\partial h^*_w}{\partial x^*}\right)^2}} \tag{7.20}$$

with $\quad g^*_w \to 0, \text{ as } X^* \to -\infty \text{ (for } U^* > 0) \quad \text{or} \quad |Y^*| \to \infty.$

The effects of the translational motion of the sphere are embedded in the dependence on time of the reaction rates. As defined in Equation (7.17), these

[1]The denominator comes from the projection of the rectangle of area $\delta X^* \delta Y^*$ on the wall onto the sphere. It is approximately equal to one near the base of the sphere and will not contribute to the leading order solution.

depend on the vertical distance $h_w^*(x^* + X^*, y^* + Y^*, t^*)$ between the ligands on the sphere and the wall and changes of h_w^* with time as the sphere moves past (see Equation (7.9)).

We now express Equation (7.20) in the reference frame of the center of the sphere, $\mathcal{R}_s$. For a given point $(x_s^*, y_s^*, 0)$ on the wall (in $\mathcal{R}_s$), let $A_{tot}^* \, g_s^*(x_s^*, y_s^*, X^*, Y^*) \, \delta x_s^* \delta y_s^* \, \delta X^* \delta Y^*$ be the number of bonds that are attached between the infinitesimal patch of area $\delta x_s^* \delta y_s^*$ at $(x_s^*, y_s^*, 0)$ on the wall and the infinitesimal patch of area

$$\frac{\delta X^* \delta Y^*}{\sqrt{1 + \left(\dfrac{\partial h_w^*}{\partial x^*}\right)^2}} \quad \text{at} \quad (x_s^* + X^*, y_s^* + Y^*, h_s^*(x_s^* + X^*, y_s^* + Y^*))$$

on the sphere. In $\mathcal{R}_s$, the center of the sphere is fixed, the sphere rotates about $\mathbf{e}_y$, and the wall slides underneath the sphere. At each point on the (x_s, y_s)-plane, the height between the wall and the sphere does not vary with time. We assume that the sphere has reached a steady state and that there are uniform and continuous distributions of adhesion molecules both on the sphere and the wall. Hence we expect the bond distribution g_s^* at each point on the (x_s, y_s)-plane to remain constant.

The evolution equation for g_s^* is obtained from a change of frame of reference from $\mathcal{R}_w$ to $\mathcal{R}_s$ in Equation (7.20). The coordinates in $\mathcal{R}_s$ of a point attached to the wall (e.g., a receptor) and the height between this point and the sphere vary in time according to the parametrization

$$y_s^* = y^*, \quad x_s^*(t^*) = x^* - V^* t^* \quad \text{and} \quad h_s^*(x_s^*(t^*), y_s^*) = h_w^*(x^*, y^*, t^*) \tag{7.21}$$

Similarly, the bond densities in each frame of reference satisfy

$$g_w^*(x^*, y^*, X^*, Y^*, t^*) = g_s^*(x_s^*(t^*), y_s^*, X^*, Y^*) \tag{7.22}$$

The time derivative in Equation (7.20), which describes nonequilibrium effects in the binding kinetics, concerns g_w^*, defined in $\mathcal{R}_w$ where the height between the wall and the sphere (on which the reaction rates depend) varies. Changing to $\mathcal{R}_s$, where this height is fixed, transforms the time-dependent problem to a purely spatial one. Applying the chain rule to Equation (7.22) with Equation (7.21b) yields:

$$\frac{\partial g_w^*}{\partial t^*} = \frac{\partial x_s^*}{\partial t^*}(t^*)\frac{\partial g_s^*}{\partial x_s^*} = -V^* \frac{\partial g_s^*}{\partial x_s^*} \tag{7.23}$$

In $\mathcal{R}_s$, nonequilibrium effects in the formation of bonds appear through the translation speed V^* of the sphere and the streamwise inhomogeneities of the bond distribution $\partial g_s^* / \partial x_s^*$.

In $\mathcal{R}_s$, using Equations (7.22) and (7.23), Equation (7.20) therefore becomes

$$-V^*\frac{\partial g_s^*}{\partial x_s^*}+\frac{\partial}{\partial X^*}[U^*g_s^*]=\frac{K_{on}^*(h_s^*(x_s^*+X^*,y_s^*+Y^*),X^*,Y^*)}{\sqrt{1+\left(\frac{\partial h_w^*}{\partial x^*}\right)^2}} - \frac{K_{off}^*(h_s^*(x_s^*+X^*,y_s^*+Y^*),X^*,Y^*)g_s^*}{\sqrt{1+\left(\frac{\partial h_w^*}{\partial x^*}\right)^2}} \tag{7.24}$$

with $g_s^* \to 0$, as $X^* \to -\infty$ (for $U^* > 0$) or $|Y^*| \to \infty$ where h_s^* has been defined in Equation (7.10). Note that Y^* and y_s^* play the roles of parameters. This is a consequence of our assumption that the sphere rotates about an axis that is always perpendicular to the streamwise direction. In what follows we consider bond densities defined in $\mathcal{R}_s$, and therefore drop the subscript s.

7.3.3 Forces and Torques

Each bond locally exerts a force on the sphere that can be broken into (1) the extensional force, which is related to the bond stretch by Hooke's law (we assume, however, that bonds do not resist compression), and (2) the torsional force, which is proportional to the angle α formed by the bond with the vertical (see Figure 7.5). These forces are defined, respectively, by

$$\mathbf{f}_E^* = \kappa^* \max(L^* - \lambda^*, 0)\mathbf{e}_r \quad \text{and} \quad \mathbf{f}_T^* = \kappa_\theta^* L^{*-1}\alpha\,\mathbf{e}_r \tag{7.25}$$

The number of bonds in the $\mathcal{O}(\lambda^* R^*)$ adhesion area is expected to scale like $A_{tot}^* K_{eq}\lambda^* R^*$. The net adhesive force $\mathbf{F}_{adh}^*$ exerted on the sphere is the

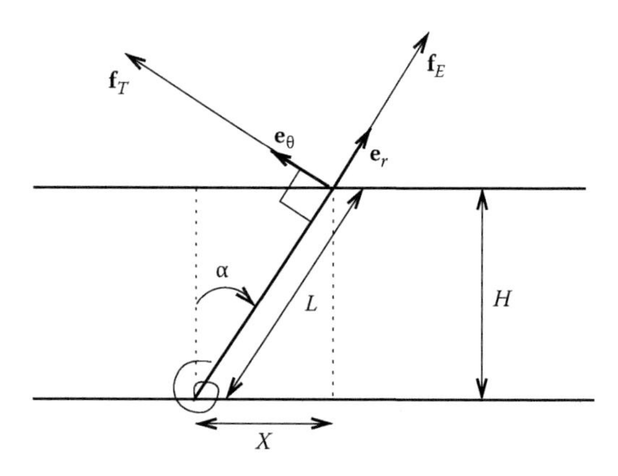

FIGURE 7.5 Schematic view of the forces exerted by an individual bond. Note that $\mathbf{e}_r = -\frac{X}{L}\mathbf{e}_x - \frac{Y}{L}\mathbf{e}_y - \frac{H}{L}\mathbf{e}_z$, $\mathbf{e}_\theta = -\frac{HX}{DL}\mathbf{e}_x - \frac{HY}{DL}\mathbf{e}_y + \frac{D}{L}\mathbf{e}_z$.

sum of forces due to bond stretching $\mathbf{F}_E^*$ and bond tilting $\mathbf{F}_T^*$, which are thus expected to scale like $A_{tot}^* K_{eq} \kappa^* \lambda^{*2} R^*$ and $A_{tot}^* K_{eq} \kappa_\theta^* R^*$, respectively:

$$\mathbf{F}_{adh}^* = \iint_{R^2} \mathbf{f}_{adh}^* \, dx_s^* \, dy_s^*, \quad \text{with} \quad \mathbf{f}_{adh}^* = \iint_{R^2} g^* \left(\mathbf{f}_E^* + \mathbf{f}_T^*\right) dX^* \, dY^* \tag{7.26}$$

being the force density exerted on the sphere per surface area on the wall.

Similarly, the adhesive torque $\mathbf{T}_{adh}^*$ about the center of the sphere can be separated into the contributions from bond stretching $\mathbf{T}_E^*$ and bond tilting $\mathbf{T}_T^*$. Furthermore, distinguishing vertical and horizontal adhesion forces, we have

$$\mathbf{T}_{adh}^* \cdot \mathbf{e}_y = \iint_{R^2} \iint_{R^2} g^* ((x_s^* + X^*)\mathbf{e}_x - (R^* + \Delta^* - h_s^*)\mathbf{e}_z) \times (\mathbf{f}_E^* + \mathbf{f}_T^*) \, dX^* \, dY^* \, dx_s^* \, dy_s^* \cdot \mathbf{e}_y \tag{7.27a}$$

$$= \iint_{R^2} \iint_{R^2} g^* (R^* + \Delta^* - h_s^*)(\mathbf{f}_E^* + \mathbf{f}_T^*) \cdot \mathbf{e}_x \, dX^* \, dY^* \, dx_s^* \, dy_s^* + \iint_{R^2} \iint_{R^2} g^* (x_s^* + X^*)(\mathbf{f}_E^* + \mathbf{f}_T^*) \cdot \mathbf{e}_z \, dX^* \, dY^* \, dx_s^* \, dy_s^* \tag{7.27b}$$

The first term on the RHS of Equation (7.27b) is the torque created by horizontal friction forces on the base of the sphere. It scales like $A_{tot}^* K_{eq} \kappa^* (\lambda^* R^*)^2$ or $A_{tot}^* K_{eq} \kappa_\theta^* R^{*2}$, depending on whether the forces arise primarily through bond stretching or bond tilting, respectively. The second term in Equation (7.27) is the torque created by the asymmetry of vertical adhesive forces between the front and the back of the sphere. Although from straightforward scaling arguments it appears to be $\mathcal{O}(\sqrt{\lambda^*/R^*})$ smaller than the first term, there are cases, as explained below, where it becomes dominant. This can arise as a consequence of the nontrivial dependence between the adhesive forces and the unknown variables U^* and V^*.

As in Equations (7.5a,b), the horizontal force and torque balance about the center of mass of an adhesive sphere moving near a wall in a shear flow, at steady state, are, respectively

$$6\pi\mu^* R^* (R^* G^* F_s + V^* |\ln \epsilon| (F_t + F_r) - U^* |\ln \epsilon| F_r) + (\mathbf{F}_E^* + \mathbf{F}_T^*) \cdot \mathbf{e}_x = 0 \tag{7.28a}$$

$$-4\pi\mu^* R^{*2} (R^* G^* T_s + 2V^* |\ln \epsilon| (T_t + T_r) - 2U^* |\ln \epsilon| T_r) + (\mathbf{T}_E^* + \mathbf{T}_T^*) \cdot \mathbf{e}_y = 0 \tag{7.28b}$$

where the coefficients with subscripts s, r, and t are the $\mathcal{O}(1)$ functions of ϵ introduced in Equation (7.2). The main difference from (7.5a,b) is that there is now no explicit relationship between V^* and Ω^*. The adhesive forces and torque are also no longer treated as unknowns, but rather as (nonlinear) functions of V^* and U^*, so that Equation (7.28) is a closed system of two equations for two unknowns. For different parameter values, it models both regimes of no adhesion (as in Section 7.2.1) and ideal adhesion (as in Section 7.2.2), as well as the transition between the two.

7.3.4 Nondimensionalization

From Equation (7.28), we have, for the horizontal force balance and torque balances

$$\left(\frac{G^*}{K^*_{off,eq}}F_s + \frac{V^*|\log\epsilon|}{R^*K^*_{off,eq}}(F_t + F_r) - \frac{U^*|\log\epsilon|}{R^*K^*_{off,eq}}F_r\right.$$
$$\left.+\frac{A^*_{tot}K^*_{eq}\left(\kappa^*\lambda^{*2}\mathbf{F}_E + \kappa^*_\theta\mathbf{F}_T\right)}{6\pi\mu^*R^*K^*_{off,eq}}\right)\cdot\mathbf{e}_x = 0 \qquad (7.29a)$$

$$\left(\frac{G^*}{K^*_{off,eq}}T_s + 2\frac{V^*|\log\epsilon|}{R^*K^*_{off,eq}}(T_t + T_r) - 2\frac{U^*|\log\epsilon|}{R^*K^*_{off,eq}}T_r\right.$$
$$\left.-\frac{A^*_{tot}K^*_{eq}\left(\kappa^*\lambda^{*2}\mathbf{T}_E + \kappa^*_\theta\mathbf{T}_T\right)}{4\pi\mu^*R^*K^*_{off,eq}}\right)\cdot\mathbf{e}_y = 0 \qquad (7.29b)$$

respectively. We then introduce the following dimensionless variables and parameters

$$h^*_s = \lambda^*H, \quad L^* = \lambda^*L, \quad V^* = V\sqrt{\epsilon}R^*K^*_{off,eq}, \quad U^* = U\epsilon R^*K^*_{off,eq} \qquad (7.30a)$$

$$G^* = GK^*_{off,eq}, \quad K^*_{off} = e_{off}K^*_{off,eq}, \quad K^*_{on} = e_{on}\frac{K^*_{on,eq}}{\lambda^*}, \quad g^* = g\frac{K_{eq}}{\lambda^{*2}} \qquad (7.30b)$$

$$\beta = \frac{\kappa^*_{ts}}{\kappa^*}, \quad \gamma = \frac{\kappa^*\lambda^{*2}}{k^*_B\mathcal{T}^*}, \quad \mathcal{C} = \frac{A^*_{tot}K_{eq}\kappa^*\lambda^{*2}}{\mu^*R^*K^*_{off,eq}}, \quad \text{and} \quad k = \frac{\kappa^*_\theta}{2\kappa^*\lambda^{*2}} \qquad (7.30c)$$

where γ and γk compare the stretching and torsion energy to thermal fluctuations, respectively, and β models the response of the bonds to extensional strain (with $\beta < 1$ for slip bonds and $\beta > 1$ for catch bonds, [6]). k therefore compares the magnitude of adhesion forces arising from bond tilting to those due to bond stretching. $\mathcal{C}$ is the visco-adhesive parameter, relating extensional bond forces to hydrodynamic forces. Typical values of the parameters are shown in Table 7.1.

Bonds are elongated horizontally by the sliding motion of the sphere with a time scale λ^*/U^* (see Figure 7.6(a)) and stretched vertically as the sphere rolls past the site where the bond is anchored to the wall with a time scale $(\lambda^*R^*)^{1/2}/V^*$ (see Figure 7.6(b)). The dimensionless ratios $U^*/K^*_{off,eq}\lambda^*R^*$

TABLE 7.1 Typical Parameter Values for Neutrophil and Microbead Rolling

Symbol	Definition	Neutrophils	Microbeads	Ref.
R^*	Cell radius	4μm	2–20 μm	[28]
λ^*	Bond length	10–300 nm	70 nm	[30,31]
G^*	Shear rate	40–2000 s^{-1}	2–2000 s^{-1}	
A^*_{tot}	Receptor density	10–10^2 μm^{-2}	0–800 μm^{-2}	[18]
$K^*_{off,eq}$	Reverse rate	1–10 s^{-1}	1–10 s^{-1}	[3,25]
$K^*_{on,eq}$	Forward rate	1–100 s^{-1}	1–10 s^{-1}	[18,25]
κ^*	Spring constant	0.01–5 dyn cm^{-1}	5 dyn cm^{-1}	[10,30]
$\kappa^*_\theta/2k^*_B\mathcal{T}^*$	Resistance to bending	0–10^3	≈ 0	
G	Dimensionless shear rate	1–10^4	1–10^4	
$\mathcal{C}$	Visco-adhesive parameter	1–10^4	1–10^6	

and $V^*/K^*_{off,eq}(\lambda^* R^*)^{1/2}$ therefore compare bonds' characteristic lifetimes with advection mechanisms occurring at the microscopic level of the bonds.

The full dimensionless problem becomes, from Equations (7.29) and (7.30),

$$G6\pi F_s + \sqrt{\epsilon}V|\log\epsilon|6\pi(F_t + F_r) - \epsilon U|\log\epsilon|6\pi F_r + \mathcal{C}(\mathbf{F}_E + k\mathbf{F}_T)\cdot\mathbf{e}_x = 0 \tag{7.31a}$$

$$G4\pi T_s + \sqrt{\epsilon}V|\log\epsilon|8\pi(T_t + T_r) - \epsilon U|\log\epsilon|8\pi T_r - \mathcal{C}(\mathbf{T}_E + k\mathbf{T}_T)\cdot\mathbf{e}_y = 0 \tag{7.31b}$$

Here, $\mathbf{F}_E$, $\mathbf{F}_T$, $\mathbf{T}_E$, and $\mathbf{T}_T$ are functions of U and V, derived from (Equations (7.26) and (7.27)) in terms of force densities as

$$\mathbf{F}_E = \iint_{R^2} \hat{\mathbf{F}}_E(x_s, y_s)\,dx_s\,dy_s, \quad \mathbf{F}_T = \iint_{R^2} \hat{\mathbf{F}}_T(x_s, y_s)\,dx_s\,dy_s \tag{7.32a}$$

$$\mathbf{T}\cdot\mathbf{e}_y = \iint_{R^2} \hat{F}_x(x_s, y_s)\,dx_s\,dy_s + \sqrt{\epsilon}\iint_{R^2} \hat{F}_z(x_s, y_s)x_s\,dx_s\,dy_s + \mathcal{O}(\epsilon) \tag{7.32b}$$

where

$$\hat{F}_x(x_s, y_s) = (\hat{\mathbf{F}}_E(x_s, y_s) + \hat{\mathbf{F}}_T(x_s, y_s))\cdot\mathbf{e}_x \tag{7.33a}$$

$$\hat{F}_z(x_s, y_s) = (\hat{\mathbf{F}}_E(x_s, y_s) + \hat{\mathbf{F}}_T(x_s, y_s))\cdot\mathbf{e}_z \tag{7.33b}$$

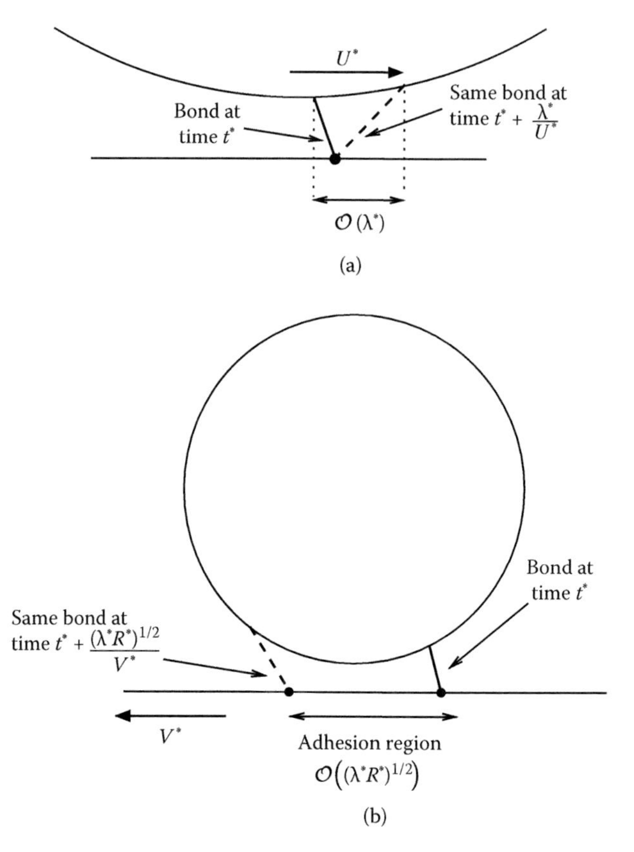

FIGURE 7.6 Schematic view of the advection time scales of consummated bonds. (a) Advection of a tiltable bond by sliding; (b) advection of a bond beneath a sphere (as viewed in $\mathcal{R}_s$).

Using Equation (7.25) and Figure 7.5, the adhesive force densities are expressed in terms of g as

$$\hat{\mathbf{F}}_E(x_s, y_s) = \iint_{R^2} -g(x_s, y_s, X, Y)\max(L-1, 0)\left[\frac{X}{L}\mathbf{e}_x + \frac{H}{L}\mathbf{e}_z\right] dX\,dY \tag{7.34a}$$

$$\hat{\mathbf{F}}_T(x_s, y_s) = \iint_{R^2} g(x_s, y_s, X, Y)\left[-\frac{HX}{DL^2}\arctan\frac{D}{H}\mathbf{e}_x + \frac{D}{L^2}\arctan\frac{D}{H}\mathbf{e}_z\right] dX\,dY \tag{7.34b}$$

where H, L, and D are functions of x_s, y_s, X, and Y, derived from Equations (7.10) through (7.12).

$$H(x_s, y_s, X, Y) = d + \tfrac{1}{2}(x_s + y_s + \sqrt{\epsilon}(X+Y))^2 \tag{7.35a}$$

$$L(x_s, y_s, X, Y) = \sqrt{H(x_s, y_s, X, Y)^2 + X^2 + Y^2} \tag{7.35b}$$

$$D(X, Y) = \sqrt{X^2 + Y^2} \tag{7.35c}$$

Following Equations (7.24) and (7.17), the bond density $g(x_s, y_s, X, Y)$ satisfies the PDE

$$-V\frac{\partial g}{\partial x_s} + U\frac{\partial g}{\partial X} = e_{on}(H(x_s, y_s, X, Y), X, Y) - e_{off}(H(x_s, y_s, X, Y), X, Y)g \tag{7.36a}$$

with errors of $\mathcal{O}(\epsilon)$. The reaction rates are given by

$$e_{off}(H(x_s, y_s, X, Y), X, Y) = \exp\left[(1-\beta)\frac{\gamma}{2}(L-1)^2\right] \tag{7.37a}$$

$$e_{on}(H(x_s, y_s, X, Y), X, Y) = \exp\left[-\gamma\left(\frac{\beta}{2}(L-1)^2 + k\arctan^2\left(\frac{D}{H}\right)\right)\right]\frac{H}{\mathcal{N}L^3} \tag{7.37b}$$

where the normalization factor $\mathcal{N}$ is defined in Equation (7.15). Note from Equation (7.36) that g depends on y_s and Y only parametrically.

We seek U and V as solutions of Equations (7.31) through (7.37) and parameterized by G, $\mathcal{C}$, ϵ, d, k, β, and γ. The solution strategy is as follows. Using Equations (7.32) through (7.37) we first determine numerically the adhesive forces and torques $\mathbf{F}_E$, $\mathbf{F}_T$, $\mathbf{T}_E$, and $\mathbf{T}_T$ as functions of U and V for a sample of parameter values. The system in Equation (7.31) then becomes an algebraic system of two equations for the two unknowns U and V, which we solve numerically. We expect that this nonlinear system has multiple solutions in some regions of parameter space. The numerical integration is undertaken, for each point on a fine grid in (U, V) space, using the subroutine `d03pcf` in the NAG library to solve the PDE (Equation (7.36)) and the subroutines `e01baf` (spline interpolation) and `e02bdf` (spline integration) for the integrals in Equations (7.34) and (7.32). The resulting data set is then fitted by 2-D spline interpolants, allowing us to implement $\mathbf{F}_E$, $\mathbf{F}_T$, $\mathbf{T}_E$, and $\mathbf{T}_T$ as smooth functions of U and V (and parameterized by ϵ, d, k, β, and γ).

7.4 Numerical Results

7.4.1 Nonlinear Force-Velocity and Torque-Velocity Relations

We first compute each component of the adhesion forces and torque for a large range of values of U and V (by solving the differential Equation (7.36) numerically, substituting in Equation (7.34), and integrating over the whole sphere using Equation (7.32)). We do so for fixed values of the parameters that describe the bonds' properties β, γ, k, for a fixed height d separating the sphere from the wall and a fixed value of ϵ.

Figure 7.7 shows a brief summary of how friction force densities between two adhesive surfaces depend on their relative sliding speed. Here the two surfaces are flat and parallel, and are separated by a constant height H_0 (so that the dependence in x_s is lost in Equation (7.36), which effectively becomes an ODE). The nonlinearity in the force–velocity relationship has been described in [24] for the 2-D case of a rigid cylinder and remains qualitatively the same for the 3-D case. When there is no sliding, the bond distribution g is symmetric and there is no net lateral force on the upper plate. At low speeds, bonds are tilted sideways, providing a frictional force opposing the motion. At high speeds, the rapid motion of the plate causes bond breakage (leading to a reduction in the magnitude of g) and a drop in the frictional force.

Figure 7.8 shows the torque exerted by the bonds about the center of the sphere, plotted as a function of V for the special case $U = 0$. Panels (a) through (c) explain qualitatively how advective effects in the bonds' formation can lead to a nonlinearity between the torque and the translation speed. When the sphere is static, the total bond density is symmetric. As the sphere rolls slowly over the wall, nonequilibrium effects delay bond formation at the front of the adhesive region, where the surfaces approach vertically, and delay bond breakage at the rear of the adhesive region, where surfaces are separating. The asymmetry in bond density creates a net torque that opposes rolling motion. At high speeds, bonds form less easily and break more readily, leading to a reduction in the adhesive torque.

Figure 7.9 summarizes these results, and shows how the friction force and the torque exerted on a sphere depend nonlinearly on both U and V. The data are computed only for $\sqrt{\epsilon}U < V$, which corresponds, physically, to $U^* < V^*$ (i.e., $\Omega^* > 0$). The remainder of the domain is not relevant for a sphere in a shear flow. The horizontal adhesive force is always negative (acting against the sliding motion, with $U > 0$) and has a minimum that scales approximately like $\mathcal{O}(\min(-1, -k))$. The torque exerted about the center of the sphere can change sign: it is negative when it comes predominantly from friction forces (and its minimum is then $\mathcal{O}(\min(-1, -k))$) and it is positive when it comes predominantly from the nonlinear advective effect described in Figure 7.8 (its maximum is then $\mathcal{O}(1)$). The asymptotic behaviors of the different components

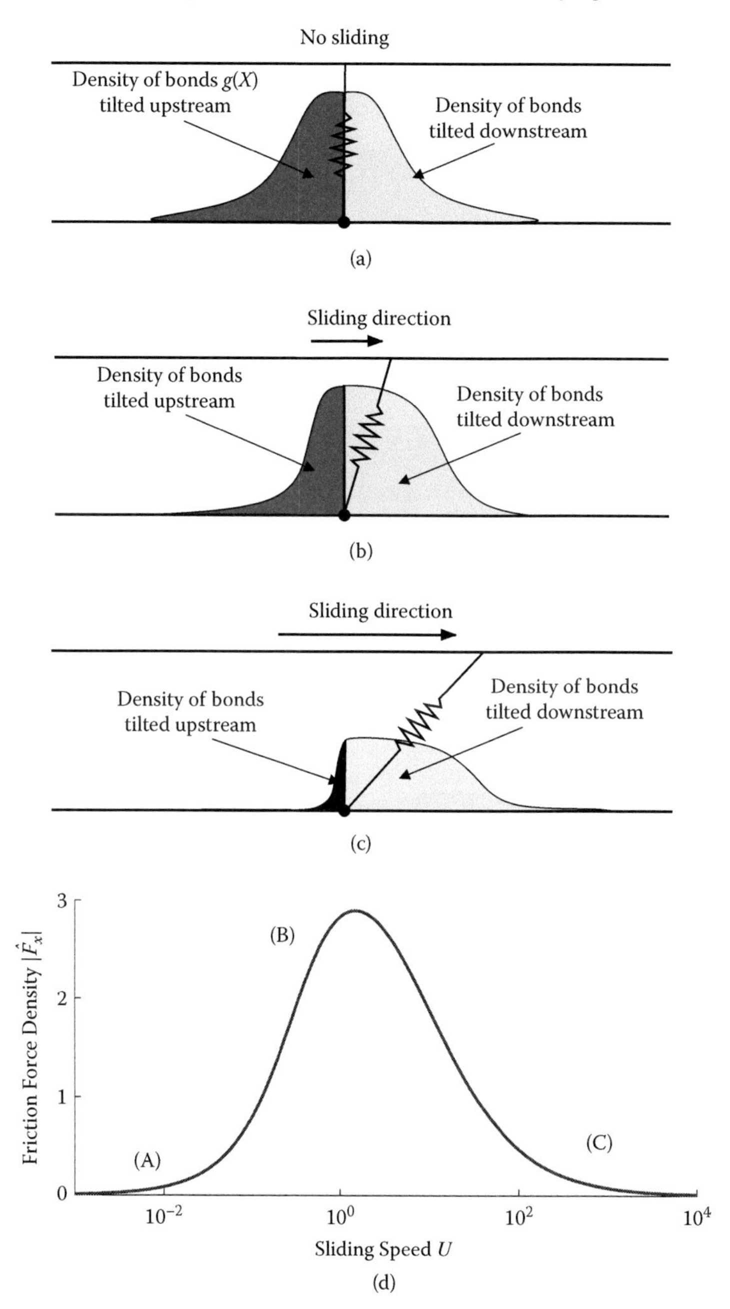

FIGURE 7.7 Schematic of the consummated bond density between two sliding surfaces. (a, b, c) Bond density between two plates for (a) $U = 0$; (b) $U = \mathcal{O}(1)$; (c) $U \gg 1$. (d) Corresponding horizontal friction force density exerted on the upper plate. Values of the parameters: $V = 0$, $H_0 = 1$, $k = 0$, $\beta = 0.9$, and $\gamma = 1$.

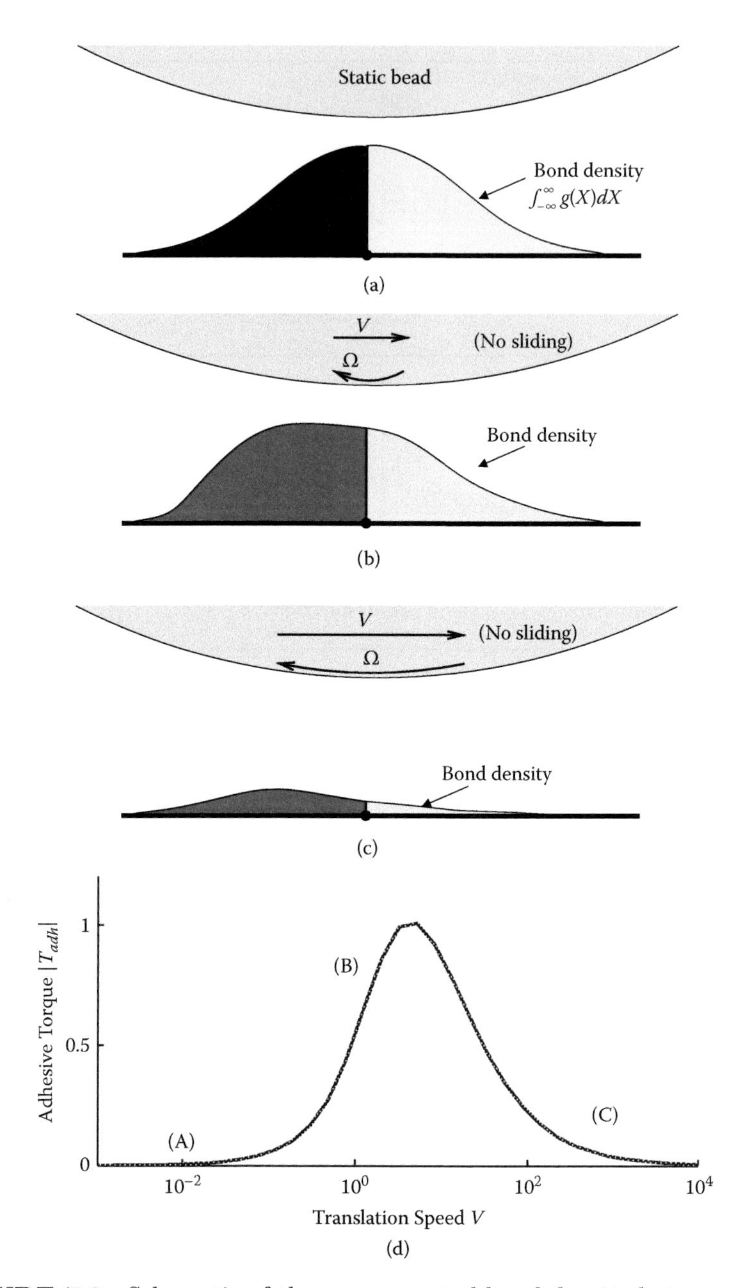

FIGURE 7.8 Schematic of the consummated bond density between a wall and a sphere rolling without sliding ($U = 0$). (a, b, c) Bond density between two plates for (a) $V = 0$; (b) $V = \mathcal{O}(1)$; (c) $V \gg 1$. (d) Corresponding total torque exerted by the bonds about the center of the sphere, as defined in Equation (7.32b). Values of the parameters: $V = 0$, $k = 0$, $\epsilon = 10^{-2}$, $\beta = 0.9$ and $\gamma = 1$.

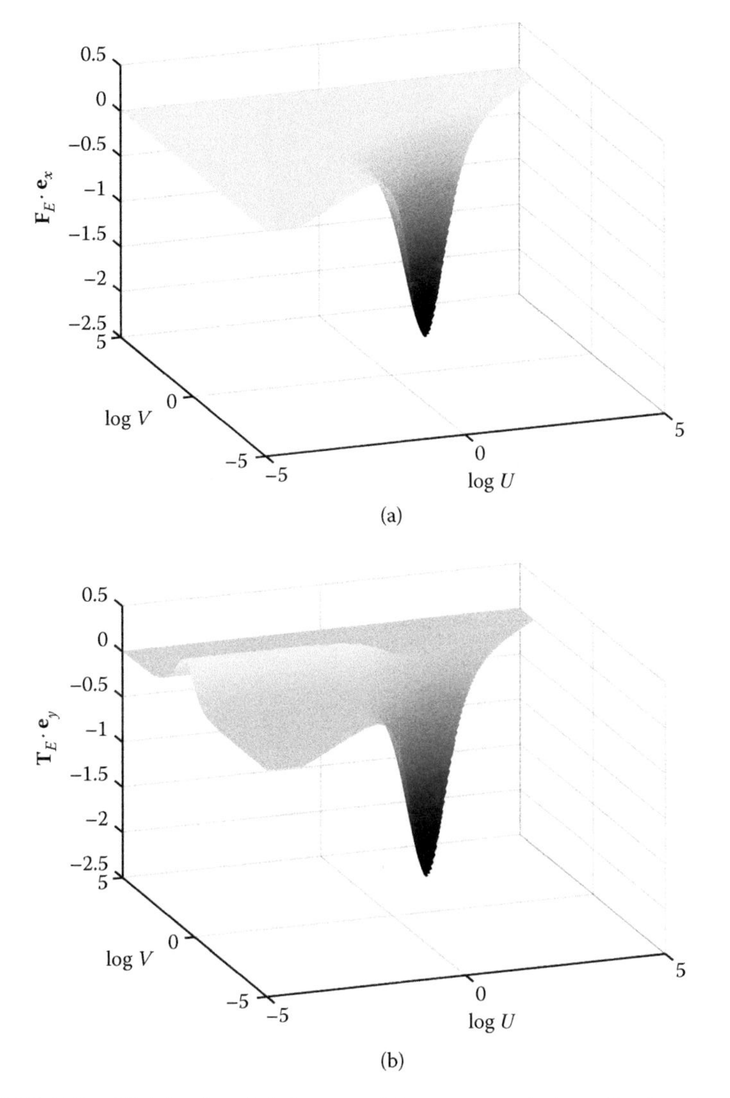

FIGURE 7.9 Net horizontal force and torque exerted by the bonds on the sphere and net adhesive torque as computed numerically by solving Equation (7.36) and integrating Equations (7.34) and (7.32). $\epsilon = 10^{-2}$, $\beta = 0.9$, and $\gamma = 1$; (a,b) $k = 0$; (c,d) $k = 10$. (Continued)

of $\mathbf{F}_E$, $\mathbf{F}_T$, $\mathbf{T}_E$, and $\mathbf{T}_T$ for $U \ll 1$, $U \gg 1$, $V \ll 1$, or $V \gg 1$ are addressed elsewhere [23].

7.4.2 Steady-State Motion of Sphere in Shear Flow

Eliminating the parameter G from the force balance in Equation (7.31) yields one equation for the two unknowns U and V, which we can solve numerically

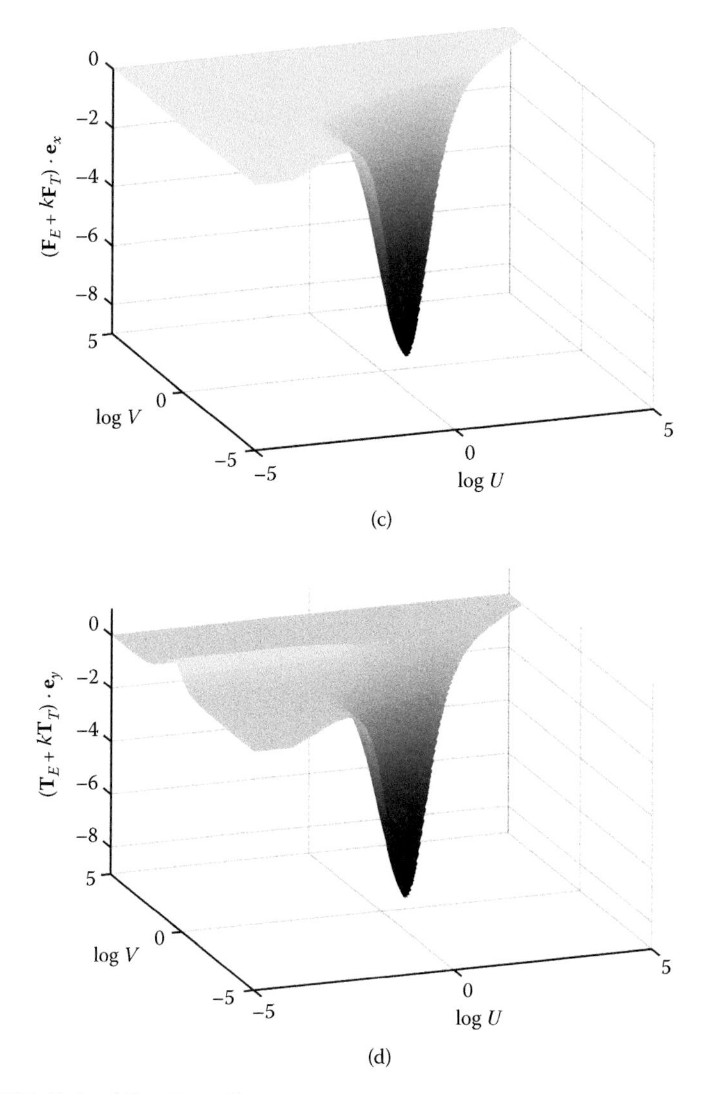

FIGURE 7.9 (*Continued*).

using the interpolated values of the adhesion forces and torque functions. To each point on this solution curve in the (U, V) space there corresponds a unique value of G that is easily determined by substituting U and V into Equation (7.31a). We therefore obtain a curve in the 3-D space (G, U, V) that describes the steady states of the sphere and we plot the projections of this curve onto the (G, U) and (G, V) planes. Results for different values of k, ϵ, and $\mathcal{C}$ are compared.

Figure 7.10 shows U and V at steady state as the shear rate G varies. The results are shown for four different values of $\mathcal{C}$, all other parameters being fixed ($\epsilon = 10^{-4}$).

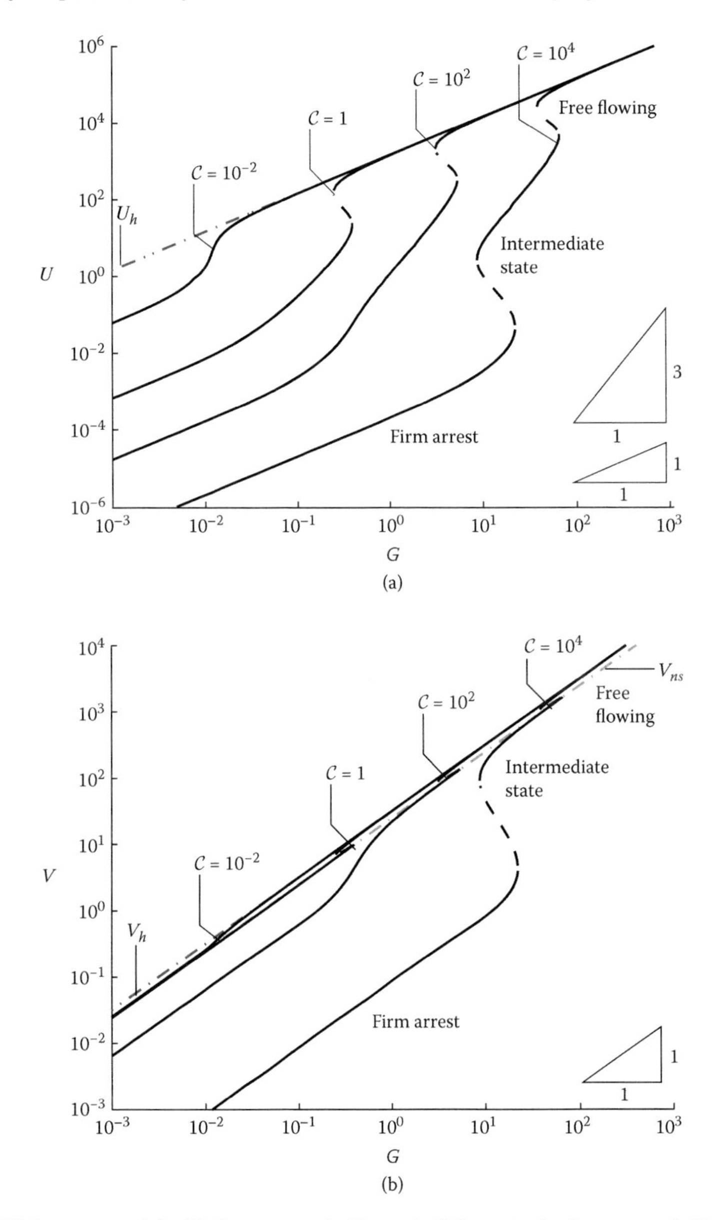

FIGURE 7.10 (a) Sliding speed U and (b) translation speed V of the sphere, solutions of Equation (7.31), versus the shear rate G for different values of the visco-adhesive parameter $\mathcal{C}$. Dashed curves represent (presumably) unstable states. The dot-dashed lines show the solutions derived in Equation (7.4) for a sphere moving at its free hydrodynamic velocity (V_h) and in Equation (7.7) for a sphere rolling without sliding (V_{ns}). The three different types of motion for an adhesive sphere are identified for the case $\mathcal{C} = 10^4$. Here $k = 10$, $d = 1$, $\epsilon = 10^{-4}$, $\beta = 0.9$, and $\gamma = 1$.

For $\mathcal{C} = 10^{-2}$ (relatively weak adhesion forces), when $G \gtrsim 0.05$, both the sliding speed and the translation speed are approximately equal to the hydrodynamic velocities U_h and V_h (see Equation (7.4)), respectively: the sphere behaves as it would in the absence of bonds and we call this state "free flowing." However, for small values of G, the sliding speed is nearly two orders of magnitude smaller than U_h and $V \approx V_{ns}$ (see Equation (7.7)); the motion is thus similar to a tank-treading sphere (i.e., one that rolls without sliding). The transition between these two states is smooth and occurs for G between 0.01 and 0.05.

For $\mathcal{C} = 1$, the behavior of the sphere is similar and the same two qualitatively distinct steady states are observed. The sliding speed of the tank-treading sphere is reduced from the case $\mathcal{C} = 10^{-2}$ by an additional two orders of magnitude, whereas the translation speed remains $V \approx V_{ns}$. However, in this case, the transition between tank-treading and free motion is abrupt and there exists a region of bistability (between $G \approx 0.25$ and $G \approx 0.4$) similar to that described in [24] for 2-D adhesive rolling of a cylinder. Increasing or decreasing the shear rate above and below the critical values makes the sphere describe a hysteresis loop between tank-treading and free flowing.

For $\mathcal{C} = 100$, the effects of adhesive forces are no longer limited to reducing the sliding speed but they also affect the translation speed. For $G \lesssim 0.8$, the latter is one order of magnitude smaller than V_{ns}, the translation speed of a tank-treading sphere. Because the sphere moves very slowly (in the frame of reference of the wall), we call this state "firm arrest." For $0.8 \lesssim G \lesssim 8$, the translation speed is close to V_{ns}, and U is much smaller than U_h. Perhaps surprisingly, the sliding speed here increases approximately as the cube of the shear rate (with slope 3 in Figure 7.10a). This contrasts with the tank-treading case discussed above (where the shear–velocity relation was linear) and indicates that different physical mechanisms might be involved (by means of asymptotic analysis, we show in [23] how this behavior emerges from Equations (7.31) through (7.37)). We call this the "intermediate state". The transition between firm arrest and intermediate state is smooth, for $G \approx 0.8$. For $3.2 \lesssim G \lesssim 5.4$, there is bistability between intermediate state and free flowing.

For $\mathcal{C} = 10^4$, the same three distinct behaviors are observed. In addition, the transition between firm arrest and the intermediate state exhibits, in this case, a region of bistability between the two states (with critical shear rates $G \approx 8.7$ and $G \approx 22$). The bistability between the intermediate state and free flowing occurs for $38 \lesssim G \lesssim 65$.

Figure 7.11 shows U and V computed for $\epsilon = 10^{-2}$ instead of 10^{-4} (all other parameters remaining unchanged). The behavior is qualitatively the same, except that the transition between the tank-treading (or intermediate) states and the free-flowing state is always smooth (regardless of the value of $\mathcal{C}$). The tank-treading behavior observed for small values of $\mathcal{C}$ and G is less significant than for $\epsilon = 10^{-4}$ (in the sense that the relative change of U compared to U_h is smaller). For $\mathcal{C} = 10^4$, the region of bistability between

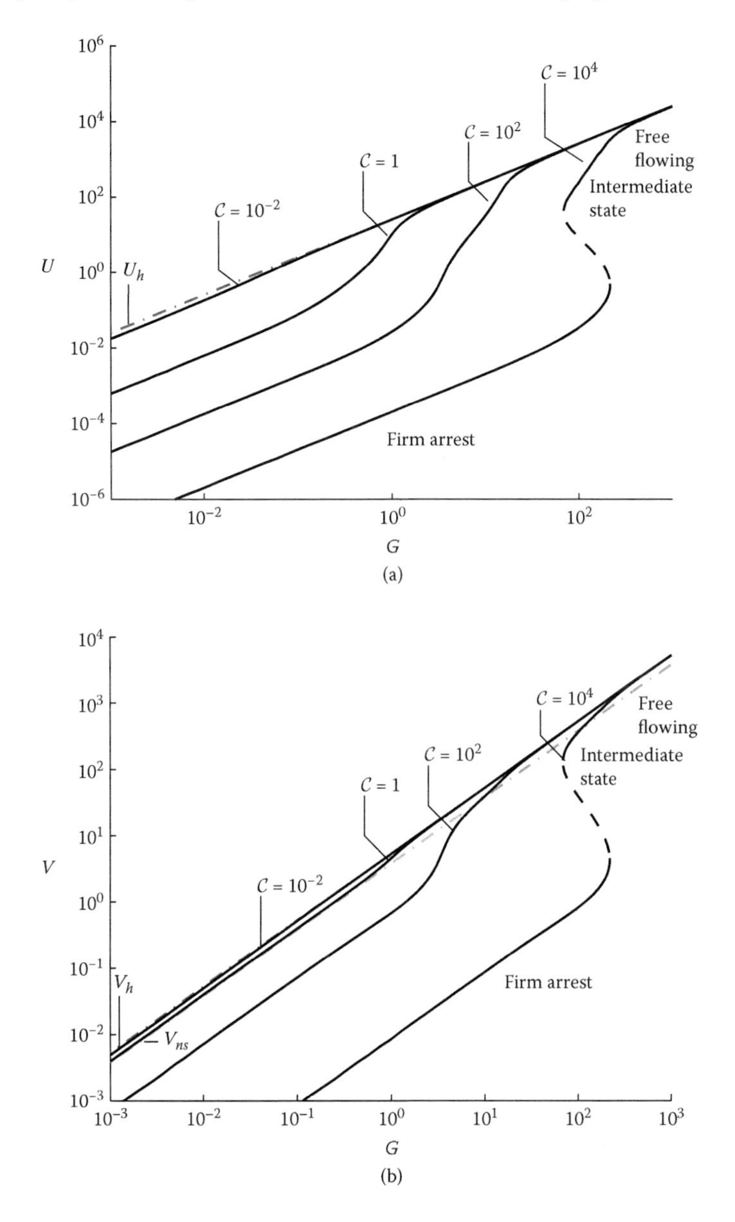

FIGURE 7.11 (a) Sliding speed U and (b) translation speed V of the sphere, solutions of Equation (7.31), versus the shear rate G for different values of the visco-adhesive parameter $\mathcal{C}$. Dashed curves represent (presumably) unstable states. The dot-dashed lines show the solutions derived in Equation (7.4) for a sphere moving at its free hydrodynamic velocity (V_h) and in Equation (7.7) for a sphere rolling without sliding (V_{ns}). The three different types of motion for an adhesive sphere are identified for the case $\mathcal{C} = 10^4$. Here $k = 10$, $d = 1$, $\epsilon = 10^{-2}$, $\beta = 0.9$ and $\gamma = 1$.

firm arrest and the intermediate state is found for $68 \lesssim G \lesssim 221$. The critical shear rates are approximately ten times larger than those found for $\epsilon = 10^{-4}$.

The effect of varying k, the parameter that compares adhesive forces due to bond tilting to that due to bond stretching, is shown in Figure 7.12, where U and V are computed for $k = 0$ ("floppy bonds") and $k = 100$ ("stiff bonds"). For comparison, the other parameters are the same as in Figure 7.10, which was obtained for $k = 10$. In all cases, the three different states are observed and so are the different regions of bistability. Qualitatively, bonds' resistance to tilting does not seem to affect the sphere's behavior. Quantitatively, we find that the sliding speed scales as $1/k$ for $k \gg 1$ in the tank-treading and firm arrest regimes. The translational speed V scarcely changes as k varies.

In summary, three different regimes can be identified: (1) the sphere adheres to the wall (the bonds preventing both sliding motion and translational motion); (2) the sphere tank-treads on the wall (the bonds preventing sliding motion); or (3) the sphere is free from adhesive forces (with most of the bonds broken). These three regimes may overlap for some values of the parameters, giving rise to regions of bistability. They are shown in Figure 7.13, where the state diagram of the sphere is reported in (G, C)-parameter space.

7.5 Discussion

In a previous theoretical study, we described the effects of sliding friction on the motion of a cylinder in a shear flow [24]; we emphasized the nonlinear relationship between sliding speed and adhesion forces. Our model has here been extended to the 3-D case of an adhesive sphere near a wall. Conceptually, the binding between the wall and a moving object occurs in a similar way in 2-D or in 3-D, but qualitative differences in the hydrodynamics induce more dramatic changes. For instance, the horizontal velocity of a cylinder scales as $\epsilon^{1/2} G^* R^*$, whereas a sphere moves faster, as $G^* R^* / |\log \epsilon|$. These velocities determine the time scale for bonds to be advected from the leading edge to the trailing edge of the rolling cell. Comparing this to the time scale for bond breakage gives a critical shear rate beyond which binding kinetics cannot be assumed to be at equilibrium (as in the so-called rapid kinetics assumption). This shear rate is $G^*_{neq} \sim K^*_{off,eq}$ in 2-D and $G^*_{neq} \sim \epsilon^{1/2} |\log \epsilon| K^*_{off,eq}$ in 3-D. For physiological parameter values (see Table 7.1) and in the limit $\epsilon \ll 1$, it makes little sense, in 3-D, to assume rapid kinetics (although it can be formally justified in 2-D). For this reason, we incorporated nonequilibrium binding kinetics effects in the present 3-D adhesion model.

In addition to the nonlinearity between adhesive friction and sliding speed (see Figure 7.7), nonequilibrium binding introduces another nonlinear relation, between the torque exerted on the sphere by adhesion molecules and the translation speed (Figure 7.8). The advection of bonds from the front to the

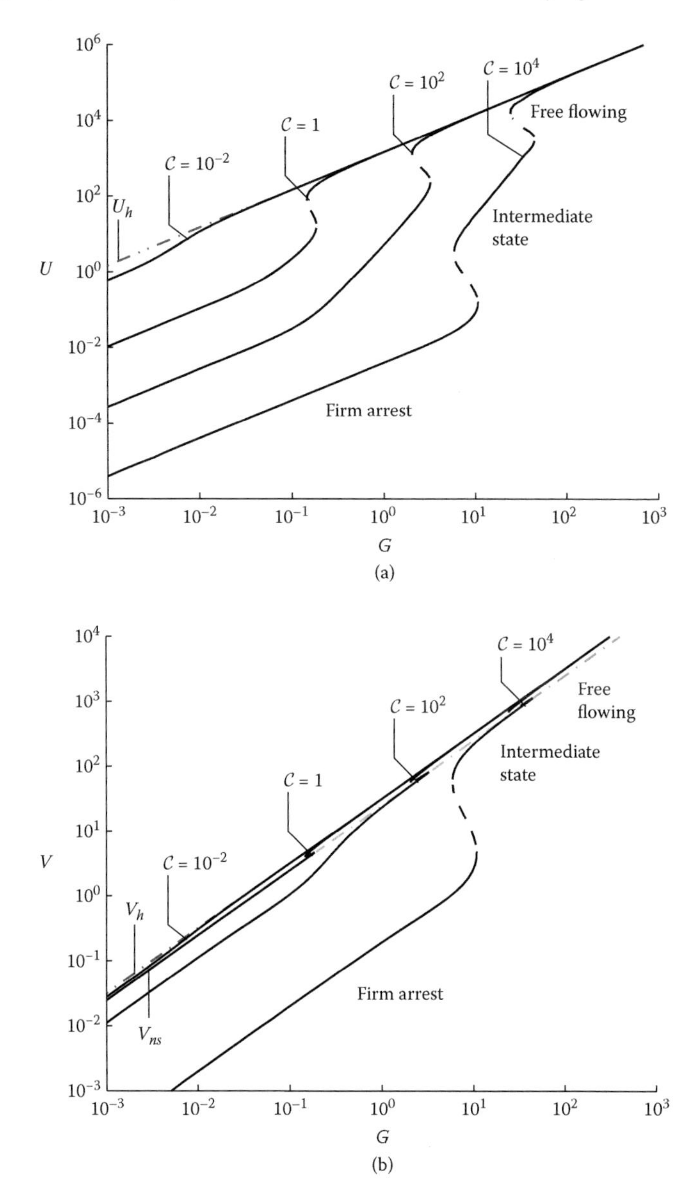

FIGURE 7.12 (a) Sliding speed U and (b) translation speed V of the sphere, solutions of Equation (7.31), versus the shear rate G for different values of the visco-adhesive parameter $\mathcal{C}$. Dashed curves represent (presumably) unstable states. The dot-dashed lines show the solutions derived in Equation (7.4) for a sphere moving at its free hydrodynamic velocity (V_h) and in Equation (7.7) for a sphere rolling without sliding (V_{ns}). The three different types of motion for an adhesive sphere are identified for the case $\mathcal{C} = 10^4$. Here $k = 0$ for (a) and (b) and $k = 100$ for (c) and (d), $d = 1$, $\epsilon = 10^{-4}$, $\beta = 0.9$, and $\gamma = 1$.

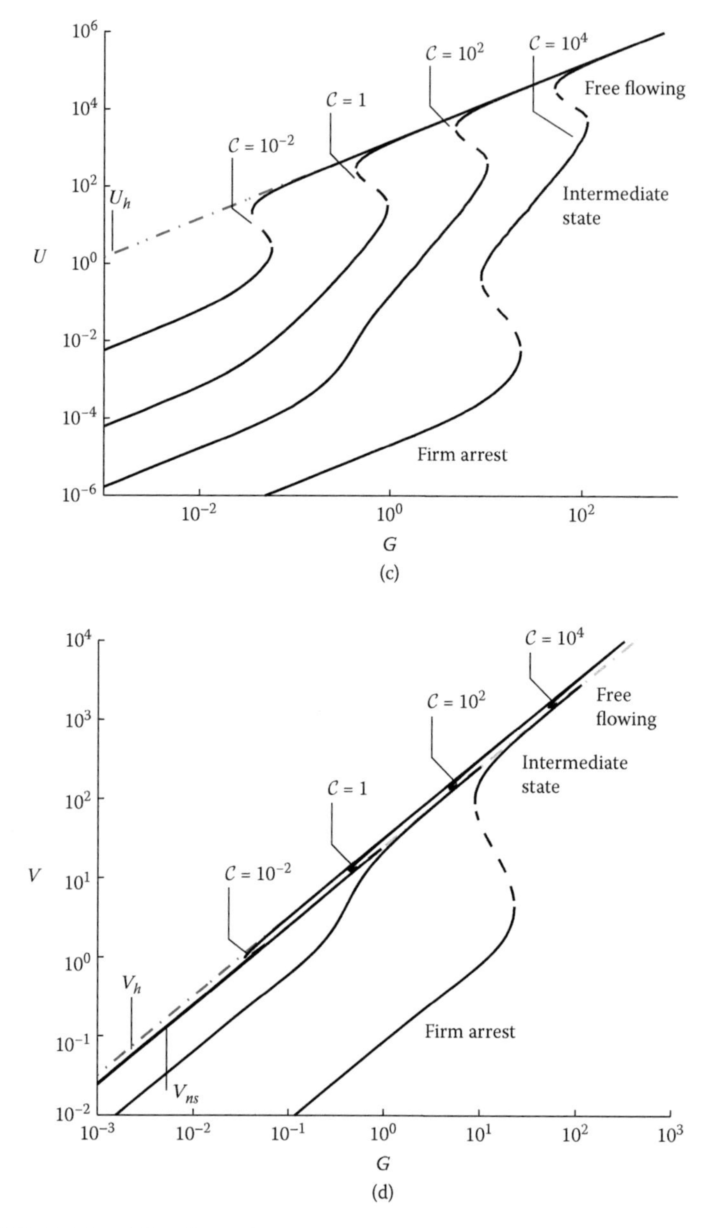

FIGURE 7.12 *(Continued)*.

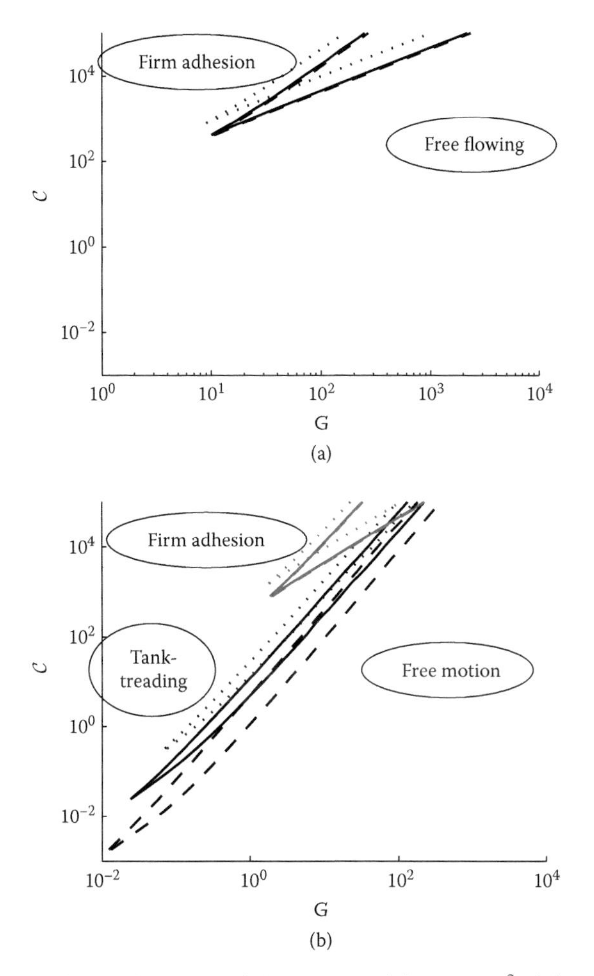

FIGURE 7.13 State diagram of the sphere. (a) $\epsilon = 10^{-2}$; (b) $\epsilon = 10^{-4}$. The wedges are the critical shear rates that bound the different bistable regions. Dotted lines: $k = 0$; solid lines: $k = 10$; dashed lines: $k = 100$. Values of the other parameters: $d = 1$, $\beta = 0.9$, and $\gamma = 1$.

back of the rolling sphere, when it occurs on a time scale comparable to (or shorter than) bonds' characteristic lifetime, causes an accumulation of consummated bonds near the back of the sphere. This asymmetry generates a torque about the center of the sphere that tends to impede the rolling motion.

As the shear rate varies, the sphere's motion at steady state exhibits a variety of possible features, characterized by the sphere's translation and sliding velocities (V and U, respectively). The behavior depends on the visco-adhesive parameter C that compares adhesion forces to viscous forces exerted on the sphere. Results also vary with the separation distance Δ^* between the sphere and the wall. Physically, this distance is determined by a vertical balance of forces on the sphere. However, the net vertical force depends strongly on the nature of nonspecific interactions between the sphere and the wall, and is therefore inherently very system dependent. For the sake of simplicity, Δ^* was chosen as a fixed parameter in our model and assumed to be comparable to the bonds' unstretched length λ^* (results are given for $\Delta^* = \lambda^*$, principally).

Typically, as the shear rate increases, we observe a transition from a regime where adhesion is important (the sphere is in the tank-treading state) to a regime where the sphere is transported freely by the flow (the free-flowing state). The transition is abrupt and there exists a bistable region where both regimes are stable (Figures 7.10 through 7.13). Depending on the bonds' physical properties and the aspect ratio λ^*/R^* between bonds' unstressed length and the radius of the sphere, the tank-treading regime can be subdivided into two qualitatively distinct regimes: (1) the intermediate state where the sphere rolls almost without sliding and at a velocity smaller than, but comparable to, the velocity it would have without adhesive binding (and in this case the translation speed does not depend on the strength of the bonds); (2) the firm arrest state where the sphere sticks to the wall and rolls much more slowly, at a velocity that scales like the inverse of (some measure of) the strength of the bonds.

To our knowledge, no experiments have been conducted that explicitly demonstrate the bistability of adhering cells in a shear flow. However, some published results suggest signs of shear-induced hysteresis. For instance, in a flow-chamber experiment, leukocytes entering an adhesive region are observed to adhere in large numbers when the shear rate is approximately 100 s^{-1} and in very small numbers for higher shear rates ($\approx$250 s^{-1}) [22]. However, tracking already adherent cells reveals that they remain bound for shear rates up to approximately 400 to 1000 s^{-1}, thus indicating a range of shear rates where both unbound and bound states are (to some extent) stable. Recently it has been argued that this hysteresis is a consequence of the catch-bond behavior of the L-selectin–PSGL-1 pair, namely that the bonds' rate of dissociation is lowered by an increase in the exerted traction [5]. The authors also propose a shear-controlled on rate as a possible explanation. Our analysis proposes an alternative explanation because a similar macroscopic behavior can be observed when using Dembo et al.'s model [6] with slip bonds, accounting also for the effects of bond tilting. Also, in our model, neither the rate of formation nor the rate of dissociation of individual bonds is directly modified by the sliding motion of the sphere.

Accounting for cell deformability is expected to lead to significant differences in the scalings of adhesion forces. The adhesion area at the base of the sphere would increase with the deformability, enhancing adhesive effects dramatically. Concurrently, asymmetries in cell deformation may break the reversibility of the Stokes equation and generate a net viscous lift force that tends to detach the cell from the wall. Further investigation is therefore required to elucidate how the state diagram in Figure 7.13 is modified by membrane deformation.

References

[1] R. Alon, S.Q. Chen, K.D. Puri, E.B. Finger, and T.A. Springer (1997). The kinetics of L-selectin tethers and the mechanics of selectin-mediated rolling. *J. Cell Biol.* 138(5):1169–1180.

[2] R. Alon, D.A. Hammer, and T.A. Springer (1995). Lifetime of the P-selectin–carbohydrate bond and its response to tensile force in hydrodynamic flow. *Nature* 374(6544):539–542.

[3] G.I. Bell (1978). Models for the specific adhesion of cells to cells. *Science* 200:618–627.

[4] K.E. Caputo and D.A. Hammer (2005). Effect of microvillus deformability on leukocyte adhesion explored using adhesive dynamics simulations. *Biophys. J.* 89(1):187–200.

[5] K.E. Caputo, D. Lee, M.R. King, and D.A. Hammer (2007). Adhesive dynamics simulations of the shear threshold effect for leukocytes. *Biophys. J.* 92(3):787–797.

[6] M. Dembo, D.C. Torney, K. Saxman, and D. Hammer (1988). The reaction-limited kinetics of membrane-to-surface adhesion and detachment. *Proc. R. Soc. Lond. B* 234:55–83.

[7] C. Dong and X.X. Lei (2000). Biomechanics of cell rolling: shear flow, cell-surface adhesion, and cell deformability. *J. Biomech.* 33(1):35–43.

[8] A.E. Filippov and V. Popov (2007). Flexible tissue with fibres interacting with an adhesive surface. *J. Phys.: Condens. Matter* 19:096012.

[9] E.B. Finger, K.D. Puri, R. Alon, M.B. Lawrence, U.H. von Andrian, and T.A. Springer (1996). Adhesion through L-selectin requires a threshold hydrodynamic shear. *Nature* 379(6562):266–269.

[10] J. Fritz, A.G. Katopodis, F. Kolbinger, and D. Anselmetti (1998). Force-mediated kinetics of single P-selectin ligand complexes observed by atomic force microscopy. *Proc. Natl. Acad. Sci.* 95(21):12283–12288.

[11] A.J. Goldman, R.G. Cox, and H. Brenner (1967). Slow viscous motion of a sphere parallel to a plane wall–I. *Chem. Eng. Sci.* 22:637–651.

[12] A.W. Greenberg, D.K. Brunk, and D.A. Hammer (2000). Cell-free rolling mediated by L-selectin and sialyl Lewis(x) reveals the shear threshold effect. *Biophys. J.* 79(5):2391–2402.

[13] D.A. Hammer (2005). Leukocyte adhesion: What's the catch? *Curr. Biology* 15(3):R96–R99.

[14] D.A. Hammer and S.M. Apte (1992). Simulation of cell rolling and adhesion on surfaces in shear-flow: General results and analysis of selectin-mediated neutrophil adhesion. *Biophys. J.* 63(1):35–57.

[15] S.R. Hodges and O.E. Jensen (2002). Spreading and peeling dynamics in a model of cell adhesion. *J. Fluid Mech.* 460:381–409.

[16] R.R. Isberg and P. Barnes (2002). Dancing with the host: flow-dependent bacterial adhesion. *Cell* 110(1):1–4.

[17] E.F. Krasik, K.L. Yee, and D.A. Hammer (2006). Adhesive dynamics simulation of neutrophil arrest with deterministic activation. *Biophys. J.* 91(4):1145–1155.

[18] M.B. Lawrence and T.A. Springer (1991). Leukocytes roll on a selectin at physiologic flow rates: distinction from and prerequisite for adhesion through integrins. *Cell* 65:859–873.

[19] L.A. Liotta (2001). Cancer: an attractive force in metastasis. *Nature* 410(6824):24–25.

[20] R.P. McEver (2001). Adhesive interactions of leukocytes, platelets, and the vessel wall during hemostasis and inflammation. *Thromb. Haemostasis* 86(3):746–756.

[21] N.W. Moore and T.L. Kuhl (2006). The role of flexible tethers in multiple ligand-receptor bond formation between curved surfaces. *Biophys. J.* 91(5):1675–1687.

[22] K.D. Puri, S. Chen, and T.A. Springer (1998). Modifying the mechanical property and shear threshold of L-selectin adhesion independently of equilibrium properties. *Nature* 392:930–933.

[23] S. Reboux (2009). Multiscale Models for Cellular Adhesion and Deformation. Ph.D. thesis.

[24] S. Reboux, G. Richardson, and O.E. Jensen (2008). Bond tilting and sliding friction in a model of cell adhesion. *Proc. R. Soc. A* 464(2090):447–467.

[25] L.J. Rinko, M.B. Lawrence, and W.H. Guilford (2004). The molecular mechanics of P- and L-selectin lectin domains binding to PSGL-1. *Biophys. J.* 86(1):544–554.

[26] J.J. Rychak, J.R. Lindner, K. Ley, and A.L. Klibanov (2006). Deformable gas-filled microbubbles targeted to P-selectin. *J. Contr. Release* 114(3):288–299.

[27] A. Saez, M. Ghibaudo, A. Buguin, P. Silberzan, and B. Ladoux (2007). Rigidity-driven growth and migration of epithelial cells on microstructured anisotropic substrates. *Proc. Natl. Acad. Sci. USA* 104(20):8281–8286.

[28] G.W. Schmidt-Schönbein, Y.C. Fung, and W. Zweifach (1975). Vascular endothelium-leukocyte interaction, sticking shear force in venules. *Circ. Res.* 36:173–184.

[29] D.W. Schmidtke and S.L. Diamond (2000). Direct observation of membrane tethers formed during neutrophil attachment to platelets or P-selectin under physiological flow. *J. Cell Biol.* 149(3):719–729.

[30] J.Y. Shao, H.P. Ting-Beall, and R.M. Hochmuth (1998). Static and dynamic lengths of neutrophil microvilli. *Proc. Natl. Acad. Sci. USA* 95(12):6797–6802.

[31] T.A. Springer (1990). Adhesion receptors of the immune-system. *Nature* 346(6283):425–434.

[32] D.F.J. Tees, R.E. Waugh, and D.A. Hammer (2001). A microcantilever device to assess the effect of force on the lifetime of selectin-carbohydrate bonds. *Biophys. J.* 80(2):668–682.

[33] M. Varenberg and S. Gorb (2007). Shearing of fibrillar adhesive microstructure: friction and shear-related changes in pull-off force. *J. R. Soc. Interface* 4:721–725.

[34] C. Zhu (2000). Kinetics and mechanics of cell adhesion. *J. Biomech.* 33(1):23–33.

Chapter 8

Understanding Adhesion Sites as Mechanosensitive Cellular Elements

Sophie Féréol, Redouane Fodil, Gabriel Pelle, Bruno Louis, Valérie M. Laurent, Emmanuelle Planus, and Daniel Isabey

Contents

8.1 State-of-the-Art on Cell Mechanosensitivity

It is now well recognized that tissue cells have the ability to sense their extracellular environment and respond by adapting their structure, internal tension, and mechanical properties [22], modulating their function without being subjected to external forces [13,48]. From soft to stiff substrates, it has been shown that: (1) cell spreading is increased and stress fibers are reinforced in epithelial cells, fibroblasts, and smooth muscle cells but not always in neutrophils [26]; and (2) cell migration is facilitated due to larger intracellular traction forces and cell spreading area is increased [34]. Moreover, the function of certain cells has been found optimal in an intermediate range of environmental stiffness. This is the case with skeletal muscle cells, which optimally differentiate when they grow on substrates with stiffness close to muscle tissue stiffness [15]. It should be emphasized that tissues have very different stiffness levels; a normal tissue might have very different elastic properties. With a Young's modulus on the order of 0.5 kPa, brain appears the softest tissue. Muscles have a Young's modulus of 10 kPa. The Young's modulus of skin is approximately 10^2 kPa while that of bones reaches up to 10^6kPa. Most importantly, pathophysiological phenomena develop in the context of altered extracellular mechanical properties. The role of these properties on the disease processes is not fully understood. This is the case for tumor progression, which is modulated by substrate and extracellular matrix (ECM) mechanical properties [49]. The definition of new therapeutic processes such as nerve tissue engineering, encapsulated cell therapies and self-renewal and differentiation of stem cells also requires considering biophysical cues in addition to biochemical cues. It has been shown that the rate of neurite extension and branching critically depends on substrate rigidity [32]. Moreover, the mechanical properties of the stem cell's microenvironment regulate its behavior [41].

Although it has been recognized for a long time that cell shape, structure, and function are controlled by external forces [6,7], the recognition that passive mechanical properties of the extracellular environment control cell shape, structure, and function raises new fundamental questions [14,34,38,50]. Several reviews have summarized how externally applied forces may trigger a cellular response [16,30,42]. The question we address in this chapter is different and can be summarized as follows: by which physicochemical mechanism do cells sense and adapt to the mechanical properties of the extracellular environment? To answer this question, a fundamental assumption to consider is that adhesion site maturation—namely, adhesion sites in their dynamic phase (i.e., not the quasi-stationary adhesion site sketched in Figure 8.1)—plays a key role [4].

Such an assumption is based on the consideration that a physicochemical coupling occurs between transmembrane proteins and actomyosin tensed cytoskeletal filaments (see Figure 8.2).

Note that adhesion site dynamics are not the only pertinent parameter for cell sensitivity. Here we explain why, in addition to adhesion site dynamics, cellular prestress must also be considered, showing that cell sensitivity to substrate stiffness results from a cellular–molecular coupling at the adhesion site level (see below and Féréol et al. [20]).

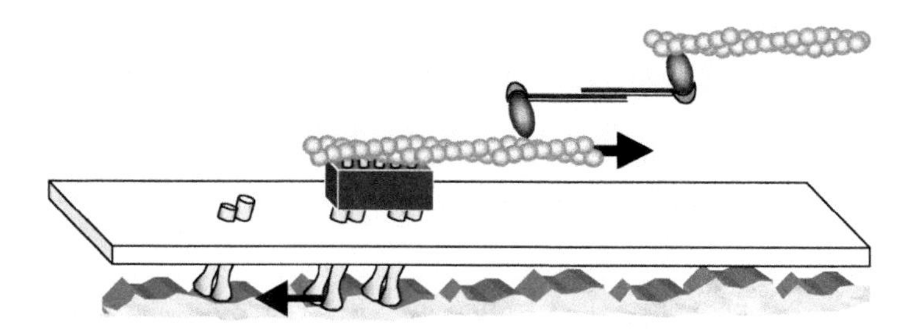

FIGURE 8.1 The stationary adhesion site model. The stationary adhesion site works in static equilibrium between intracellular traction forces generated by actomyosin coupling and the integrated reaction force raised by the assumed constant viscoelastic properties of the underlying substrate. (Modified from [3, 39].)

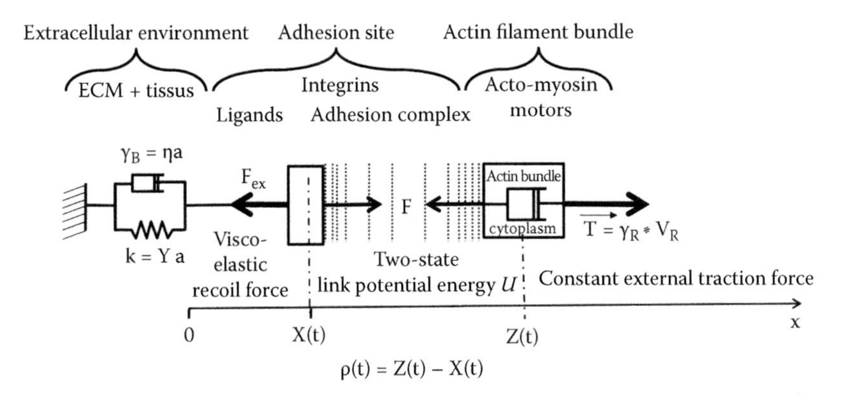

FIGURE 8.2 The dynamic adhesion site model (extracellular environment + adhesion site + actin bundle filament). The dynamic adhesion site takes into account the nanoscale motion $\rho(t) = Z(t) - X(t)$ of the adhesion site relative to the actin bundle on which a constant traction force T is exerted. Z(t) is the coordinate of the displacement of the actin bundle and X(t) is the coordinate of the adhesion site. The link force between adhesion site and actin bundle derives from a two-state (or multistate) potential energy U. F may be much smaller than T at the onset of adhesion site maturation. In response to the binding force, a reaction force F_{ex} increases proportionally with time. This reaction force depends on the spring constant k of the extracellular environment and on V_R (the speed of actin retrograde motion in response to intracellular tension). The other important mechanical parameters are the friction coefficient between the actin bundle and cytoplasm γ_R, and the friction coefficient within the extracellular environment, which is supposed to behave as a simple Voigt solid model. (Modified from [20].)

It has been clearly established that adhesion sites are clusters of membrane-associated proteins constituting a discrete physicochemical link between the cytoskeleton (CSK) and the extracellular environment [45]. The physical link between extracellular matrix, transmembrane receptor, CSK, and nucleus constitutes the mechanotransduction pathways [29,35,37,45]. These specific pathways transmit to the nucleus the force signal initiated extracellularly or intracellularly secondary to actomyosin motor protein located along the F-actin filaments, allowing the cell to continuously adapt to its extracellular matrix [25]. Most simple models of the cell–substrate interaction—typically, the model depicted in Figure 8.1—are based on the static mechanical equilibrium between contractile forces and substrate reaction forces [8,26]. More sophisticated models are based on the complex interplay of physical and biochemical signals in the feedback of matrix stiffness on contractility and cell signaling [13,40]. More precisely, while sensitivity to external forces involves outside-in signaling pathways [39], triggering passive mechanical properties of the extracellular matrix supposes a bi-directional transmission of the mechanical information through inside-out and outside-in signaling pathways [13,40]. However, these models fail to provide a clear view of the mechanisms implicated in cell sensitivity.

The alternative approach we propose to delineate cell sensitivity considers that adhesion sites pass through different stages of development (e.g., initial adhesion (IA) [27], focal complex (FC) [24], focal adhesion (FA) [51]) characterized by the recruitment of an increasing number of constituent components resulting in molecular reinforcement of the links between CSK and extracellular environments. Then, as adhesion sites gain in molecular complexity and strength (without necessarily increasing their area), they lose their dynamic character and become more stationary, providing an evolutionary cell signaling that contributes to cell adaptation. Indeed, it can be shown that Newton's action–reaction principle governing the static force equilibrium at a given adhesion site does not leave room for fully mature adhesion sites (FAs) to exhibit such a substrate stiffness sensitivity [20]. Dynamic adhesion sites behave in a way such that mechanical relaxation of the extracellular environment tends to slow down drastically the "instantaneous" biochemical process of receptor–ligand binding and thereby provides an adhesion site force regulation that depends on the mechanical properties of the substrate. The same theory shows that intracellular tension would also be able to slow down the biochemical process of adhesion receptor binding. Classical experiments by Choquet et al. using optical tweezers as a calibrated spring have shown that nascent or dynamic adhesion sites match the force they exert to the stiffness of the substrate [9,38]. Thus, different cells would produce different responses that adapt to the wide variety of extracellular mechanical environments. We presently use two models (i.e., alveolar epithelial cells (AECs) and alveolar macrophages (AMs)), exhibiting, respectively, stationary and dynamic adhesion sites, and compare their sensitivity to substrate stiffness with theoretical predictions.

8.2 Rationale for Cell Mechanosensitivity

8.2.1 Force Regulation of Surface Adhesion Molecules

The effect of mechanical stress on molecular adhesion binding motivates an entire research field on the dynamic response of single molecular interactions (i.e., protein mechanics). A striking application is the understanding of adhesive interactions of leukocytes and blood vessel walls, which involves a competition between bond formation and breakage [36]. It is now known that specific interactions of P-selectin, expressed by endothelial cells or platelets with PSGL-1 (P-selectin glycoprotein ligand-1), enable leukocytes to roll on vascular surfaces during inflammatory response by transient interruption of cell transport (through a tethering process) in blood flow under sustained wall shear stress [44]. The applied force (i.e., the mechanical work) lowers the free-energy barrier to bond rupture and thus shortens bond lifetimes [2]. However, based on theoretical considerations, it has been hypothesized that force could also prolong bond lifetimes by deforming the adhesion complexes into an alternative locked or bound state [12]. These two distinct dynamic responses to external forces are referred to as slip and catch bonds [12]. Using experimental devices (e.g., atomic force microscopy and flow chamber experiments), it has been recently shown that increasing forces first prolonged (catch bond) and, beyond a critical force, shortened (slip bond) the lifetime of P-selectin complexes with P-selectin glycoprotein ligand-1 [36]. Note that the force-dependent reinforcement of catch bonds is susceptible to explain the paradox of leukocyte rolling whose "stability" was found to increase as hydrodynamic forces increase [23]. This well-known phenomenon illustrates the counterintuitive behavior of the mechanical resistance of receptor-ligand binding.

From a theoretical point of view, the model proposed by Bell for surface adhesion molecules is based on the knowledge of binding properties of solutions formed by the same molecules [2]. In the initial approach of Bell, the receptor–ligand complex randomly oscillates due to Brownian thermal excitation, providing some probability of bond rupture P described by the Boltzmann factor $P = \exp[(-\Delta E)/k_B T]$, where k_B is the Boltzmann constant, T is the absolute temperature, and ΔE is the activation energy required for bond dissociation. Note that $k_B T$, the thermal energy, is close to 4 pN·µm at biological temperatures (~300K). If a force F is applied to the adhesion bond, the energy required to break the bond will decrease because the force brings to the bond a mechanical energy F·d*, where d* is a length scale (in the nanometer range) characterizing the molecular displacement associated with the deformation of the adhesion complex. The effect of the force on the dissociation rate k_{off} is classically given by $k_{off} = k_0 \exp(F \cdot d^*/k_B T)$, known as Bell's law in the biological field. On the other hand, because thermodynamic principles imply a decrease in receptor–ligand affinity (k_a) when a dissociation force is applied [12], $k_{off}(= k_a/k_{on})$ does not necessarily increase but could decrease, such as in the case of catch bonds.

The limit of these early approaches comes from the complexity of molecular interactions considering that there are not only two states such as slip and catch bonds, but an infinity of states related to a continuum in the mechanochemical energy landscape that governs physical strength and kinetics of molecular bonds under stress. Under external forces, barriers in the energy landscape are lowered and bond lifetime shortens because the chemical energy barrier at distance x decreases by $F \cdot x$ brought by the mechanical work. When isolated bonds are ruptured under steady ramps of force, barriers diminish in time and thus the rupture force depends on the loading rate (= force/time). This has been demonstrated by experiments in which weak bonds were probed with ramps of forces over an enormous range of loading rate (10 to 10^5 pN·s^{-1}) [17]. The most frequent forces for failure plotted versus

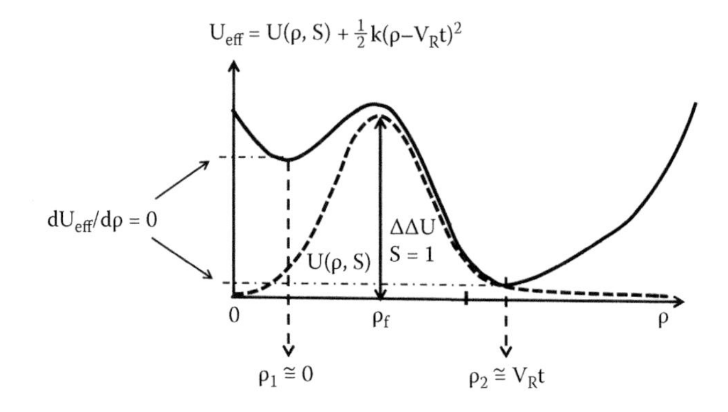

FIGURE 8.3 The potential energy function characterizing the link between dynamic adhesion site and actin bundle is plotted versus the relative distance ρ. A simple typical shape (dotted line) is chosen for $U(\rho, S)$. S is the state variable of the adhesion site (see Figure 8.4). $\Delta\Delta U$ is the activation energy to "escape out from the well" for $S = 1$. The effective potential energy $U_{eff} = U(\rho, S) + 0.5k(\rho - V_R t)^2$ is plotted versus ρ (continuous line). U_{eff}, which includes the substrate elastic energy, behaves as a double-well potential energy with two characteristic values of ρ that correspond to the minima of U_{eff} ($\delta U_{eff}/\delta\rho = 0$) obtained for (1) $\rho_1 \approx 0$, which corresponds to adhesion site binding; and (2) $\rho_2 \approx V_R t$, which corresponds to adhesion site dissociation.

a logarithmic scale of the loading rate establish a dynamic spectrum of bond strength that images the prominent energy barriers that characterize each specific molecular interaction. One of the simplest forms of the energy–distance curves is shown by the dotted line in Figure 8.3. This means that bond rupture supposes to cross over a succession of barriers (actually only one in the simplified form is shown in Figure 8.3).

The Bell model consists of estimating the rupture frequency of the molecular link as the product of a "frequency to escape out from the well" times "the probability to have the available activation energy ΔE." For instance, in the PSGL-1/L-selectin model [18], energy barriers were found 0.06 nm and 0.4 nm from equilibrium position while forces above 75 pN created bond dissociation in less than 10 ms.

8.2.2 Force Regulation of Adhesion Sites

8.2.2.1 Stationary Adhesion Sites

It is assumed that the adhesion site is in static equilibrium between the traction force (T) transmitted to the adhesion site via a bundle of actin filaments and an opposed reaction force generated within the viscoelastic substrate in response to the traction force (see Figure 8.1). The traction force is generated

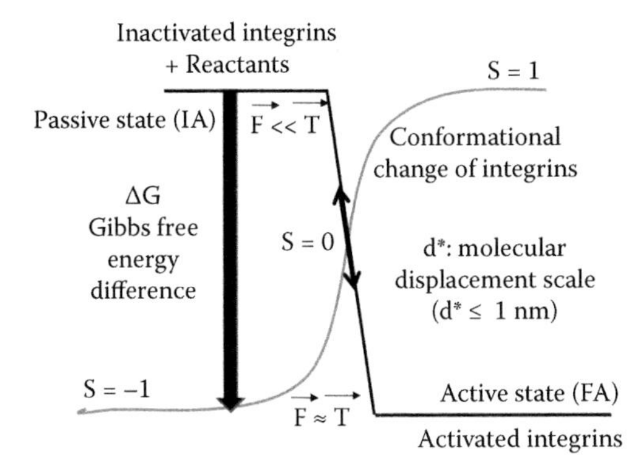

FIGURE 8.4 The mechanochemical response of the adhesion site is characterized by a state variable S given by $\langle S \rangle_F = \tanh \frac{1}{k_B T}(-\Delta G + F \cdot d^*)$. $S = -1$ is the passive state and $S = 1$ is the active state. $S = 0$ corresponds to $\Delta G = F \cdot d^*$, usually called degenerated state. During integrin activation, the adhesion site passes from passive to active states due to the conformational change of integrins. If a force F is exerted on the adhesion site, the mechanical work $F \cdot d^*$ contributes to reduce the activation energy required for adhesion site activation.

intracellularly by actomyosin motors, which also contribute to internal tension. The traction force generates a more or less local stress distribution inside the substrate whose area integral is equal to the traction force (T). Using linear elasticity theory, it can be shown, in 1-D and 3-D continuous mechanical models, that the substrate stress field cannot depend on the elasticity properties of the substrate [4,20]. The reaction forces exerted by the substrate on a stationary adhesion site are therefore independent of substrate rigidity. Hence, the activation of stationary adhesion sites, which contain force-sensitive proteins at the cell–substrate interface, cannot be regulated by substrate rigidity. This is not inconsistent with the idea that regulation of mature adhesion sites is still determined by mechanical forces. These mechanical forces are normally traction forces generated intracellularly by myosin II motor proteins coupled to the actin filaments that are connected to the site (Figure 8.1), but external forces can also stimulate adhesion site development [3,39]. It is indeed well-known that the size of mature focal adhesion sites reversibly increases and decreases as a function of the applied force [25], but the lack of sensitivity of mature adhesion sites for mechanical properties of extracellular environment constitutes a rather new idea [4,20].

8.2.2.2 Dynamic Adhesion Sites

In contrast to the stationary case, the dynamic adhesion site already used by [4,20] is assumed to be connected to the actin filament bundle through a link potential energy $U(\rho, S)$ that describes the variable mechanochemical linkage between the actin filament bundle and the adhesion site. Thus, compared to the stationary adhesion site, the filament bundle in a dynamic adhesion site is similarly exposed to a traction force T directed along the x-axis in Figure 8.2, but the difference is that the force F exerted on the site (which has direction T) could be only a fraction of the actomyosin traction force T. The link potential energy $U(\rho, S)$ depends on two variables: (1) $\rho(t)(= Z(t) - X(t))$, which is the relative displacement (at the nanometer scale) between the actin filament bundle and the linked adhesion site (see Figure 8.2); and (2) the variable S, which characterizes the variable state of a given site changing from passive ($S = -1$) to active ($S = 1$).

The feature of the link potential energy U and the variable distance ρ between adhesion site and actin filament resembles that found for surface adhesion molecules (see curve with dotted line in Figure 8.3 and the paragraph above): (1) $U(\rho, S)$ has an absolute minimum near $\rho = 0$, (2) $U(\rho, S)$ tends toward 0 as ρ increases toward large values, (3) the activation energies to escape out of the potential well (barriers) are denoted ΔU for $S = -1$ and $\Delta\Delta U$ for $S = 1$. These maxima of $U(\rho, S)$ are supposed to occur at $\rho = \rho_f$ (see Figure 8.3). $\Delta U/\rho_f$ is supposed to be much smaller than T, meaning that the force of the link potential is much weaker than the traction force (passive state). In the active state, the activation level is supposed to be much higher than in the passive state: $\Delta U/\rho_f < \Delta\Delta U/\rho_f \sim T$.

The state of the dynamic adhesion site is supposed to reversibly vary between an inactivated (or passive) state ($S = -1$) and an activated sate ($S = 1$), which takes into account integrin activation induced by phosphatase activity triggered by a force-induced conformational change of adhesion site integrins. Under chemical equilibrium conditions, the likelihood value of the site variable S is obtained by the expression:

$$\langle S \rangle_F = \tanh \frac{1}{k_B T}(-\Delta G + F \cdot d^*) \tag{8.1}$$

where ΔG, the Gibbs free energy difference between these two states (in the absence of applied force), depends on the number of integrins at the site and on the concentration of reactants involved in the activation reaction. $F \cdot d^*$ is the thermodynamic work performed by the force F to the link between the adhesion site and the cytoskeleton; d^* is the characteristic length scale (in the nanometer range) of the molecular displacement during the conformational change of integrins. The typical evolution of the state variable S is shown in Figure 8.4.

The main feature of the dynamic adhesion site model presented in Figure 8.2 is to integrate microscale and nanoscale effects in a unique set of equations. In response to the force F derived from the link potential energy and applied to the adhesion site—from the cellular side—a recoil force resulting from the viscoelastic reaction of the substrate is exerted on the adhesion site. This recoil force consists of the viscous drag $\gamma_B \frac{dX}{dt}$ proportional to substrate friction coefficient $\gamma_B(\approx \eta a)$ added to the substrate recoil force $-kX$ proportional to the spring constant $k(\approx E_{sub} \cdot a)$, where a is the size of the dynamic adhesion site supposed to be unchanged during the maturation. $f(t)$ is the thermal random noise exerted on the site and is provided by fluctuation–dissipation theorem. The dynamic adhesion site motion is described by the equation:

$$\gamma_B \frac{dX}{dt} + kX = \frac{dU(\rho, S)}{d\rho} + f(t) \tag{8.2}$$

The motion of the actin filament bundle depends on the equilibrium of the sum of forces exerted on the bundle on the cellular side, that is, the constant traction force (T), the link force, the thermal fluctuation random force $f^*(t)$, and the viscous retarding force $\gamma_R \frac{dZ}{dt}$, the latter resulting from the friction between actin filaments and the cytoplasm (friction coefficient: γ_R).

$$\gamma_R \frac{dZ}{dt} = \frac{dU(\rho, S)}{d\rho} + T + f^*(t) \tag{8.3}$$

T $(= \gamma_R \cdot V_R)$ is also responsible for the retrograde motion of the actin bundle at a constant velocity V_R. V_R, the velocity of actin retrograde motion, can thus be used to characterize the basic activity of actomyosin motors in the cell, which is close to the concept of internal tension [47]. Note that when the link force F is cancelled i.e., under conditions such that the link potential energy function $U(\rho, S)$ reaches a minimum $(dU/d\rho \approx 0)$, the filament bundle is under the unique effect of T, not arrested by the adhesion site linkage, and therefore the filament bundle is drawn intracellularly at the speed V_R.

Equations (8.2) and (8.3) can be combined to provide the equation of motion—at the nanoscale—of the adhesion site relative to the actin filament bundle. Moreover, assuming times much shorter than the mechanical relaxation time (i.e., $t \ll \tau = (\gamma_B + \gamma_R)/k (= \gamma_R/k$ because $\gamma_B \ll \gamma_R$, a condition experimentally verified when polyacrylamide gels are used [43]), the relative motion of the adhesion site of coordinate $\rho(t)$ can be described by the following equation:

$$\gamma_B \frac{d\rho}{dt} + k\rho - kV_R t \cong -\frac{dU(\rho, S)}{d\rho} + \gamma_B V_R + f(t) \tag{8.4}$$

Equation (8.4) is a Langevin equation that describes the nanoscale motion of a particle of coordinate $\rho(t)$, moving in a link potential $U(\rho, S)$, subjected to a small fraction of the constant traction force $(\gamma_B/\gamma_R)T$ and responsible for a time-dependent reaction force $F_{ex}(t)(= kV_R t)$, the latter depending on the global mechanical properties of both the substrate and the cell. The effective potential energy is given by $U_{eff} = U(\rho, S) + 0.5k(\rho - V_R t)^2$ and its variation with the relative distance ρ is shown in Figure 8.3. The term 0.5 $k(\rho - V_R t)^2$ corresponds to the elastic energy accumulated in the deformable substrate (giving a time-dependent parabola not plotted in Figure 8.3). U_{eff} has the form of a double well potential with a first minimum ρ_1 near $\rho = 0$

and a second minimum ρ_2 near $\rho = V_R t$. The first minimum corresponds to an adhesion site bound to the filament bundle while the second minimum corresponds to a dissociated site. Note that a few percent of the traction force $(\gamma_B/\gamma_R)T$ transmitted to the adhesion site is sufficient to initiate the reaction force $F_{ex}(t)(= kV_R t)$ and thereby the adhesion site development. According to Equation 8.4, this reaction force can even be viewed as the sole opportunity for the adhesion site to maturate with time. The slope (kV_R) of the time-dependent reaction force plays a role equivalent to the "loading rate" described above by limiting (thus controlling) the adhesion site maturation. The two limiting factors are k and V_R. The spring constant k represents the passive and global elasticity properties of the substrate while the velocity of actin retrograde motion V_R can be related to the global intracellular level of activation of actomyosin motors.

At the instant of dissociation, the most likely force $\langle F(k)\rangle$ exerted on the adhesion site can be estimated by applying Kramers' method to Equation 8.4 [28]. This leads to a logarithmic dependence on kV_R, which has the same physical origin as the logarithmic dependence on force loading rate of the well-known Bell-Evans expression for surface adhesion molecules [17] (see above):

$$\frac{\langle F(k)\rangle}{f_\beta} \approx ln\left\{\frac{kV_R}{Jf_\beta}\right\} + \frac{1}{f_\beta}\frac{\langle \Delta U\rangle}{\rho_f} \quad with \quad f_\beta = k_B T/\rho_f \tag{8.5}$$

Equation (8.5) is transformed into a nondimensional equation (see Equation (8.6)) after appropriate normalization of force by f_β, energy by $k_B T$, distance by ρ_f. Moreover, after considering that $\langle \Delta U\rangle$, the likelihood value of activation energy of the potential during dissociation, or thermal average of the passive $(S = -1)$ and active $(S = 1)$ activations energies, can be written in terms of activation energies $\Delta U(S = -1)$ and $\Delta\Delta U(S = 1)$; that is, $\langle \Delta U\rangle = \Delta U + 1/2\Delta\Delta U(1+\langle S\rangle_F)$.

$$\tilde{F}(\tilde{k}) \approx ln\tilde{k} + \left[\Delta\tilde{U} + \frac{\Delta\Delta\tilde{U}}{2}\left\{1 + \tanh\left[(\tilde{F}(\tilde{k}) - \tilde{F}_c)\frac{d^*}{\rho_f}\right]\right\}\right] \tag{8.6}$$

Results of Equation 8.6 are plotted in Figure 8.5.

The logarithmic dependence on substrate rigidity occurs up to a force level $\tilde{F}_c$ corresponding to the threshold rigidity $\tilde{k}_c$ where the energies of active and passive states degenerate $(F_c = \Delta G/d^*)$ and the dissociation force level diverges (see curve in Figure 8.5). These critical conditions correspond to $S = 0$ in Equation (8.1). For $-1 < S < 0$ (i.e., $\tilde{k} < \tilde{k}_c$), the dynamic adhesion site follows a substrate-dependent reinforcement, meaning that the dissociation force will increase logarithmically with the normalized substrate stiffness $\tilde{k}$. Results in Figure 8.5 also suggest that extremely soft substrates and/or totally immature adhesion sites $(S = -1)$ will not be able to reinforce even partially. This case would correspond to fully slipping adhesion sites. On the other hand, in the case of too rigid substrates and/or fully mature adhesion

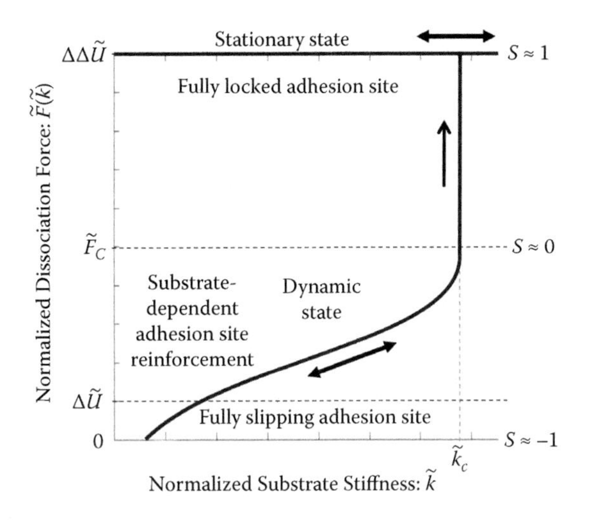

FIGURE 8.5 Graph of the function given by Equation (8.6). The normalized dissociation force $\tilde{F}(\tilde{k})$ for a given adhesion site is plotted versus the normalized substrate stiffness $\tilde{k}$. The characteristic stiffness used for such a normalization is given by: $k_0(= Jf_\beta/V_R)$, a stiffness scale which characterizes the dynamic properties of the adhesion site at the molecular level. $J[= J_0 \exp(-\frac{\Delta U}{k_B T})]$ is the attempt rate for "escape out of the well" or rate of dissociation (i.e., close to k_{off} in the Bell's law above) which intricately depends on (1) local curvature of $U(\rho)$ in the vicinity of site dissociation, (2) friction coefficients, and (3) temperature. Values of the state parameter S increase from -1 to 1 as the energy level required for adhesion site dissociation increases. $S \sim -1$ corresponds to the lowest level of dissociation energy ($\Delta\tilde{U} = \Delta U/k_B T$) and thus to a fully slipping adhesion site which could never maturate. Dynamic adhesion sites would correspond to an intermediate range of S values (-1; 0) corresponding to the transition between logarithmic dependence and the increasing contribution of elastic energy associated to the substrate elasticity (i.e., $0.5k(\rho - V_R t)^2$). Substrate-dependent adhesion site reinforcement is a reversible process. Beyond a critical value of substrate rigidity $\tilde{k}_c$ (also called degenerated state), the adhesion site is reinforced and can fully transmit the actomyosin traction force toward the substrate. Stationary adhesion sites pertain to the upper (nonlogarithmic) zone of the diagram (S varies in the range (0; +1) in which adhesion sites are stationary and irreversibly reinforced, fully locked once the level of dissociation energy ($\Delta\Delta\tilde{U} = \Delta\Delta U/k_B T$) expected for $S \sim 1$ is reached. (Modified from [20].)

sites, cells would also remain insensitive to substrate stiffness, following the upper horizontal line with no backward evolution, as previously indicated by Bruinsma [4]. This case corresponds to fully locked adhesion sites, which can be described by the stationary adhesion site model treated above.

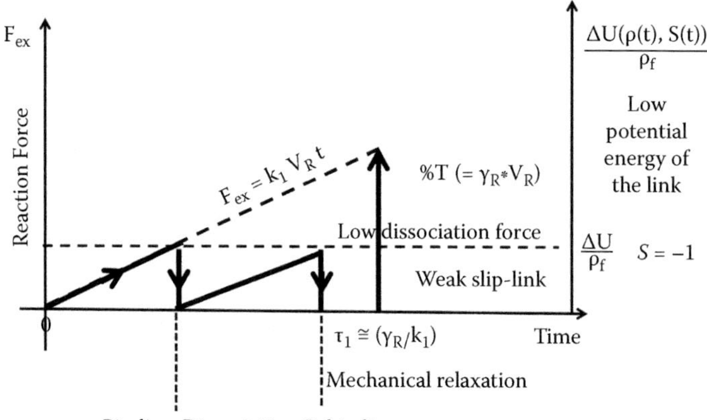

FIGURE 8.6 In response to the traction (or only a weak part of the traction), the reaction force F_{ex} increases linearly as time increases. The slope of the reaction force (kV_R) constitutes the loading rate. The present diagram sketches the case of a rather soft substrate (small k_1) and rather low tensed cell (small V_R). Because the dissociation force level is low (S close to -1), the reaction force will rapidly reach the dissociation threshold and the adhesion linkage will break, leading to a succession of binding, dissociation, and rebinding. Note that dissociation always occurs for an instantaneous local minimum of the link potential energy function and a time much shorter than the mechanical relaxation time.

To summarize the main message brought by the dynamic adhesion site model and to synthesize the knowledge brought by Equations (8.4) through (8.6), we have plotted in Figure 8.6 and Figure 8.7 the time evolution in the two typical cases described below.

1. For soft substrates (k_1) and/or low tensed cell, the loading rate ($k_1 V_R$) is low and the reaction force starts to increase in response to even a small fraction of intracellular traction transmitted. Due to the low dissociation force level (close to $\Delta U/\rho_f$), the reaction force rapidly reaches the dissociation level, the adhesion site breaks, and one can expect a succession of binding, dissociation, and rebinding. Note that dissociation occurs much before the time for mechanical relaxation, which is approached by $\tau_1 = \gamma_R/k_1$ (see Figure 8.6).
2. For stiff substrates (k_1) and/or highly tensed cell, the loading rate ($k_2 V_R$) is higher and the reaction force increases more rapidly in response to intracellular traction. Due to the higher dissociation force level associated with stiffer substrate (see Figure 8.5 and Equations (8.5) and (8.6)), the dissociation force increases, which allows sufficient time for the reaction force to reach the degenerated state ($S = 0$), integrin

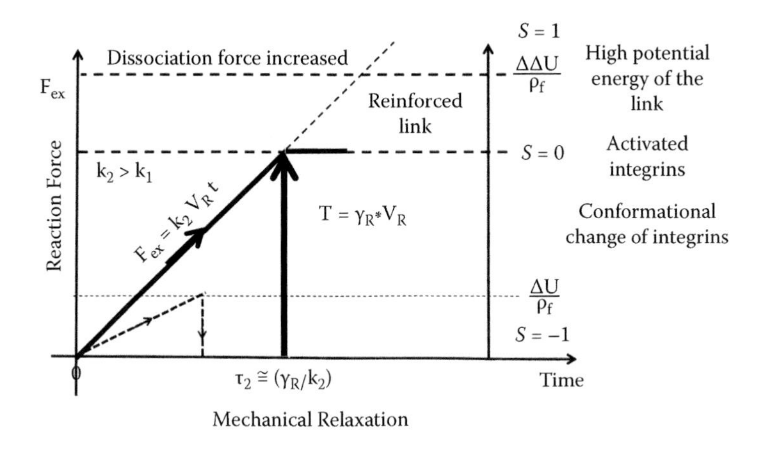

FIGURE 8.7 The same graph as in Figure 8.6 but for a stiffer substrate ($k_2 > k_1$) and/or a more tensed cell. Because the loading rate is increased, the reaction force will reach more rapidly a higher level while, at the same time, the dissociation force level is increased secondary to the increase in substrate rigidity (see Figure 8.5 and Equations (8.5) or (8.6)). Therefore, the adhesion site has more chances to become reinforced, thus impeding breaking. Hence, after a time equivalent to the new mechanical relaxation time τ_2, the whole traction force can be transmitted to the adhesion site, and the actin filament bundle is arrested. Once reinforced, the adhesion site behaves as a stationary adhesion site.

activation, and thus adhesion site reinforcement. Then the adhesion site has a chance to maturate toward a fully locked and fully reinforced link. The traction force is fully transmitted to the adhesion site and the filament bundle is arrested. The adhesion site could not dissociate before the mechanical relaxation time $\tau_2 = \gamma_R/k_2$ even though $\tau_2 < \tau_1$ (see Figure 8.7).

In conclusion, a key assumption in understanding cell sensitivity to substrate mechanical properties is to consider that the activation of substrate-sensitive cell elements must be mechanically limited by substrate relaxation and not by chemical reaction between active and passive states, which is definitely much faster.

8.3 Experiments on Cell Mechanosensitivity

We recently characterized cell sensitivity to substrate stiffness using two cellular models to represent two different systems of adhesion [20]: (1) alveolar epithelial cells (AECs) grown at confluence or subconfluence representative of a stationary adhesion system and (2) resting alveolar macrophages (AMs) representative of a dynamic adhesion system. We used: (1) rigid substrates of plastic or glass (i.e., Young's modulus $E_{sub} \geq 3$ MPa), (2) stiff polyacrylamide gel substrate (i.e., $E_{sub} \approx 60$ kPa), and (3) soft polyacrylamide gel substrate (i.e., $E_{sub} \approx 20$ kPa). AMs were also cultured on epithelial cell monolayers (i.e., $E_{sub} \approx 0.5$-1 kPa) [19]. The results are summarized in Figures 8.8 and 8.9.

The two cellular models—AECs and AMs—were chosen because they are, respectively, representative of stationary (Figure 8.8(d)) and dynamic (Figure 8.8(c)) adhesion systems. Alveolar epithelial cells grown at confluence or subconfluence (AECs) are representative of a focal adhesion system (bottom right of Figure 8.8(d)), which is also a stationary adhesion site system [1,5]. Alveolar macrophages (AMs) provide a cellular model representative of the

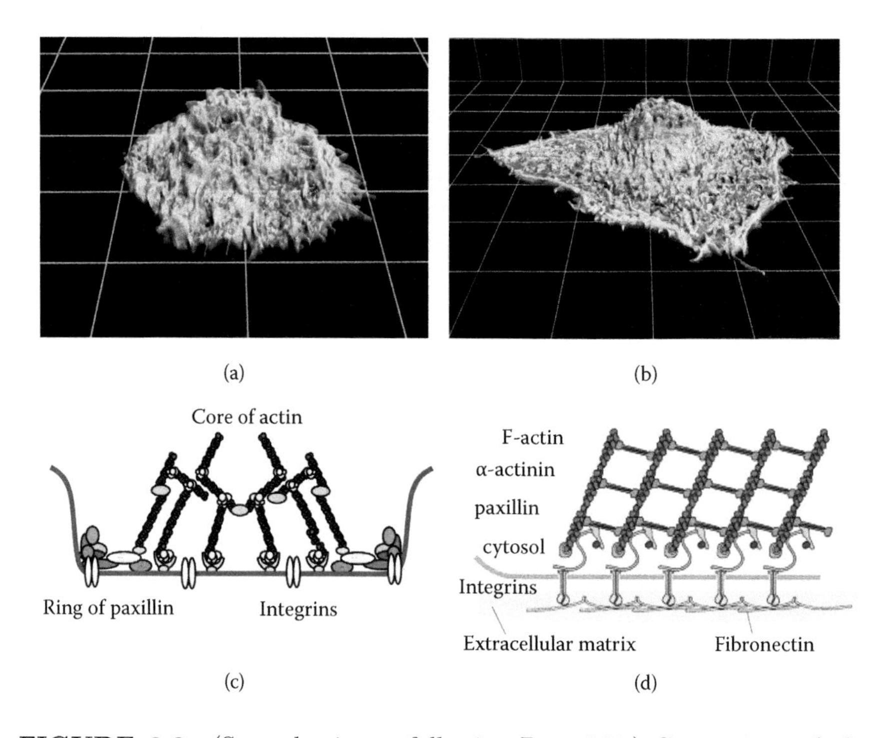

FIGURE 8.8 (See color insert following Page 398.) Comparison of alveolar macrophages (AMs) and epithelial cells (AECs) for cell structures and adhesion sites. (a) and (b): 3-D reconstructions of F-actin structures and 3-D localizations of paxillin, an adhesion molecule present in adhesion sites of both cell types. (c) and (d): The ultrastructure of adhesion sites in AMs (c): the podosome-like structure with a core of actin and a ring of paxillin. The ultrastructure of adhesion sites in AECs (d): the well-known focal adhesion sites made of bundles of parallel filaments connected to integrins through numerous specific molecules including paxillin.

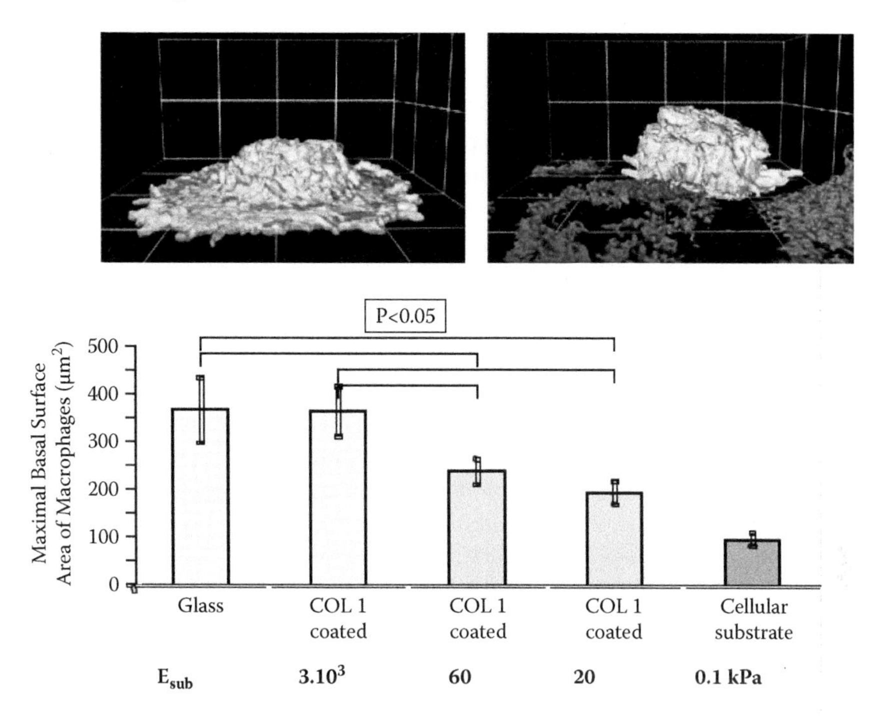

FIGURE 8.9 The effect of substrate stiffness of rigid (3 MPa), stiff gel (60 kPa), soft gel (20 kPa), epithelial cell monolayer (0.5 to 1 kPa) on the shapes of alveolar macrophages (AMs). The maximal basal surface area of AMs decreases as substrate stiffness decreases. This result is confirmed by the 3-D reconstructions of actin structure of AMs shown for glass substrate (top left) and cell substrate (top right). (Modified from [19,21,33].)

podosome-like system (sketched on Figure 8.8(c)), which can be seen as a dynamic system of adhesion [3,11]. Spatial reconstruction of paxillin and F-actin after immunostaining (see reconstructed images in Figures 8.8(a) and (b)) has allowed us to verify that the two cellular models are both physically linked to the substrate through adhesion sites containing paxillin. Moreover, the cellular and molecular structures existing between the cell and the substrate are markedly different in terms of F-actin organization and adhesion site structure. In AECs (Figures 8.8(b) and (d)), the dense F-actin network near the basal plane covers the entire cell due to multiple interconnections between short and long dense actin stress fibers. The latter generally end at focal adhesion plaques visible through reconstructed aggregates of paxillin. These plaques are notably located at the cell periphery in AECs. In AMs (Figures 8.8(a) and (c)), the dense F-actin network does not extend throughout the entire cell, forming either punctuated structured or only local networks, the exception may be at the origin of the circumferential lamellipodium.

We concluded that AECs correspond to adhesion sites with the highest maturation (or affinity; i.e., $0 < S < 1$) which means highest dissociation forces. Energy values of such a reinforced link could approach $100k_BT$ and might correspond to dissociation forces in the range 50 to 100 pN. Single-bond mechanics of macrophage adhesion via a variety of specific ligands has been investigated by Knöner et al. [31], leading to a maximum dissociation force of 10 pN at 10 pN s^{-1} loading rate. Such data confirm that AM adhesion sites have rupture forces markedly smaller than AECs (see Figures 8.5 through 8.7). The present theory predicts that AMs probably adapt their shape to substrate properties in a stiffness-dependent manner (Figure 8.5). Results in Figure 8.9 indeed confirm that AMs are able to adapt their shape to substrate stiffness. This is evidenced by the passage from a flattened shape to a more rounded cell with higher cell height as substrate stiffness decreases [21]. In contrast, AECs grown at confluence or subconfluence do not really modify their shape when cultured on different substrates (results not shown in this chapter). Surprisingly, this AM adaptation occurs despite an evident lack of stress fibers, the latter elements classically considered the key mechanotransduction pathways in tissue cells [8,10,46]. This suggests that AECs and AMs fundamentally differ in terms of mechanosensitive pathways, as suggested in a previous study [19].

Note that assuming a 1-μm-wide site, for the three substrates tested in this study (plastic, stiff gel, soft gel), typical values of spring constant are k = 3000 pN nm^{-1}, 58 pN nm^{-1}, and 23 pN nm^{-1}, respectively. Corresponding values of γ_B for the polyacrylamide gels tested remain within the range 3.10^{-1} to 3.10^{-4} pN s nm^{-1} [43]. γ_R is on the order of 0.1 pN s nm^{-1} per filament, that is, 10 to 100 pN s nm^{-1} for 10^2 to 10^3 filaments per actin bundle. Thus, $\gamma_B \ll \gamma_R$ is a condition generally verified by the values given above for polyacrylamide gels (a condition required for the validity of Equation (8.4)). V_R is in the micrometer-per-minute (μm/min) range and characterizes actin dynamics in the lamellipodia. Values of kV_R were 48000 pN s^{-1}, 928 pN s^{-1}, and 368 pN s^{-1} for the three substrates tested.

8.4 Conclusions

As time or force increases, adhesion sites reinforce from initial adhesion supporting the piconewton range to adhesion complex supporting the nanonewton range. Fully mature focal adhesion can support up to a hundred nanonewtons. Fully mature adhesion sites are mainly stationary while non mature adhesion sites are more dynamic, leading to different sensitivities to substrate or environmental stiffness.

Fundamental principles of continuum mechanics predict that reaction forces exerted by the substrate on a stationary adhesion site are independent of substrate rigidity. Hence, activation of stationary adhesion sites (containing force-sensitive proteins) at the cell–substrate interface could not be regulated by substrate rigidity.

Considering the nanoscale displacement of the dynamic adhesion site relative to an actin bundle on which actomyosin exerts a constant traction force reveals that, in response to this intracellular traction, a time-dependent reaction force generated extracellularly by the substrate and intracellularly by actomyosin motors raises and dramatically slows the physicochemical linkage between adhesion site and actin filament. It also reveals that the dissociation force depends on the logarithm of substrate stiffness. This mechanism limits the time response to integrin activation and, moreover, makes cell sensitivity to substrate stiffness dependent on more global intracellular and extracellular factors such as intracellular tension or mechanical properties of the extracellular environment.

References

[1] N.Q. Balaban, U.S. Schwarz, D. Riveline, P. Goichberg, G. Tzur, I. Sabanay, D. Mahalu, S. Safran, A. Bershadsky, L. Addadi, and B. Geiger (2001). Force and focal adhesion assembly: a close relationship studied using elastic micropatterned substrates. *Nat. Cell. Biol.* 3:466–472.

[2] G.I. Bell (1978). Models for the specific adhesion of cells to cells. *Science* 200:618–627.

[3] M.R. Block, C. Badowski, A. Millon-Fremillon, D. Bouvard, A.P. Bouin, E. Faurobert, D. Gerber-Scokaert, E. Planus, and C. Albiges-Rizo (2008). Podosome-type adhesions and focal adhesions, so alike yet so different. *Eur. J. Cell Biol.* 87:491–506.

[4] R. Bruinsma (2005). Theory of force regulation by nascent adhesion sites. *Biophys. J.* 89:87–94.

[5] K. Burridge and M. Chrzanowska-Wodnicka (1996). Focal adhesions, contractility, and signaling. *Annu. Rev. Cell Dev. Biol.* 12:463–518.

[6] C.S. Chen and D.E. Ingber (1999). Tensegrity and mechanoregulation: from skeleton to cytoskeleton. *Osteoarthritis and Cartilage* 7:81–94.

[7] M.E. Chicurel, C.S. Chen, and D.E. Ingber (1998). Cellular control lies in the balance of forces. *Curr. Opin. Cell Biol.* 10:232–239.

[8] M. Chiquet, A.S. Renedo, F. Huber, and M. Fluck (2003). How do fibroblasts translate mechanical signals into changes in extracellular matrix production? *Matrix Biol.* 22:73–80.

[9] D. Choquet, D.P. Felsenfeld and M.P. Sheetz (1997) Extracellular matrix rigidity causes strengthening of integrin-cytoskeleton linkages. *Cell* 88:39–48.

[10] F. Chowdhury, S. Na, O. Collin, B. Tay, F. Li, T. Tanaka, D.E. Leckband, and N. Wang (2008). Is cell rheology governed by nonequilibrium-to-equilibrium transition of noncovalent bonds? *Biophys. J.* 95:5719–5727.

[11] O. Collin, P. Tracqui, A. Stephanou, Y. Usson, J. Clement-Lacroix, and E. Planus (2006). Spatiotemporal dynamics of actin-rich adhesion microdomains: influence of substrate flexibility. *J Cell Sci.* 119:1914–1925.

[12] M. Dembo, D.C. Torney, K. Saxman, and D. Hammer (1988). The reaction-limited kinetics of membrane-to-surface adhesion and detachment. *Proc. R. Soc. London B. Biol. Sci.* 234:55–83.

[13] D.E. Discher, P. Janmey, and Y.L. Wang (2005). Tissue cells feel and respond to the stiffness of their substrate. *Science* 310:1139–1143.

[14] H.G. Dobereiner, B.J. Dubin-Thaler, G. Giannone, and M.P. Sheetz (2005). Force sensing and generation in cell phases: analyses of complex functions. *J. Appl. Physiol.* 98:1542–1546.

[15] A.J. Engler, M.A. Griffin, S. Sen, C.G. Bonnemann, H.L. Sweeney, and D.E. Discher (2004). Myotubes differentiate optimally on substrates with tissue-like stiffness: pathological implications for soft or stiff microenvironments. *J. Cell Biol.* 166:877–887.

[16] B.T. Estes, J.M. Gimble, and F. Guilak (2004). Mechanical signals as regulators of stem cell fate. *Curr. Top. Dev. Biol.* 60:91–126.

[17] E. Evans (2001). Probing the relation between force lifetime and chemistry in single molecular bonds. *Annu. Rev. Biophys. Biomol. Struct.* 30:105–128.

[18] E. Evans, A. Leung, D. Hammer, and S. Simon (2001). Chemically distinct transition states govern rapid dissociation of single L-selectin bonds under force. *Proc. Natl. Acad. Sci. USA* 98:3784–3789.

[19] S. Féréol, R. Fodil, B. Labat, S. Galiacy, V.M. Laurent, B. Louis, D. Isabey, and E. Planus (2006). Sensitivity of alveolar macrophages to substrate mechanical and adhesive properties. *Cell. Motil. Cytoskeleton* 63:321–340.

[20] S. Féréol, R. Fodil, V.M. Laurent, M. Balland, B. Louis, G. Pelle, S. Hénon, E. Planus, and D. Isabey (2009). Prestress and adhesion site dynamics control cell sensitivity to extracellular stiffness. *Biophys. J.* 96:2009–2022.

[21] S. Féréol, R. Fodil, V.M. Laurent, E. Planus, B. Louis, G. Pelle, and D. Isabey (2008). Mechanical and structural assessment of cortical and

deep cytoskeleton reveals substrate-dependent alveolar macrophage remodeling. *Biomed. Mater. Eng.* 18:S105–S118.

[22] S. Féréol, R. Fodil, G. Pelle, B. Louis, and D. Isabey (2008). Cell mechanics of alveolar epithelial cells (AECs) and macrophages (AMs). *Respir. Physiol. Neurobiol.* 163:3–16.

[23] E.B. Finger, K.D. Puri, R. Alon, M.B. Lawrence, U.H. von Andrian, and T.A. Springer (1996). Adhesion through L-selectin requires a threshold hydrodynamic shear. *Nature* 379:266–269.

[24] C.G. Galbraith, K.M. Yamada, and M.P. Sheetz (2002). The relationship between force and focal complex development. *J. Cell Biol.* 159:695–705.

[25] B. Geiger and A. Bershadsky (2001). Assembly and mechanosensory function of focal contacts. *Curr. Opin. Cell Biol.* 13:584–592.

[26] P.C. Georges and P.A. Janmey (2005). Cell type-specific response to growth on soft materials. *J. Appl. Physiol.* 98:1547–1553.

[27] G. Giannone, G. Jiang, D.H. Sutton, D.R. Critchley, and M.P. Sheetz (2003). Talin1 is critical for force-dependent reinforcement of initial integrin-cytoskeleton bonds but not tyrosine kinase activation. *J. Cell Biol.* 163:409–419.

[28] P. Hänngi, P. Talkner, and M. Borkovec (1990). Reaction rate theory: 50 years after Kramers. *Rev. Mod. Phys.* 62:25–341.

[29] D.E. Ingber (1997). Integrins, tensegrity, and mechanotransduction. *Gravit. Space Biol. Bull.* 10:49–55.

[30] P.A. Janmey and D.A. Weitz (2004). Dealing with mechanics: mechanisms of force transduction in cells. *Trends Biochem. Sci.* 29:364–370.

[31] G. Knoner, B.E. Rolfe, J.H. Campbell, S.J. Parkin, N.R. Heckenberg, and H. Rubinsztein-Dunlop (2006). Mechanics of cellular adhesion to artificial artery templates. *Biophys. J.* 91:3085–3096.

[32] J.B. Leach, X.Q. Brown, J.G. Jacot, P.A. Dimilla, and J.Y. Wong (2007). Neurite outgrowth and branching of PC12 cells on very soft substrates sharply decreases below a threshold of substrate rigidity. *J. Neural. Eng.* 4:26–34.

[33] S. Linder and M. Aepfelbacher (2003). Podosomes: adhesion hot-spots of invasive cells. *Trends Cell Biol.* 13:376–385.

[34] C.M. Lo, H.B. Wang, M. Dembo, and Y.L. Wang (2000). Cell movement is guided by the rigidity of the substrate. *Biophys. J.* 79:144–152.

[35] A.J. Maniotis, C.S. Chen, and D.E. Ingber (1997). Demonstration of mechanical connections between integrins, cytoskeletal filaments, and nucleoplasm that stabilize nuclear structure. *Proc. Natl. Acad. Sci. USA* 94:849–854.

[36] B.T. Marshall, M. Long, J.W. Piper, T. Yago, R.P. McEver, and C. Zhu (2003). Direct observation of catch bonds involving cell-adhesion molecules. *Nature* 423:190–193.

[37] S. Na, O. Collin, F. Chowdhury, B. Tay, M. Ouyang, Y. Wang, and N. Wang (2008). Rapid signal transduction in living cells is a unique feature of mechanotransduction. *Proc. Natl. Acad. Sci. USA* 105:6626–6631.

[38] R.J. Pelham, Jr. and Y. Wang (1997). Cell locomotion and focal adhesions are regulated by substrate flexibility. *Proc. Natl. Acad. Sci. USA* 94:13661–13665.

[39] D. Riveline, E. Zamir, N.Q. Balaban, U.S. Schwarz, T. Ishizaki, S. Narumiya, Z. Kam, B. Geiger, and A.D. Bershadsky (2001). Focal contacts as mechanosensors: externally applied local mechanical force induces growth of focal contacts by an mDia1-dependent and ROCK-independent mechanism. *J. Cell Biol.* 153:1175–1186.

[40] K. Rottner, A. Hall, and J.V. Small (1999). Interplay between Rac and Rho in the control of substrate contact dynamics. *Curr. Biol.* 9:640–648.

[41] K. Saha, A.J. Keung, E.F. Irwin, Y. Li, L. Little, D.V. Schaffer, and K.E. Healy (2008). Substrate modulus directs neural stem cell behavior. *Biophys. J.* 95:4426–4438.

[42] F.H. Silver and L.M. Siperko (2003). Mechanosensing and mechanochemical transduction: how is mechanical energy sensed and converted into chemical energy in an extracellular matrix? *Crit. Rev. Biomed. Eng.* 31:255–331.

[43] C. Verdier. Personal communication. 2008.

[44] D. Vestweber and J.E. Blanks (1999). Mechanisms that regulate the function of the selectins and their ligands. *Physiol. Rev.* 79:181–213.

[45] N. Wang, J.P. Butler, and D.E. Ingber (1993). Mechanotransduction across the cell surface and through the cytoskeleton [see comments]. *Science* 260:1124–1127.

[46] N. Wang and Z. Suo (2005). Long-distance propagation of forces in a cell. *Biochem. Biophys. Res. Commun* 328:1133–1138.

[47] N. Wang, I.M. Tolic-Norrelykke, J. Chen, S.M. Mijailovich, J.P. Butler, J.J. Fredberg, and D. Stamenovic (2002). Cell prestress. I. Stiffness and prestress are closely associated in adherent contractile cells. *Am. J. Physiol. Cell Physiol.* 282:C606–C616.

[48] J.Y. Wong, J.B. Leach, and X.Q. Brown (2004). Balance of chemistry, topography, and mechanics at the cell–biomaterial interface: issues and challenges for assessing the role of substrate mechanics on cell response. *Surf. Sci.* 570:119–133.

[49] M.A. Wozniak, R. Desai, P.A. Solski, C.J. Der, and P.J. Keely (2003). ROCK-generated contractility regulates breast epithelial cell differentiation in response to the physical properties of a three-dimensional collagen matrix. *J. Cell Biol.* 163:583–595.

[50] T. Yeung, P.C. Georges, L.A. Flanagan, B. Marg, M. Ortiz, M. Funaki, N. Zahir, W. Ming, V. Weaver, and P.A. Janmey (2005). Effects of substrate stiffness on cell morphology, cytoskeletal structure, and adhesion. *Cell Motil. Cytoskeleton* 60:24–34.

[51] E. Zamir and B. Geiger (2001). Molecular complexity and dynamics of cell-matrix adhesions *J. Cell Sci.* 114:3583–3590.

Chapter 9

Cancer Cell Migration on 2-D Deformable Substrates

Valentina Peschetola, Claude Verdier, Alain Duperray, and Davide Ambrosi

Contents

9.1 Introduction

Cell migration is an important feature of many biological processes such as tissue invasion, the immunologic response, etc. and involves sophisticated mechanisms such as the development of focal adhesions, possible extracellular matrix (ECM) degradation, and activation of the actin–myosin complex to develop forces necessary for traction. Depending on their type and the environment, cells can move according to different types of locomotion, either in 2-D or 3-D. For example, fibroblasts move on a 2-D substrate in the most conventional way, as described by Sheetz [34]: they form a lamellipodium at the front, and develop adhesion complexes so that they can pull on such anchors to detach the rear part and move forward. On the other hand, fish keratocytes migrating on 2-D substrates take the form of a crescent [6,22]. In 3-D, two types of motions have been observed for cancer cells migrating in 3-D collagen, the amoeboid motion and the mesenchymal one [15,16].

2-D migration can depend on cell type and ligand matrix density, as shown in previous works on substrates covered by different densities of ECM [27].

The migration velocity versus ligand density curve usually exhibits a bell-shaped curve with a maximum velocity at an intermediate ligand density, which is the signature of an optimum affinity between receptor and ligand molecules. This is the basis for understanding the complex machinery provided by cells developing focal adhesions at the front while removing them at the back in the conventional five-step motion [34]. To better understand these features, fluorescence (confocal) microscopy is a very useful tool to show the precise location of adhesion complexes involved during cell migration. The formation of such focal contacts [4] allows the development of important forces that the cell uses to move forward, usually called "traction forces." They are difficult to determine by straightforward analysis. Therefore, many attempts have been made in the past 10 years to measure such forces in an indirect way and different techniques have been developed:

- The original technique of Harris and co-workers [18], consisting of observing the magnitude and shape of wrinkles on a soft elastic film below a cell
- The most classical method of Dembo and co-authors [9,10] measures the motion of fluorescent beads embedded in a gel deformed by a cell
- Micro-patterned substrates [4,12,35] with thin poles whose deflection directly gives the forces exerted by the cell

All these techniques have been tested *in vitro* and were quite effective for the study of the motion of cells on 2-D rigid substrates. The first technique [18] was initially qualitative, but further developments by Burton et al. [6], in particular devoted to the study of keratocyte locomotion, revealed them as quite quantitative. The last one, introduced by Balaban et al. [4], recently has received a lot of interest [17] and important results concerning cell migration of 3T3 fibroblasts on rigid substrates have been obtained regarding the influence of substrate rigidity. Indeed, as earlier understood by Lo et al. [24], cells usually exert larger traction forces on more rigid substrates, this behavior correlated with an increased contact surface and a reduced migration velocity. The traction force increases for larger stiffnesses until a critical rigidity is reached (around 80 kPa), where the force exhibits a plateau [17]. Although very elegant, the technique of Balaban et al. still poses questions such as whether cells behave similarly on micropatterned substrates as compared to classical polyacrylamide gels or when migrating *in vivo*.

In this chapter we focus on the technique developed by Dembo and Wang [10], and its further recent developments [2,3,25,30]. The basic idea is to measure the displacements of fluorescent beads embedded in rigid substrates of polyacrylamide (a mixture of acrylamide and bis-acrylamide to be more precise). Component concentrations can be tuned to obtain the desired gel rigidity, usually ranging between 1 and 100 kPa. As cells are laid onto the functionalized substrate, they adhere and start to exert tractions that deform the substrate so that fluorescent beads located in the underlying gel are displaced.

Based on the hypothesis that the gel is isotropic and elastic, and displacements are small enough so that linear elasticity applies, the gel depth being large enough so that it can be approximated by a half space, the Boussinesq equations can be solved in the whole gel. Usually a thickness of 100 μm is enough, but a recent work argues about the advantages of using thinner gels [25], so that one can compute the traction stresses exactly at the gel surface. This is an inverse problem, because the particle displacement vectors $\mathbf{u}(\mathbf{x}, t)$ are known at certain locations (with an unavoidable experimental error) and the traction stresses $\mathbf{T}(\mathbf{x}, t)$ are to be computed everywhere on the gel surface. The solution of the Boussinesq problem for an elastic gel with Young's modulus E and Poisson coefficient ν reads:

$$\mathbf{u}(\mathbf{r}) = \int \mathbf{G}(\mathbf{r} - \mathbf{r}')\, \mathbf{T}(\mathbf{r}')\, d\mathbf{r}' \tag{9.1}$$

where the kernel $\mathbf{G}$ is a Green's function tensor [21]:

$$\mathbf{G}(\mathbf{r}) = \frac{1+\nu}{\pi E}\left((1-\nu)\,\frac{\mathbf{1}}{r} + \nu\frac{\mathbf{r}\otimes\mathbf{r}}{r^3}\right) \tag{9.2}$$

while r denotes the modulus of $\mathbf{r}$.

In the initial approach of Dembo and Wang [10], bead displacements are first determined. Bilinear shape functions are defined on an unstructured quadrilateral mesh. The optimal stress field is obtained minimizing the error by an iterative procedure under force magnitude penalization. Because the method is rather complicated, other methods have been introduced [7] that make use of the particular structure of the equations. Noticing that Equation (9.1) is a convolution, one can compute the Fourier transform $\tilde{\mathbf{F}}(\mathbf{k}) = \int_{-\infty}^{\infty} \mathrm{e}^{-i\mathbf{k}.\mathbf{r}}\, \mathbf{F}(\mathbf{r})\, d\mathbf{r}$ on both sides of Equation (9.1), $\mathbf{k}$ being the wave vector:

$$\tilde{\mathbf{u}}(\mathbf{k}) = \tilde{\mathbf{G}}(\mathbf{k})\tilde{\mathbf{T}}(\mathbf{k}) \tag{9.3}$$

Following the approach of [7], the expression of $\tilde{\mathbf{G}}(\mathbf{k})$ is given by a 2×2 matrix, assuming that vertical displacements can be neglected. This is usually the case as long as gels are stiff enough and are not deformed by vertical tractions exerted by cells, which is a good enough hypothesis in this case. Finally Equation (9.3) can be inverted to obtain $\tilde{\mathbf{T}}(\mathbf{k})$ and then going back into the real space gives the traction stress field $\mathbf{T}(\mathbf{r})$:

$$\mathbf{T}(\mathbf{r}) = FT_2^{-1}(\tilde{\mathbf{G}}^{-1}(\mathbf{k})\, \tilde{\mathbf{u}}(\mathbf{k})) \tag{9.4}$$

where FT_2^{-1} denotes the two–dimensional inverse Fourier transform and $\tilde{\mathbf{G}}^{-1}(\mathbf{k})$ is the inverse of $\tilde{\mathbf{G}}(\mathbf{k})$, which can be computed explicitly thanks to simple manipulations [7]. The main drawback of this method is that despite its simplicity, it requires a periodic displacement field to compute the fast Fourier transforms.

One further work [33] deals with pointwise traction force reconstruction and the effect of noise, which are well known to affect the results. Due to the ill posedness of the problem, it is necessary to use a regularization algorithm to obtain significant data. Other useful information is the location of focal adhesions. Another possible hypothesis [25] is to use thin substrates (typically 10 μm thick), which can help get explicit solutions of the Boussinesq equations. This allows for increased accuracy in the determination of these traction stresses. To conclude, let us mention the recent work of Sabass et al. [30] who combined three approaches to determine traction forces, one based on the integral boundary element method (BEM) [10], one using Fourier transform traction cytometry (FTTC) [7], and the last one using traction recovery with point forces (TRPF) [33]. All methods can be improved when proper regularizations are used and can lead to an increased resolution. The TRPF procedure seems to be the most promising one, as long as focal adhesions are well developed, but requires more sophisticated experiments. These experiments are needed in any case, as one wants to clearly understand how forces are connected with focal adhesions, and also how the cell cytoskeleton plays a role in the different motility regimes.

In this chapter a different method [2,3] is used as a promising tool for studying the migration of cancer cells. It is based on a formulation arising from the minimization of an energy, combined with the use of a penalty parameter. The method is presented in the next section. Then detailed experimental methods are given, followed by the results showing displacements and stresses in the case of migration T24 cancer cells. The discussions concern the effect of gel rigidity, as well as comparisons with other cell types.

9.2 Adjoint Method for Cell Traction

An alternative approach to obtain the pattern of the shear stresses exerted by a cell on a flat substrate is the adjoint method proposed by Ambrosi [2,3]. The mathematical model is based on the classical functional analysis framework due to Lions [23]; the general theory is applied to the specific problem of small deformation of a homogeneous elastic material subjected to body forces only.

Let Ω be the whole domain and $\mathbf{u}(\mathbf{x})$ the displacement vector field, $\mathbf{x} \in \Omega \subset \mathbb{R}^3$. The displacement is known only in a subset $\Omega_0 \subset \Omega$ where beads are located; the related function $\mathbf{u}_0(\mathbf{x})$ has support in Ω_0. Let $\Omega_c \subset \Omega$ be the region covered by the cell and where the shear stress is applied. As explained before, the traction forces are generated through the actin–myosin interactions and act on the underlying substrate through focal adhesion sites. These areas are not localized precisely in our experiments and we do not restrict the force support to these sites as is done in the algorithm of Schwarz et al. [33].

Consider the following elastic problem in the whole domain Ω:

$$-\mu\Delta\mathbf{u} - (\mu + \lambda)\nabla(\nabla\cdot\mathbf{u}) = \mathbf{f}, \quad \mathbf{u}|_{\partial\Omega} = 0 \tag{9.5}$$

where μ and λ are the Lamé constants that characterize the material. The problem can be rewritten in the form $\mathrm{A}\mathbf{u} = \mathbf{f}$, where A is a linear operator that is deduced easily from the above equation. The aim is to obtain the force field $\mathbf{f}$, which is inferred by a known displacement (inverse problem). If we try to invert directly the equation, we find that the problem is ill posed because the displacement is known only in a subset Ω_0 of Ω. It is necessary to introduce the projector P, $P: \Omega \to \Omega_0$, and a functional $J(\mathbf{f})$, $J: L^2(\Omega) \to \mathbb{R}$, defined as

$$J(\mathbf{f}) = \int_{\Omega_0} |\mathbf{u} - \mathbf{u}_0|^2\, dV + \varepsilon \int_{\Omega} |\mathbf{f}|^2 dV \tag{9.6}$$

where ε is a real positive number. This functional measures the difference between the displacement field produced by $\mathbf{f}$ and the experimental one defined by $\mathbf{u}_0$ under penalization of the square norm of the force field itself. We look for $\mathbf{g}$ minimizing J:

$$J(\mathbf{g}) \leq J(\mathbf{f}), \qquad \forall \mathbf{f} \in V_c \tag{9.7}$$

where $V_c \subset L^2(\Omega)$ is the space of the finite energy functions with support in Ω_c. The minimization of J accomplishes the minimization of the distance of the solution from the measured value $\mathbf{u}_0$ under penalization of the magnitude of the associated force $\mathbf{f}$ per unit surface. The penalty parameter ε balances the two requirements. An equivalent condition of inequality (9.7) is given by $\mathrm{J}'(\mathbf{g})[\mathbf{f} - \mathbf{g}] \geq 0$; making the Gateaux derivative explicit suggests the introduction of the adjoint equations, $A^* : L^2(\Omega) \to \Omega_0$

$$A^*\mathbf{q} = P\mathbf{u}(\mathbf{g}) - \mathbf{u}_0, \qquad \mathbf{q}|_{\partial\Omega} = 0 \tag{9.8}$$

Substituting back in the functional derivative allows us to obtain the solution of inequality (9.7) that represents the optimal body force:

$$\mathbf{g} = -\frac{\chi_c}{\varepsilon}\mathbf{q}$$

χ_c is the characteristic function of the domain Ω_c and $\mathbf{q}$ is a volume force.

In short, the set of equations that we need to solve is given by two elliptic partial differential equations, one for the displacement $\mathbf{u}$ and the adjoint one for the volume force $\mathbf{q}$.

On the basis of dimensional arguments, it is possible to reduce this 3-D system (in space) into a 2-D one. In fact, vertical averaging along an effective thickness h allows us to introduce two parameters $\hat{\mu}$ and $\hat{\lambda}$:

$$\hat{\mu} = h\frac{E}{2(1+\nu)}, \qquad \hat{\lambda} = h\frac{E\nu}{1-\nu^2}$$

E and ν are the Young's modulus and the Poisson ratio, respectively; h is the averaging height fixed by the depth of field of the microscope, that is, 1.5 μm in our case. Below this depth, the beads are not in focus and their positions are not measured; the displacement $\mathbf{u}$ should be understood as the average displacement along h, which is nearly the displacement of the center of the beads. Finally, the 2-D system becomes

$$\begin{aligned} -\hat{\mu}\Delta\mathbf{u} - (\hat{\mu} + \hat{\lambda})\nabla(\nabla\cdot\mathbf{u}) &= -\frac{\chi_c}{\varepsilon}\mathbf{p}, \qquad \mathbf{u}|_{\partial\Omega} = 0 \\ -\hat{\mu}\Delta\mathbf{p} - (\hat{\mu} + \hat{\lambda})\nabla(\nabla\cdot\mathbf{p}) &= \chi_o\mathbf{u} - \mathbf{u}_0, \qquad \mathbf{p}|_{\partial\Omega} = 0 \end{aligned} \tag{9.9}$$

where χ_c and χ_0 are the characteristic functions related to Ω_c and Ω_0, respectively. Note that the structure of the two equations is the same because the problem is self-adjoint. $\mathbf{p}$ now represents the traction stresses (N/m^2) previously introduced as $\mathbf{T}$ and will be determined once the two equations are solved.

In the ill-posed problem, the penalty parameter, ε in our case, plays an important role. In order to fix ε, it is necessary to re-interpret the system in Equation (9.9) on the basis of arguments suggested by modal analysis. Suppose that $\Omega_0 = \Omega_c = \Omega$ under periodic boundary conditions; the previous system of equations rewrites like a Tikhonov filter. The amplitude of the Fourier components of the solution u_k, p_k satisfies the algebraic relations

$$\begin{aligned} hEk^2 u_k &\simeq -\frac{1}{\varepsilon}p_k \\ hEk^2 p_k &\simeq u_k - u_{0,k} \end{aligned} \tag{9.10}$$

that is,

$$u_k \simeq \frac{u_{0,k}}{1 + \varepsilon h^2 E^2 k^4} \tag{9.11}$$

where $u_{0,k}$ represents the amplitude of the kth Fourier component of u_0. According to Equation (9.11), if the data are known all over the domain, the system of Equations (9.9) is a filter damping the modes corresponding to wave numbers $k > \varepsilon^{-1/4}h^{-1/2}E^{-1/2}$. The choice of ε can be interpreted in terms of filtering modes falling below the experimental accuracy. Equation (9.11) shows that the key parameter of the inversion procedure is actually εh^2 and the solution does not change for combinations of the averaging layer h and penalty parameter ε that preserve this quantity.

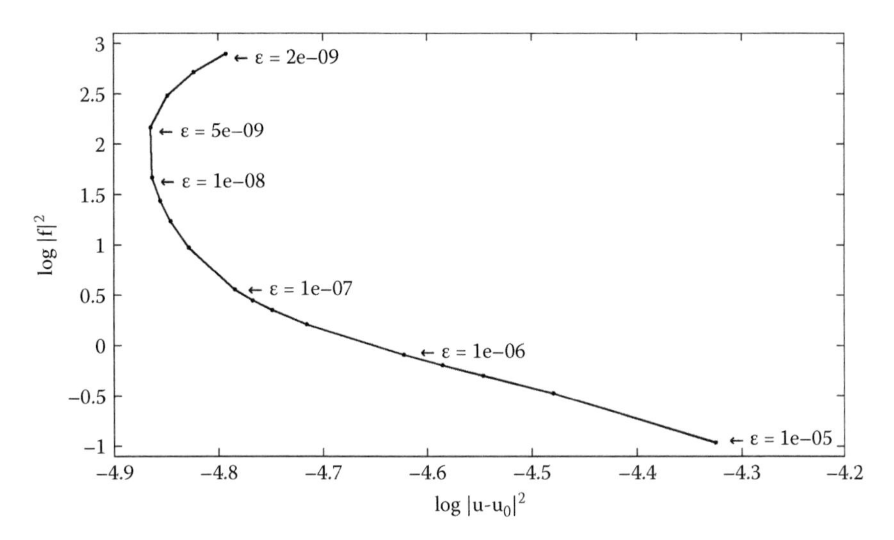

FIGURE 9.1 Discrete L-curve with 17 values of ε obtained for our model in the case of the stiff gel ($E = 10$ kPa). The corner corresponds to the optimal balance between data agreement and regularization.

A convenient tool for the analysis of discrete ill-posed problems is the L-curve criterion, which is a log-log plot, for all valid regularization parameters, of the norm of the regularized solution $||f||_2$ versus the corresponding residual norm $||\mathbf{u} - \mathbf{u}_0||_2$. In this way, the L-curve displays the minimization of these two quantities and its corner, which is intrinsic to the data, corresponds to the optimal balance between data agreement and regularization. Figure 9.1 shows an example of the discrete L-curve obtained for 17 values of the ε parameter for our model in the case of the stiff gel ($E = 10$ kPa). In our work we take the minimum value of ε that does not yield erratic results in the displacement, that is, the corner of the L-curve.

The system of Equations (9.9) has been discretized by a finite element method using linear basis functions on an unstructured mesh. To avoid any iterative coupling between the two equations in (9.9), a global conjugate method has been used to solve the resulting system of linear equations numerically. Figure 9.2 shows an example of the computational setting: the computational mesh, the bead displacement, and the cell contour. The triangular mesh satisfies two constraints: it has a node in every point where displacements are known (bead locations) and a sequence of element sides coincides with the cell contour. In particular, the boundary Ω_c of the cell, is described as a piecewise linear curve following the shape of the cell and represents the boundary between the intracellular domain over the substrate and the rest of the domain where the forces are not applied.

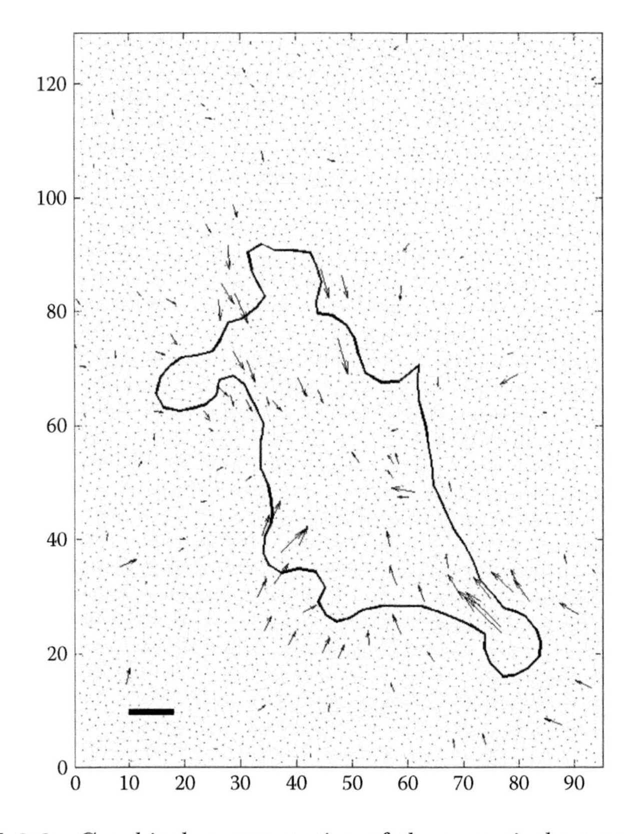

FIGURE 9.2 Graphical representation of the numerical set-up. The computational mesh, made of triangles, is represented in light gray. The mesh satisfies two constraints: it has one node at every point where the displacement is known and a sequence of element sides coincides with the boundary of the cell. The arrows indicate the experimental bead displacement. The scale bar at the bottom left corner is 0.5 μm long.

9.3 Experimental Methods

In this work we characterize cell migration on gels with different rigidities. T24 epithelial cells from a bladder cancer line are used. This type of cancer cell is known to be invasive. Regarding gel preparation, the following experimental procedure is used:

- *Gel preparation and stiffness.* To prepare gels with different stiffnesses, we change the ratio between polyacrylamide and bis-acrylamide components. Three different gels have been prepared containing 5, 7.5, and 10% polyacrylamide and the bis-acrylamide percentage is 0.03%. The mechanical properties have been measured using conventional dynamic shear rheometry tests on a Malvern rheometer (Gemini 150, stress-controlled). Sinusoidal oscillations at a known shear deformation $\gamma = \gamma_0 \sin(\omega t)$ are applied within the linear regime (small enough deformation $\gamma_0 \sim 0.01$) at different angular frequencies ω. The stress response $\sigma = \sigma_0 \sin(\omega t + \phi)$ (where σ_0 is a constant stress and ϕ is the phase angle) is measured and the elastic (G') and viscous moduli (G'') are deduced. These gels are quite elastic, as long as the acrylamide concentration is large enough, which is the case. Indeed, experiments show a constant elastic modulus G' when the frequency f (related to ω by $\omega = 2\pi f$) ranges from 0.1 to 10 Hz. The loss modulus G'' is usually lower by two orders of magnitude. Therefore we can assume that $\nu = 0.5$, for an incompressible gel, thus $E = 3G'$. Note that the hypothesis that $E = 3G'$ is relevant here in comparison to the work of Boudou et al. [5] showing that $\nu \sim 0.48$ in such polyacrylamide gels, as determined using micropipettes. This leads to the typical gel Young's moduli 2, 6.3, and 10 kPa for the soft, medium, and hard gels, respectively, used in this study. An example of the influence of the polyacrylamide content (for the three gels) is shown in Figure 9.3. The typical slope of ~2 is

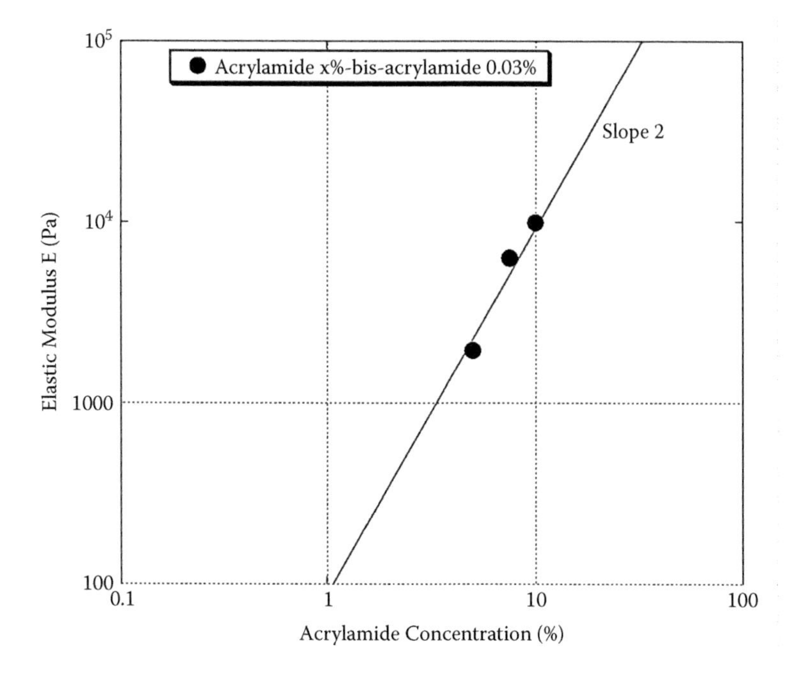

FIGURE 9.3 Elastic moduli E (Pa) as a function of the acrylamide concentration. Bis-acrylamide concentration is fixed at 0.03%.

current in such biological networks as observed, for example, in actin and fibrin gels [20]. Comparisons have been made previously [3] with other authors [5,13,14,37] and revealed good qualitative agreement.

- *Methodology.* Gels were prepared on a silanated square cover glass 22 mm × 22 mm, covered with a circular cover glass (35 mm diameter) and treated with NaOH (0.1 M), (3-aminopropyl)trimethoxysilane (APTMS, 10 min), and 0.5% glutaraldehyde (30 min). Fluorescent beads (Molecular Probes, 0.2 μm diameter) were seeded before addition of the crosslinker. Gels were laid onto the square cover glass, and the circular cover glass was brought carefully to capture the gel by capillarity, to avoid flipping the preparation. Indeed, beads need to sediment fast so that they are located closer to the gel upper surface where measurements are made. The gel volumes were chosen to have a thickness of about 70 μm, as described in other works [28]. The gels were left to polymerize for nearly 90 min.
- *Gel functionalization.* After polymerization, the square cover glass was gently removed and sulfo-Sanpah 1 mM mixed with dimethylsulfoxide (DMSO) and phosphate buffered saline (PBS) was added to functionalize the gels (15 min under UV). This procedure was repeated twice; then the surface was rinsed with PBS. Finally a 20 μg/ml fibronectin solution was used overnight to bind the above surface.
- *Cell seeding.* Cancer cells of epithelial bladder type (T24) were then seeded at a low density. They usually adhered rapidly and spread. The cover glass was attached at the bottom of a 35-mm culture dish (containing medium) in order to carry out microscopic observations. Two types of images were made: a phase contrast one to observe the cell and its contour, and a fluorescent one focused on the beads (at a slightly different vertical position). The depth of field of the images was around 1.5 μm. All operations were carried out automatically in order to take one set of images at regular time steps (2 min, for example).
- *Microscopy.* Images were then collected and treated using ImageJ software [19] to determine trajectories and/or displacements with respect to the initial position, as explained in the following section. The initial bead position was determined at the end of the experiment by adding distilled water to detach the cells, allowing a few minutes for the gel to go back to its initial position.

The basic idea in this work is to follow cells as they move to determine displacements, traction stresses, and velocities of migration during a significant duration (usually around 1 hour or more). This feature is indeed important because most other studies focus on stresses at a given time, but it will be shown here that stresses can vary significantly as cell migration proceeds. Then displacements and tractions will be studied as functions of gel rigidity.

9.4 Determination of Displacements

The basic step for the calculation of the traction field is the determination of the bead displacements. Essential data for the computation were the cell boundary, the initial bead position, and the in-plane displacement of the markers. The displacement field was extracted from a stack of images: the first one showed the beads in the undeformed position and the other images showed the gel deformed under cell traction. Usually a stack of 30 images was used and the times between frames were chosen to be 2 or 3 min. This enabled us to capture sufficient changes in the bead positions so that beads could be followed along their trajectories.

The processing of these images began with the correction for relative translational shifts. Here we used the ImageJ software and, in particular, the "Align Slice" plug-in to perform a recursive alignment of a stack of images. The alignment proceeded by propagation: each image was used as a template with respect to which the next slice was aligned.

From the corrected images sequences, we localized cells and divided images into small areas, typically 100 μm $\times$ 100 μm, each containing one isolated cell. Bead detection was made using "Particle Tracker," another ImageJ plug-in. The plug-in implemented point detection and a tracking algorithm as described in [32]; it performed two different steps: first the detection of the bead positions in each image and then the bead link into trajectories. The estimation of the bead center location was done by finding the maximum local intensity in the image. The point locations were refined under the assumption that the bead local intensity maxima were near the true geometric centers of the beads, and finally spurious detections such as dust or particle aggregates were rejected. The linking algorithm identified centers corresponding to the same physical particle in subsequent frames, using a graph technique theory, and linked these positions into trajectories.

An example of the image processing technique is provided in Figure 9.4; in particular, Figure 9.4(a) represents the fluorescent beads as recorded with the microscope; this configuration is related to the undisturbed position of the markers. Figure 9.4(b) shows the trajectories of these fluorescent markers under cell traction after 70 min. They are obtained after treatment of a stack containing 30 images.

9.5 Determination of Traction Stresses

To investigate the traction field generated by T24 cancer cells, we observed cells on different substrates, with Young's modulus (E) ranging from 2 to 10 kPa. After extracting the bead displacements as described in the previous

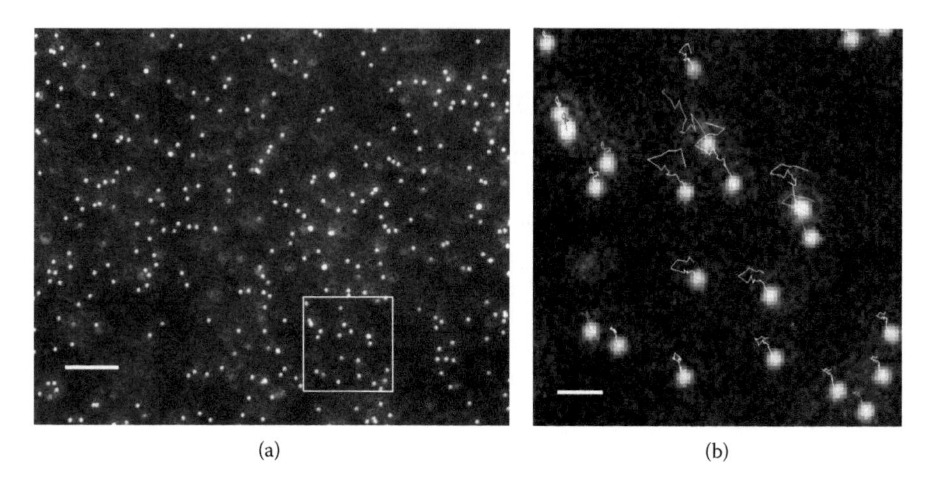

FIGURE 9.4 (a) Images of undisturbed positions of the fluorescent beads as recorded with the microscope. The scale bar is 10 μm. (b) Bead trajectories detected with ImageJ after processing a stack containing 30 images. The scale bar is 2 μm. (b) is a zoom of the square area delineated in (a).

section, the traction stress field was obtained by solving the system in Equation (9.9).

Figure 9.5(a) shows a T24 cancer cell on the soft gel, $E = 2$ kPa. The cell was observed for 2 hours and the image was recorded 30 min after the beginning of the experiment. The tumor cell is moving on the surface toward the bottom right part of the figure; it is quite elongated on such a surface. There are many beads whose displacements are significant, usually around 0.5 μm; they occur

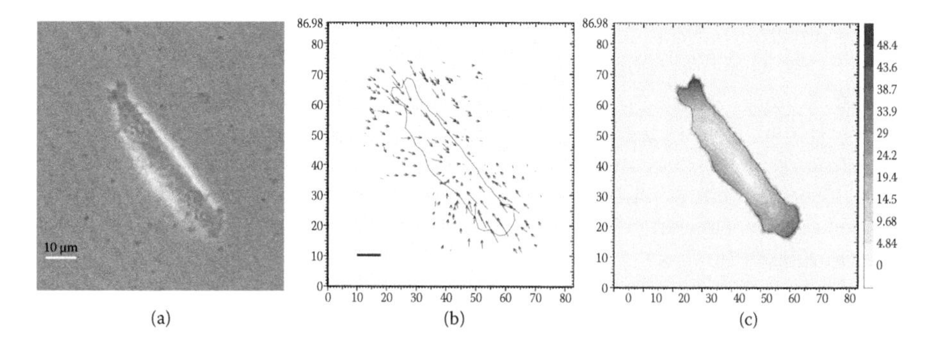

FIGURE 9.5 (a) T24 cell adhering on a soft polyacrylamide substrate ($E =$ 2 kPa). The cell is elongated on the surface. (b) The bead displacements under and around the cell are detected with ImageJ. (c) The traction stresses are shown as a gray-level map for the magnitude. The maximum value is around 50 pN μm^{-2}. The scale bar for displacements is 0.5 μm long. The grayscale map is in pN μm^{-2}.

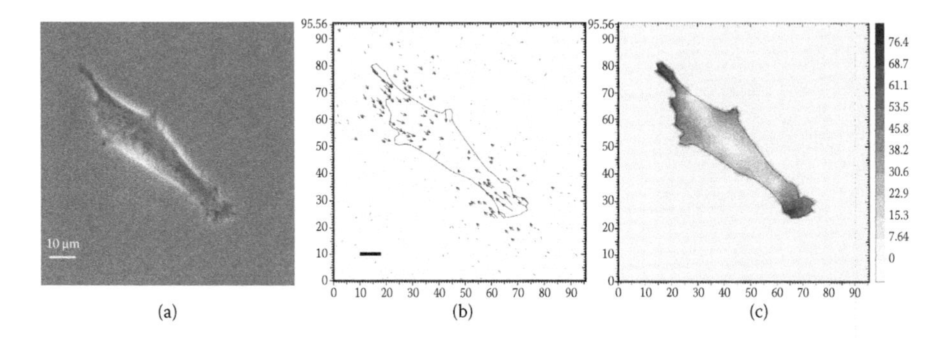

FIGURE 9.6 (a) T24 cell adhering on a medium stiffness polyacrylamide substrate ($E = 6.3$ kPa). The cell looks elongated, with small shape protrusion compared to Figure 9.5(a). (b) The displacement field has roughly the same magnitude as Figure 9.5. (c) The traction field has a maximum magnitude of about 80 pN μm^{-2}. The scale bar for displacements is 0.5 μm long. The grayscale map is in pN μm^{-2}.

essentially at the cell edges, as seen in Figure 9.5(b). In this same region, the maximum value of the traction stress is obtained, that is, around 50 pN μm^{-2} as seen in the gray-scale map of Figure 9.5(c), representing the traction stress levels.

A T24 cell is shown next on the medium stiff gel, $E = 6.3$ kPa in Figure 9.6(a). After 46 min, the cancer cell is quite elongated and exhibits a less symmetric shape than in the previous case. The beads have moved by about 0.5 μm maximum (Figure 9.6(b)). The stresses exerted by the tumor cell are shown in Figure 9.6(c): the traction stresses are small along the cell edges with peaks at the front and tail where the maximum magnitude is around 80 pN μm^{-2}.

Finally, Figure 9.7(a) shows a T24 cancer cell moving on the stiffer gel, $E = 10$ kPa, as well as the related numerical traction field in Figure 9.7(c). The picture was taken 10 min after the beginning of the experiment; the cell exhibits a shape with protrusions and it is anchored by the tail at the bottom right side where the shear stress reaches its maximum value, around 145 pN μm^{-2}. In this case, the beads are located close to the cell contour and the bead maximum displacements are still around 0.5 μm, as shown in Figure 9.7(b).

From the above, the maximum displacements in these three experiments seem to be in the same range (around 0.5 μm), independent of the rigidity of the substrate.

It is important to study the maximum value of the traction stresses at each time step. For the T24 cancer cell adhering on the rigid substrate, values of the traction stresses were recorded and vary between 90 and 190 pN μm^{-2} as presented in Figure 9.8. Cells do not move in a simple way, but their motion requires the development of protrusions until corresponding stable

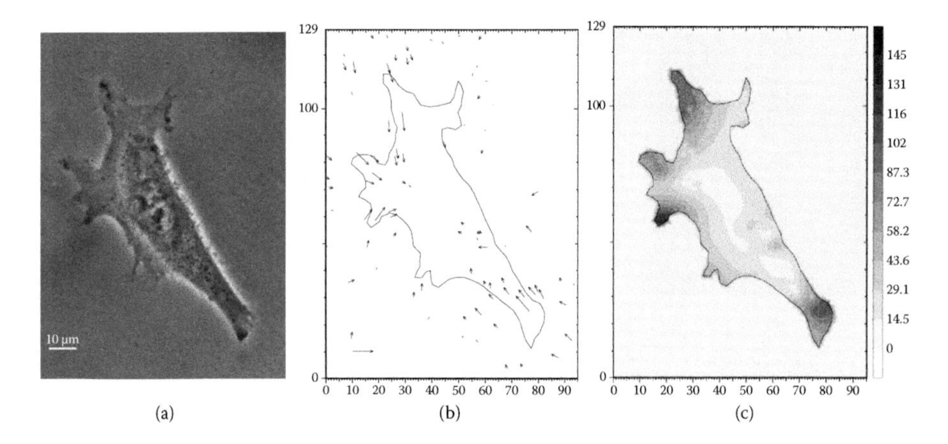

FIGURE 9.7 (a) T24 cell adhering on a stiff polyacrylamide substrate (E = 10 kPa). The cell has a less symmetric shape. (b) The displacement field is of the same magnitude as in Figures 9.5 and 9.6. (c) The traction field has a maximum magnitude around 145 pN μm^{-2}. The reference vector for displacements is 0.5 μm long. The grayscale map is in pN μm^{-2}.

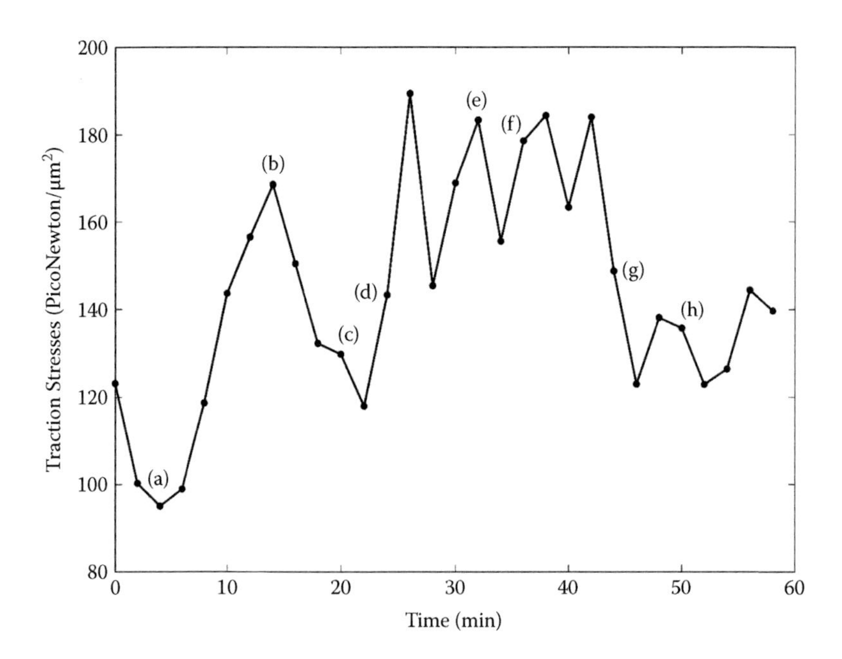

FIGURE 9.8 Evolution of the maximum value of the traction stresses during T24 cell migration on the rigid substrate (E = 10 kPa).

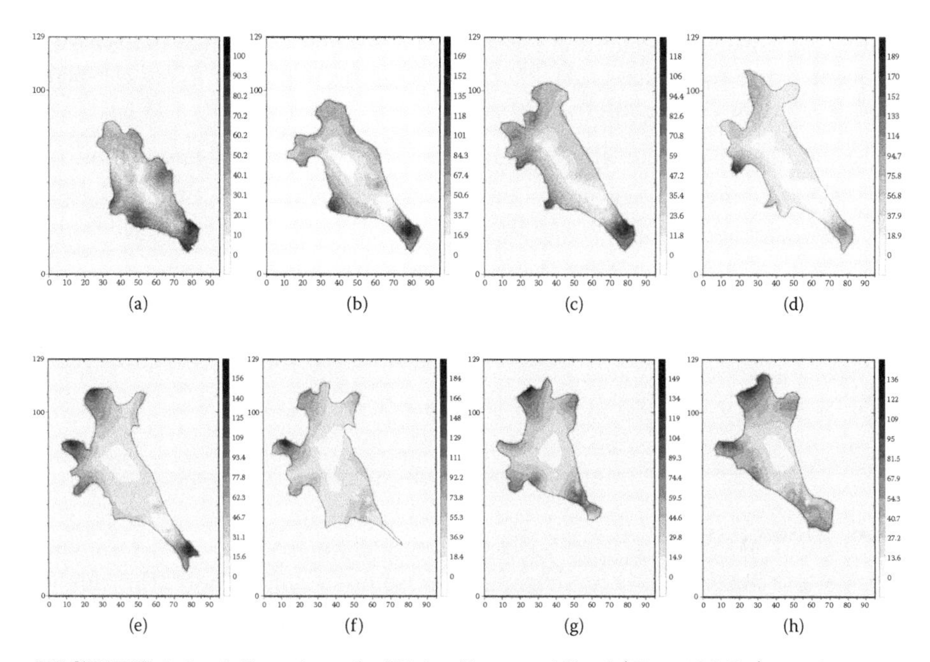

FIGURE 9.9 Migration of a T24 cell on a stiff gel ($E = 10$ kPa), at time $t = 2, 14, 22, 26, 34, 38, 44, 50$ min. The cell first adheres strongly (dark region) at its lower right part (a), then starts to exhibit protrusions in (b) through (e) until it eventually forms new adhesion sites on the upper left of the picture as in (e) and (f). At this precise time, it is able to pull on these adhesion sites, contract and detach its rear (see (f) and (g)); then it starts to move back toward the right direction (h). Note that the grayscale is reset to a range between minimum and maximum values in each frame.

focal adhesions can be formed. After this is achieved, cells can retract their uropods by pulling on focal adhesions; thus they move in a noncontinuous way [34]. This is precisely the significance of Figure 9.8 where an unsteady regime of traction stress is observed changing abruptly from one value to another due to rapid pulling on focal adhesion sites. Clearly such mechanisms involved during migration require tight spatial and temporal regulation; we can study this process more accurately in terms of traction stress maps, as proposed in Figure 9.9 on the stiff gel.

The cell first adheres and binds to the lower right part (a), then it develops protrusions (b) through (e), as explained, with the formation of stable adhesion sites (d) through (f). This corresponds to the dark regions (large stresses). The T24 starts to pull on these new adhesion sites until it moves (f) and (g). The disassembly of adhesions at the rear of the cell and the retraction of the tail complete the migration picture and enable cell translocation.

This mechanism of locomotion for the T24 cancer cell is comparable with the four- or five-step picture [1,34] that describes the motion of many cell types and is summarized as follows:

- Formation of a lamellipodium at the front by actin polymerization
- Development of new focal adhesions coupled with the actin cytoskeleton
- Force traction—cell contraction
- Release of bonds at the rear—actin and protein recycling

In the next section the observed features are discussed and compared with previous works.

9.6 Discussion

The study presented here demonstrates some advantages of the adjoint method [2] when combined with relevant experiments [3]. The major interest is that, instead of solving the CPU-consuming integral equation [10], it directly forces the 2-D averaged problem of partial differential equations to be solved by finite element method. In this respect, the resolution becomes easier and can be performed on a personal computer in a few seconds. The longest part is the data processing of images for the determination of the displacements, whatever the method [7,10,33]. The results strongly depend on the quality of the images. Further refinements have been made recently as compared to our initial work [3]: pictures are now taken at different heights separated by 0.2 μm; stacks of about eight to ten pictures are taken at each time step, and the best focused images are selected. Thus beads are localized more easily and fewer beads are lost during the collection of trajectories. Then a good resolution is obtained, as shown by the squared error in Figure 9.1, which is around 10^{-5}. More precisely, we obtain an error $||\mathbf{u} - \mathbf{u}_0||_2 \sim 3.0\ 10^{-3}$ μm, which is small compared to displacements in the range of 0.05 to 0.5 μm. This clearly indicates a good resolution as compared to experimental data.

Nevertheless, improvements can still be made; in particular, recent work by two groups [25,30] suggests how the resolution can be further enhanced. Merkel et al. [25] use thinner substrates, allowing an analytical resolution of the problem. Sabass et al. [30] illustrate the advantage of a combination of complementary fluorescent microscopy to capture the location and evolution of focal adhesions in relation to traction forces, a method also proposed earlier in [4]. This enables us to show an increased accuracy in all techniques, in all cases (boundary element method [10], Fourier transform traction cytometry [7], traction recovery with point forces [33]). In our case, we limit ourselves to the usual traction stress determination but it is possible to combine fluorescence and observe focal adhesions as well as the actin cytoskeleton

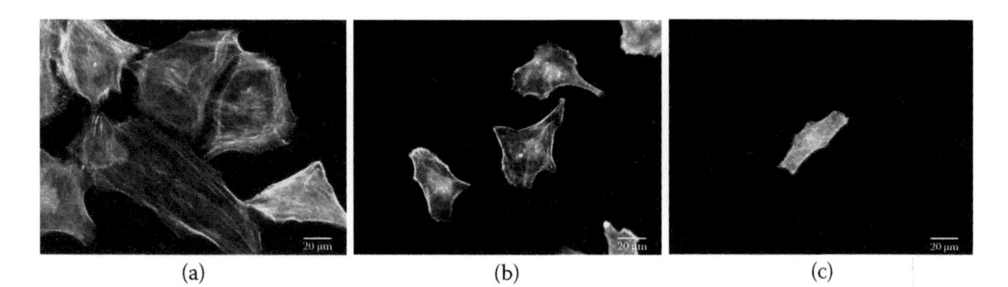

FIGURE 9.10 (See color insert following Page 398.) T24 cells adhering on substrates with various stiffness $E = 10 - 6.3 - 2$ kPa (decreasing from left to right). The actin cytoskeleton has been marked with GFP. Note the formation of stress fibers and a large cell area on the more rigid substrate. Conversely, in the case of the very soft gel, the actin structure is less developed and the cell barely adheres to the substrate, showing a very small contact area. The scale bar represents 20 μm.

development. As an example, Figure 9.10 shows the actin cytoskeleton organized in stress fibers, in the case of T24 cells adhering on substrates with different rigidities (respectively the ones used earlier, $E = 10-6.3-2$ kPa). It is known that stress fibers are colocalized with focal adhesions [1]. In Figure 9.10, most of the stress fibers come close to the cell edges where focal adhesion complexes are usually localized. These photographs clearly indicate that T24 cells form many stress fibers on rigid substrates ($E = 10$ kPa) but fewer on soft ones ($E = 2$ kPa). The intermediate case also shows stress fibers but they are less numerous as compared to the case of the rigid substrate.

To continue the discussion of substrate rigidity further, let us concentrate on the stresses exerted by cells on such substrates, a subject already discussed by other authors [3,11,17,24]. This current work emphasizes the ability of cells to adapt to a different environment. T24 cancer cells also obey this principle; they develop similar strains when adhering to various rigid substrates. Indeed, the level of deformation does not seem to be affected by substrate rigidity in the relevant range, as shown in Figures 9.5, 9.6, and 9.7 where it was mentioned that the displacements are very similar (of the order 0.5 μm). Additionally, the ratio between maximum stress and Young's modulus is almost a constant (strain) on the order 0.02 [3]. This is in agreement with previous work [11,31]. It is likely that, as the substrate rigidity increases, such a (stress-rigidity) linear behavior will fail and there should be a saturation [17] because one expects that cells cannot reinforce their integrin–cytoskeleton links indefinitely [8]. Note that such behaviors can be related to the cell microrheological properties [36], in particular their ability to modulate the growth of stress fibers.

It has been also shown that migration of cancer cells on 2-D rigid substrates follows a rather classical picture divided into four or five steps. Indeed,

Figure 9.8 and Figure 9.9 provide direct evidence of the motion, which can be decomposed into these different steps. First, the cancer cell adheres and is not very active in Figure 9.9(a); thus stresses are low. Then it searches for new adhesion sites until it reaches some places that will allow it to establish larger traction stresses. This is precisely where the traction maxima are obtained in Figures 9.8(e) and (f). Thanks to these forces, the rear (uropod) can be detached. The cell again relaxes its stresses in Figure 9.9(g) until it starts again to try and move in another direction (Figure 9.9(h)). Therefore, a clear connection has been made between the stress diagram in Figure 9.8 and the migration patterns showing stress maps in Figure 9.9.

Finally, our study also shows that cell contact area with the substrate increases with substrate rigidity, in close correspondence with the development of stress fibers (see Figure 9.10). This fact was only observed previously regarding cell area [24]. The velocity of migration also varies in the usual way [3]: cells with more stable focal adhesions on rigid substrates definitely move less rapidly as compared to soft substrates, where they move faster because of the difficulties in developing stable adhesions [11].

The order of magnitude of tractions can finally be discussed. T24 cells seem to develop smaller stresses (typically 0.05 to 0.2 kPa here) in the range of stresses found for human airway smooth muscle (HASM) cells [7] and much lower than the ones usually found with fibroblasts [10] or endothelial cells [29]. This may provide a possible explanation of why cancer cells move rapidly to form metastases and therefore use small traction stresses to migrate faster. Still, this hypothesis needs to be confirmed in more realistic situations such as *in vitro* 3-D migration.

9.7 Conclusions

The study developed here has revealed some interesting features of cancer cell migration on rigid substrates. It is important to note that although many studies have already focused on the determination of traction forces, the current method is very efficient and shows interesting time-dependent features. This method is promising and should allow us to investigate new problems related to cancer cells. The ability to distinguish cell types is of course one perspective. Second, attempts should be made to differentiate invasive from noninvasive cancer cells [26]. This could lead to early diagnosis. Finally, substrate specificity may also be used to compare cell migration characteristics such as their velocity or exerted traction stresses. But the most promising follow-up of this work will probably rely on the ability to correlate precisely the cytoskeleton and focal adhesion dynamics with the development of forces exerted by cancer cells.

References

[1] B. Alberts et al. (1994). *Molecular Biology of the Cell, Third Edition*, New York, Garland.

[2] D. Ambrosi (2006). Cellular traction as an inverse problem. *SIAM J. Appl. Math.* 66:2049–2060.

[3] D. Ambrosi, A. Duperray, V. Peschetola, and C. Verdier (2009). Traction patterns of tumor cells. *J. Math. Biol.* 58:163–181.

[4] N.Q. Balaban, U.S. Schwarz, D. Riveline, et al. (2001). Force and focal adhesion assembly: a close relationship studied using elastic micropatterned substrates. *Nat. Cell Biol.* 3:466–472.

[5] T. Boudou, J. Ohayon, C. Picart, and P. Tracqui (2006). An extended relationship for the characterization of Young's modulus and Poisson's ratio of tunable polyacrylamide gels. *Biorheology* 43:721–728.

[6] K. Burton, J.H. Park, and D.L. Taylor (1999). Keratocytes generate traction forces in two phases. *Mol. Biol. Cell* 10:3745–3769.

[7] J.P. Butler, I.M. Tolic-Norrelykke, B. Fabry, and J.J. Fredberg (2002). Traction fields, moments, and strain energy that cells exert on their surroundings. *Am. J. Physiol. Cell Physiol.* 282:C595–C605.

[8] D. Choquet, D.P. Felsenfeld, and M.P. Sheetz (1997). Extracellular matrix rigidity causes strengthening of integrin-cytoskeleton linkages. *Cell* 88:39–48.

[9] M. Dembo, T. Oliver, A. Ishihara, and K. Jacobson (1996). Imaging the traction stresses exerted by locomoting cells with elastic substratum method. *Biophys. J.* 70:2008–2022.

[10] M. Dembo and Y.L. Wang (1999). Stresses at the cell-to-substrate interface during locomotion of fibroblasts. *Biophys. J.* 76:2307–2316.

[11] D.E. Discher, P. Janmey, and Y. Wang (2005). Tissue cells feel and respond to the stiffness of their substrate. *Science* 310:1139–1143.

[12] O. du Roure, A. Saez, A. Buguin, et al. (2005). Force mapping in epithelial cell migration. *Proc. Natl. Acad. Sci. USA* 102:2390–2395.

[13] A. Engler, L. Bacakova, C. Newman, A. Hategan, M. Griffin, and D. Discher (2004). Substrate compliance versus ligand density in cell on gel responses. *Biophys. J.* 86:617–628.

[14] S. Féréol, R. Fodil, B. Labat, et al. (2006). Sensitivity of alveolar macrophages to substrate mechanical and adhesive properties. *Cell Motil. Cytoskeleton* 63:321–340.

[15] P. Friedl, S. Borgmann, and E.B Bröcker (2001). Amoeboid leukocyte crawling through extracellular matrix: lessons from the Dictyostelium paradigm of cell movement. *J. Leukoc. Biol.* 70:491–509.

[16] P. Friedl and K. Wolf (2003). Tumour-cell invasion and migration: diversity and escape mechanisms. *Nature* 3:362–374.

[17] M. Ghibaudo, A. Saez, L. Trichet, et al. (2008). Traction forces and rigidity sensing regulate cell functions. *Soft Matter* 4:1836–1843.

[18] A.K. Harris, P. Wild, and D. Stopak (1980). Silicone rubber substrata: a new wrinkle in the study of cell locomotion. *Science* 208:177–179.

[19] ImageJ, by Rasband, W.S. 1997–2000. Bethesda, Maryland, USA, U.S. National Institutes of Health. http://rsb.info.nih.gov/ij/.

[20] P.A. Janmey, U. Euteneuer, P. Traub, and M. Schliwa (1991). Viscoelastic properties of vimentin compared with other filamentous biopolymer networks. *J. Cell Biol.* 113:155–160.

[21] L. Landau and E. Lisfchitz (1967). *Théorie de l'Élasticité*, Moscow, Editions Mir.

[22] J. Lee, M. Leonard, T. Oliver, A. Ishihara, and K. Jacobson (1994). Traction forces generated by locomoting keratocytes. *J. Cell Biol.* 127:1957–1964.

[23] J.L. Lions (1968). *Contrôle Optimal de Systèmes Gouvernés par des Équations aux Dérivées Partielles*, Paris, Dunod et Gauthier-Villard.

[24] C.M. Lo, H.B. Wang, M. Dembo, and Y.L. Wang (2000). Cell movement is guided by the rigidity of the substrate. *Biophys. J.* 79:144–152.

[25] R. Merkel, N. Kirchgessner, C.M. Cesa, and B. Hoffmann (2007). Cell force microscopy on elastic layers of finite thickness. *Biophys. J.* 93:3314–3323.

[26] C.T. Mierke, D.P. Zitterbart, B. Kollmannsberger, et al. (2008). Breakdown of the endothelial barrier function in tumor cell transmigration. *Biophys. J.* 94:2832–2846.

[27] S.P. Palecek, J.C. Loftus, M.H. Ginsberg, D.A. Lauffenburger, and A.F. Horwitz (1997). Integrin-ligand binding properties govern cell migration speed through cell-substratum adhesiveness. *Nature* 385:537–540.

[28] R.J. Pelham and Y. Wang (1997). Cell locomotion and focal adhesions are regulated by substrate flexibility. *Proc. Natl. Acad. Sci. USA* 94:13661–13665.

[29] C.A. Reinhart-King, M. Dembo, and D.A. Hammer (2005). The dynamics and mechanics of endothelial cell spreading. *Biophys. J.* 89:676–689.

[30] B. Sabass, M.L. Gardel, C.M. Waterman, and U.S. Schwarz (2008). High resolution traction force microscopy based on experimental and computational advances. *Biophys. J.* 94:207–220.

[31] A. Saez, A. Buguin, P. Silberzan, and B. Ladoux (2005). Is the mechanical activity of epithelial cells controlled by deformations or forces? *Biophys. J.* 89:L52–L54.

[32] I.F. Sbalzarini and P. Koumoutsakos (2005). Feature point tracking and trajectory analysis for video imaging in cell biology. *Struct. Bio. J.* 151:182–195.

[33] U.S. Schwarz, N.Q. Balaban, D. Riveline, A. Bershadsky, B. Geiger, and S.A. Safran (2002). Calculation of forces at focal adhesions from elastic substrate data: the effect of localized force and the need for regularization. *Biophys. J.* 83:1380–1394.

[34] M.P. Sheetz, D. Felsenfeld, C.G. Galbraith, and D. Choquet (1999). Cell migration as a five-step cycle. *Biochem. Soc. Symp.* 65:233–243.

[35] J.L. Tan, J. Tien, D.M. Pirone, D.S. Gray, K. Bhadriraju, and C.S. Chen (2003). Cells lying on a bed of microneedles: an approach to isolate mechanical force. *Proc. Natl. Acad. Sci. USA* 100:1484–1489.

[36] C. Verdier (2003). Rheological properties of living materials: from cells to tissues. *J. Theor. Med.* 5:67–91.

[37] T. Yeung, P.C. Georges, L.A. Flanagan, et al. (2005). Effects of substrate stiffness on cell morphology, cytoskeletal structure, and adhesion. *Cell Motil. Cytoskeleton* 60:24–34.

Chapter 10

Single-Cell Imaging of Calcium in Response to Mechanical Stimulation

Tae-Jin Kim and Yingxiao Wang

Contents

10.1 Introduction

Calcium ions (Ca^{2+}) play an important role in the regulation of many aspects of cellular activities, such as muscle contraction, embryogenesis, cell differentiation, proliferation, gene expression, secretion, learning and memory, and apoptosis [3,6,14,23,26,29,33,38,46,49,56]. Recent evidence indicates that mechanical stimulation plays an important role in regulating various cellular functions, including Ca^{2+}signaling [17,21,22,50]. For example, mechanical factors such as substrate stiffness play important roles in determining the differentiation lineage and commitment of human mesenchymal stem cells [9]. Substrate stiffness also regulates Ca^{2+} oscillatory signals via the RhoA pathway in human mesenchymal stem cells [22]. Furthermore, capacity Ca^{2+} entry channels and mechanosensitive channels can be activated by mechanical stimulation, which results in an increase in intracellular calcium concentration and consequently the alteration of a variety of cellular activities [17,47,58].

The application of fluorescent dyes for imaging of intracellular calcium dynamics and the improvement in fluorescence microscopy systems have led to a revolution in our understanding of the ubiquitous roles of Ca^{2+} in the physiological functions of cells over the past two decades. The introduction of fluorescent proteins (FPs) with a variety of spectra and the advancement of genetically encoded calcium indicators have made possible the visualization of calcium signaling in a living single cell at subcellular levels with high spatiotemporal resolutions. These calcium imaging techniques in combination with modern technologies that are able to create diverse mechanical environments have allowed us to explore calcium signaling of a single cell in response to mechanical stimulation. In this chapter we introduce design strategies on how to approach the single cell imaging of calcium in response to mechanical stimulation. In particular, we provide a detailed discussion of the integration of genetically encoded calcium FRET biosensors and stiffness matrix with different elasticity and optical laser tweezers.

10.2 Ca^{2+} Signaling

10.2.1 Fundamental Mechanism of Ca^{2+} Signaling

Most cells maintain their cytosolic Ca^{2+} concentration at a low level (approximately 100 nM) by major players such as the Na^+/Ca^{2+} exchanger (NCX), the plasma-membrane Ca^{2+}-ATPase (PMCA), and the sarco(endo)plasmic reticulum Ca^{2+}-ATPase (SERCA) for calcium homeostasis [4,5]. When cells are stimulated, the intracellular Ca^{2+} concentration ($[Ca^{2+}]i$) rapidly increases and reaches the micromolar range, which subsequently impacts many different cellular processes. In general, Ca^{2+} entry from the extracellular pool is mediated by three main entry channels: (1) voltage-operated channels (VOCs), (2) receptor-operated channels (ROCs), and (3) store-operated channels (SOCs) [20,24,25]. Ca^{2+} release from internal stores is mediated by two families of Ca^{2+} channels—ryanodine receptors (RyRs) and 1,4,5-trisphophate receptors (IP3Rs)—each of which has three major isoforms identified [16,28,30,42,55]. Restoration of Ca^{2+} homeostasis is achieved primarily by pumping across the plasma membrane by NCX and PMCA, and uptaking into the endoplasmic reticulum (ER) and sarcoplasmic reticulum (SR) by SERCA [37,48].

10.2.2 Ca^{2+} Entry Mechanisms

Ca^{2+} entry is driven primarily by a large electrochemical gradient across the plasma membrane. Cells utilize the external Ca^{2+} by activating various Ca^{2+} entry channels that have very different kinetic properties. Voltage-operated Ca^{2+} channels (VOCCs) have been studied extensively in excitable cells in brain, skeletal, cardiac and smooth muscle endocrine glands, and other tissues. Fundamentally, VOCCs respond to changes in membrane potential through a voltage-sensor domain integral to the channel pore forming protein, which results in a conformation change that opens the channels to allow for the influx of extracellular Ca^{2+}. VOCCs can be activated very rapidly in response to depolarization and generate rapid Ca^{2+} increases in the cytoplasm to regulate fast cellular processes such as muscle contraction, excitability, and exocytosis. ROCs open in response to the binding of extracellular ligands. For instance, NMDA (N-methyl-D-aspartate) receptors (NMDARs) are NMDA-gated ion channels that are permeable to Ca^{2+}. When glutamate binds to the NMDARs, $Ca^{2+}$and Na^{+} enter the cell and K^{+} leaves through NMDA-gated channels [5]. Belonging to ATP receptors and Ach receptors, P_2X_7 and nicotinic acetylcholine (nACh) are also permeable to Ca^{2+} upon ligand binding [13,36]. SOCs are regulated by filling or depleting internal stores through a mechanism known as capacitative calcium entry, later called store-operated Ca^{2+} entry (SOCE) [24,44,45]. When the internal Ca^{2+} stores are empty, Ca^{2+} channels are activated in the plasma membrane to aid in refilling the Ca^{2+} store. Most SOC channels appear to belong to the transient receptor protein (TRP) channels associated with mechanosensitive channels, such as thermosensors and stretch-activated channels. These channels have been largely classified into three groups—canonical TRPC, vanilloid TRPV, and melastatin TRPM—which are all involved in mechanical transduction. In particular, TRP channels have low conductance and play a crucial role in controlling slow cellular processes such as smooth-muscle contractility and cell proliferation. To date, the detailed mechanism of SOCE remains unclear because the identity of the entry channels is not clear. However, the discovery of stromal interaction molecules (STIM) sheds new light on our understanding of the molecular mechanism of SOCE. This STIM1 is localized mainly in the ER membrane. STIM1 senses ER Ca^{2+} to activate Orai1, a pore-forming subunit of the Ca^{2+} release-activated Ca^{2+} channel (CRCA) in the plasma membrane for SOCE [11]. Thus, STIM1 can activate SOCE and help refill the ER Ca^{2+} store.

10.2.3 Ca^{2+} Release from Internal Stores

There are two main channels responsible for releasing Ca^{2+} from the internal stores: (1) IP3 receptors (IP3Rs) and (2) ryanodine receptors (RyRs). Activation of phospholipase $C_\beta(PLC_\beta)$ by G protein coupled receptors (GPCRs) and PLC_γ by receptor tyrosine kinases can induce cleavage of phosphatidylinositol

4, 5 bisphosphate (PIP2) into 1,4,5-inositol trisphosphate (IP3) and diacylglycerol (DAG). IP3 can bind to the IP3 receptors located in ER membrane to trigger Ca^{2+} from ER stores. The RyRs are also intracellular calcium channels located primarily in the ER/SR of various excitable cells, particularly muscle and neuronal cells. RyRs are similar to IP3Rs, which can be stimulated to transport Ca^{2+} into the cytosol by recognizing the low level of Ca^{2+} on its cytosolic side. Both channels are very sensitive to calcium. Therefore, it appears that the process of calcium-induced calcium release (CICR) contributes to the increase in intracellular calcium concentration, and as such CICR contributes to the generation of calcium spikes and calcium waves.

10.3 Genetically Encoded FRET-Based Ca^{2+} Biosensor

Fluorescence resonance energy transfer (FRET)-based Ca^{2+} biosensors for single cell imaging have allowed researchers to quantitatively monitor Ca^{2+} dynamics and their signaling cascades in live cells with high spatiotemporal resolution. FRET is a phenomenon of quantum mechanics and describes an energy transfer mechanism between two chromophores. When one chromophore (donor) and another (acceptor) are in proximity and the emission spectrum of the donor overlaps the excitation spectrum of the acceptor, the excitation of the donor will cause a sufficient energy transfer to the acceptor and result in emission from the acceptor. This FRET efficiency depends on the relative orientations and the distance between these two chromophores [52,54]. Hence, the conformational change in orientation and distance between the two chromophores can alter the FRET efficiency and change the acceptor/donor emission ratio. A wide range of FRET-based Ca^{2+} biosensors has been developed to visualize calcium dynamics and activities [41]. In general, there are three different categories, classified according to properties such as binding moiety and strategy for Ca^{2+} sensing:

1. Calmodulin/FRET-based Ca^{2+} biosensors
2. Troponin C/FRET-based Ca^{2+} biosensors
3. Single fluorophore biosensors

These genetically encoded Ca^{2+} biosensors have been applied for targeting various organelles such as the nucleus, endoplasmic reticulum (ER), mitochondria, Golgi, and plasma membrane in living single cells in an effort to visualize dynamic Ca^{2+} signals [12,15,18,19,27,32,34,40,41]. Here we briefly describe the calmodulin/FRET based Ca^{2+} biosensor, which is one of the most popular and commonly used among the various kinds of FRET-based Ca^{2+} indicators. Calmodulin (CaM) is a calcium binding protein and is ubiquitously expressed in all eukaryotic cells. As a monomer with an approximate molecular weight of

17,000, CaM is located primarily in the cytosol, but can be translocated to the nucleus where it regulates transcription and gene expression. CaM mediates numerous fundamental cellular processes, such as inflammation, apoptosis, cell cycle progression, metabolism, and calcium transport. Structurally, CaM has four EF-hand motifs, each of which binds a Ca^{2+}ion and undergoes a conformational change. These CaMs can be applied to the development of calcium sensors. In fact, a variant of CaM and a CaM-binding peptide connected by two FPs was developed as a FRET-based Ca^{2+} biosensor by Tsien and co-workers; this was the first generation of genetically encoded calcium sensors called "Cameleons" [32]. The sensor originally consisted of blue and green FPs (BFP as the donor and GFP as the acceptor), flanking *Xenopus* calmodulin (XcaM) and peptide M13. Upon Ca^{2+} binding, Ca^{2+}/calmodulin wraps around the neighboring M13 peptide, thus causing a conformational change and increasing the energy transfer efficiency from BFP to GFP [32]. To overcome the weak fluorescence brightness and relatively poor photobleaching property of BFP, this BFP-GFP FRET pair was switched to cyan FP (CFP) and yellow FP (YFP) in Yellow Cameleon 2.0 (YC 2.0), which provided better signal-to-noise ratios and more stable FRET signals in live cells. Further efforts for making better Cameleons were carried out afterward, resulting in the improvement of signal strength of the biosensors and a reduction in perturbation to endogenous cell signaling cascades. At present, various improved versions of Cameleons have been developed, ranging from YC2.0 to YC6.1 [15,31,32,34,51].

10.4 Single Cell Imaging of Calcium in Response to Mechanical Stiffness

To date, it is clear that all organisms from bacteria to mammals are mechanically sensitive. They possess highly specialized and widely expressed mechanosensitive systems that are involved in a wide range of physiological functions. Recently, emerging evidence indicates that mechanical stimulation such as fluid shear stress and mechanical strain regulate the proliferation and differentiation of stem cells by means of various signaling pathways. Interestingly, it appears that many cells sense and respond to the dynamic and static characteristics of mechanical stimulation, including the elasticity of the extracellular microenvironment. In fact, there is a wide range of tissues in the human body with varying degrees of stiffness, ranging from soft tissues like the brain with a Young's modulus of tenths of a kilopascal (kPa) to hard tissues like bone with a Young's modulus of hundreds of kilopascals.

10.4.1 Cell Culture Protocol on Substrate Stiffness Gel

At present, polyacrylamide gels with varying acrylamide and bis-crosslinker concentrations can be applied to control the substrate rigidity of the extracellular environment with defined elastic moduli to mimic the mechanical properties of *in vivo* systems [43]. In our experiment, 200 μL of a 0.1 N NaOH solution was dropped on the surface of a glass cover slip and left overnight to air dry. 3-Aminopropyltrimethoxysilane (3-APTMS) was then smeared over the surface for 6 min. After washing twice with water for 15 min, 100 μl of 0.5% gluteraldehyde was applied to the glass surface for 30 min. The glass cover slip was then rinsed with distilled water. Polyacrylamide gel solutions were then prepared with 40% w/v acrylamide stock solution (5%) and 2% w/v bis-acrylamide stock solution (0.03–0.3%). 10% w/v ammonium persulfate (APS), and N,N,N9,N9-tetramethylethylenediamine (TEMED) were added to catalyze the polymerization of the solutions. After rinsing with 100 nM HEPES, sulfo-SANPAH was used to crosslink the extracellular matrix proteins onto the gel surface and 200 μL of freshly made SANPAH solution (1 mM solution of SANPAH with DMSO in 100 mM HEPES) was applied to each surface before it was exposed to UV photoactivation for 6 min. After rinsing off the SANPAH solutions, the photoactivation procedure was repeated one more time. The gels were then washed for 3 min with 100 mM HEPES and coated with fibronectin (FN). An elastic gel with controlled stiffness and FN-coated surface was developed and ready for use. Human mesenchymal stem cells (HMSCs) were ultimately cultured on top of these coated gels in human mesenchymal stem cell growth medium (MSCGM) containing 10% fetal bovine serum, 2 mM L-glutamine, 100 U/mL penicillin, and 100 μg/mL streptomycin in a humidified incubator of 95% O_2 and 5% CO_2 at 37°C.

10.4.2 Ca^{2+} Imaging Utilizing FRET-Based Ca^{2+} Biosensor in Response to Mechanical Stiffness

A FRET biosensor for reporting Ca^{2+} dynamics in live cells (Cameleon) was initially developed by Atsushi Miyawaki in Roger Y. Tsien's lab [32]. This Ca^{2+} biosensor has been modified into new versions, improving the properties of the biosensor in terms of its sensitivity and other factors. We have also developed a calcium biosensor, pairing an enhanced cyan FP (ECFP) with a newly developed yellow FP (Ypet) [35]. This Ca^{2+} biosensor provides a high dynamic range in monitoring intracellular Ca^{2+} concentrations [39]. The application of this FRET-based Ca^{2+} biosensor in HMSCs has allowed us to successfully monitor the spontaneous Ca^{2+} oscillation at subcellular locations inside HMSCs, as shown in our previous work [22] and Figure 10.1. We were also able to monitor ER calcium signaling using an ER-targeted calcium biosensor that has an ER retention sequence to anchor the biosensor inside the lumen of ER (Figure 10.1D) [22,40].

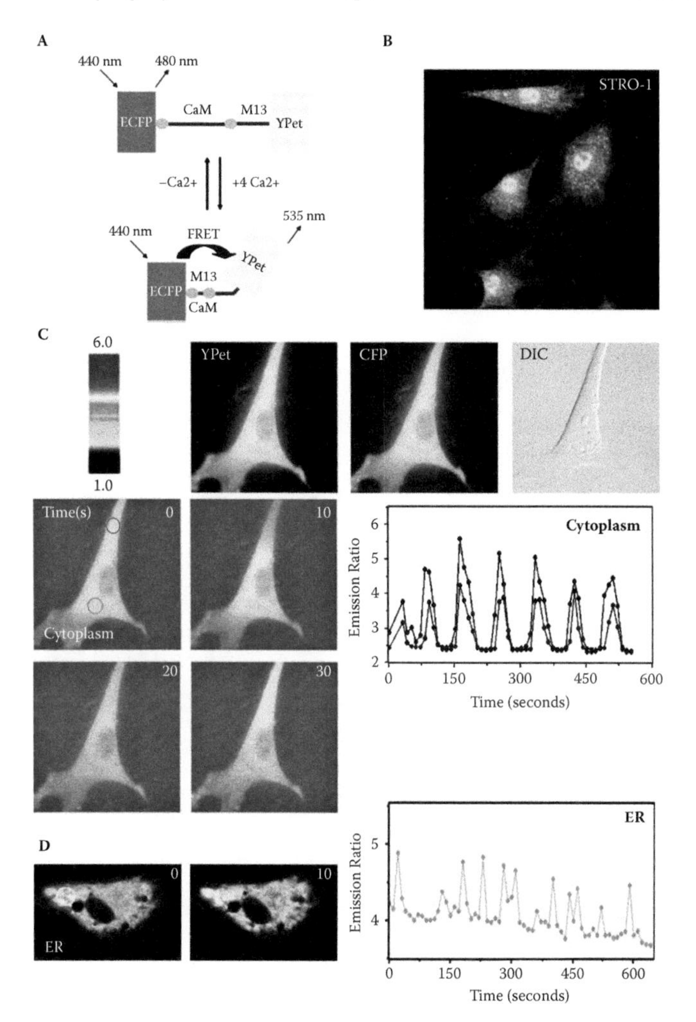

FIGURE 10.1 (See color insert following Page 398.) The application of FRET-based Ca^{2+} biosensors in human mesenchymal stem cells (HMSCs). A: A schematic drawing of the activation mechanism of the Ca^{2+} FRET biosensor. B: HMSCs were immunostained by monoclonal antibody against STRO-1, a MSCs marker. C: The FRET change of the Ca^{2+} biosensor targeted at cytoplasm. HMSCs were transfected with the cytoplasmic Ca^{2+} biosensor, which visualizes a spontaneous Ca^{2+} oscillation. Color images represent the YPet/ECFP emission ratio of the cytoplasmic Ca^{2+} biosensor. The color scale bar represents the YPet/ECFP emission ratio, with cold and hot colors indicating low and high levels of Ca^{2+} concentration, respectively. The time course curves represent the YPet/ECFP emission ratio averaged over regions on (blue) and distal (red) to the triggering corner. D: The color images and time course curves represent the oscillatory FRET changes of the Ca^{2+} biosensor targeted inside ER. (From Kim, T.J., *J Cell Physiol.*, 218, 287, 2009. With permission.)

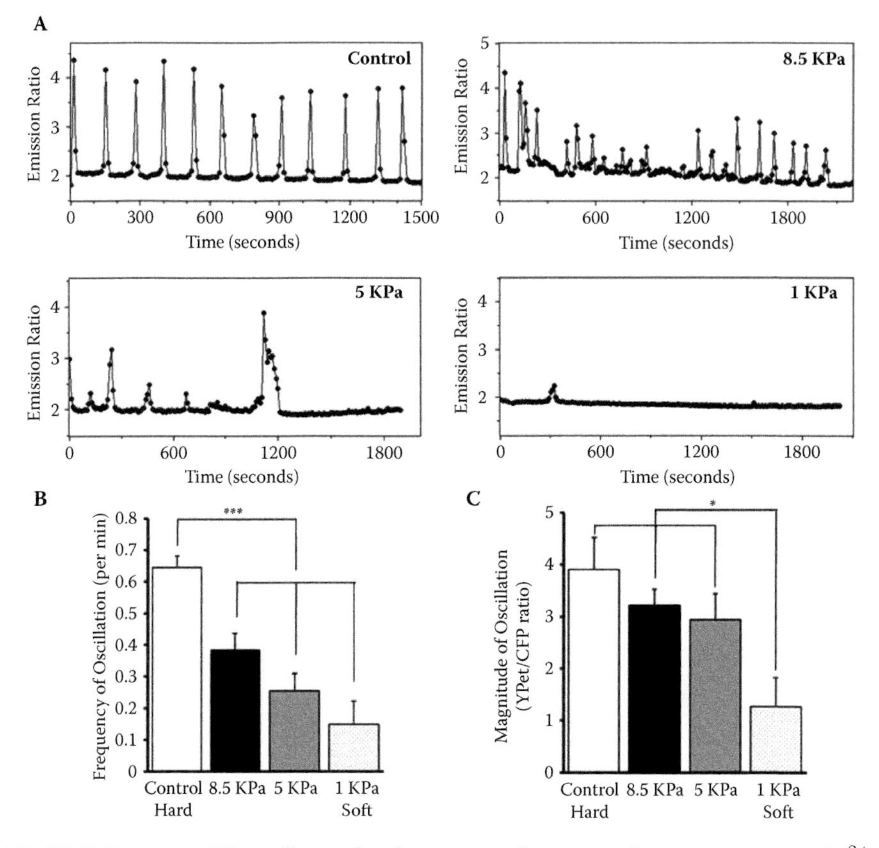

FIGURE 10.2 The effect of substrate stiffness on the spontaneous Ca^{2+} oscillation. A: The time courses represent the cytoplasmic Ca^{2+} concentrations in cells cultured on gels with different rigidity, as indicated. Bar graphs (mean $\pm$ SEM) represent the (B) frequency and (C) magnitude of spontaneous Ca^{2+} oscillations in cells cultured on gels with different rigidity, as indicated. Error bars indicate standard errors of mean; $*P < 0.05$; $***P < 0.001$. (From Kim, T.J. et al., *J. Cell Physiol.*, 218, 289, 2009. With permission.)

10.4.3 Regulation of Ca^{2+} Signaling by Substrate Rigidity of Extracellular Environment

Because both mechanical stiffness and Ca^{2+} oscillatory signals play a crucial role in regulating the stem cell commitment for differentiation [7–9], we investigated whether mechanical stiffness can affect spontaneous Ca^{2+} oscillation in HMSCs. As shown in Figure 10.2, gels with lower elasticity reduced the spontaneous Ca^{2+} oscillation activity in HMSCs [22].

Interestingly, we found that a 1-kPa gel inhibited both the frequency and magnitude of Ca^{2+} oscillation whereas 5-kPa and 8.5-kPa gels only affected the frequency. Therefore, these results demonstrate that the substrate rigidity of the extracellular environment can affect Ca^{2+} signaling in HMSCs, with the frequency of Ca^{2+} oscillation possibly more sensitive to mechanical stiffness than the magnitude [22]. It remains unclear as to the detailed molecular mechanism by which the substrate rigidity affects Ca^{2+} oscillation. The substrate rigidity of the extracellular environment was shown to affect cell morphology, cytoskeletal structure, and cell adhesion [57]. Another study postulated that substrates with different rigidities may regulate cellular functions by differentially altering signaling molecules and cytoskeletal structures involved in cell adhesion and force sensing [10]. Thus, we have further examined whether cytoskeletal elements and intracellular tension mediate the oscillatory Ca^{2+} signals in sensing the substrate rigidity of the extracellular environment [22]. As shown in Figure 10.3, the spontaneous Ca^{2+} oscillation of HMSCs was found to be independent of cytoskeleton and MLCK/myosin.

However, RhoA and its downstream molecule, Rho-associated kinase (ROCK), were demonstrated to be involved in the regulation of Ca^{2+} oscillation of HMSCs in response to substrate rigidity. Moreover, we found that RhoA is not sufficient to restore the Ca^{2+} oscillation modulated by substrate rigidity [22]. This suggests that other factors, independent of RhoA, possibly participate in regulating the Ca^{2+} oscillation in response to mechanical stiffness.

10.5 Ca^{2+} Imaging and Mechanobiology in Live Cells with Optical Laser Tweezers

Since Ashkin trapped a bead with a focused laser [1], optical laser tweezers have become a powerful tool for force measurement and mechanical stimulation in mechanobiology. Optical laser tweezers are based on the transfer of mechanical momentum change from the trapping beam to the particle. Using this method, beads, spherical particles, or microspheres can be trapped at the focal plane of a high numerical aperture (NA) objective lens within an inverted microscope. When a bead is mechanically coupled to a cell surface, the forces generated in the bead by laser tweezers can be adjusted and calibrated, and transmitted into the cell. Recently, FRET technology has been used in combination with optical laser tweezers to study mechanotransduction [53]. In these experiments, beads were coated with fibronectin (FN). Because of adhesion between FN and cell membrane receptors, integrins can induce the formation of adhesion complexes to mechanically couple the bead to the cytoskeleton. The FN–integrin–cytoskeleton coupling can transmit the laser-induced force from the bead into the cell. When the FN beads are bound to integrins on the plasma membrane of human umbilical vein endothelial cells (HUVECs) pre-transfected with a FRET-based Src biosensor, a series of

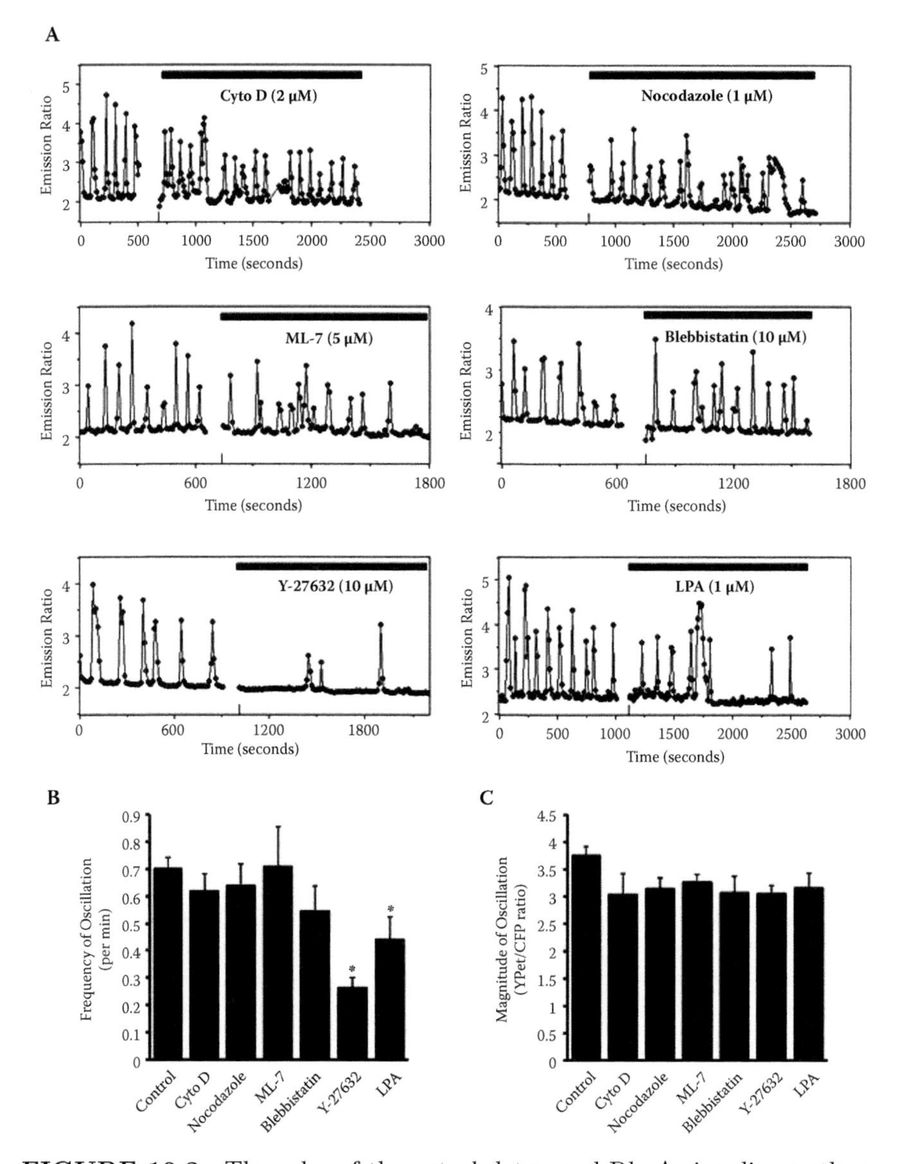

FIGURE 10.3 The roles of the cytoskeleton and RhoA signaling pathway in spontaneous cytoplasmic Ca^{2+} oscillation. A: Representative time courses of the cytoplasmic Ca^{2+} concentration in cells pre-treated with Cyto D, Noc, ML-7, blebbistatin, Y-27632, and LPA. Bar graphs (mean ± SEM) represent the (B) frequency and (C) magnitude of spontaneous Ca^{2+} oscillations in cells pretreated with different reagents, as indicated. Error bars indicate standard errors of mean; $^{*}P < 0.05$. (From Kim, T.J. et al., *J. Cell Physiol.*, 218, 290, 2009. With permission.)

signal transduction events at the cell membrane can be observed in response to mechanical force in the form of wave propagations of the FRET signals [53]. Recently, it has also been shown that the mechanical force induced by pulling FN-coated beads on the membrane of vascular smooth muscle cells can trigger Ca^{2+} sparks, followed by global Ca^{2+} mobilization [2]. More recently, it has been reported that direct mechanical stimulation of an actin stress fiber using optical laser tweezers can activate mechanosenstive channels and subsequently cause a local intracellular calcium increase near focal adhesions in HUVECs [17]. However, when different beads that are not linked to cytoskeleton were used, mechanical force did not induce calcium signaling events. These results indicate that actin filaments and cytoskeleton play important roles in transducing mechanical force into intracellular Ca^{2+} signaling events.

10.6 Future Directions

The combination of genetically encoded FRET Ca^{2+} biosensors with various technologies, including elastic gels with different rigidities and optical laser tweezers capable of applying mechanical stimulation, has proven powerful for the study of molecular mechanisms by which cells perceive and transduce mechanical cues into biochemical signals, that is, mechanobiology. The development of and improvement in new FRET-based Ca^{2+} biosensors targeted at subcellular compartments of individual cells (such as nucleus, endoplasmic reticulum, mitochondria, Golgi complex, and the plasma membrane) are undergoing intensive investigation. Hence, it is envisioned that different kinds of sensitive Ca^{2+} biosensors localized at any subcellular compartment will be available soon such that Ca^{2+} signaling in response to mechanical stimulation can be monitored at global levels and local sites in live cells with high spatiotemporal resolution.

References

[1] A. Ashkin, J.M. Dziedzic, and T. Yamane (1987). Optical trapping and manipulation of single cells using infrared laser beams. *Nature* 330:769–771.

[2] L. Balasubramanian, A. Ahmed, C.M. Lo, J.S. Sham, and K.P. Yip (2007). Integrin-mediated mechanotransduction in renal vascular smooth muscle cells: activation of calcium sparks. *Am. J. Physiol. Regul. Integr. Comp. Physiol.* 293:R1586–R1594.

[3] M.J. Berridge (1995). Calcium signalling and cell proliferation. *Bioessays* 17:491–500.

[4] M.J. Berridge (1997). Elementary and global aspects of calcium signalling. *J. Exp. Biol.* 200:315–319.

[5] M.J. Berridge, M.D. Bootman, and H.L. Roderick (2003). Calcium signalling: dynamics, homeostasis and remodelling. *Nat. Rev. Mol. Cell Biol.* 4:517–529.

[6] D.E. Clapham (2007). Calcium signaling. *Cell* 131:1047–1058.

[7] E. den Dekker, D.G. Molin, G. Breikers, R. van Oerle, J.W. Akkerman, G.J. van Eys, and J.W. Heemskerk (2001). Expression of transient receptor potential mRNA isoforms and Ca(2+) influx in differentiating human stem cells and platelets. *Biochim. Biophys. Acta* 1539:243–255.

[8] S.J. D'Souza, A. Pajak, K. Balazsi, and L. Dagnino (2001). Ca^{2+} and BMP-6 signaling regulate E2F during epidermal keratinocyte differentiation. *J. Biol. Chem.* 276:23531–23538.

[9] A.J. Engler, S. Sen, H.L. Sweeney, and D.E. Discher (2006). Matrix elasticity directs stem cell lineage specification. *Cell* 126:677–689.

[10] A.J. Engler, H.L. Sweeney, D.E. Discher, and J.E. Schwarzbauer (2007). Extracellular matrix elasticity directs stem cell differentiation. *J. Musculoskelet. Neuronal Interact.* 7:335.

[11] S. Feske, Y. Gwack, M. Prakriya, S. Srikanth, S.H. Puppel, B. Tanasa, P.G. Hogan, R.S. Lewis, M. Daly, and A. Rao (2006). A mutation in Orai1 causes immune deficiency by abrogating CRAC channel function. *Nature* 441:179–185.

[12] L. Filippin, P.J. Magalhaes, G. Di Benedetto, M. Colella, and T. Pozzan (2003). Stable interactions between mitochondria and endoplasmic reticulum allow rapid accumulation of calcium in a subpopulation of mitochondria. *J. Biol. Chem.* 278:39224–39234.

[13] S. Fucile (2004). Ca^{2+} permeability of nicotinic acetylcholine receptors. *Cell Calcium* 35:1–8.

[14] M. Gomez, E. De Castro, E. Guarin, H. Sasakura, A. Kuhara, I. Mori, T. Bartfai, C.I. Bargmann, and P. Nef (2001). Ca^{2+} signaling via the neuronal calcium sensor-1 regulates associative learning and memory in *C. elegans*. *Neuron* 30:241–248.

[15] O. Griesbeck, G.S. Baird, R.E. Campbell, D.A. Zacharias, and R.Y. Tsien (2001). Reducing the environmental sensitivity of yellow fluorescent protein. Mechanism and applications. *J. Biol. Chem.* 276:29188–29194.

[16] S.L. Hamilton (2005). Ryanodine receptors. *Cell Calcium* 38:253–260.

[17] K. Hayakawa, H. Tatsumi, and M. Sokabe (2008). Actin stress fibers transmit and focus force to activate mechanosensitive channels. *J. Cell Sci.* 121:496–503.

[18] N. Heim and O. Griesbeck (2004). Genetically encoded indicators of cellular calcium dynamics based on troponin C and green fluorescent protein. *J. Biol. Chem.* 279:14280–14286.

[19] K. Ishii, K. Hirose, and M. Iino (2006). Ca^{2+} shuttling between endoplasmic reticulum and mitochondria underlying Ca^{2+} oscillations. *EMBO Rep.* 7:390–396.

[20] S. Jagannathan, S.J. Publicover, and C.L. Barratt (2002). Voltage-operated calcium channels in male germ cells. *Reproduction* 123:203–215.

[21] G.S. Kassab (2006). Biomechanics of the cardiovascular system: the aorta as an illustratory example. *J. Roy. Soc. Interface* 3:719–740.

[22] T.J. Kim, J. Seong, M. Ouyang, J. Sun, S. Lu, J.P. Hong, N. Wang, and Y. Wang (2009). Substrate rigidity regulates Ca^{2+} oscillation via RhoA pathway in stem cells. *J. Cell Physiol.* 218:285–293.

[23] J.W. Landsberg and J.X. Yuan (2004). Calcium and TRP channels in pulmonary vascular smooth muscle cell proliferation. *News Physiol. Sci.* 19:44–50.

[24] R.S. Lewis (2007). The molecular choreography of a store-operated calcium channel. *Nature* 446:284–287.

[25] S.W. Li, J. Westwick, and C.T. Poll (2002). Receptor-operated Ca^{2+} influx channels in leukocytes: a therapeutic target? *Trends Pharmacol. Sci.* 23:63–70.

[26] Y. Maeda, Y. Nitanai, and T. Oda (2007). From the crystal structure of troponin to the mechanism of calcium regulation of muscle contraction. *Adv. Exp. Med. Biol.* 592:37–46.

[27] M. Mank, D.F. Reiff, N. Heim, M.W. Friedrich, A. Borst, and O. Griesbeck (2006). A FRET-based calcium biosensor with fast signal kinetics and high fluorescence change. *Biophys. J.* 90:1790–1796.

[28] A.R. Marks (2001). Ryanodine receptors/calcium release channels in heart failure and sudden cardiac death. *J. Mol. Cell Cardiol.* 33:615–624.

[29] T.A. McKinsey, C.L. Zhang, and E.N. Olson (2002). MEF2: a calcium-dependent regulator of cell division, differentiation and death. *Trends Biochem. Sci.* 27:40–47.

[30] K. Mikoshiba (2007). The IP3 receptor/Ca^{2+} channel and its cellular function. *Biochem. Soc. Symp.* 74:9–22.

[31] A. Miyawaki, O. Griesbeck, R. Heim, and R.Y. Tsien (1999). Dynamic and quantitative Ca^{2+} measurements using improved cameleons. *Proc. Natl Acad. Sci. USA* 96:2135–2140.

[32] A. Miyawaki, J. Llopis, R. Heim, J.M. McCaffery, J.A. Adams, M. Ikura, and R.Y. Tsien (1997). Fluorescent indicators for Ca^{2+} based on green fluorescent proteins and calmodulin. *Nature* 388:882–887.

[33] L. Munaron (2002). Calcium signalling and control of cell proliferation by tyrosine kinase receptors (review). *Int. J. Mol. Med.* 10:671–676.

[34] T. Nagai, S. Yamada, T. Tominaga, M. Ichikawa, and A. Miyawaki (2004). Expanded dynamic range of fluorescent indicators for Ca(2+) by circularly permuted yellow fluorescent proteins. *Proc. Natl Acad. Sci. USA* 101:10554–10559.

[35] A.W. Nguyen and P.S. Daugherty (2005). Evolutionary optimization of fluorescent proteins for intracellular FRET. *Nat. Biotechnol.* 23:355–360.

[36] R.A. North (2002). Molecular physiology of P2X receptors. *Physiol. Rev.* 82:1013–1067.

[37] D. Oceandy, P.J. Stanley, E.J. Cartwright, and L. Neyses (2007). The regulatory function of plasma-membrane Ca(2+)-ATPase (PMCA) in the heart. *Biochem. Soc. Trans.* 35:927–930.

[38] I. Ohtsuki (1999). Calcium ion regulation of muscle contraction: the regulatory role of troponin T. *Mol. Cell. Biochem.* 190:33–38.

[39] M. Ouyang, J. Sun, S. Chien, and Y. Wang (2008). Determination of hierarchical relationship of Src and Rac at subcellular locations with FRET biosensors. *Proc. Natl Acad. Sci. USA* 105:14353–14358.

[40] A.E. Palmer, C. Jin, J.C. Reed, and R.Y. Tsien (2004). Bcl-2-mediated alterations in endoplasmic reticulum Ca^{2+} analyzed with an improved genetically encoded fluorescent sensor. *Proc. Natl Acad. Sci. USA* 101:17404–17409.

[41] A.E. Palmer and R.Y. Tsien (2006). Measuring calcium signaling using genetically targetable fluorescent indicators. *Nat. Protoc.* 1:1057–1065.

[42] R.L. Patterson, D. Boehning, and S.H. Snyder (2004). Inositol 1,4,5-trisphosphate receptors as signal integrators. *Annu. Rev. Biochem.* 73:437–465.

[43] R.J. Pelham, Jr. and Y. Wang (1997). Cell locomotion and focal adhesions are regulated by substrate flexibility. *Proc. Natl Acad. Sci. USA* 94:13661–13665.

[44] J.W. Jr. Putney (1986). A model for receptor-regulated calcium entry. *Cell Calcium* 7:1–12.

[45] J.W. Putney, Jr. (1990). Capacitative calcium entry revisited. *Cell Calcium* 11:611–624.

[46] Y. Rong and C.W. Distelhorst (2008). Bcl-2 protein family members: versatile regulators of calcium signaling in cell survival and apoptosis. *Annu. Rev. Physiol.* 70:73–91.

[47] R.V. Sharma, M.W. Chapleau, G. Hajduczok, R.E. Wachtel, L.J. Waite, R.C. Bhalla, and F.M. Abboud (1995). Mechanical stimulation increases intracellular calcium concentration in nodose sensory neurons. *Neuroscience* 66:433–441.

[48] A.A. Sher, P.J. Noble, R. Hinch, D.J. Gavaghan, and D. Noble (2008). The role of the Na^+/Ca^{2+} exchangers in Ca^{2+} dynamics in ventricular myocytes. *Prog. Biophys. Mol. Biol.* 96:377–398.

[49] T.R. Soderling (1993). Calcium/calmodulin-dependent protein kinase II: role in learning and memory. *Mol. Cell Biochem.* 127–128:93–101.

[50] J. Solon, I. Levental, K. Sengupta, P.C. Georges, and P.A. Janmey (2007). Fibroblast adaptation and stiffness matching to soft elastic substrates. *Biophys. J.* 93:4453–4461.

[51] K. Truong, A. Sawano, H. Mizuno, H. Hama, K.I. Tong, T.K. Mal, A. Miyawaki, and M. Ikura (2001) FRET-based in vivo Ca^{2+} imaging by a new calmodulin-GFP fusion molecule. *Nat. Struct. Biol.* 8:1069–1073.

[52] R.Y. Tsien (1998). The green fluorescent protein. *Annu. Rev. Biochem.* 67:509–544.

[53] Y. Wang, E.L. Botvinick, Y. Zhao, M.W. Berns, S. Usami, R.Y. Tsien, and S. Chien (2005). Visualizing the mechanical activation of Src. *Nature* 434:1040–1045.

[54] Y. Wang and S. Chien (2007). Analysis of integrin signaling by fluorescence resonance energy transfer. *Meth. Enzymol.* 426:177–201.

[55] X.H. Wehrens, S.E. Lehnart, and A.R. Marks (2005). Intracellular calcium release and cardiac disease. *Annu. Rev. Physiol.* 67:69–98.

[56] A.E. West, W.G. Chen, M.B. Dalva, R.E. Dolmetsch, J.M. Kornhauser, A.J. Shaywitz, M.A. Takasu, X. Tao, and M.E. Greenberg (2001). Calcium regulation of neuronal gene expression. *Proc. Natl Acad. Sci. USA* 98:11024–11031.

[57] T. Yeung, P.C. Georges, L.A. Flanagan, B. Marg, M. Ortiz, M. Funaki, N. Zahir, W. Ming, V. Weaver, and P.A. Janmey (2005). Effects of substrate stiffness on cell morphology, cytoskeletal structure, and adhesion. *Cell Motil. Cytoskeleton* 60:24–34.

[58] S.H. Young, H.S. Ennes, J.A. McRoberts, V.V. Chaban, S.K. Dea, and E.A. Mayer (1999). Calcium waves in colonic myocytes produced by mechanical and receptor-mediated stimulation. *Am. J. Physiol.* 276:G1204–1212.

Part IV

From Cellular to Multicellular Models

Chapter 11. Mathematical Framework to Model Migration of Cell Population in Extracellular Matrix

Cell migration is an essential feature of physiologic and pathologic phenomena in biology, such as embryonic development, wound healing, and tumor invasion. According to the local micro-environment and the function of the migrating cell, the characteristics of migration may vary considerably. In connective tissue, cells (cancer cells, fibroblasts, etc.) interact both with other cells and with the surrounding tissue (ECM, extracellular matrix), which provides them with a natural complex scaffold to which to adhere in order to migrate. Recently much attention has been devoted to the description of the mechanics of cell motion as a result of their interactions with the ECM. Experiments have evidenced two types of motion, amoeboid and mesenchymal, that relate to different migration strategies. The amoeboid motion corresponds to a "path finding" strategy involving morphological adaptation of cells, while the mesenchymal motion corresponds to a "path generating" strategy involving proteolytic activity of cells to degrade the fibers of the ECM. This chapter takes a closer look at the individual interaction mechanisms to develop a model for amoeboid cell migration that includes both a preferential movement of cells along the collagen fibers of the ECM—a phenomenon called "contact guidance"—and a randomly oriented migration due to interactions among cells in denser areas. A modeling framework is derived at the mesoscopic (kinetic) scale, and a continuous (macroscopic) model is deduced through a diffusive limit of the kinetic one. The response of the cells to external stimuli (taxis), capable of influencing and biasing the motion, is also included. Finally, numerical simulations are presented to illustrate the ability of the model to account for the influence of (1) the heterogeneity and/or the anisotropy of the ECM medium and (2) various sorts of taxes (chemotaxis, haptotaxis, repellent behavior).

Chapter 12. Mathematical Modeling of Cell Adhesion and Its Applications to Developmental Biology and Cancer Invasion

Cellular adhesion is a key factor in many biological processes. Interactions of adhesion molecules at the molecular scale lead to cell rearrangements at the cellular scale and these can generate macroscopic patterns at the tissue scale. A multitude of discrete and continuous models of cell adhesion have been proposed that take into account the effects at the various scales. Such models are reviewed and then a continuous model of cell adhesion [N.J. Armstrong et al. (2006), *J. Theor. Biol.* 243:98–113] is discussed in more detail. This model captures molecular and cellular scale effects in an integral (nonlocal) term defining a cell velocity due to adhesive effects. This velocity is then employed to drive rearrangements of cell densities at the tissue scale in an advection–diffusion–reaction system. The application of this framework to successfully model effects as observed in cell sorting experiments and cancer cell invasion demonstrates the suitability and generality of the approach. Analytical and numerical challenges of the framework are discussed and possible extensions are outlined.

Chapter 13. Bridging Cell and Tissue Behavior in Embryo Development

Embryological tissues undergo massive morphological changes that present a challenge to a mechanistic understanding of developmental biology. One outstanding issue concerns the mechanisms through which molecular information leads to the individual or collective movement of cells that robustly shape tissues. Recent imaging and computational advances now allow us to track thousands of cells and monitor their shapes and reorganization over time. Here we present and discuss recent progress to measure and decompose tissue deformations into the relevant cellular and multicellular components, that is, cell shape change and cell–cell slippage. This multiscale approach works with unprecedented spatial and temporal resolution to dramatically extend the scope of phenotypic descriptions available to biologists. It thus provides a suitable framework for extracting representative features and for quantitatively comparing mutant phenotypes or species. This opens up new opportunities to understand the cellular mechanisms underlying tissue deformation and to identify the physical and biological parameters controlling embryo morphogenesis.

Chapter 14. Modeling Steps from Benign Tumor to Invasive Cancer: Examples of Intrinsically Multiscale Problems

The step from benign tumors to invasive cancer is characterized by neovascularization, detachment of cells from the main tumor, and eventually invasion of cells into the surrounding tissue and blood vessels, leading to distant metastases. We will for each of these steps show how experimental observations can be explained by the interplay of processes on the molecular and the cellular scale within a framework using individual-based models. The representation of the cell in the models permits us to analyze physical, particularly biomechanical, constraints. We first study how a neoformation of blood vessels can affect the development of tumor size and shape if the nutrients transported in the vessels control the growth rates of the individual cells. Cell detachment is often triggered by a malfunctioning of the beta-catenin-degrading apparatus in the cytosol. We demonstrate how an elevated beta-catenin concentration in one cell can trigger a cascade of other cells stepwise detaching as well, and migrating freely into the surrounding tissue. Before the cells can form distant metastases, they need to invade blood vessels. We show how the competition between N-CAM and VE-cadherin bonds can facilitate invasion of a cancer cell into a blood vessel if the involved pathways have defects.

Chapter 15. Delaunay Object Dynamics for Tissues Involving Highly Motile Cells

Biomechanics is now recognized as a major organization principle in biological pattern formation. Therefore, mathematical modeling must incorporate biomechanical parameters beyond the level of imposing physical constraint. Many tools are now available to tackle different aspects of biomechanical phenomena from single cells to tissues. The method of Delaunay object dynamics allows us to investigate a large number of individual cells, each having its own phenotype dynamics and mechanical properties. The simulation framework is, in particular, applicable to studies of highly dynamic systems such as fast migrating cells. The method is illustrated with the example of secondary lymphoid tissue organization. With a small set of local interactions, the typical morphology of this immune tissue can be reproduced *in silico*. The model demonstrates the appearance of stable patterns built from motile cells in a flow equilibrium. An experimentally well-known intermediate state of the pattern formation process can be traced back to the biomechanics of lymphocytes without the need to impose any additional regulatory mechanism to the system.

Chapter 11

Mathematical Framework to Model Migration of Cell Population in Extracellular Matrix

Arnaud Chauvière and Luigi Preziosi

Contents

11.1 Introduction and Biological Background

Cell migration is an essential feature of both normal and pathological biological phenomena. Tissue formation in embryonic development requires cell movements and coordination among cells. Migration of cells plays a fundamental role in immune response and tissue homeostasis in mature multicellular organisms. It is also the main process of metastasis dissemination and tumor invasion in cancer.

The characteristics of migration may vary considerably, being either intrinsic properties of the cells or resulting from their adaptation to the environment. Cell movement is partially regulated by external factors that may include diffusive (such as chemoattractant) and nondiffusive (like ligands bound to the extracellular matrix or ECM) chemicals. Physical interactions of cells and the ECM also play an important role in cell movement.

The ECM is the defining feature of connective tissue and serves many functions, such as providing support and anchorage for cells. It regulates the cell's dynamic behavior and has additional regulatory functions for apoptosis and proliferation. The ECM is composed of an interlocking mesh of fibrous proteins as collagen and fibronectin, and provides directional information either through matrix-bound ligands (a process called *haptotaxis*) or directly through the fibers along which cells tend to align. This last process is known as *contact guidance* and is illustrated in Figure 11.1.

From recent experimental studies, much has been learned about cell movement in fibrous tissues [16]. Various cell migratory behaviors in the ECM have

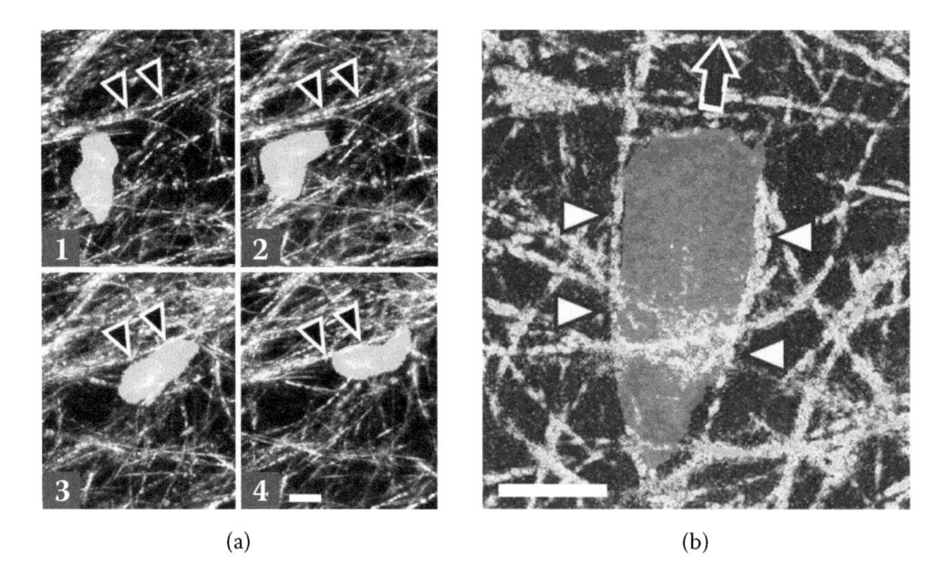

FIGURE 11.1 (See color insert following Page 398.) Contact guidance process. Amoeboid T-cell migration, cell–fiber interaction, and contact guidance within 3-D collagen matrix. (a) Alignment along fiber strands (black arrowheads). Image sequence 1–4 is a time series spanning 8 min of observation time. (b) T-cell alignment is parallel to matrix fibers (white arrowheads) upon forward migration (black arrow). Scale bars are 5 μm. From K. Wolf, R. Müller, S. Borgmann, E.-B. Bröcker, and P. Friedl (2003). Amoeboid shape change and contact guidance: T-lymphocyte crawling through brillar collagen is independent of matrix remodeling by MMPs and other proteases. *Blood* 102(9):3262–3269. (With permission).

been identified [15]. Cells can migrate while only interacting briefly with other migrating cells (individual migration) or develop adhesive bonds to form clusters (collective migration). Additionally, individual cell migration in the ECM can be split into the *amoeboid* and *mesenchymal* types. The mesenchymal migration relates to the cell strategy to generate space to move by secreting matrix degrading enzymes (MMPs). In the amoeboid case, cells migrate using the ECM as a scaffold and squeeze into free spaces of the matrix, establishing brief contacts with fibers and frequently changing direction. They generate only minor fiber bending that does not permanently alter the ECM.

In this chapter the focus is on individual amoeboid cell migration. We propose a didactic approach to derive our modeling framework. The structure of the chapter is the following. In Section 11.2 we introduce two different ways to represent both the migrating cell population and the ECM, namely statistical and continuum descriptions. We present the generic mesoscopic framework of transport equations for the velocity-jump processes that we use. In Section 11.3 we review various models based on velocity-jump processes that have been used for cell migration modeling. We present first the basic approach for random migration and then extend this further to account for contact guidance. We show how simple cell–cell interactions can also be included and conclude with the extension of the earlier models to account for the influence of environmental factors that we specialize for chemotaxis, haptotaxis, and self-repellent behavior. Then in Section 11.4 we show how macroscopic continuum models can be derived from the mesoscopic transport equations and focus on the so-called diffusive approximation. We derive further our generic macroscopic model of cell migration in the ECM. The governing drift-diffusion equation will be solved numerically. In particular in Section 11.5 we present numerical simulations to illustrate the ECM effects on cell migration. We include evidence of the influence of heterogeneity and anisotropy, and conclude with the effects of environmental factors previously introduced in the derivation of our generic macroscopic model.

11.2 Mathematical Descriptions of System

11.2.1 Statistical Description

We consider a cell population moving within a d-dimensional domain $\mathcal{D} \subseteq \mathbb{R}^d$. Each cell of the population moves with its own velocity $\mathbf{v} \in V \subseteq \mathbb{R}^d$. The cell population is thus described by the distribution function $p = p(t, \mathbf{x}, \mathbf{v})$, which depends on time $t > 0$, location $\mathbf{x} \in \mathcal{D}$, and velocity $\mathbf{v} \in V$. Here, $d \geq 1$ represents the spatial dimension of interest, with *in vivo* motion corresponding to $d = 3$ and planar motion on a substrate to $d = 2$. We assume that the space V of allowed velocities is radially symmetric and can be written $V = |V| \times S^{d-1}$, where $|V|$ denotes the range of possible speeds and S^{d-1} is the unit sphere in $\mathbb{R}^d$. We furthermore introduce the unit vector $\hat{\mathbf{v}} = \mathbf{v}/v \in S^{d-1}$ in direction of velocity $\mathbf{v}$, where $v = |\mathbf{v}| \in |V|$ denotes the modulus of the velocity.

The fibers of the extracellular matrix are described by the distribution function m. In this chapter we focus on describing the amoeboid cell motion. Thus we neglect alteration of the ECM by the cells and assume a time independence of m. Thanks to the observation that ECM fibers are symmetrical along their axis, both fiber directions are identical and we finally write the distribution function $m = m(\mathbf{x}, \mathbf{n})$, where $\mathbf{n} \in S_+^{d-1}$ is a unit vector that represents the fiber orientation defined over the half unit sphere S_+^{d-1}. However, it will be useful to extend m to S^{d-1} by introducing the distribution function

$$m^e(\mathbf{x}, \mathbf{n}) = \begin{cases} m(\mathbf{x}, \mathbf{n}) & \text{for } \mathbf{n} \in S_+^{d-1} \\ m(\mathbf{x}, -\mathbf{n}) & \text{for } \mathbf{n} \in S_-^{d-1} \end{cases} \tag{11.1}$$

The modeling framework is formulated in the form of a transport equation in which changes in cell velocity are described through an operator that models the characteristic properties of cell migration. This continuous transport equation approach uses a microscopic description for cell motion; however, it provides an output at the level of a cell population described by the distribution function. This approach is therefore commonly referred to as a *mesoscopic description.* In this chapter our attention focuses on velocity-jump processes to describe cell motion and is formulated as follows:

$$\frac{\partial p}{\partial t}(t, \mathbf{x}, \mathbf{v}) + \mathbf{v} \cdot \nabla p(t, \mathbf{x}, \mathbf{v}) = \mathcal{M}(t, \mathbf{x}, \mathbf{v}) \tag{11.2}$$

where the operator ∇ denotes the spatial gradient and $\mathcal{M}$ is an integral operator describing peculiar cell motion with velocity-jump processes. The nature of this operator is illustrated in Section 11.3 for various examples from the literature and specialized for movement into the ECM.

We remark that when the matrix is remodeled or degraded by cells, Equation (11.2) can be coupled to an evolution equation for the distribution function m as proposed in [8, 21, 31].

11.2.2 Continuum Description

A so-called continuum description of the system can be derived from the above mesoscopic description by means of averaging processes. These processes relate to the classical notion of moments used in the kinetic theory framework. In this section we introduce these moments for both the cell and fiber populations, and give an interpretation of each of them. The cell (number) density ρ is defined by

$$\rho(t, \mathbf{x}) = \int_V p(t, \mathbf{x}, \mathbf{v})\, d\mathbf{v} \tag{11.3}$$

The next-order moment gives the cell population flux (or momentum) $\mathbf{j} = \rho\mathbf{U}$, where $\mathbf{U}$ denotes the mean cell velocity, defined as

$$\mathbf{U}(t, \mathbf{x}) = \frac{1}{\rho(t, \mathbf{x})} \int_V p(t, \mathbf{x}, \mathbf{v}) \, \mathbf{v} \, d\mathbf{v} \tag{11.4}$$

The second-order moments relate to pressure and internal energy. Because the focus of this chapter is the dynamical behavior of the cell population, all energetic considerations are neglected. Thus we introduce only the notion of the pressure tensor $\mathbb{P}$. This tensor is also known as the variance–covariance matrix of the velocity distribution and is defined by

$$\mathbb{P}(t, \mathbf{x}) = \int_V p(t, \mathbf{x}, \mathbf{v}) \, [\mathbf{v} - \mathbf{U}(t, \mathbf{x})] \otimes [\mathbf{v} - \mathbf{U}(t, \mathbf{x})] \, d\mathbf{v} \tag{11.5}$$

where the operator $\otimes$ denotes the tensorial product. It measures the statistical deviations of the cell velocities $\mathbf{v}$ from the local mean velocity $\mathbf{U}$.

Regarding now the moments of the fiber distribution, we define the fiber density by

$$M(\mathbf{x}) = \int_{S_+^{d-1}} m(\mathbf{x}, \mathbf{n}) \, d\mathbf{n} = \frac{1}{2} \int_{S^{d-1}} m^e(\mathbf{x}, \mathbf{n}) \, d\mathbf{n} \tag{11.6}$$

Due to the symmetry property (11.1), of the distribution function m^e, its first moment is null which corresponds to

$$\int_{S^{d-1}} m^e(\mathbf{x}, \mathbf{n}) \, \mathbf{n} \, d\mathbf{n} = \mathbf{0} \tag{11.7}$$

For this reason, the orientation of the fiber network must be described by the so-called orientation tensor $\mathbb{D}$ (see [32]). This tensor is also known as the variance–covariance matrix of the angle distribution and is defined by

$$\mathbb{D}(\mathbf{x}) = \frac{d}{M(\mathbf{x})} \int_{S_+^{d-1}} m(\mathbf{x}, \mathbf{n}) \, \mathbf{n} \otimes \mathbf{n} \, d\mathbf{n} \tag{11.8}$$

This tensor is symmetric, positive definite, and trace-invariant with $tr(\mathbb{D}) = d$. It also provides a visualization tool, when one refers to the ellipsoid $\mathbf{x} \cdot (\mathbb{D}^{-1})\mathbf{x} = 1$. Indeed, the mean quantity of fibers projected in a certain direction $\mathbf{n}$ is proportional to the square root of the distance between the center of the ellipsoid and the intersection of the line directed along $\mathbf{n}$ with the ellipsoid. This ellipsoid has its axes identified by the eigenvectors of the tensor $\mathbb{D}$ and gives the principal directions of the orientation of the fiber network. More precisely, the main direction is given by the eigenvector associated with the principal eigenvalue and corresponds to the largest axis of the ellipsoid. In the case of fibers all aligned along the same direction $\mathbf{n}$, the ellipsoid degenerates into a segment with direction $\mathbf{n}$. Other examples detailed below are illustrated for three cases in Figure 11.2 in a 2-D configuration for simplicity reasons. Through them we aim to provide a better understanding of the meaning of the orientation tensor.

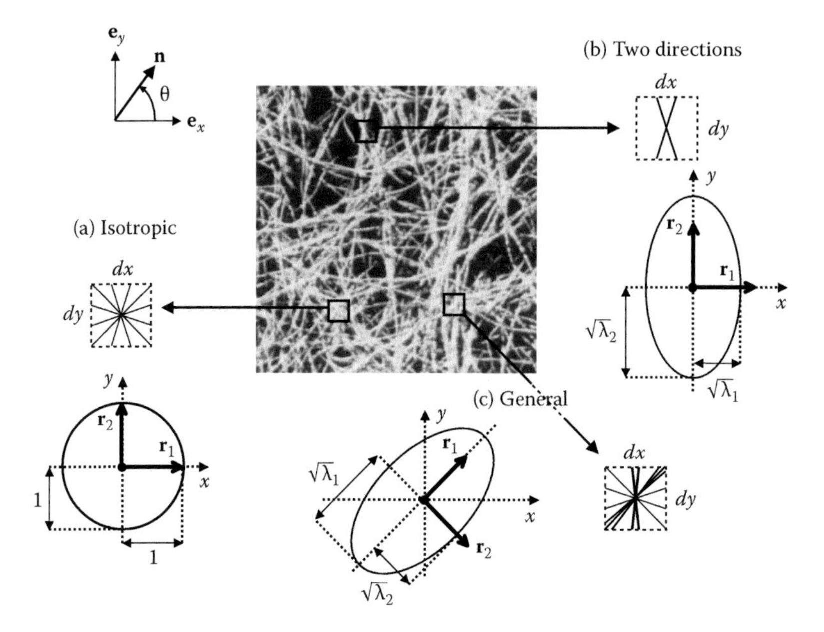

FIGURE 11.2 Local angular fiber distributions in the ECM. (a) Isotropic configuration. (b) Bi-directional configuration. (c) General fiber distribution. We graphically represent the orientation tensor $\mathbb{D}$ by the ellipse $\mathbf{x} \cdot (\mathbb{D}^{-1})\mathbf{x} = 1$ whose eigenvectors $\mathbf{r}_1$ and $\mathbf{r}_2$, respectively, associated with the eigenvalues λ_1 and λ_2, give locally the principal orientation of the fibers.

- We consider a locally isotropic distribution of fibers in the element $d\mathbf{x} = dx\,dy$ centered at one point $\mathbf{x}$ of the domain $\mathcal{D}$. At this point the fiber distribution is modeled by

$$m(\mathbf{x}, \mathbf{n}) = \frac{1}{\pi} M(\mathbf{x})$$

that results in the orientation tensor being the identity tensor. Its graphical representation is simply the circle of unit radius as shown in Figure 11.2a.

- We consider now an ECM having locally only two fiber directions given by the unit vectors $\mathbf{n}_i = \cos\theta_i\,\mathbf{e}_x + \sin\theta_i\,\mathbf{e}_y$ with $\theta_i \in [0, \pi]$ for $i = \{1, 2\}$ and assume that these two angles satisfy $\theta_1 + \theta_2 = \pi$. This configuration is described by

$$m(\mathbf{x}, \mathbf{n}) = \frac{1}{2} M(\mathbf{x})[\delta(\mathbf{n} - \mathbf{n}_1) + \delta(\mathbf{n} - \mathbf{n}_2)]$$

where δ denotes the Dirac function. The corresponding orientation tensor at $\mathbf{x}$ is diagonal and written

$$\mathbb{D}(\mathbf{x}) = 2\begin{pmatrix} \cos^2\theta_1 & 0 \\ 0 & \sin^2\theta_1 \end{pmatrix}$$

The two eigenvalues are respectively $\lambda_1 = 2\cos^2\theta_1$ (with eigenvector $\mathbf{r}_1 = \mathbf{e}_x$) and $\lambda_2 = 2\sin^2\theta_1$ (with eigenvector $\mathbf{r}_2 = \mathbf{e}_y$). The direction of the eigenvector associated with the maximum eigenvalue gives the main fiber direction, which is the largest axis of the ellipse. In the schematic representation of Figure 11.2(b), the maximum eigenvalue is λ_2 and the main fiber direction is therefore given by $\mathbf{r}_2 = \mathbf{e}_y$.

- Finally we illustrate the general case that we describe by the distribution function

$$m(\mathbf{x}, \mathbf{n}) = M(\mathbf{x})\frac{f(\mathbf{n})}{\displaystyle\int_{S_+^{d-1}} f(\mathbf{n})\,d\mathbf{n}}$$

We write the unit vector $\mathbf{n} = \cos\theta\,\mathbf{e}_x + \sin\theta\,\mathbf{e}_y$ with $\theta \in [0, \pi]$ so that the angle distribution can be performed through the function $f(\theta)$. The corresponding orientation tensor at $\mathbf{x}$ is then

$$\mathbb{D}(\mathbf{x}) = \frac{d}{\displaystyle\int_0^\pi f(\theta)\,d\theta}\begin{pmatrix} \displaystyle\int_0^\pi f(\theta)\cos^2\theta\,d\theta & \displaystyle\int_0^\pi f(\theta)\cos\theta\sin\theta\,d\theta \\ \displaystyle\int_0^\pi f(\theta)\cos\theta\sin\theta\,d\theta & \displaystyle\int_0^\pi f(\theta)\sin^2\theta\,d\theta \end{pmatrix}$$

A sample angular distribution modeling the configuration (c) in Figure 11.2 may be

$$f(\theta) = 1 + \cos\left(2\left[\theta - \frac{\pi}{4}\right]\right)$$

that will lead to

$$\mathbb{D}(\mathbf{x}) = \begin{pmatrix} 1 & \frac{1}{2} \\ \frac{1}{2} & 1 \end{pmatrix} \quad \text{with} \quad \begin{cases} \lambda_1 = \frac{3}{2} & \text{and} \quad \mathbf{r}_1 = \mathbf{e}_x + \mathbf{e}_y \\ \lambda_2 = \frac{1}{2} & \text{and} \quad \mathbf{r}_2 = \mathbf{e}_x - \mathbf{e}_y \end{cases}$$

Therefore the orientation of the fiber network at $\mathbf{x}$ can be graphically represented by the rotated ellipse of axes $\mathbf{e}_x + \mathbf{e}_y$ and $\mathbf{e}_x - \mathbf{e}_y$ as shown in Figure 11.2(c). Because the highest eigenvalue is λ_1, the main fiber orientation is given by $\mathbf{r}_1 = \mathbf{e}_x + \mathbf{e}_y$.

In Section 11.4 we present methods to derive evolution equations for the macroscopic variables that characterize cell migration, such as the cell density ρ or the cell mean velocity $\mathbf{U}$. We explain how to link the statistical and continuum descriptions introduced above. In particular, we focus on scaling methods and detail the so-called diffusive approximation that leads to an evolution equation for the cell density ρ in the ECM described by the fiber density M and the orientation tensor $\mathbb{D}$.

11.3 Mesoscopic Modeling of Cell Migration in ECM

In this section we review various models based on velocity-jump processes that have been used for cell migration modeling. We start with the simplest modeling approach and show further on how ECM effects can be included. We conclude by showing how environmental factors can be accounted for in this modeling framework.

11.3.1 Random Migration

A basic behavioral mode of motion modeled by a velocity-jump process is the so-called *run and tumble* motion, in which cells move by smooth runs interrupted at discrete times by an instantaneous random reorientation. In general, the choice of a new velocity and the average runtime between reorientations would depend on environmental factors. This kind of movement has been modeled within the mesoscopic framework (Equation (11.2)) by the turning operator

$$\mathcal{M}(t,\mathbf{x},\mathbf{v}) = -\mu p(t,\mathbf{x},\mathbf{v}) + \mu \int_V T(\mathbf{v}';\mathbf{v}) p(t,\mathbf{x},\mathbf{v}')\, d\mathbf{v}' \tag{11.9}$$

The first term of the right-hand side describes turning of cells away from velocity $\mathbf{v}$ with a frequency μ that may depend on environmental factors. The integral term calculates the rate at which cells reorient into velocity $\mathbf{v}$ given previous velocity $\mathbf{v}'$. The reorientation function $T(\mathbf{v}';\mathbf{v})$ defines a probability distribution for a cell with previous velocity $\mathbf{v}'$ to choose the new velocity $\mathbf{v}$, and satisfies

$$\int_V T(\mathbf{v}';\mathbf{v})\, d\mathbf{v} = 1 \tag{11.10}$$

dictated by cell number conservation, which yields

$$\int_V \mathcal{M}(t,\mathbf{x},\mathbf{v})\, d\mathbf{v} = 0 \tag{11.11}$$

The simplest illustration of velocity-jump process is a uniform reorientation probability associated with the turning operator

$$T(\mathbf{v}'; \mathbf{v}) \equiv T_R(\mathbf{v}'; \mathbf{v}) = \frac{1}{\mathcal{V}_d}\psi(v) \tag{11.12}$$

In Equation (11.12), $\mathcal{V}_d = \int_{S^{d-1}} d\hat{\mathbf{v}}$ is the surface of the unit sphere in $\mathbb{R}^d$, and ψ gives the distribution of newly chosen speeds that, due to Equation (11.10), satisfies

$$\int_{|V|} \psi(v) v^{d-1}\, dv = 1 \tag{11.13}$$

This simple mechanism leads to the turning operator

$$\mathcal{M}(t, \mathbf{x}, \mathbf{v}) \equiv \mathcal{M}_R(t, \mathbf{x}, \mathbf{v}) = \mu \left(\rho(t, \mathbf{x}) \frac{\psi(v)}{\mathcal{V}_d} - p(t, \mathbf{x}, \mathbf{v}) \right) \tag{11.14}$$

Examples of the speed distribution function are proposed below for:

- Cells moving with a constant speed U:

$$\psi(v) = \frac{1}{\mathrm{U}^{d-1}} \delta(v - \mathrm{U}) \tag{11.15}$$

- Cells moving with speed randomly chosen within a range $[\mathrm{U}_1, \mathrm{U}_2]$:

$$\psi(v) = \frac{d}{\mathrm{U}_2^d - \mathrm{U}_1^d} \tag{11.16}$$

The basic turning operator proposed in Equation (11.12) is based on completely random reorientation that assumes neither memory effect of the prior velocity to tumble (therefore no orientational persistence) nor information to orient the motion toward a particular direction. Such a description assumes then that inertia in cell movement is out of resolution. We should mention here that polarization processes that lead to orientational persistence in motion are observed for many cells. To model persistence mechanisms, memory effect can easily be introduced by adapting the turning operator T as in [19,28], including a dependence on the prior direction of motion (that is the vector $\hat{\mathbf{v}}'$). Representation of persistence can also be absorbed in the turning frequency, that could depend on environmental factors such as chemical fields [1] in order to avoid an increased complexity of the model. A more detailed presentation of effects of environmental factors is given in Section 11.3.4. In the next section we focus rather on the directional effect of ECM fibers.

11.3.2 Contact Guidance

To model the movement of a cell in a given fiber network, it is commonly assumed that the dominant process is contact guidance. The matrix or tissue gives a selection of preferred directions along which a cell can move. This alignment process has been modeled in [21] by neglecting cell–cell interactions. This type of model refers to the class of alignment models that has also been developed for reorientation of actin filaments [17]. In [21] the turning operator that models cell alignment along the fibers depends then on the angular distribution m^e of fiber and, in our notation, reads

$$T(\mathbf{v}'; \mathbf{v}) \equiv T_m(\mathbf{v}'; \mathbf{v}) = \frac{m^e(\mathbf{x}, \hat{\mathbf{v}})}{2M(\mathbf{x})}\psi(v) \tag{11.17}$$

We notice that in Equation (11.17) the spatial dependence in m^e allows the description of heterogeneous ECM, which was not originally possible in [21]. From Equation (11.17) one can derive

$$\mathcal{M}(t, \mathbf{x}, \mathbf{v}) \equiv \mathcal{M}_m(t, \mathbf{x}, \mathbf{v}) = \mu \left(\rho(t, \mathbf{x})\psi(v)\frac{m^e(\mathbf{x}, \hat{\mathbf{v}})}{2M(\mathbf{x})} - p(t, \mathbf{x}, \mathbf{v}) \right) \tag{11.18}$$

In [31], Painter assumes that 2-D random reorientation is biased by contact guidance and that a parameter $b \in [0, 1]$ reflects the degree of bias. This assumption leads the author to write the associated turning operator as a linear combination of Equations (11.12) and (11.17), which we extend to

$$T(\mathbf{v}'; \mathbf{v}) \equiv T_b(\mathbf{v}'; \mathbf{v}) = \left[\frac{1-b}{\mathcal{V}_d} + \frac{b\, m^e(\mathbf{x}, \hat{\mathbf{v}})}{2M(\mathbf{x})} \right] \psi(v) \tag{11.19}$$

This leads to

$$\mathcal{M}(t, \mathbf{x}, \mathbf{v}) \equiv \mathcal{M}_b(t, \mathbf{x}, \mathbf{v}) = \mu \left(\rho(t, \mathbf{x})\psi(v) \left[\frac{1-b}{\mathcal{V}_d} + \frac{b\, m^e(\mathbf{x}, \hat{\mathbf{v}})}{2M(\mathbf{x})} \right] - p(t, \mathbf{x}, \mathbf{v}) \right) \tag{11.20}$$

In the models mentioned above, we reformulated the corresponding operators in our notation and considered potential heterogeneity of the ECM. However, this form does not enable us to account truly for ECM heterogeneity because the spatially dependent ECM density $M(\mathbf{x})$ is only used as a normalization factor. This simplification can be readily illustrated when considering the angular fiber configurations in Section 11.2.2. In contrast, heterogeneity influence was introduced in [8] in the context of kinetic theory, by also incorporating cell-cell interactions. In the following we introduce this kinetic theory framework and start with the general expression of the migration operator

$$\mathcal{M}(t, \mathbf{x}, \mathbf{v}) = -\mathcal{L}(t, \mathbf{x}, \mathbf{v}) + \mathcal{G}(t, \mathbf{x}, \mathbf{v}) \tag{11.21}$$

where $\mathcal{L}$ is a *loss* term (that is, the rate at which cells turn away from velocity $\mathbf{v}$) and $\mathcal{G}$ is a *gain* term giving the rate at which cells reorient into velocity $\mathbf{v}$ (see [6] for more details). The cell number conservation dictates that property (11.11) has to be satisfied also for the above expression of $\mathcal{M}$. We first focus on describing the contact guidance process and assume that realignment along

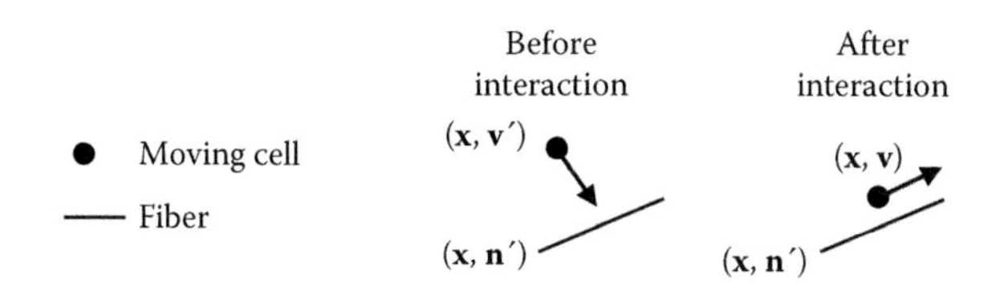

FIGURE 11.3 Schematic representation of cell–fiber interaction. At location $\mathbf{x}$ the moving cell with velocity $\mathbf{v}'$ aligns along the fiber with orientation $\mathbf{n}'$, choosing its new velocity direction into $\hat{\mathbf{v}} = \mathbf{n}'$.

the fibers does not appear at a turning frequency, but is caused by interactions between cells and fibers of the ECM that occur with rate η_m. A schematic representation of the interaction is shown in Figure 11.3. Then we can state the following expressions:

$$\mathcal{L}(t, \mathbf{x}, \mathbf{v}) \equiv \mathcal{L}_m(t, \mathbf{x}, \mathbf{v}) = p(t, \mathbf{x}, \mathbf{v}) \int_{S_+^{d-1}} \eta_m \, m(\mathbf{x}, \mathbf{n}') \, d\mathbf{n}' \tag{11.22}$$

$$\mathcal{G}(t, \mathbf{x}, \mathbf{v}) \equiv \mathcal{G}_m(t, \mathbf{x}, \mathbf{v}) = \iint_{V \times S_+^{d-1}} \eta_m \, \psi_m(\mathbf{v}', \mathbf{n}'; \mathbf{v}) p(t, \mathbf{x}, \mathbf{v}') m(\mathbf{x}, \mathbf{n}') \, d\mathbf{v}' d\mathbf{n}' \tag{11.23}$$

The function $\psi_m(\mathbf{v}', \mathbf{n}'; \mathbf{v})$ defines a transition probability distribution for a cell with given velocity $\mathbf{v}'$ to chose a new velocity $\mathbf{v}$ when interacting with a fiber oriented toward $\mathbf{n}'$, and satisfies

$$\int_V \psi_m(\mathbf{v}', \mathbf{n}'; \mathbf{v}) \, d\mathbf{v} = 1 \tag{11.24}$$

which, similar to Equation (11.10), is dictated by cell number conservation. In general the encounter rate η_m would depend on microscopic quantities to account for the cell velocity or for the fiber size. For example, in classical kinetic theory, the encounter rates are proportional to the relative velocity modulus raised to some power (see [3], for example). For simplicity, we assume that η_m is constant. Additionally, we assume that the alignment process along a fiber is independent of the prior velocity $\mathbf{v}'$. Using the generic speed distribution function ψ introduced previously, we write

$$\psi_m(\mathbf{v}', \mathbf{n}'; \mathbf{v}) = \psi(v) \frac{1}{2} [\delta(\hat{\mathbf{v}} - \mathbf{n}') + \delta(\hat{\mathbf{v}} + \mathbf{n}')] \tag{11.25}$$

which leads to

$$\mathcal{M}(t, \mathbf{x}, \mathbf{v}) \equiv J_m(t, \mathbf{x}, \mathbf{v}) = \eta_m M(\mathbf{x}) \left(\rho(t, \mathbf{x}) \psi(v) \frac{m^e(\mathbf{x}, \hat{\mathbf{v}})}{2M(\mathbf{x})} - p(t, \mathbf{x}, \mathbf{v}) \right) \tag{11.26}$$

We first observe the similarity between J_m (Equation (11.26)) and $\mathcal{M}_m$ (Equation (11.18)). The two approaches lead to the same expression, assuming that the turning frequency μ is locally proportional to the ECM density, *i.e.* $\mu \equiv \mu(\mathbf{x}) = \eta_m M(\mathbf{x})$. This dependence states that the turning frequency increases for an increasing ECM density. Then the average movement (that could be characterized by the mean squared displacement) is slowed down for an increasing ECM density, which is in agreement with numerical results of [7]. The second observation is that by substituting into Equation (11.26)

$$\eta_m \to \frac{\mu}{M}, \qquad m^e \to b\, m^e + (1-b)\frac{2M}{\mathcal{V}_d} \tag{11.27}$$

we find the operator (11.20) that models random walk biased by contact guidance. This enables us to work with the operator J_m to model contact guidance, knowing that it can be straightforwardly extended to account for random motion by substitutions (11.27). This random effect leads to a partial alignment along the fibers, which could also be modeled by considering a smooth alignment distribution ψ_m as in [17,25] rather than the Dirac function used in Equation (11.25).

11.3.3 Influence of Cell–Cell Interactions

Interactions between cells are very complex and of many different types: contact inhibition, adhesion, and repulsion are well-known examples. Here we focus on dynamical aspects and consider only the orientational effect that results from the interaction between two moving cells. We also assume that realignment processes are dominated by fiber guidance. In a similar manner, as in the previous section, we now focus on cell–cell interactions for which we state the following loss and gain terms:

$$\mathcal{L}(t,\mathbf{x},\mathbf{v}) \equiv \mathcal{L}_c(t,\mathbf{x},\mathbf{v}) = p(t,\mathbf{x},\mathbf{v}) \int_V \eta_c\, p(t,\mathbf{x},\mathbf{v}'_*)\, dv'_* \tag{11.28}$$

$$\mathcal{G}(t,\mathbf{x},\mathbf{v}) \equiv \mathcal{G}_c(t,\mathbf{x},\mathbf{v}) = \iint_V \eta_c\, \psi_c(\mathbf{v}',\mathbf{v}'_*;\mathbf{v}) p(t,\mathbf{x},\mathbf{v}') p(t,\mathbf{x},\mathbf{v}'_*)\, d\mathbf{v}' d\mathbf{v}'_* \tag{11.29}$$

Here, the function $\psi_c(\mathbf{v}',\mathbf{v}'_*;\mathbf{v})$ defines a transition probability distribution for a moving cell with given velocity $\mathbf{v}'$ to choose the new velocity $\mathbf{v}$ when interacting with a *field* cell, that is, a surrounding cell with velocity $\mathbf{v}'_*$, and satisfies

$$\int_V \psi_c(\mathbf{v}',\mathbf{v}'_*;\mathbf{v})\, d\mathbf{v} = 1 \tag{11.30}$$

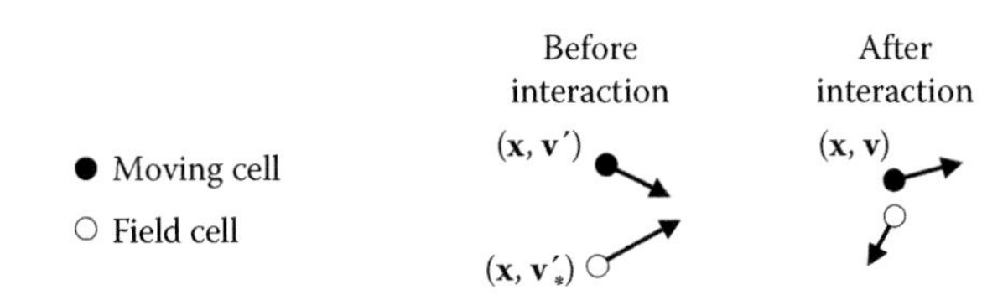

FIGURE 11.4 Schematic representation of cell–cell interaction. The moving cell collides with another cell and randomly reorients its movement by changing its velocity from $\mathbf{v}'$ to $\mathbf{v}$.

For coherence with the previous paragraph, we assume that:

1. Cell–cell interactions occur with rate η_c that is constant.
2. Collision between two migrating cells leads to random reorientation of their movement independently of pre-collision velocities (see Figure 11.4).

We model this process by the uniform transition probability

$$\psi_c(\mathbf{v}', \mathbf{v}'_*; \mathbf{v}) = \frac{1}{\mathcal{V}_d}\psi(v) \tag{11.31}$$

More complex phenomena could be accounted for, for example, alignment caused by cell–cell interaction described in [27], that would lead to other expressions of ψ_c. We remark that the formation of cell clusters mentioned in the introduction (which relates to collective migration) could also be modeled but would lead to much more sophisticated aggregation models that we do not present here. For didactical reasons, we maintain the description of the simplest type of interaction through Equation (11.31) and derive accordingly

$$\mathcal{M}(t, \mathbf{x}, \mathbf{v}) \equiv J_c(t, \mathbf{x}, \mathbf{v}) = \eta_c\rho(t, \mathbf{x})\left(\rho(t, \mathbf{x})\frac{\psi(v)}{\mathcal{V}_d} - p(t, \mathbf{x}, \mathbf{v})\right) \tag{11.32}$$

which models random reorientation of colliding cells. Expression (11.32) can be compared to Expression (11.14). Both are identical when assuming that the frequency μ is proportional to the cell density, *i.e.* $\mu \equiv \mu(t, \mathbf{x}) = \eta_c\rho(t, \mathbf{x})$. Similar to cell–fiber interaction, density effects are considered by our approach and result in the deceleration of the average movement when cell density increases.

At this point, we have briefly reviewed some mathematical models of different cell migration types in the mesoscopic framework of transport equations. We introduced the basic requirements to model random motion, alignment processes, and cell–cell interactions. In the next section, we show how to incorporate the influence of environmental factors.

11.3.4 Influence of Environmental Factors

Environmental factors can be of various natures. They can be diffusible chemicals secreted into the environment that will trigger a chemotactic movement of the cell in response. This response results from the external detection of a signal that is transducted to internal pathways. Signaling molecules can also be released and felt by the same population in order to communicate, enabling a single cell to sense its surrounding density (a phenomenon called *quorum sensing* [36]). They can also be any other exogenous chemical within the environment, like fibronectin present in the ECM [9].

In this section we propose a mathematical description of the cell response to environmental signals, a phenomenon called *taxis*, and focus in particular on chemotactic and haptotactic cues, the latter associated to nondiffusive signals. A common feature of taxis models is a reduction of the signaling pathway complexity into a sensitivity function that could depend on the signal itself. Mathematical modeling of chemotaxis has been widely developed, and the interested reader can find a review in [24]. In the context of transport equations that we present, chemotaxis is introduced as a bias of the main movement, which is often assumed to be random motion [29]. Different structures of the bias lead to particular sensitivity functions that can also account for internal mechanisms [13,18]. The derivation of the bias is performed according to the cell-sensing strategy, which depends on the cell nature. Mathematical modeling usually refers to strictly local, local average, neighbor-based, and gradient-based as listed in [2], or more sophisticated sampling radius-based models [23]. A few models have been developed for taxis in which the structure of the surrounding tissue impacts also on cell movement (see [10,34], for example).

In the following we use the formalism introduced in previous sections and extend it to account for taxis modeling in a generic way based on the approach developed in [5]. We start from the description of cell migration in the ECM as a combination of random motion, contact guidance, and cell–cell interaction. Therefore we propose to integrate the influence of signaling as a bias of the main motion described by the operator

$$\mathcal{M}(t, \mathbf{x}, \mathbf{v}) = J_m(t, \mathbf{x}, \mathbf{v}) + J_c(t, \mathbf{x}, \mathbf{v}) \tag{11.33}$$

where J_m and J_c are given by Equations (11.26) and (11.32), respectively. We assume that cells are able to detect or measure environmental properties of their local neighborhood, for example, by detecting signaling molecules that are released into the environment, and adapt their behavior on the basis of this information. The direction of movement $\hat{\mathbf{v}}$ is then chosen according to a signal with a probability that depends on the signal's nature and intensity. In [5] the approach is based on the assumption that one cell measures the density S of signaling molecules in its neighborhood described by a sampling radius, compares the values in every direction, and finally evaluates the signal direction thanks to this comparison. Both random reorientation and alignment processes along the fibers are affected; hence, a cell may choose to follow a fiber up or down according to the external signal. Therefore J_m and J_c are extended to

$$J_m^{\mathcal{B}}(t, \mathbf{x}, \mathbf{v}) = \eta_m M(\mathbf{x}) \left(\rho(t, \mathbf{x})\psi(v) \frac{m^e(\mathbf{x}, \hat{\mathbf{v}})}{2M(\mathbf{x})} [1 + \mathcal{B}(t, \mathbf{x}, \hat{\mathbf{v}})] - p(t, \mathbf{x}, \mathbf{v}) \right) \tag{11.34}$$

$$J_c^{\mathcal{B}}(t, \mathbf{x}, \mathbf{v}) = \eta_c \rho(t, \mathbf{x}) \left(\rho(t, \mathbf{x}) \frac{\psi(v)}{\mathcal{V}_d} [1 + \mathcal{B}(t, \mathbf{x}, \hat{\mathbf{v}})] - p(t, \mathbf{x}, \mathbf{v}) \right) \tag{11.35}$$

In Equations (11.34) and (11.35), the bias $\mathcal{B}$ accounts for an external stimulus that modifies the rate at which a cell reorients into $\hat{\mathbf{v}}$. Under the assumption of a small sampling radius, in [5], the simplest expression is derived, leading to the following gradient-based bias:

$$\mathcal{B}(t, \mathbf{x}, \hat{\mathbf{v}}) = \pm\Gamma \frac{\nabla \mathcal{S}(t, \mathbf{x}) \cdot \hat{\mathbf{v}}}{\beta_{\mathcal{S}} + \mathcal{S}(t, \mathbf{x})} \tag{11.36}$$

In Equation (11.36) the parameter Γ reflects the cell sensitivity to the signal and represents the small sampling radius around each cell. The $\pm$ sign is associated with repellent $(-)$ or attractive $(+)$ effect in the direction of the gradient $\nabla\mathcal{S}$ of signaling molecule density $\mathcal{S}$, while $\beta_{\mathcal{S}} > 0$ is introduced to avoid a singular behavior when $\mathcal{S} = 0$. The meaning of $\mathcal{B}$ is illustrated in Figure 11.5 for a 2-D configuration. The gradient of signaling molecules can be written as $\nabla\mathcal{S} = |\nabla\mathcal{S}|\mathbf{e}_{\mathcal{S}}$ where $\mathbf{e}_{\mathcal{S}} = \cos\theta_{\mathcal{S}}\,\mathbf{e}_x + \sin\theta_{\mathcal{S}}\,\mathbf{e}_y$ is the unit vector in the direction of the gradient. Similarly we write the velocity direction $\hat{\mathbf{v}} = \cos\theta\,\mathbf{e}_x + \sin\theta\,\mathbf{e}_y$. Both $\theta_{\mathcal{S}}$ and θ belong to the interval $[-\pi, \pi]$. Thus the bias can be rewritten as

$$\mathcal{B}(t, \mathbf{x}, \theta) = \pm\Gamma \frac{|\nabla \mathcal{S}(t, \mathbf{x})|}{\beta_{\mathcal{S}} + \mathcal{S}(t, \mathbf{x})} \cos(\theta - \theta_{\mathcal{S}}) \tag{11.37}$$

and is illustrated in Figure 11.5. In Equation (11.37) the sign is respectively associated with attractive and repellent signals. In Figure 11.5(b) the attractive signal increases the probability that direction $\mathbf{e}_{\mathcal{S}}$ of the gradient is chosen. In Figure 11.5(a) the repellent signal has the opposite effect: the lowest reorientation probability corresponds to the direction $\mathbf{e}_{\mathcal{S}}$ of the gradient while the highest one is in the opposite direction $\theta_{\mathcal{S}} - \pi$.

The modified probability distribution that accounts for the influence of signaling molecules may be used for different types of stimulus $\mathcal{S}$. A diffusive chemical substance $\mathcal{C}$ yields the *Chemotactic bias*:

$$\mathcal{B}(t, \mathbf{x}, \hat{\mathbf{v}}) \equiv \mathcal{B}_{\mathcal{C}}(t, \mathbf{x}, \hat{\mathbf{v}}) = \pm\,\Gamma\, \frac{\nabla \mathcal{C}(t, \mathbf{x}) \cdot \hat{\mathbf{v}}}{\beta_{\mathcal{C}} + \mathcal{C}(t, \mathbf{x})} \tag{11.38}$$

Usually $\mathcal{C}$ refers to a chemoattractant that corresponds to the positive sign in Equation (11.38). However, chemorepulsion has also been evidenced lately in [35] for glioma cells, which would require the negative sign in the bias.

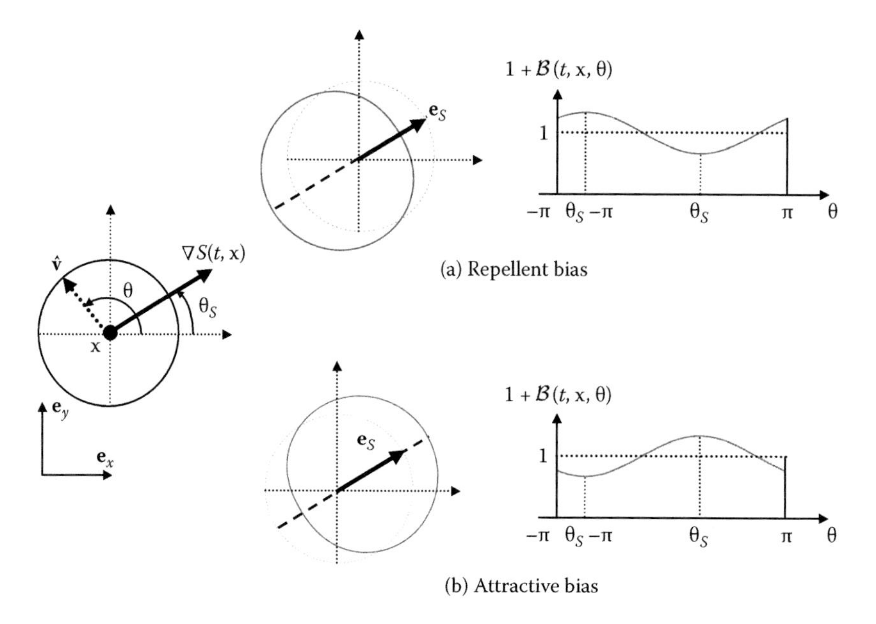

FIGURE 11.5 Schematic representation of the generic bias $\mathcal{B}$. On the left we represent geometrically the velocity direction $\hat{\mathbf{v}}$ and the gradient of chemical density $\mathcal{S}$. On the right we illustrate the typical probability distributions to choose θ as the new direction of movement when sensing (a) a repellent chemical and (b) an attractive one. In the center, these probability distributions are also illustrated in polar coordinates: the dotted circles represent the uniform distributions while the solid lines are the biased ones.

In a similar manner we model haptotaxis, that is, the movement of cells toward a gradient of ligands into the ECM. We assume that these nondiffusive ligands are proportional to the ECM density and write the corresponding *Haptotaxis bias*:

$$\mathcal{B}(t, \mathbf{x}, \hat{\mathbf{v}}) \equiv \mathcal{B}_{\mathcal{H}}(\mathbf{x}, \hat{\mathbf{v}}) = +\Gamma \, \frac{\nabla M(\mathbf{x}) \cdot \hat{\mathbf{v}}}{\beta_{\mathcal{H}} + M(\mathbf{x})} \tag{11.39}$$

We already introduced repellent chemotaxis evidenced in [35]. In a more recent article [12], Eckerisch et al. proposed that contactin is an interesting counter-adhesive molecule that may contribute to repulsion between adjacent tumor cells. We assume that this repellent auto-inducer is proportional to the cell density and call this assumption *Repellent quorum sensing bias*:

$$\mathcal{B}(\mathrm{t}, \mathbf{x}, \hat{\mathbf{v}}) \equiv \mathcal{B}_{\mathcal{R}}(t, \mathbf{x}, \hat{\mathbf{v}}) = -\Gamma \, \frac{\nabla \rho(t, \mathbf{x}) \cdot \hat{\mathbf{v}}}{\beta_{\mathcal{R}} + \rho(t, \mathbf{x})} \tag{11.40}$$

The dependence on the density field could also be more complicated.

It is straightforward to consider a combination of chemotaxis, haptotaxis, and repellent quorum sensing by deriving the associated bias as a linear combination of Equations (11.38), (11.39), and (11.40). Additionally, other mechanisms may also be included using our framework, with the remaining assumption that these mechanisms influence the main movement as a bias of the initial operators (11.26) and (11.32).

Throughout this section we introduced mesoscopic modeling of cell motion and reviewed various models of biological phenomena of importance that can be accounted for in the mesoscopic framework. Specifically, we presented the basic approaches to model random motion and how contact guidance is included in this context, thus accounting for anisotropy of the ECM. Then we extended existing models using the description provided by the kinetic theory framework and evidenced density effects arising from cell–ECM and cell–cell interactions. We finally derived a generic formulation to account for environmental factors in the ECM. This concludes the presentation of mesoscopic modeling of cell motion in the ECM. In the next section we show how macroscopic models can be derived from a mesoscopic description.

11.4 From Mesoscopic to Macroscopic Modeling

11.4.1 Introduction to Methods

The derivation of macroscopic equations from the transport Equation (11.2) was introduced in the context of fluid dynamics. We start with a short review of the methods that have been developed in the literature.

The first class of methods is based on the direct derivation of equations for the moments ρ and $\mathbf{U}$ introduced in Section 11.2.2. Note that second-order moments associated with pressure or energy can also be considered. This approach leads to an open system in which, typically, an equation for the ith order moment requires knowledge of the $(i+1)$th order moment. The moment closure method is a fast and easy method to obtain a closed moment system. Generally, there are many ways to close a moment system. From a modeling point of view it is sensible to assume that the pressure term is dominated by the equilibrium distribution of the system. However, using this closure, it is not obvious that the solution of the macroscopic system is an approximation of the moments of the solution p. Other physically motivated moment closure techniques use entropy or energy minimization approaches to justify the moment closure (see, for example, [20,26]).

The second class relates to asymptotic methods that require the existence of a small parameter ε for scaling processes. They rely on the scaling choice and may give rise to the so-called hydrodynamical and diffusive limits. An excellent overview of the hydrodynamical limits, including Hilbert's and Chapman-Enskog's methods, can be found in [33]. In [22], Hillen and Othmer detail the diffusion limit of transport equations derived from velocity-jump processes.

These methods have been often used for chemotaxis [11,14,30] and contact guidance [8,21] modeling, and gave rise to a wide variety of sophisticated models.

In the next section we use Hilbert's method to derive the diffusive limit for the type of transport models introduced throughout Section 11.3. To this end we need to identify the small parameter ε that is required for asymptotic methods. We start then with the nondimensionalization of the kinetic equation

$$\frac{\partial p}{\partial t}(t,\mathbf{x},\mathbf{v}) + \mathbf{v}\cdot\nabla p(t,\mathbf{x},\mathbf{v}) = J_m^{\mathcal{B}}(t,\mathbf{x},\mathbf{v}) + J_c^{\mathcal{B}}(t,\mathbf{x},\mathbf{v}) \tag{11.41}$$

for the following expressions of the r.h.s. operators that model biased migration in the ECM:

$$J_m^{\mathcal{B}}(t,\mathbf{x},\mathbf{v}) = \eta_m M(\mathbf{x})\left(\rho(t,\mathbf{x})\psi(v)\frac{m^e(\mathbf{x},\hat{\mathbf{v}})}{2M(\mathbf{x})}[1+\mathcal{B}(t,\mathbf{x},\hat{\mathbf{v}})] - p(t,\mathbf{x},\mathbf{v})\right) \tag{11.42}$$

$$J_c^{\mathcal{B}}(t,\mathbf{x},\mathbf{v}) = \eta_c\rho(t,\mathbf{x})\left(\rho(t,\mathbf{x})\frac{\psi(v)}{\mathcal{V}_d}[1+\mathcal{B}(t,\mathbf{x},\hat{\mathbf{v}})] - p(t,\mathbf{x},\mathbf{v})\right) \tag{11.43}$$

We introduce reference values ρ_0 and M_0 to scale the cell and fiber densities, and scale velocity by a maximal speed U_0 that occurs in $|V|$. Thus the appropriate reference values are, respectively, $p_0 = \rho_0/\mathrm{U}_0^d$ for the distribution function p and $\psi_0 = 1/\mathrm{U}_0^d$ for the speed distribution function ψ. We introduce now a typical time t_0 associated with the reorientation processes that occur with frequency $\lambda_c \equiv \eta_c\rho$ due to cell–cell interaction and $\lambda_m \equiv \eta_m M$ for cell–fiber interaction. Thus we introduce the typical microscopic length $\ell_0 = \mathrm{U}_0 t_0$ that corresponds to the mean free path, that is, the distance covered by a moving cell during a time interval $[t, t+t_0]$ without reorientation. Finally, we introduce macroscopic scalings T_0 for time and X_0 for length, and assume that $\mathrm{X}_0 \gg \ell_0$ so that we can introduce a small dimensionless parameter $\varepsilon > 0$ defined by $\varepsilon = \ell_0/\mathrm{X}_0$ that is equivalent to a Knudsen number. This assumption gives rise naturally to two time scales corresponding, respectively, to drift and diffusive macroscopic processes. The first one is associated with a typical time $\mathrm{T}_0 = \mathrm{T}_{\text{drift}} = \mathrm{X}_0/\mathrm{U}_0 = t_0/\varepsilon$, while the second one relates to a diffusive time $\mathrm{T}_0 = \mathrm{T}_{\text{diff}}$ characterized, for one cell, by a diffusion coefficient $D = \mathrm{U}_0^2 t_0/d$ in $\mathbb{R}^d$. Because $\mathrm{T}_{\text{diff}} = \mathrm{X}_0^2/D$, it is straightforward to calculate $\mathrm{T}_{\text{diff}} = d\, t_0/\varepsilon^2$. We use now the above reference values to perform the following nondimensionalization:

$$\begin{aligned} p &= \frac{\rho_0}{\mathrm{U}_0^d}\,\tilde{p}\,, \quad \rho = \rho_0\,\tilde{\rho}\,, \quad m = M_0\,\tilde{m}\,, \quad M = M_0\,\tilde{M}\,, \\ t &= \mathrm{T}_0\,\tilde{t}\,, \quad \mathbf{x} = \mathrm{X}_0\,\tilde{\mathbf{x}}\,, \quad \mathbf{v} = \mathrm{U}_0\,\tilde{\mathbf{v}}\,, \quad \psi = \frac{1}{\mathrm{U}_0^d}\,\tilde{\psi} \end{aligned} \tag{11.44}$$

According to the choice of the time scaling, that is, either $\mathrm{T}_0 = \mathrm{T}_{\text{drift}}$ or $\mathrm{T}_0 = \mathrm{T}_{\text{diff}}$, the change of variables will lead to different approximations of the transport equation associated with different time scales.

While both scalings are interesting, we focus on the diffusive one and use the reference time scale $T_0 = T_{\text{diff}}$ to rewrite the initial Equation (11.41) as

$$\varepsilon^2 \frac{\partial \tilde{p}}{\partial \tilde{t}}(\tilde{t}, \tilde{\mathbf{x}}, \tilde{\mathbf{v}}) + \varepsilon d \, \tilde{\mathbf{v}} \cdot \tilde{\nabla} \tilde{p}(\tilde{t}, \tilde{\mathbf{x}}, \tilde{\mathbf{v}}) = d \, \tilde{J}_m^{\mathcal{B}}(\tilde{t}, \tilde{\mathbf{x}}, \tilde{\mathbf{v}}) + d \, \tilde{J}_c^{\mathcal{B}}(\tilde{t}, \tilde{\mathbf{x}}, \tilde{\mathbf{v}}) \quad (11.45)$$

where $\tilde{\nabla}$ denotes the gradient associated with the rescaled space variable. We assume that contact guidance remains the mechanism that drives cell migration in the ECM and take the typical microscopic time $t_0 = 1/(\eta_m M_0)$. We remark that assuming dominant cell–cell interactions would simply lead to $t_0 = 1/(\eta_c \rho_0)$ and would not change the dimensionless structure. The right-hand-side operators of Equation (11.45) are thus rewritten as:

$$\tilde{J}_m^{\mathcal{B}}(\tilde{t}, \tilde{\mathbf{x}}, \tilde{\mathbf{v}}) = \tilde{M}(\tilde{\mathbf{x}}) \left(\tilde{\rho}(\tilde{t}, \tilde{\mathbf{x}}) \psi(\tilde{v}) \frac{\tilde{m}^e(\tilde{\mathbf{x}}, \hat{\mathbf{v}})}{2\tilde{M}(\tilde{\mathbf{x}})} [1 + \varepsilon \tilde{\mathcal{B}}(\tilde{t}, \tilde{\mathbf{x}}, \hat{\mathbf{v}})] - \tilde{p}(\tilde{t}, \tilde{\mathbf{x}}, \tilde{\mathbf{v}}) \right) \quad (11.46)$$

$$\tilde{J}_c^{\mathcal{B}}(\tilde{t}, \tilde{\mathbf{x}}, \tilde{\mathbf{v}}) = \alpha \, \tilde{\rho}(\tilde{t}, \tilde{\mathbf{x}}) \left(\tilde{\rho}(\tilde{t}, \tilde{\mathbf{x}}) \frac{\psi(\tilde{v})}{\mathcal{V}_d} [1 + \varepsilon \tilde{\mathcal{B}}(\tilde{t}, \tilde{\mathbf{x}}, \hat{\mathbf{v}})] - \tilde{p}(\tilde{t}, \tilde{\mathbf{x}}, \tilde{\mathbf{v}}) \right) \quad (11.47)$$

In Equation (11.47) we introduced $\alpha = (\eta_c \rho_0)/(\eta_m M_0)$, which reflects the influence of cell–cell interaction compared to cell–fiber interaction. Additionally, the bias $\mathcal{B}$ has been normalized using a reference value $\mathcal{S}_0$ for the generic density $\mathcal{S}$ and assuming that the coefficient Γ, which represents the typical sampling radius, is rescaled as $\Gamma = \ell_0 \tilde{\Gamma}$ with $\tilde{\Gamma} = \mathcal{O}(1)$. This means that the surrounding neighborhood in which a cell feels a signal is, at most, on the order of the microscopic length scale ℓ_0.

Starting from the rescaled governing Equation (11.45), we now present the derivation of its diffusive approximation.

11.4.2 Formal Limit of Diffusive Approximation

In this section we derive the formal diffusive limit of Equation (11.45), for which the interaction operators $\tilde{J}_m^{\mathcal{B}}$ and $\tilde{J}_c^{\mathcal{B}}$ are of the form discussed in the previous section. We recall that, as dictated by the cell number conservation in Equation (11.11), these operators satisfy

$$\int_V \tilde{J}_m^{\mathcal{B}}(\tilde{t}, \tilde{\mathbf{x}}, \tilde{\mathbf{v}}) \, d\tilde{\mathbf{v}} = \int_V \tilde{J}_c^{\mathcal{B}}(\tilde{t}, \tilde{\mathbf{x}}, \tilde{\mathbf{v}}) \, d\tilde{\mathbf{v}} = 0 \quad (11.48)$$

For the sake of clarity in this section, we suppressed the tilde notation and the dependence on time and space variables, which do not play a role in the calculations. Solutions for the distribution function p is sought in terms of the Hilbert expansion $p^{(\varepsilon)} = p^{(0)} + \varepsilon p^{(1)} + \mathcal{O}(\varepsilon^2)$. The right-hand-side operators are expanded in a similar manner as $J_m^{\mathcal{B}(\varepsilon)} = J_m^{\mathcal{B}(0)} + \varepsilon J_m^{\mathcal{B}(1)} + \varepsilon^2 J_m^{\mathcal{B}(2)} + \mathcal{O}(\varepsilon^3)$ and $J_c^{\mathcal{B}(\varepsilon)} = J_c^{\mathcal{B}(0)} + \varepsilon J_c^{\mathcal{B}(1)} + \varepsilon^2 J_c^{\mathcal{B}(2)} + \mathcal{O}(\varepsilon^3)$. We also introduce the density

$$\rho^{(0)} = \int_V p^{(0)}(\mathbf{v}) \, d\mathbf{v} \quad (11.49)$$

and recall that the vector $\mathbf{v}$ is, throughout this section, the normalized velocity that belongs to the normalized velocity domain. Substituting these expansions into the transport equation and equating the coefficients of ε^n gives rise to the following equations:

In ε^0:

$$0 = J_m^{\mathcal{B}(0)}(\mathbf{v}) + J_c^{\mathcal{B}(0)}(\mathbf{v}) \tag{11.50}$$

In ε^1:

$$\nabla \cdot [p^{(0)}(\mathbf{v})\mathbf{v}] = J_m^{\mathcal{B}(1)}(\mathbf{v}) + J_c^{\mathcal{B}(1)}(\mathbf{v}) \tag{11.51}$$

In ε^2:

$$\frac{\partial p^{(0)}(\mathbf{v})}{\partial t} + d\,\nabla \cdot [p^{(1)}(\mathbf{v})\mathbf{v}] = d\,J_m^{\mathcal{B}(2)}(\mathbf{v}) + d\,J_c^{\mathcal{B}(2)}(\mathbf{v}) \tag{11.52}$$

The solution $p^{(0)}$ to Equation (11.50), also called the equilibrium solution, can be determined explicitly as:

$$p^{(0)}(\mathbf{v}) = \frac{\rho^{(0)}\,\psi(v)}{M + \alpha\rho^{(0)}}\left(\frac{1}{2}m^e(\hat{\mathbf{v}}) + \frac{\alpha}{\mathcal{V}_d}\rho^{(0)}\right) \tag{11.53}$$

which leads to the associated mean cell velocity $\mathbf{U}^{(0)} = \mathbf{0}$. We summarize below the methodology that is detailed in [5] to derive the diffusive approximation:

1. We find the expression of $p^{(1)}$ from Equation (11.51).
2. We integrate Equation (11.52) over the velocity domain V, which yields

$$\frac{\partial \rho^{(0)}}{\partial t} + d\,\nabla \cdot \int_V p^{(1)}(\mathbf{v})\mathbf{v}\,d\mathbf{v} = \mathcal{O}(\varepsilon) \tag{11.54}$$

   thanks to property (11.48) that is satisfied for any order of the expansions $J_m^{\mathcal{B}(\varepsilon)}$ and $J_c^{\mathcal{B}(\varepsilon)}$.
3. We evaluate the first moment of $p^{(1)}$ that appears in Equation (11.54) and find

$$\int_V p^{(1)}(\mathbf{v})\mathbf{v}\,d\mathbf{v} = \pm\frac{\kappa}{d}\,\frac{\mathbb{T}^{(0)}\nabla\mathcal{S}}{\beta_S + \mathcal{S}}\,\rho^{(0)} - \frac{1}{d}\,\frac{\nabla \cdot \mathbb{P}^{(0)}}{M + \alpha\rho^{(0)}} \tag{11.55}$$

   where

$$\kappa = \Gamma \int_{|V|} \psi(v) v^d\,dv \tag{11.56}$$

   On the right-hand side of Equation (11.55) the first term relates to the bias accounting for external factors with normalized signaling molecule density $\mathcal{S}$. Additionally we introduce the tensor

$$\mathbb{T}^{(0)} = \frac{M\mathbb{D} + \alpha\rho^{(0)}\,\mathbb{I}}{M + \alpha\rho^{(0)}} \tag{11.57}$$

where $\mathbb{D}$ is the orientation tensor defined in Equation (11.8) to describe macroscopically the angular fiber distribution. The last term in Equation (11.55) relates to the main cell motion in the ECM. There, $\mathbb{P}^{(0)}$ corresponds to the pressure tensor defined in Equation (11.5) that is evaluated at equilibrium:

$$\mathbb{P}^{(0)} = \int_V p^{(0)}(\mathbf{v})\mathbf{v} \otimes \mathbf{v}\, d\mathbf{v} = \sigma \rho^{(0)} \mathbb{T}^{(0)} \quad \text{with} \quad \sigma = \int_{|V|} \psi(v) v^{d+1}\, dv \tag{11.58}$$

4. We substitute Expression (11.55) into (11.54) and take the limit $\varepsilon \to 0$ to obtain the diffusive approximation:

$$\frac{\partial \rho^{(0)}}{\partial t} \pm \kappa \nabla \cdot \left[\rho^{(0)} \frac{\mathbb{T}^{(0)} \nabla \mathcal{S}}{\beta_{\mathcal{S}} + \mathcal{S}} \right] = \sigma \nabla \cdot \left[\frac{\nabla \cdot \left[\mathbb{T}^{(0)} \rho^{(0)} \right]}{M + \alpha \rho^{(0)}} \right] \tag{11.59}$$

Equation (11.59) is our governing macroscopic model of cell migration within the ECM. It is a drift-diffusion equation for the cell density $\rho^{(0)}$ that is the leading order of the approximation $\rho^{(\varepsilon)}$ assumed to converge to ρ. We use this assumption and cancel the superscript to reformulate our macroscopic model equation in the generic form:

$$\frac{\partial \rho}{\partial t} = \nabla \cdot \left[\sigma \left(\frac{\mathbb{T} \nabla \rho}{M + \alpha \rho} + \frac{\rho \nabla \cdot \mathbb{T}}{M + \alpha \rho} \right) \mp \rho \mathbb{X} \nabla \mathcal{S} \right] \tag{11.60}$$

In Equation (11.60), the first term on the right-hand side relates to anisotropic and nonlinear diffusion, while the second term may be seen as a drift term that accounts for spatial variation of the ECM anisotropy. The last term is also a drift term, caused by the gradient of signaling molecule density $\mathcal{S}$ with a coefficient called sensitivity [29]. The positive sign refers to a drift in direction of the gradient (attractive signal) while the negative sign refers to the opposite direction (repellent signal). In our model the sensitivity is actually a tensor that reads

$$\mathbb{X} = \frac{\kappa \mathbb{T}}{\beta_{\mathcal{S}} + \mathcal{S}} \tag{11.61}$$

whose signal dependence comes from the structure of Equation (11.36) that we proposed as the simplest choice of the bias. In Equation (11.61), the tensor $\mathbb{T}$ accounts for the local ECM orientation in taxis modeling. Therefore, sensitivity Expression (11.61) belongs to the set of signal-dependent sensitivity models, but extended to ECM effects. When the ECM is isotropic, the drift velocity $\mathbb{X}\nabla\mathcal{S}$ becomes $\chi\nabla\mathcal{S}$, where the sensitivity χ is the extension of the logistic model [24] that corresponds to $\beta_{\mathcal{S}} = 0$. In the next section we illustrate the effect of this drift term and put in evidence its role for the phenomena introduced in Section 11.3.4, that is, chemotaxis, haptotaxis, and repellent quorum sensing.

11.5 Numerical Illustrations

We consider the 2-D square domain $\mathcal{D}$ of side length X_0 containing N_c cells and describe the ECM with N_m fibers whose angular distribution is described later in this section. We take the reference values $\rho_0 = N_c/X_0^2$ and $M_0 = N_m/X_0^2$. Therefore we deal throughout this section with the dimensionless configuration in the domain $[0, 1] \times [0, 1]$ in which we have

$$\int_{[0,1]\times[0,1]} \rho(t, \mathbf{x})\, d\mathbf{x} = \int_{[0,1]\times[0,1]} M(\mathbf{x})\, d\mathbf{x} = 1 \tag{11.62}$$

where $\mathbf{x} = (x, y)$ are the normalized space variables.

To perform numerical simulations, we use a mass-preserving finite volume method to solve numerically the generic Equation (11.60) with a symmetric splitting scheme for both directions and operators. The nonlinear parabolic component is solved with a Crank-Nicholson scheme, in which the implicit nonlinear term is treated by a Beam and Warming scheme. The hyperbolic component is solved through a high-resolution wave-propagation algorithm developed for spatially varying flux, using Van-Leer limiter. Periodic boundary conditions are used for the sake of simplicity.

To avoid additional complexity, we also assume that all cells are moving with the same velocity. We already presented this assumption that leads to the particular Expression (11.15) for the speed redistribution function ψ. This assumption yields $\sigma = 1$ in Equation (11.60) and $\kappa = \Gamma$ in Equation (11.61), where the dimensionless parameter κ may be seen as a Peclet number that compares advective processes to diffusive ones.

In the following we illustrate the effects of fiber guidance on cell migration in an inhomogeneous and anisotropic ECM. We focus first on density and anisotropy effects and do not consider chemical factors. We will consider their influence in Section 11.5.2.

11.5.1 Spreading in Heterogeneous and Anisotropic ECM

In this section we illustrate the influence of ECM heterogeneity and anisotropy on cell motion and do not consider taxis. Thus we provide some numerical results for the following equation:

$$\frac{\partial \rho}{\partial t} = \nabla \cdot \left[\frac{\nabla \cdot [\mathbb{T}\rho]}{M + \alpha\rho} \right] \tag{11.63}$$

In Equation (11.63) the only parameter is $\alpha = (\eta_c \rho_0)/(\eta_m M_0) = (\eta_c N_c)/(\eta_m N_m)$ because we took $\rho_0 = N_c/X_0^2$ and $M_0 = N_m/X_0^2$ as reference values. We use the value $\alpha = 0.01$ and will maintain it for all simulations.

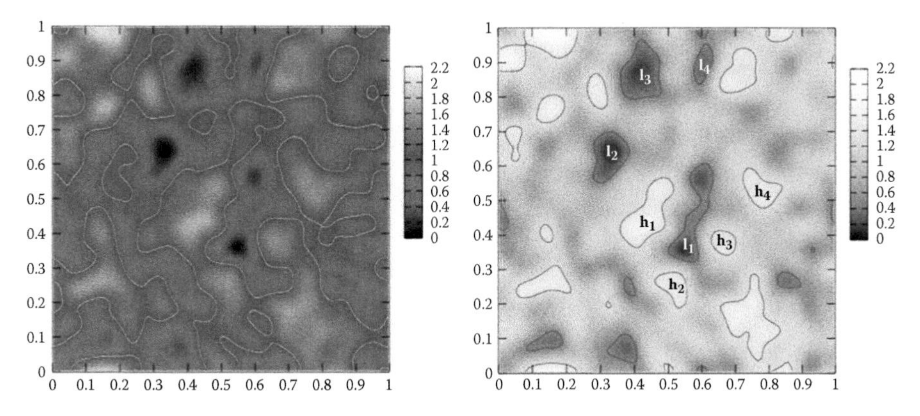

FIGURE 11.6 Contour levels of the fiber density $M(x, y)$. (Left) The plain line corresponds to $M(x, y) = 1$. (Right) Areas of high ($M(x, y) \geq 1.4$) and low ($M(x, y) \leq 0.5$) fiber density. Areas that impact on motion are indicated and respectively named $\mathbf{h}_i$ for high-density regions and $\mathbf{l}_i$ for low-density ones.

This value corresponds to a "dilute" configuration in which the surrounding ECM effects are dominant.

To account for the inhomogeneity of the ECM, we choose a randomly distributed fiber density $M(x, y)$. The characteristics of this distribution are presented in Figure 11.6. To account for the anisotropy of the ECM, we choose now a particular orientation tensor $\mathbb{D}$, namely the diagonal configuration in Figure 11.2(b). In this case the anisotropy of the matrix is completely described by the first element of the diagonal denoted by D while the second element is $2 - D$. Therefore the anisotropy of the ECM is described by a randomly space-dependent coefficient $D(x, y)$ later referred to as the *anisotropy coefficient*. The corresponding random distribution and areas associated with high anisotropy are presented in Figure 11.7.

We choose the initial condition $\rho(t = 0, x, y)$ arbitrarily. It is represented in the first row of Figure 11.8 by its contour levels. The influence of the ECM is shown in the columns (b) and (c) of Figure 11.8, which must be compared with the homogeneous and isotropic case presented in column (a).

In column (b) of Figure 11.8, only heterogeneity is considered. The circular shape of the cell density contour levels changes due to the spatial variations of the fiber density. While the areas of high fiber density slightly affect the evolution, an important effect of the low ones is observed. An accelerated spreading is observed early in zones $\mathbf{l}_1$ and $\mathbf{l}_2$, which deforms the contour levels. When area $\mathbf{l}_3$ is later reached (see last snapshot), the movement is again affected and accelerated. Concerning the weak effect of high fiber density, it has already been shown in [7] that the effect exists but needs larger density values to be clearly observed. Attention now focuses on anisotropy effects, that is, column (c) of Figure 11.8. The elongated shape of the central part is caused by the horizontal fiber orientation in $\mathbf{x}_1$, while vertical deformations of the

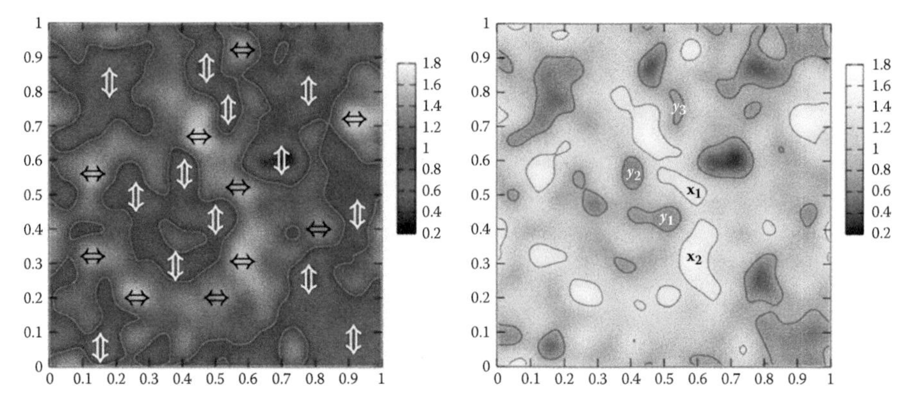

FIGURE 11.7 Contour levels of the anisotropy coefficient $D(x, y)$. (Left) The plain line corresponds to $D(x, y) = 1$ (i.e., a locally isotropic fiber distribution) and the arrows evidence the main local orientation of the fiber network. (Right) Areas of high anisotropy of the ECM. The two contour levels indicate respectively zones with preferentially horizontal orientation ($D(x, y) \geq 1.33$) of the fiber network and preferentially vertical orientation ($D(x, y) \leq 0.77$). Areas that impact on motion are indicated and respectively named $\mathbf{x}_i$ when the fiber orientation is mainly horizontal and $\mathbf{y}_i$ when mainly vertical.

contour levels appear in areas $\mathbf{y}_1$ and $\mathbf{y}_2$ due to the vertical fiber orientation. Additionally, an acceleration of the front is also observed horizontally in $\mathbf{x}_2$ and vertically in $\mathbf{y}_3$.

To illustrate in the next section the effects of various types of external factors, we introduce the reference configuration of a population spreading in a heterogeneous and anisotropic ECM configuration. We consider both heterogeneity through the nonuniform fiber density M shown in Figure 11.6 and anisotropy through the anisotropic coefficient D shown in Figure 11.7. Therefore the simulation of Figure 11.9 is actually a combination of the effects, respectively, evidenced in columns (b) and (c) of Figure 11.8. We use it as a reference in what follows to illustrate the effects of the environmental factors previously introduced.

11.5.2 Illustration of Taxis Effects

In the previous section we evidenced the effects of ECM heterogeneity and anisotropy on the spreading of a cell population. In this section we aim to illustrate taxis mechanisms that we modeled through Equation (11.60) by the drift term associated with the gradient of a generic signaling molecule density $\mathcal{S}$.

We start with the illustration of positive chemotaxis and take $\mathcal{S} \equiv \mathcal{C}$, where $\mathcal{C}$ denotes the normalized density of an exogenous chemoattractant. We consider a time-independent Gaussian distribution of $\mathcal{C}$ centered at $x = 0.3$,

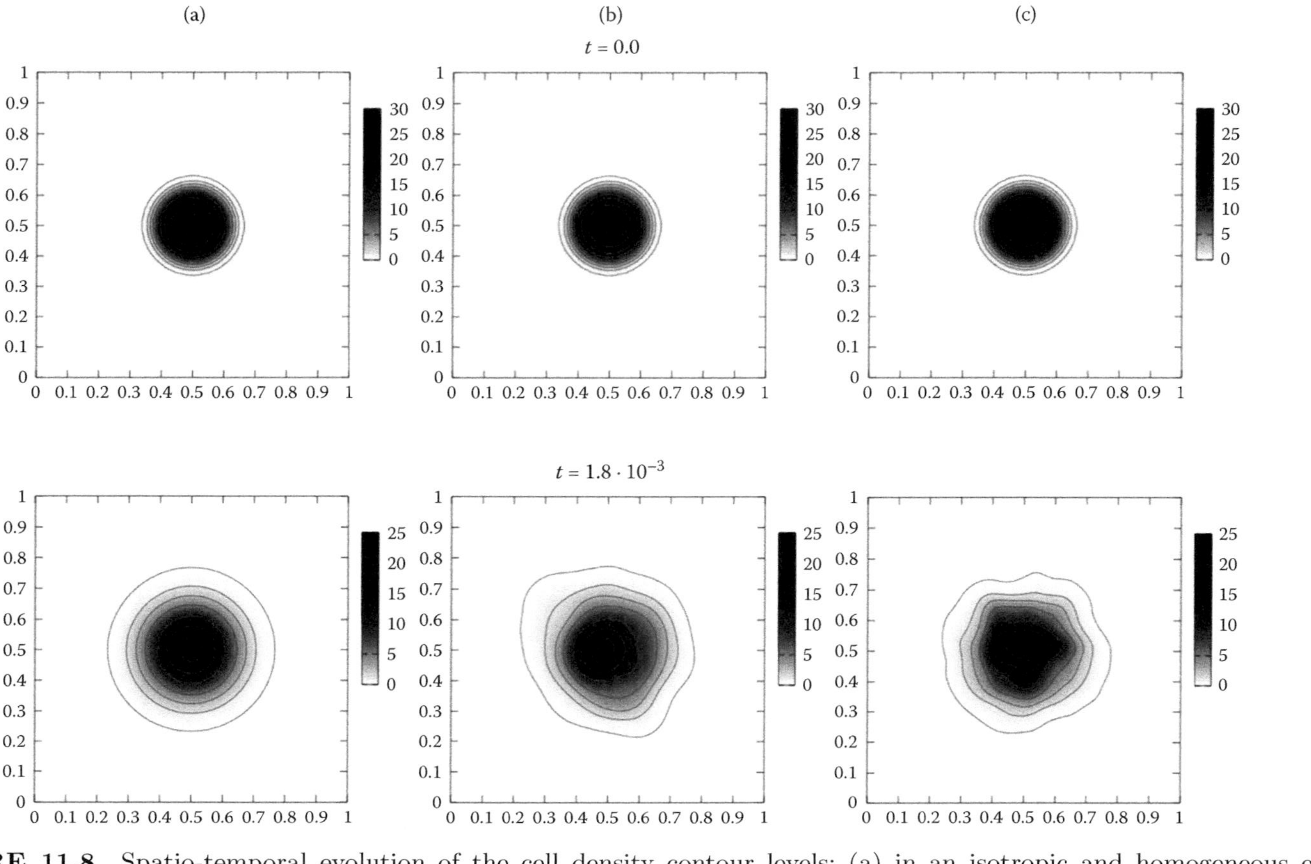

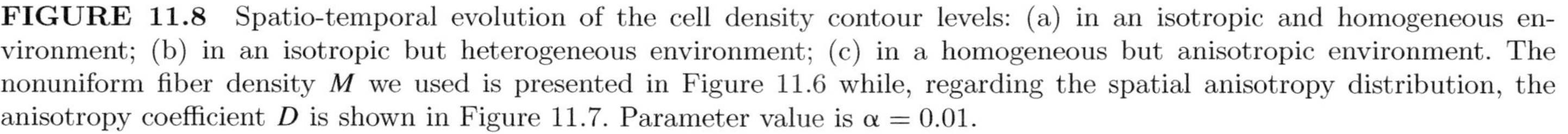

FIGURE 11.8 Spatio-temporal evolution of the cell density contour levels: (a) in an isotropic and homogeneous environment; (b) in an isotropic but heterogeneous environment; (c) in a homogeneous but anisotropic environment. The nonuniform fiber density M we used is presented in Figure 11.6 while, regarding the spatial anisotropy distribution, the anisotropy coefficient D is shown in Figure 11.7. Parameter value is $\alpha = 0.01$.

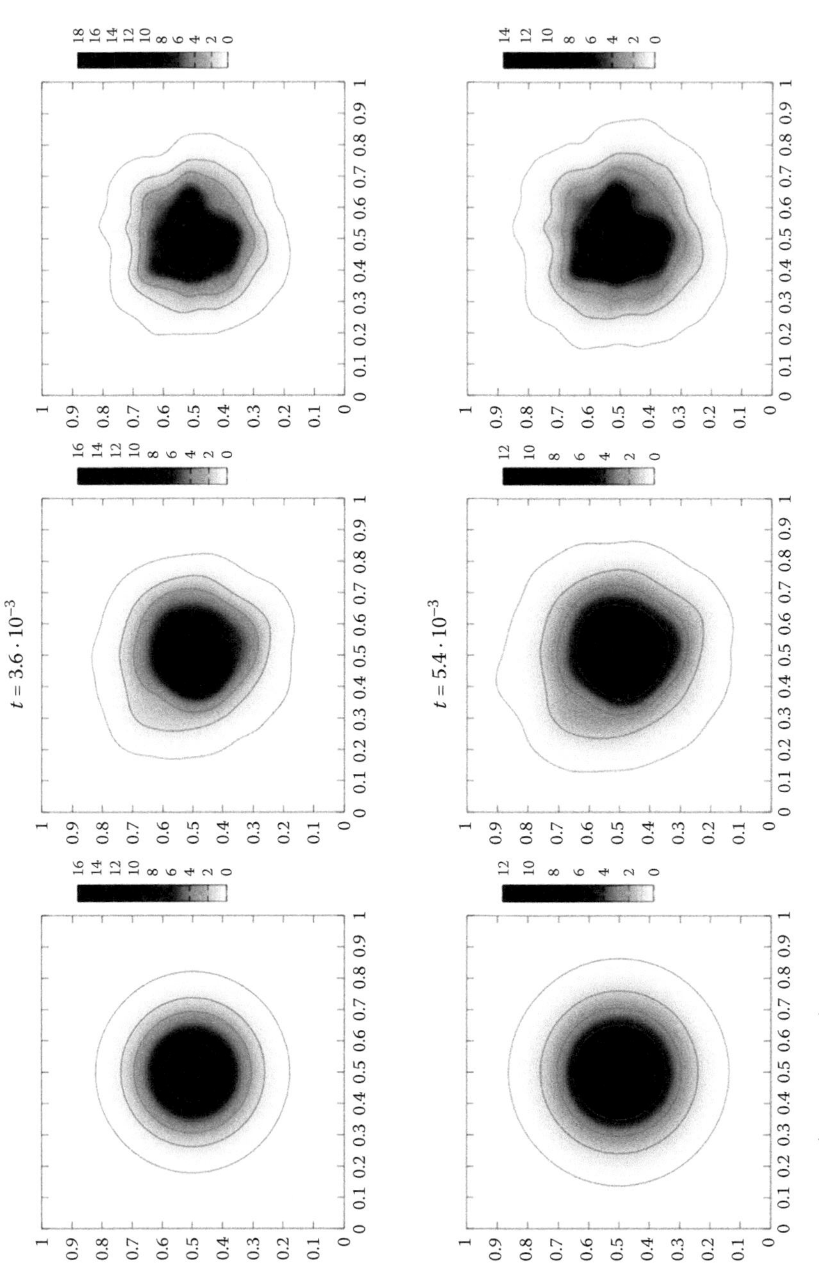

FIGURE 11.8 *(Continued).*

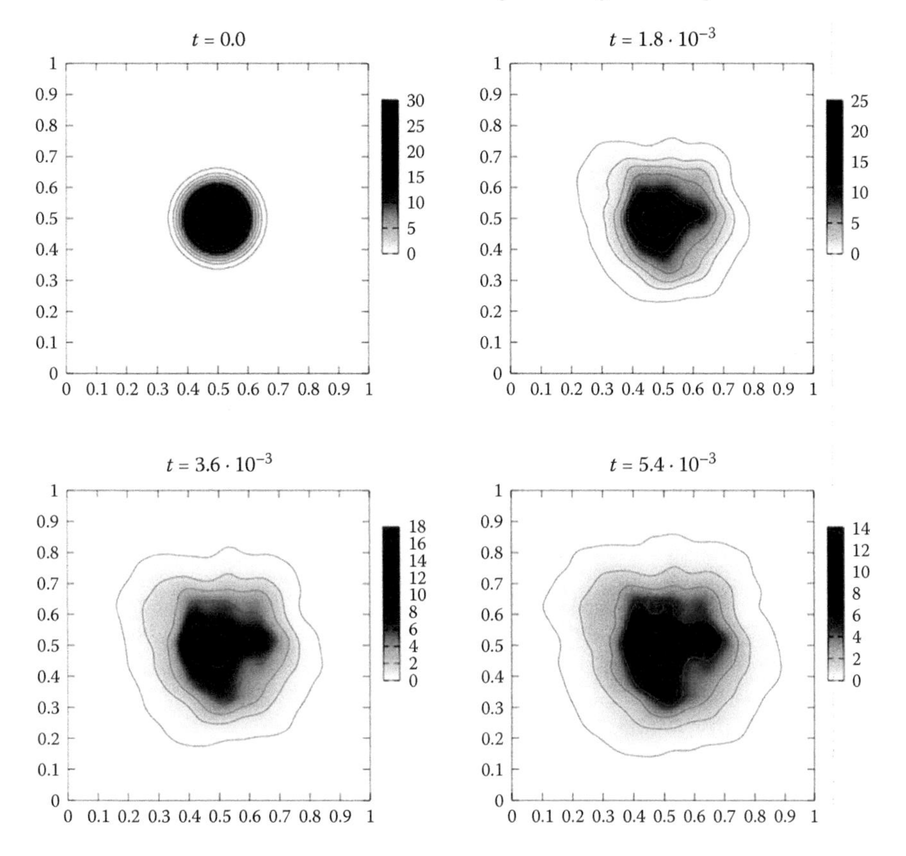

FIGURE 11.9 Reference configuration. Spatio-temporal evolution of the cell density contour levels in an anisotropic and heterogeneous environment. The corresponding fiber density M and anisotropy coefficient D we used are shown in Figures 11.6 and 11.7, respectively.

$y = 0.3$ that, thanks to the adapted reference value used to nondimensionalise our problem, satisfies

$$\int_{[0,1]\times[0,1]} \mathcal{C}(\mathbf{x})\, d\mathbf{x} = 1 \tag{11.64}$$

The extension to a time-dependent distribution of a diffusive endogenous attractant can be easily performed by introducing an evolution equation for $\mathcal{C}$ to form a coupled system of the type of Keller-Segel as generally used (see [4], for example). However, here we simply illustrate in Figure 11.10 the effect of a stationary nonuniform distribution of chemoattractant (that is, neither

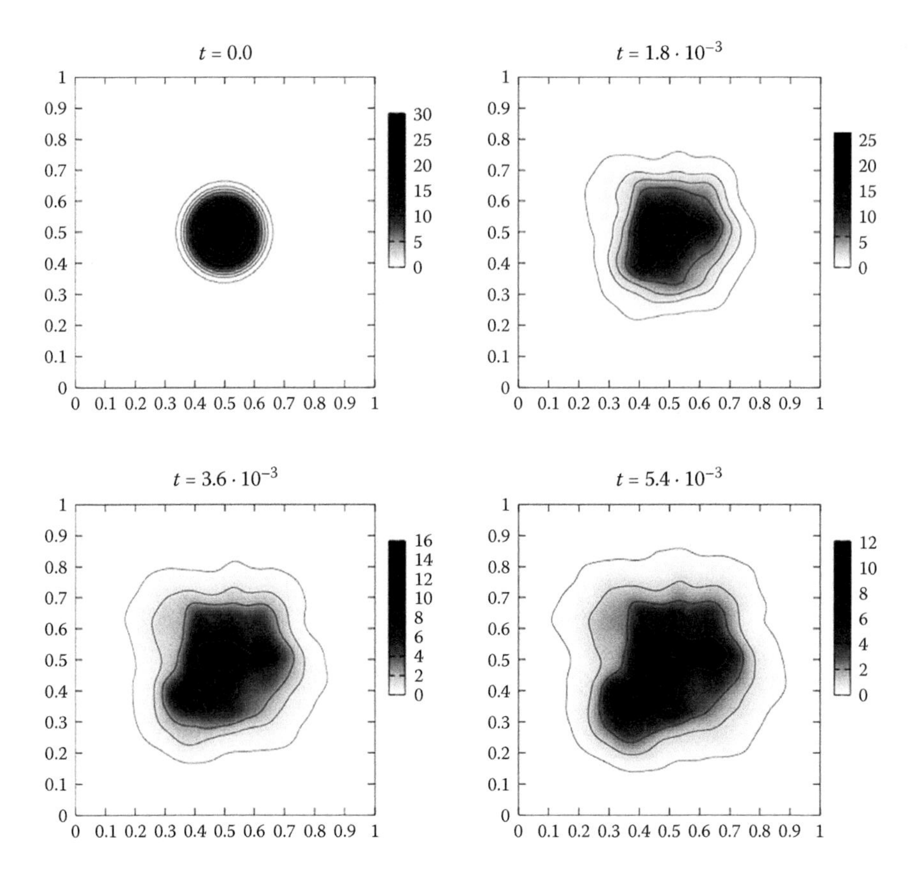

FIGURE 11.10 Chemotaxis effects. Spatio-temporal evolution of the cell density contour levels in an anisotropic and heterogeneous environment biased by a time-independent chemoattractant $\mathcal{C}$ spatially distributed as a Gaussian centered at $x = 0.3$, $y = 0.3$ with variance 0.005. Parameter values are $\Gamma = 1$ and $\beta_{\mathcal{C}} = 1$.

decay nor diffusion of the chemical) and present a numerical solution to our drift-diffusion model equation for chemotactic migration in the ECM:

$$\frac{\partial \rho}{\partial t} + \Gamma \nabla \cdot \left[\rho \frac{\mathbb{T} \nabla \mathcal{C}}{\beta_{\mathcal{C}} + \mathcal{C}} \right] = \nabla \cdot \left[\frac{\nabla \cdot [\mathbb{T} \rho]}{M + \alpha \rho} \right] \tag{11.65}$$

The effect of the chemoattractant is straightforward: the cell density expansion is strongly modified in the lower-left area of the domain, and cells accumulate toward high chemical concentration values. The expansion in other areas remains almost unaffected due to the limited operating range of the chemoattractant. It is worth mentioning that although we only provide a

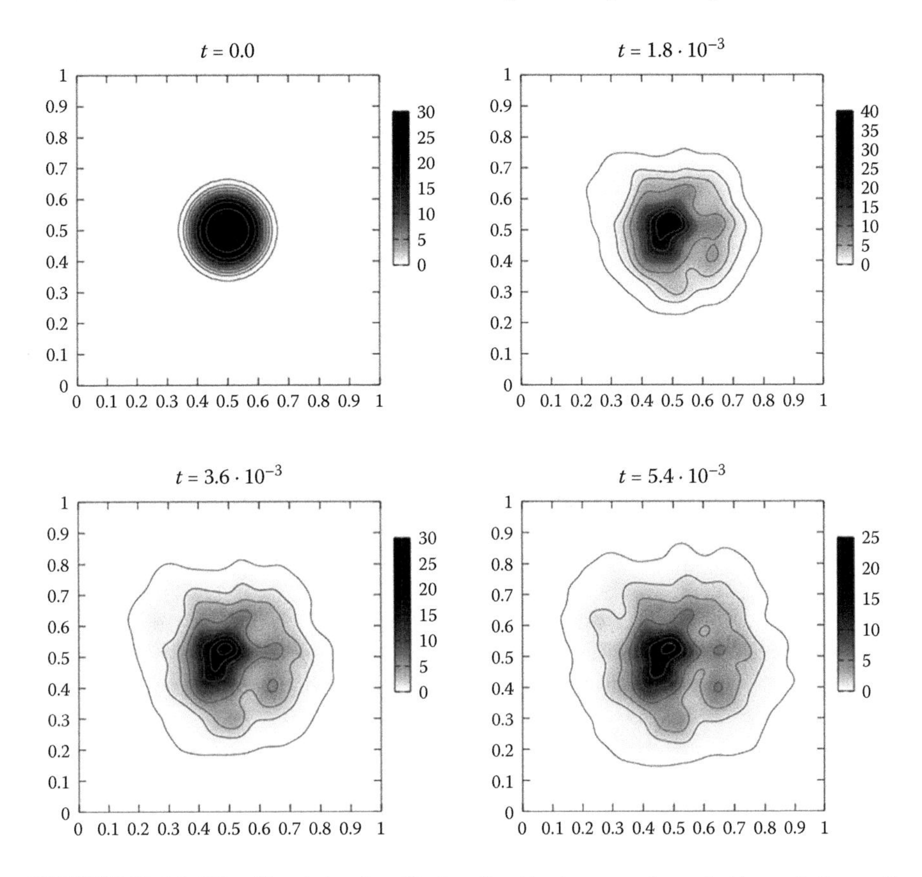

FIGURE 11.11 Haptotaxis effects. Spatio-temporal evolution of the cell density contour levels in an anisotropic and heterogeneous environment biased by haptotactic cues. Parameter values are $\Gamma = 1$ and $\beta_{\mathcal{H}} = 0$.

simple illustration here, Equation (11.65) is the only continuum representation of chemotactic effect in an anisotropic environment.

The second application of our model is also a significant improvement to former approaches. We illustrate in Figure 11.11 the effects of haptotactic cues by taking $\mathcal{S} \equiv M$ and presenting a numerical solution to our Drift-diffusion model equation for haptotactic migration in the ECM:

$$\frac{\partial \rho}{\partial t} + \Gamma \, \nabla \cdot \left[\rho \, \frac{\mathbb{T} \nabla M}{\beta_{\mathcal{H}} + M} \right] = \nabla \cdot \left[\frac{\nabla \cdot [\mathbb{T} \rho]}{M + \alpha \rho} \right] \tag{11.66}$$

The effects of a drift velocity in the direction of the fiber density gradient are not surprising but particularly significant. The spatio-temporal evolution shown in Figure 11.11 differs strongly from the reference one in Figure 11.9.

The areas $\mathbf{l}_i$ of low fiber density defined in Figure 11.6 are clearly circumvented, while aggregation processes are triggered toward the areas $\mathbf{h}_i$ of high fiber density. We suggest that haptotactic cues may increase heterogeneities in the expansion of a cellular population. By this example we also demonstrate the importance of our model. This last example has the ability to strongly reflect asymmetrical spreading when various effects of the ECM are accounted for, in particular fiber density, anisotropy and haptotaxis.

We conclude with the application of our framework to repellent quorum sensing and take $\mathcal{S} \equiv \rho$ to state our Diffusion model equation for repellent quorum sensing migration in the ECM:

$$\frac{\partial \rho}{\partial t} = \nabla \cdot \left[\frac{\nabla \cdot [\mathbb{T}\rho]}{M + \alpha\rho} + \frac{\Gamma\,\rho}{\beta_{\mathcal{R}} + \rho} \mathbb{T}\nabla\rho \right] \tag{11.67}$$

The mathematical structure of governing Equation (11.67) adds to the original unbiased term a second diffusion one with a nonlinear coefficient that results from the self-repellent mechanism. We remark that $\beta_{\mathcal{R}}$ is not necessary to avoid singularities in the coefficient and take the value $\beta_{\mathcal{R}} = 0$. In this case, the nonlinearity of the additional diffusion coefficient simplifies and provides linear diffusion whose intensity is reflected by Γ. However, the influence of the anisotropy remains present through the tensor $\mathbb{T}$. Thus, repellent quorum sensing acts as an additional diffusion that does not take into account density effects. Therefore the main effects observed in Figure 11.12, compared with the reference configuration of Figure 11.9, are diffusion related and namely: a faster spreading of the cells from the center of the domain and a faster decrease of the central mass. Additionally, a smoothing effect is observed, caused by weaker density effects while anisotropy ones remain significant.

We provided throughout this last section numerical examples of our model for various phenomena of importance that occur during cell migration in the ECM. We qualitatively compared results obtained for the spreading of a cell population in a heterogeneous or anisotropic ECM, to a uniform and isotropic one. We put in evidence significant differences, which lead us to point out the pertinence of the type of model we derived. Indeed, current modeling of complex biological phenomena involving cell migration in the ECM does not usually take both heterogeneity and anisotropy ECM properties into account. We believe that our framework can provide significant improvements to the current modeling when ECM description is required. We finally conclude here by pointing out the didactical way with which we chose to present our work in this chapter. Some assumptions or simplifications could have seemed naïve to an expert reader. However, we wanted to contain the complexity level of such approaches to promote their development. We hope that the interested reader will have found enough material within this chapter to acquire a good overview and develop his or her own modeling.

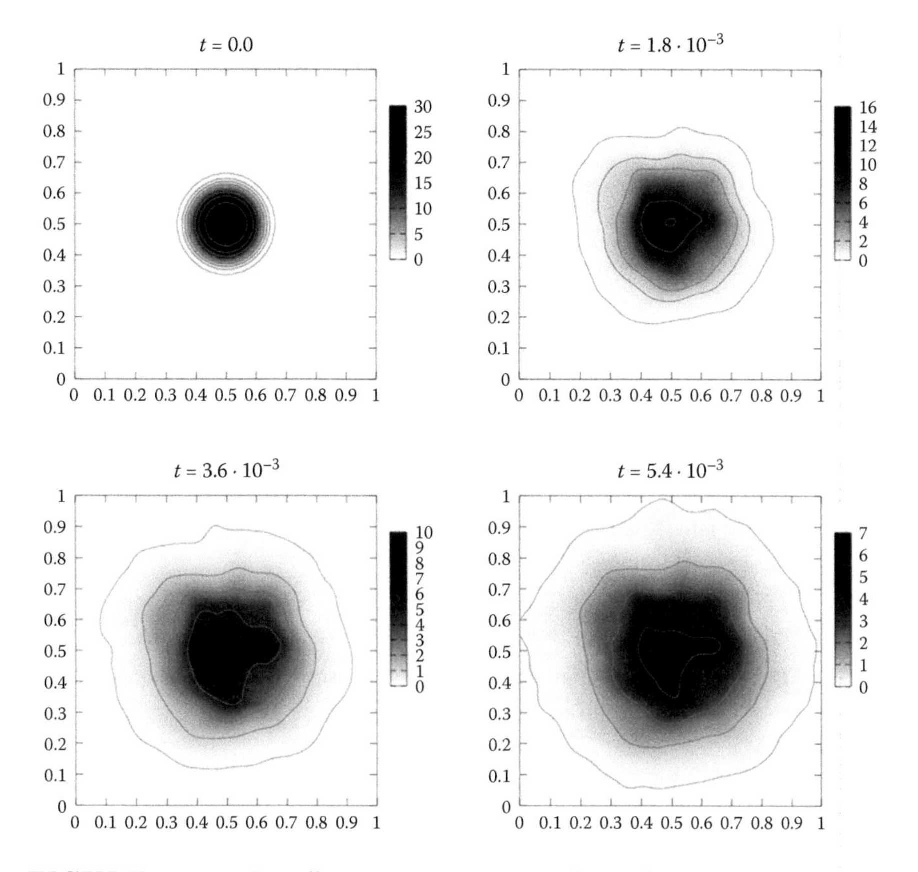

FIGURE 11.12 Repellent quorum sensing effects. Spatio-temporal evolution of the cell density contour levels affected by repellent quorum sensing in an anisotropic and heterogeneous environment. Parameter values are $\Gamma = 1$ and $\beta_{\mathcal{R}} = 0$.

Acknowledgments

We are very grateful to Katarina Wolf for the wonderful images she provided to illustrate the contact guidance mechanism. We further acknowledge Haralambos Hatzikirou for insightful discussions, and Edward Flach for fruitful comments concerning the contents of this chapter.

Finally, we thank Andreas Deutsch for his support during the preparation of this book and, in particular, for this chapter.

This work has been partially supported by a grant Biologistics of the Daimler-Benz foundation.

References

[1] W. Alt (1980). Biased random walk models for chemotaxis and related diffusion approximations. *J. Math. Biol.* 9:147–177.

[2] A.Q. Cai, K.A. Landman, and B.D. Hughes (2006). Modelling directional guidance and motility regulation in cell migration. *Bull. Math. Biol.* 68:25–52.

[3] C. Cercignani, R. Illner, and M. Pulvirenti (1994). *The Mathematical Theory of Dilute Gases*, New York, Springer.

[4] F.A.C.C. Chalub, P.A. Markowich, B. Perthame, and C. Schmeiser (2004). Kinetics models for chemotaxis and their drift-diffusion limits. *Monatsh. Math.* 142:123–141.

[5] A. Chauviere (2009). Derivation of taxis-biased models for cell migration in the extracellular matrix. Submitted.

[6] A. Chauviere and I. Brazzoli (2006). On the discrete kinetic theory for active particles. Mathematical tools. *Math. Comp. Mod.* 43:933–944.

[7] A. Chauviere, T. Hillen, and L. Preziosi (2007). Modeling the motion of a cell population in the extracellular matrix. *Discr. Cont. Dyn. Syst.* Suppl.:250–259.

[8] A. Chauviere, T. Hillen, and L. Preziosi (2007). Modeling cell movement in anisotropic and heterogeneous network tissues. *Networks Heterogeneous Media* 2(2):333–357.

[9] J.M. Davis, J. St-John, and H. Tak Cheung (1990). Haptotactic activity of fibronectin on lymphocyte migration *in vitro*. *Cell Immunol.* 129(1):67–79.

[10] R.B. Dickinson (2000). A generalized transport model for biased cell migration in an anisotropic environment. *J. Math. Biol.* 40:97–135.

[11] Y. Dolak and C. Schmeiser (2005). Kinetic models for chemotaxis: hydrodynamic limits and spatio-temporal mechanisms. *J. Math. Biol.* 51:595–615.

[12] C. Eckerich, S. Zapf, U. Ulbricht, S. Müller, R. Fillbrandt, M. Westphal, and K. Lamszus (2006). Contactin is expressed in human astrocytic gliomas and mediates repulsive effects. *Glia* 53(1):1–12.

[13] R. Erban and H.G. Othmer (2006). Taxis equations for amoeboid cells. *J. Math. Biol.* 54:847–885.

[14] F. Filbet, P. Laurençot, and B. Perthame (2005). Derivation of hyperbolic models for chemosensitive movement. *J. Math. Biol.* 50:189–207.

[15] P. Friedl (2004). Prespecification and plasticity: shifting mechanisms of cell migration. *Curr. Opin. Cell Biol.* 16:14–23.

[16] P. Friedl and E.-B. Bröcker (2000). The biology of cell locomotion within three dimensional extracellular matrix. *Cell Motility Life Sci.* 57:41–64.

[17] E. Geigant, K. Ladizhansky, and A. Mogilner (1998). An integrodifferential model for orientational distributions of F-actin in cells. *SIAM J. Appl. Math.* 59(3):787–809.

[18] D. Grünbaum (2000). Advection-diffusion equations for internal state-mediated random walks. *SIAM J. Appl. Math.* 61(1):43–73.

[19] N.A. Hill and D.-P. Häder (1997). A biased random walk model for the trajectories of swimming micro-organisms. *J. Theor. Biol.* 186(4):503–526.

[20] T. Hillen (2005). On the L^2-closure of transport equations: the general case. *Discr. Cont. Dyn. Syst. B* 5:299–318.

[21] T. Hillen (2006). M^5-mesoscopic and macroscopic models for mesenchymal motion. *J. Math. Biol.* 53(4):585–616.

[22] T. Hillen and H.G. Othmer (2000). The diffusion limit of transport equations derived from velocity jump processes. *SIAM J. Appl. Math.* 61(3):751–775.

[23] T. Hillen, K. Painter, and C. Schmeiser (2007). Global existence for chemotaxis with finite sampling radius. *Discr. Cont. Dyn. Syst. B* 7(1):125–144.

[24] T. Hillen and K. Painter (2009). A user's guide to PDE models for chemotaxis. *J. Math. Biol.* 58:183–217.

[25] K. Kang, B. Perthame, A. Stevens, and J.J.L Velazquez (2009). An integro-differential equation model for alignment and orientational aggregation. *J. Differ. Equations* 246(4):1387–1421.

[26] C.D. Levermore (1996). Moment closure hierarchies for kinetic theories. *J. Stat. Phys.* 83:1021–1065.

[27] A. Mogilner and L. Edelstein-Keshet (1995). Selecting a common direction I. How orientational order can arise from simple contact responses between interacting cells. *J. Math. Biol.* 33(6):619–660.

[28] H.G. Othmer, S.R. Dunbar, and W. Alt (1988). Models of dispersal in biological systems. *J. Math. Biol.* 26(3):263–298.

[29] H.G. Othmer and A. Stevens (1997). Aggregation, blowup, and collapse: the ABC's of taxis in reinforced random walks. *SIAM J. Appl. Math.* 57(4):1044–1081.

[30] H.G. Othmer and T. Hillen (2002). The diffusion limit of transport equations II: chemotaxis equations. *SIAM J. Appl. Math* 62(4):1222–1250.

[31] K.J. Painter (2009) Modelling cell migration strategies in the extracellular matrix. *J. Math. Biol.* 58:511–543.

[32] R.T. Tranquillo and V.H. Barocas (1997). A continuum model for the role of fibroblast contact guidance in wound contraction, in *Dynamics of Cell and Tissue Motion*, Eds. W. Alt, A. Deutsch, and G. Dunn, Basel, Birkhauser-Verlag, pp. 159–164.

[33] C. Villani (2002). Limites hydrodynamiques de l'équation de Boltzmann, in *Séminaire Bourbaki Volume 2000/2001*, Paris, Société Mathématique de France, 893-1–893-39.

[34] M.A. Wagle and R.T. Tranquillo (2000). A self-consistent cell flux expression for simultaneous chemotaxis and contact guidance in tissues. *J. Math. Biol.* 41:315–330.

[35] T.E. Werbowetski, R. Bjerkvig, and R.F. Del Maestro (2004). Evidence for a secreted chemorepellent that directs glioma cell invasion. *J. Neurobiol.* 60(1):71–88.

[36] P. Williams, K. Winzer, W.C. Chan, and M. Camara (2007). Look who's talking: communication and quorum sensing in the bacterial world. *Phil. Trans. R. Soc. B* 362:1119–1134.

[37] K. Wolf, R. Müller, S. Borgmann, E.-B. Bröcker, and P. Friedl (2003). Amoeboid shape change and contact guidance: T-lymphocyte crawling through fibrillar collagen is independent of matrix remodeling by MMPs and other proteases. *Blood* 102(9):3262–3269.

Chapter 12

Mathematical Modeling of Cell Adhesion and Its Applications to Developmental Biology and Cancer Invasion

Alf Gerisch[1] **and Kevin J. Painter**[2]

Contents

12.1 Introduction

From the earliest embryonic stages to the complexity of the adult, the ability of cell populations to adhere to each other or the surrounding extracellular matrix (ECM) is of critical importance to the survival of the organism.

[1]A.G. gratefully acknowledges financial support by the Division of Mathematics, University of Dundee during a long-term visit in 2007 introducing him to the topic.

[2]K.J.P. is supported in part by the Integrative Cancer Biology Program Grant CA113004 from the U.S. National Institutes of Health, by a BBSRC grant BB/D019621/1 to the Centre for Systems Biology at Edinburgh, and by the Mathematical Biosciences Institute, Ohio State University, Columbus, Ohio.

During embryonic development, carefully regulated adhesion plays a fundamental role directing the various cell populations into the developing organs while maintaining strong adhesive contacts is essential in preserving the integrity and structure of the adult tissues. The manifest importance of cellular adhesion is exposed by its abnormal functioning in a wide variety of pathological conditions, including malignant cancer growth (e.g., [49]) and cardiovascular diseases (e.g., [40]).

Adhesion can generally be classified into two principal forms: *cell–cell adhesion* and *cell–matrix adhesion*. The former defines the direct binding between cells through the creation of transmembrane protein–protein complexes, the prototype examples of which are the strong contacts maintaining epithelial structures such as the epidermal skin layer. The latter describes the attachment of cells to the surrounding ECM, the scaffold support surrounding cells and composed of a variety of molecules including collagens, fibronectins, and laminins. While the ECM is present in all tissues, its prevalence in connective tissues such as the dermal skin layer makes cell–matrix adhesion particularly important for stromal populations such as fibroblasts and immune cells.

The control of cell–cell and cell–matrix adhesion is fundamentally determined through the expression and regulation of a wide variety of membrane-based proteins, the cell adhesion molecules (CAMs); for a general review, see [1]. Four principal families of CAMs have been classified: the cadherins (e.g., E-cadherin, N-cadherin); the immunoglobin superfamily (e.g., NCAM, EpCAM); the integrins; and the selectins. Members of these families generally consist of transmembrane molecules with an intracellular domain linking to intracellular signaling pathways and an extracellular domain connecting to other cells or the matrix. Adhesion is achieved through protein–protein coupling of the extracellular domain to form either *homophilic* interactions (i.e., binding between two proteins of the same type, such as E-cadherin–E-cadherin) or *heterophilic* interactions (binding between two molecules of different types).

The cadherins form a large family of transmembrane adhesion molecules widely recognized for their capacity to mediate direct cell–cell adhesion, although their function extends to a host of other cellular processes, ranging from apoptosis to signaling (for reviews on the behavior and function of cadherins, see [36,57]). Classic cadherins tend to form homophilic interactions in the intermembrane space separating two cells, although heterophilic interactions can also occur (e.g., E-cadherin–P-cadherin), albeit with different adhesive intensity [25]. The transmembrane binding fastens cells in a zipper-like manner, conferring a key role to cadherins in all aspects of an organism's lifespan, from coordinating multicellular tissue movements during development to maintaining the tissue structure of the adult. A wide variety of cadherins have been identified, distributed across different cell populations. For example, the E-cadherins are mainly associated with epithelial cell populations, while more migratory mesenchymal cells (e.g., fibroblasts) tend to favor N-cadherins [79].

The integrins form the dominant CAMs regulating adhesion to the extracellular matrix [10]. The extracellular domain couples the cells to ligands of the

ECM to create various types of cell–matrix adhesion structures that, in turn, modulate the intracellular component to interact with intracellular signaling. These adhesion structures have the capacity to recruit additional molecules (e.g., matrix proteases) and therefore locally alter the structure of the ECM. Dynamic control of cell–matrix adhesion is crucial to the migration of cells in ECM-rich environments, such as connective tissue, where migration proceeds through a continuous cycle of attachment at the leading edge, extension and translocation of the cell body, and detachment at the cell rear (e.g., see Friedl and Wolf [29]). Consequently, the structure of the ECM plays a significant role in directing migration: certain cells may migrate toward ligand-dense (i.e., more adhesive) regions of the matrix, a process termed *haptotaxis*; toward more rigid regions, called *durotaxis* [46]; or even along the aligned collagen fibers, called *contact guidance* [26].

12.1.1 Cell Adhesion during Pattern Formation and Development

In a series of classical experiments, Townes and Holtfreter [74] demonstrated the intrinsic capacity for certain embryonic cell populations, when dissociated and randomly mixed, to spontaneously reorganize into their original embryonic relationship, a process attributed at the time to tissue affinity. The underlying mechanism(s) governing this cell sorting have been subject to a significant degree of speculation and experimentation over the years, with the differential adhesion hypothesis (DAH) of Steinberg (see the reviews [27,68] of Foty and Steinberg) at the forefront of theories. The series of experiments by Steinberg in the 1960s [65–67] demonstrated that embryonic cell types obey strict rules. Whatever the initial distribution for two separate populations was, the cells always rearranged into the same configuration; see Figure 12.1(a). Furthermore, populations formed hierarchical relationships. If cells of type B are engulfed by cells of type A and cells of type C are engulfed by cells of type B, then C will always be engulfed by A; see Figure 12.1(b).

Based on these observations, the DAH employs thermodynamic principles, proposing that cell sorting derives from variation in cell surface tensions that, in turn, depend on the different adhesive properties of the cell types; Cells are assumed to rearrange in a manner to minimize their free adhesive energy, analogous to the behavior of two immiscible liquids. Through these arguments, a mixture of two cell populations, A and B, can be predicted to rearrange into four basic configurations according to the relative strengths of self-adhesion (i.e., the binding between two cells of the same type, S_{AA} and S_{BB}) and cross-adhesion (i.e., binding between two cells of different type, C_{AB}): mixing, engulfment, partial engulfment, and complete sorting; see Figure 12.1(c).

Over the past decade or so, a series of thorough experiments have substantiated the DAH for sorting (see reviews [27,68] for further details). Experiments with two cell lines expressing different levels of cadherins (and hence varying degrees of adhesiveness) resulted in the population expressing higher cadherin

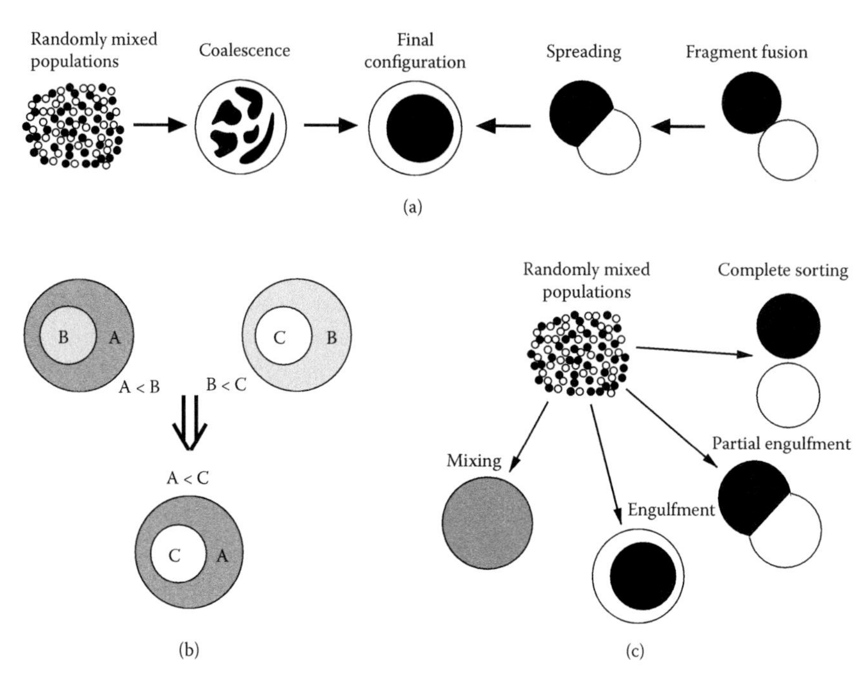

FIGURE 12.1 Sketches showing the behavior of two adhesive cell populations, as predicted by the DAH. (a) The same populations always approach the same final configuration, regardless of initial distribution. Starting from the left, populations of mixed and dissociated cells coalesce before evolving to a final configuration (shown here as "engulfment"). Starting from the right, the same two populations, when placed together as fragments, spread over one another before reaching the same pattern. (b) Hierarchical relationships in adhesive populations. (c) Two populations, A and B, evolve into various final configurations according to their self-adhesion S_{AA}, S_{BB} (between A and A, between B and B) and cross-adhesion C_{AB} (between A and B) strengths. For two populations, the observed patterns are *mixing* (in which the populations are uniformly distributed, requiring dominant cross-adhesion $C_{AB} > \frac{S_{AA}+S_{BB}}{2}$); *engulfment* (in which the more cohesive population is engulfed by the less cohesive population, requiring $S_{BB} < C_{AB} < S_{AA}$ or $S_{AA} < C_{AB} < S_{BB}$); *partial engulfment* (for which the cross-adhesion strength is less than both the self-adhesion strengths—$C_{AB} < S_{AA}$ and $C_{AB} < S_{BB}$); and *complete sorting* (for which $C_{AB} = 0$ and the two populations form separate aggregations). Adapted from R.A. Foty and M.S. Steinberg (2004). Cadherin-mediated cell–cell adhesion and tissue segregation in relation to malignancy. *Int. J. Dev. Biol.* 48:397–409.

levels aggregating to the center, with the other line confined to the periphery, consistent with the predictions of the DAH [25,69]. Recent experiments of Foty and Steinberg [28] have directly linked the surface tensions underlying sorting of tissues to differing strengths of cell–cell adhesion.

The capacity for differential adhesion to spatially sort out different populations implies an important role during the morphogenetic patterning of the embryo; indeed, examples of spatio-temporally controlled alterations to the adhesive properties of cells and the matrix include whole-embryo tissue movements during gastrulation, formation of the boundaries during segmentation of the hindbrain, and the precisely controlled movements of differentiated cell types during the patterning of the insect compound eye (for these and further examples, see [13,36,44,68,72,73]). In segmentation, the embryo is subdivided into a number of discrete blocks along the anterior–posterior axis, laying a blueprint for future development. For a number of organisms, including birds, fish, and mammals, this proceeds through *somitogenesis*, in which parallel stripes of mesenchymal tissue (the paraxial mesoderm) metamerically pinch off to form somitic pairs on either side of the developing neural tube. Compaction from the mesenchyme into epithelialised somites is thought to arise through increases to cell adhesion [16], with a number of studies indicating roles for both matrix molecules such as fibronectin (e.g., [24,53]) and cadherins (e.g., [45,71]). Somites undergo subdivision, first into distinct anterior and posterior portions (e.g., [59]) before they subsequently sort into further embryonic subpopulations [8]. Differential cell–cell adhesion has been suggested to pattern somites into their anterior and posterior segments [70], a theory strengthened by the distribution of various cadherins in the developing somite (e.g., [24,42]).

12.1.2 Cell Adhesion in Cancer Invasion

Understanding the processes that regulate the control of adhesion during tissue development and homeostasis is crucial when it comes to determining the factors that lead to tumor progression. The transition from a benign, compact tumor to an invasive, spreading tumor capable of forming metastases is a pivotal moment for prognosis, and it is now widely accepted that modifications to the adhesive properties of the cells and surrounding ECM correlate with malignant development for a wide range of cancer types (e.g., [14,15,17,49]).

For many tumors of epithelial origin, a link between increased malignancy and progressive loss of function in the cell–cell adhesion molecule E-cadherin has been observed [17], with forced expression of E-cadherin in cultures resulting in a reversal from an invasive to a benign phenotype (e.g., [11]). In a number of cancers, the loss of E-cadherin is accompanied by a gain in N-cadherin expression, a "cadherin-switching" mechanism [79] similar to those seen in various embryonic processes—for example, ingression of cells through the primitive streak. Such transitions are believed to give rise to the evolution of a more invasive migratory form.

To infiltrate surrounding healthy tissue it is necessary for the tumor cells to interact with the surrounding ECM, a structure that can both provide a substrate through which cells can move as well as a physical barrier against migration. To migrate, cells must attach to the matrix through the formation of focal adhesions, mediated through the integrin family of CAMs. These focal adhesions provide a site for recruiting matrix proteases (e.g., MMPs) that degrade the ECM and hence provide space for tumor invasion and expansion to occur [29]. A wide number of *in vitro* and *in vivo* studies have investigated the importance of integrins and MMPs for cancer cell invasion, yet the precise impact on invasion (e.g., promoting or inhibiting) varies widely according to cancer origin.

The form of the invasive front is also variable, with different tumors displaying diverse patterns of invasion, often resulting in an indistinct and diffuse tumor–host tissue interface [29]. Certain tumors (e.g., lymphomas, glioblastomas) tend to invade as individual cells, occasionally forming single-file cell chains known as "Indian chains" (e.g., in breast carcinomas). Other tumor types, particularly those of epithelial origin, tend to invade in a collective fashion in which multicellular strands of tumor cells known as "fingers" protrude into the host tissue or cell clusters migrate out from the tumor while maintaining close contacts. Once again, these distinct patterns of invasion correspond to different patterns of CAM expression, with the individual cell migration phenotypes, typified by mesenchymal and amoeboid cell types, expressing high levels of integrins and proteases while collective cell invasion is characterized by epithelial cell types with strong cell–cell adhesion.

12.1.3 Chapter Outline

Clearly, cellular adhesion plays a crucial role in many biological processes. While a wide range of models have incorporated adhesion at the discrete level, the incorporation into continuous models has received relatively little attention, a fact that can be attributed primarily to a lack of models able to replicate the characteristic behaviors of adhesive populations. In this chapter we first explore the history of modeling in this fundamental process. We review the derivation of the continuous model for cell–cell adhesion developed in [6] and show how it captures the fundamental properties of aggregation and cell sorting. In Section 12.5 we consider an application of this model to tumor invasion [33,64]. Finally, we raise a number of biological, modeling, analytical, and numerical challenges stimulated by these works.

Supplementary material for this chapter, including color figures and simulation movies, is available online at

`http://sim.mathematik.uni-halle.de/gerisch/2009/GerischPainter09`

12.2 Mathematical Modeling of Cell Adhesion

The recognition of cellular adhesion as a major driving force behind various biological processes has led to the development of a variety of modeling approaches and models. Naturally, the structure of a model will inevitably depend on the precise biological question to be addressed. However, it is reasonable to expect that any model of cell–cell adhesion at a *population* level should capture core properties, such as an ability to predict the aggregation/coalescence of a population as the "adhesivity" of the cells is increased and, when expanded to include multiple populations, the various sorting properties predicted by the DAH. The mechanism of cell–cell adhesion—a nonlocal interaction between two cells through transmembrane receptor binding—naturally suggests the usage of discrete cell (i.e., individual cell or agent-based) approaches, which retain the finite cell size and permit relatively straightforward incorporation of the molecular interactions and/or forces that act between cells. Weighed against such advantages, however, are the significant computational times required to simulate large populations and difficulties in obtaining analytical insight. Consequently, it is desirable to augment such methodologies with continuous models that capture the dynamics of population-level behavior.

12.2.1 Discrete Models for Cell Adhesion

The past decade has witnessed the development of a wide variety of discrete models that incorporate cell adhesion and are of increasing sophistication. Generally, such models can be classified into two major classes: lattice-based and lattice-free approaches. We start this section with a brief discussion of a number of discrete lattice-based models (the book by Deutsch and Dormann [21] reviews these models in greater detail with a specific focus on cellular adhesion in its Chapter 7) and consider cellular automata, the discrete-continuum technique, and cellular Potts models.

In lattice-based approaches, the morphology of a cell is restricted according to some underlying discretization of space, which can be either regular (e.g., rectangular or hexagonal in two dimensions) or irregular (e.g., a Voronoi tessellation). These approaches can generally be further subclassified into those for which one cell correlates to one lattice site and spatially extended approaches, with a cell defined by a connected set of sites. Examples of the former class include many cellular automata models: for the evolution of cells under the influence of differential adhesion, see, for example [20]; in [50], a similar approach was employed to demonstrate how different adhesive properties can generate zebrafish pigmentation stripes. A second example of the single-site class is the discrete-continuum technique developed by Anderson and co-workers [2,4]. Here, the discrete cells interact with each other and surrounding continuous fields representing extracellular matrix densities and growth factor concentrations. Movement probabilities are derived from these interactions, which include adhesion of cells to the extracellular matrix, and drive the reorganization of the cell pattern in space and time. The primary application of this technique has been in models of tumor cell invasion.

A prime example of a spatially extended approach is the cellular Potts model (or Glazier-Graner-Hogeweg model). Originating in theoretical physics, it was adapted and applied to cell populations by Graner and colleagues in the 1990s (see [34,35]). Here, each (biological) cell is of a certain cell type and represented as a number of sites (vertices) of a regular lattice. For a given state of the system, a Hamiltonian function is defined based on the surface energy along the cell boundaries and deviations of cell sizes from typical values. The evolution of the system is then driven by a Monte Carlo-like scheme that aims to reduce the value of the Hamiltonian by changing the cell association of a randomly chosen lattice site to that of one of its neighboring sites. The surface energy depends on the cell types on either side of the cell surface and consequently accounts for self- and cross-adhesive effects. A background medium (e.g., representing the extracellular matrix) can also be included in the model. The generic model structure of this Potts model has been elaborated by various authors to make it suitable for particular application areas, for example, cellular slime mold morphogenesis [48], vertebrate development [58], epidermal homeostasis [62], solid tumor growth [75], and angiogenesis [9].

The artificiality of the imposed grid can be countered through the adoption of a lattice-free approach in which individual cells are allowed to move freely through continuous space. In a number of models of this type, cells are given variable, yet predefined, shapes such as deformable ellipsoids of fixed volume in a model for cell movement of *Dictyostelium discoideum* [19,55,56]. Another option, which allows for cell growth and division, is that the *average* cell shape at any point in the life cycle of a cell is predefined while the *actual* cell shapes are reconstructed from that by taking neighboring cells into account. This approach, introduced by Drasdo et al. [23], is followed in models of tumor growth, epidermal homeostasis, and early development; for a brief review and further references, we refer the reader to [30]. One recent extension of this approach has been the incorporation of intracellular and transmembrane molecular interactions, courtesy of an ordinary differential equation system for each cell that describes the regulation of E-cadherin through the β-catenin signaling pathway [60]. In both these and the deformable ellipsoid model described above, movement of individual cells is driven by equilibrating forces, including adhesive ones; alternatively, as in [23], movement is governed by a Monte Carlo algorithm based on a suitable interaction potential.

A number of further lattice-free models provide even greater flexibility to the manner in which cells refine their shape. The model of Schaller and Meyer-Hermann [63] adopts a Voronoi-Delaunay method, permitting cells to shift smoothly between spherical and polyhedral with increasing tissue density, thereby providing greater control over the amount of cell–cell contact. The subcellular element model of Newman [51] provides additional intracellular structure through subdividing each cell into a set of continuously deforming elements, giving high malleability to the shape of a cell according to its interactions with neighbors and the environment. Finally, in the immersed boundary models for individual cells [22,61], each cell is described as a fluid-elastic structure in which its membrane is represented by a deformable boundary immersed in a fluid. Force balances again are used to represent the adhesive forces that describe the movement and deformation of cells while channels at the membrane permit the influx of fluid required for growth into the cell.

12.2.2 Continuous Models Incorporating Cellular Adhesion

While discrete models for cells permit the straightforward incorporation of many intra-, extra-, and intercellular processes, they also have their drawbacks. Of particular concern is that the transition from the cellular to the tissue scale can require a formidable number of cells, which in many models—and certainly for the more detailed ones—is computationally infeasible. In addition, discrete models often resist a thorough analytical investigation that can shed light on generic properties of the system under study. Both of these issues can be relaxed by considering continuum-scale (PDE) models where cells are represented through their density at the tissue level, and events at the cellular level are accounted for by the particular choice of terms and parameter functions in those models. In the following we briefly review some continuous models that account for cellular adhesion.

The modeling of cell–extracellular matrix adhesion has been phenomenologically captured in a number of models through the idea of haptotactic migration (e.g., [3]). Here, cells are assumed to migrate up gradients in the density of an extracellular matrix through the incorporation of an advective-flux type term qualitatively the same as those traditionally employed in continuous chemotaxis models (e.g., [39]).

Incorporation of cell–cell adhesion, however, has proved generally problematic at a continuous level. One approach, adopted in a number of models (e.g., [41]) has been to include cell–cell adhesion through a density-dependent cell diffusion coefficient. While this phenomenologically captures one aspect of adhesion (i.e., the restricted movement of cells in regions of high density), its capacity to describe more complex phenomena such as self-aggregation and sorting of multiple populations is unknown. Byrne and Chaplain [12] presented a model of cancer growth and invasion that accounts for cell–cell adhesion through the incorporation of a surface tension force at the tumor surface controlling the evolution of the tumor shape during growth. This idea has been taken up and extended in recent models [18,47]. The single-phase approach in these models has been broadened to multiphase using a diffuse interface framework in [80]. This model accounts for cell–cell and cell–matrix adhesive effects by incorporating them into a system energy that drives the system following an energy variation scheme. The nonlocal energy term is assumed to be sufficiently localized and the corresponding truncated expansion of that term leads to a fourth-order PDE model of Cahn-Hilliard type.

The modeling approach of Armstrong et al. [6], which is the focus of this chapter, also employs nonlocal terms to account for adhesive effects. In contrast to [80], no expansion of these terms is performed so that the resulting model equations are nonlocal or integro-partial differential equations of second order. This approach has been employed to show that upregulated adhesion can drive both the formation and subsequent anterior–posterior compartmentalization of somites [7] and, as we expand on below, incorporated into models for tumor invasion [33,64].

A highly desirable objective is to develop continuous models for cellular adhesion as the appropriate limit from an underlying individual model for cell movement; in the case of chemotactic cell movement; this has been studied in detail (see [37] for a review) and an obvious advantage lies in the determination of the macroscopic parameters (such as diffusion coefficients and chemotactic sensitivities) in terms of measurable microscopic parameters (e.g., cell velocities, turning rates). A number of recent attempts have been made to approach this problem. In [77], a 1-D representation of a cellular Potts model incorporating adhesion was taken, under specific scaling arguments, to its continuous limit, yet the resulting model is relatively unwieldy and it has not been shown whether sorting properties can be captured. Another approach adopted is to consider the evolution of a particle executing one-step jumps on a discrete lattice (e.g., [5,54]). While these models can capture self-aggregation of a population, the ill-posed nature of the resultant continuum equations can create singular behavior. Finally, [52] considers the limit of a Langevin-based individual model. Interestingly, the resulting continuum model incorporates nonlocal terms similar to those of the phenomenological model of Armstrong et al. [6], described below.

12.3 Derivation of a Nonlocal Model for Cell Adhesion

We begin by reviewing and extending the phenomenological derivation for an integro-partial differential equation model for cell–cell adhesion first developed in [6]. Here, a mass conservation approach was employed in which the cell density for an adhesive cell population $u(\mathbf{x}, t)$ ($\mathbf{x} \in \mathbb{R}^n$) was proposed to be governed by:

$$\frac{\partial u(\mathbf{x}, t)}{\partial t} = -\nabla \cdot \mathbf{J} + h(\cdot) \tag{12.1}$$

where $\mathbf{J}$ represents the cell flux and $h(\cdot)$ describes cell kinetics. A multitude of factors are known to dictate cell movement *in vivo*, ranging from long-range chemoattractants to local cell-cell and cell-ECM interactions, indicating a flux of the form

$$\mathbf{J} = \mathbf{J}_{random} + \mathbf{J}_{adhesion} + \mathbf{J}_{taxis} \tag{12.2}$$

where $\mathbf{J}_{random}$ is the flux due to random cell movement (typically modeled as a Fickian diffusion, $\mathbf{J}_{random} = -D_u \nabla u$, where D_u is the cell diffusion coefficient), $\mathbf{J}_{adhesion}$ is the flux due to adhesion, and $\mathbf{J}_{taxis}$ is the flux due to long-range substances such as chemoattractants. For the latter, the classical assumption is to take $\mathbf{J}_{taxis} = u\chi(u, c)\nabla c$, where c represents the chemoattractant concentration and the function χ is referred to as the chemotactic sensitivity [39,43].

To model the contribution of adhesion to the cell flux, $\mathbf{J}_{adhesion}$, we assume that movement occurs due to the forces generated when cells bind with other cells or the surrounding matrix, the density of which we denote by $m(\mathbf{x}, t)$. For a cell at $\mathbf{x}$, binding with a cell at $\mathbf{x} + \mathbf{r}$ will create a *local force* $\mathbf{f}$ in the direction $\mathbf{r}$ (equally, the cell at $\mathbf{x} + \mathbf{r}$ experiences the opposite force). To describe adhesion-based movement, we assume that the size of this local force depends on the adhesivity of this site, namely the numbers and types of adhesion molecules. Rather than explicitly modeling the concentrations of such molecules, the adhesivity is taken to simply depend on the cell density (indicating the likelihood of forming a cell–cell bond) and the matrix density (indicating the likelihood of forming a cell–matrix bond) at $\mathbf{x}+\mathbf{r}$ through the function $g(u(\mathbf{x} + \mathbf{r}, t), m(\mathbf{x} + \mathbf{r}, t))$. Note that the density of additional cell types can be included here, allowing for cross-adhesion between cell types. The possibility of a cell at $\mathbf{x}$ forming a bond at $\mathbf{x} + \mathbf{r}$ is further expected to depend on the distance between the two sites: cells establish adhesive bonds at the membrane-substrate interface, yet their capacity to change shape (e.g., become elongated) or extend thin cell protrusions ranging from shorter range lamellipodia to longer range filopodia (occasionally up to 100 μm in length, [81]) suggests that the probability of forming bonds may vary with distance.

Together, these assumptions lead us to propose the local force generated at $\mathbf{x}$ via adhesive binding at $\mathbf{x} + \mathbf{r}$ to be

$$\mathbf{f}(\mathbf{x}, \mathbf{r}) = \frac{\mathbf{r}}{|\mathbf{r}|}\Omega(|\mathbf{r}|)g(u(\mathbf{x} + \mathbf{r}, t), m(\mathbf{x} + \mathbf{r}, t)) \tag{12.3}$$

where the right-hand side terms break down into the direction of the force (a vector), the dependence of the force magnitude on the distance at which bonds are formed, Ω (a scalar), and the dependence of the force on the adhesivity, g (a scalar). We discuss various functional forms for these terms below.

The *total* force exerted at $\mathbf{x}$, $\mathbf{F}(\mathbf{x})$, will be the sum of all local forces $\mathbf{f}(\mathbf{x}, \mathbf{r})$, where $\mathbf{r}$ ranges over a finite volume V indicating the sensing region: the space over which the cell at $\mathbf{x}$ can make adhesion bonds. As described above, this V is minimally determined by the mean cell volume, yet is likely to be significantly larger due to cell shape change and protrusions. Thus, we compute the total force to be

$$\mathbf{F}(\mathbf{x}) = \int_V \frac{\mathbf{r}}{|\mathbf{r}|}\Omega(|\mathbf{r}|)g(u(\mathbf{x} + \mathbf{r}, t), m(\mathbf{x} + \mathbf{r}, t))\, d\mathbf{r} \tag{12.4}$$

To incorporate the above into the mass balance Equation (12.1), we note that at the low speeds of eukaryotic cell migration (typically 0.1 to 10 μm/min, according to cell type) we can reasonably expect inertia to be negligible and drag proportional to velocity and the cell radius R (Stokes law for a ball of radius R in a laminar flow). The adhesive flux will then be proportional to the cell density and the forces between them and therefore we take

$$\mathbf{J}_{adhesion} = \frac{\phi u}{R}\mathbf{F} \tag{12.5}$$

where ϕ is a constant of proportionality. Finally, we substitute Equation (12.5) with $\mathbf{F}$ as given in Equation (12.4) into Equation (12.2), and assume Fickian diffusion and a generic taxis cue $c(\mathbf{x}, t)$ to obtain the following cell density evolution equation:

$$\begin{aligned}\frac{\partial u(\mathbf{x},t)}{\partial t} = {} & \overbrace{D_u \nabla^2 u}^{\textit{Random movement}} \quad \overbrace{- \nabla \cdot (u\chi(u,c)\nabla c)}^{\textit{Taxis movement}} \\ & \overbrace{- \nabla \cdot \left[\frac{\phi u}{R}\int_V \frac{\mathbf{r}}{|\mathbf{r}|}\Omega(|\mathbf{r}|)g(u(\mathbf{x}+\mathbf{r},t), m(\mathbf{x}+\mathbf{r},t))\, d\mathbf{r}\right]}^{\textit{Adhesive movement}} + \overbrace{h(\cdot)}^{\textit{Cell kinetics}}\end{aligned} \tag{12.6}$$

The above forms our basic model for cell adhesion and, when combined with appropriate dynamics for matrix and chemical signaling, can be applied to a wide range of biological processes; a version of the above equation was first considered in [6] to model the basic properties of an adhesive population and, through the incorporation of an extra adhesive population, extended to model cell sorting (see Section 12.4). An amalgamation of Equation (12.6) into a chemical signaling system has been developed to model somite formation during embryonic development (see [7]), and the incorporation into the modeling of tumor invasion has been considered in [33] and [64] (see Section 12.5).

12.3.1 Cohesion through Adhesion

A fundamental test for any model for cell–cell adhesion is to determine its capacity to predict the organization of a population of dispersed cells into aggregations. Populations of cell lines aggregate rapidly into large and cohesive clumps with increasing cadherin expression (e.g., [28]). To demonstrate the ability of Equation (12.6) to allow this basic phenomenon, we neglect any effects from cell–matrix adhesion and chemoattractants and ignore cell kinetics (i.e., cell growth is assumed to be negligible on the time scale of adhesion-driven movement) to derive:

$$\frac{\partial u(\mathbf{x},t)}{\partial t} = D_u \nabla^2 u - \nabla \cdot \left[\frac{\phi u}{R}\int_V \frac{\mathbf{r}}{|\mathbf{r}|}\Omega(|\mathbf{r}|)g(u(\mathbf{x}+\mathbf{r},t))\, d\mathbf{r}\right] \tag{12.7}$$

It remains to define appropriate functional forms for the various components in the nonlocal term. In the simulations that follow, we restrict to two spatial dimensions and take the cell sensing region V to be a circle. The function Ω defines the dependence on the distance from $\mathbf{x}$. The simplest assumption is to assume that Ω is constant throughout the sensing region; however, a form in which Ω decreases due to the diminished likelihood of forming a bond with distance from the cell may be more appropriate. For the purposes here, we adopt the simplest form and take $\Omega(|\mathbf{r}|) = \text{constant}$; the impact of other forms has been considered in [64] for a 1-D version of the model.

For the adhesivity component, with *attractive* interactions we expect g to (at least initially) increase with cell density u due to the increased likelihood of forming bonds within areas of higher cell densities (and hence more adhesion receptors). Yet at even higher cell densities, it is reasonable to expect the attractive force magnitude to either saturate (e.g., due to all receptors becoming bound) or even decrease (due to an impedance against migrating into "crowded" regions). To explore the impact from different forms of g, we consider respectively linear, saturating, and logistic forms, all depending on an adhesion parameter α:

$$g(u) = \alpha u, \quad g(u) = \frac{\alpha u}{K + u}, \quad g(u) = \alpha u \max\left\{0, 1 - \frac{u}{U_{\max}}\right\} \tag{12.8}$$

We have solved Equation (12.7) for each functional form of g from Equations (12.8) on a square spatial domain $(0, 10)^2 \subset \mathbb{R}^2$ with periodic boundary conditions. The initial cell density $u(\mathbf{x}, 0) = 0.1 + \mathcal{U}(\mathbf{x})$ is constant with a uniformly distributed perturbation $\mathcal{U}(\mathbf{x}) \in 10^{-2}[-0.5, 0.5]$. The sensing region V is a circle of radius 1 and the other parameters used are

$$D_u = \phi = R = 1\,, \quad \Omega(|\mathbf{r}|) = 1 \quad \text{for } |\mathbf{r}| \in [0, 1]\,, \quad \alpha = 30\,, \quad K = U_{\max} = 2 \tag{12.9}$$

The numerically computed cell density $u(\mathbf{x}, t)$ at three output times t is shown in Figure 12.2. With the setting described above, we observe aggregation of cells for all three functional forms of g given in Equations (12.8). With the linear form of g, we obtain a very fast aggregation process leading to many small cell clusters with large cell density up to 20. As time proceeds, some of these clusters coalesce, leading to a further increase in cell density; see Figure 12.3 (top). The diffusion in the model prevents a further increase (also the finite grid width contributes to this; on finer grids, the maximum solution value becomes even larger). With the saturating form of function g, the onset of aggregation becomes visible only much later than with the other two forms. This can be understood from observing that at the low initial cell densities (≈ 0.1), the saturating form gives $g \sim \alpha u/2$, whereas $g \sim \alpha u$ for the other two functional forms; the adhesive pull driving aggregation is therefore much lower. Once the clusters have formed, a slow but steady increase in the maximum density occurs, which only flattens off as the density increases above 10 and the impact of the saturation in g takes hold. Finally, the logistic form for function g leads, like the linear form, to a quick formation of cell

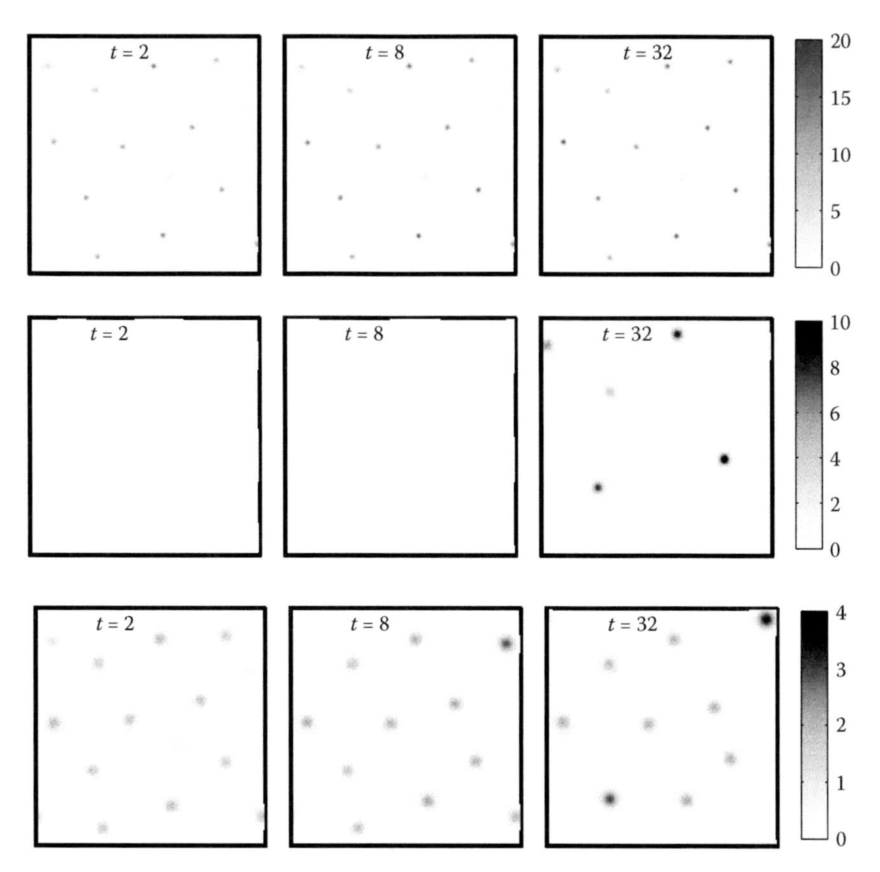

FIGURE 12.2 Simulation results $u(\mathbf{x}, t)$ for Equation (12.7) with linear (top row), saturating (middle row), and logistic (bottom row) form of function g; see Equations (12.8), at three output time points t. Note the different color scalings for the three functional forms of g. (A color version of this figure can be found in the online supplementary material.)

aggregates. However, unlike in the linear case, the maximum cell density is much smaller here and appears to be bounded by ≈ 4. This value is larger than the parameter $U_{\max} = 2$; for the dependence of the maximum cell density on the value of α, see Figure 12.3 (bottom). In a reduction of the 2-D case to a quasi-1-D problem, Sherratt et al. [64] have shown that the density is bounded by $U_{\max} = 2$, provided the adhesion parameter α is below some critical value; consequently, this result appears either not to generalize to the genuinely 2-D setting or imposes additional constraints on the size of α for boundedness by $U_{\max}$.

Based on the reasonably fast aggregation and the capacity to bound cell densities at lower levels, the choice of the logistic form for function g is recommended and will be considered in the remainder of this chapter.

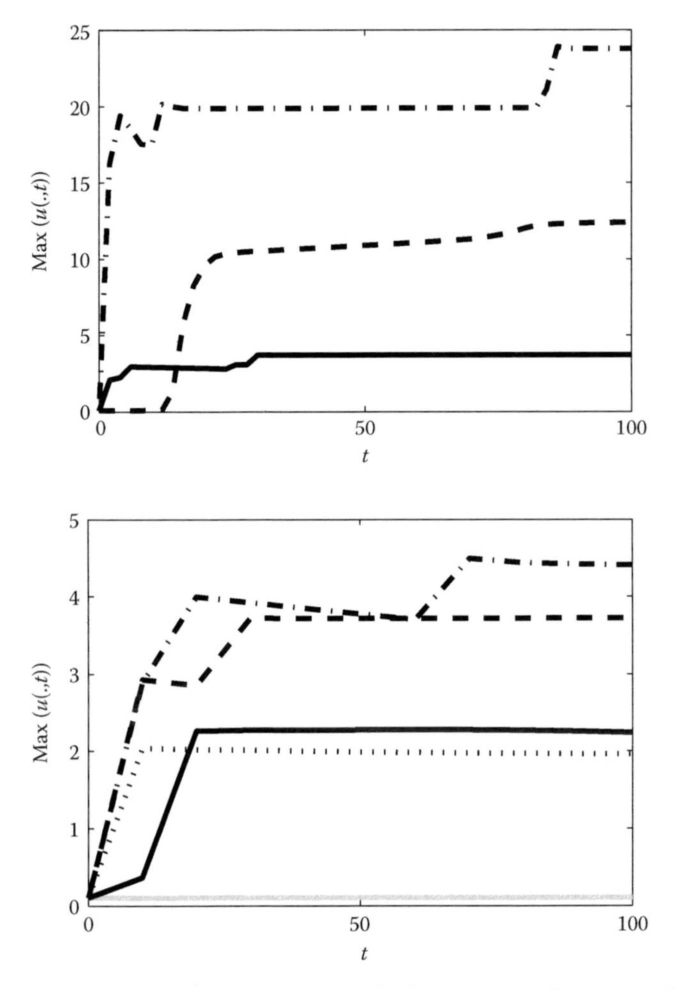

FIGURE 12.3 Top: The maximum cell density as a function of time for the numerical experiments shown in Figure 12.2 using $\alpha = 30$ and a function g that is of linear (dot-dashed), saturating (dashed), or logistic (solid) type. Bottom: The maximum cell density as a function of time for the numerical experiment shown in Figure 12.2 (bottom), i.e., with logistic type function g, but with different α values: no aggregation for $\alpha = 12$ (solid gray line) and aggregation for $\alpha = 16$ (solid), $\alpha = 20$ (dotted), $\alpha = 30$ (dashed), and $\alpha = 40$ (dot-dashed).

12.4 Modeling Cell–Cell Sorting

In this section we aim to demonstrate whether the continuous framework developed in Section 12.3 can replicate the predictions of the DAH for cell sorting; cf. Figure 12.1. The prototypical setting here is to consider two cell populations that differ only in their adhesive properties. Initially, the two cell populations are distributed more or less arbitrarily and one is interested in the long-term configuration of the system; see Figure 12.1. We denote the densities of the two cell populations by $u_A(\mathbf{x}, t)$ and $u_B(\mathbf{x}, t)$. It is reasonable to assume that cell proliferation is negligible on the time scale of cell sorting and we further assume that the random motility coefficient is approximately the same for each population. Under these simplifications we obtain the following set of PDEs describing the spatio-temporal evolution of the system:

$$\frac{\partial u_i(\mathbf{x}, t)}{\partial t} = D\nabla^2 u_i - \nabla \cdot \left[\frac{\phi u_i}{R}\int_V \frac{\mathbf{r}}{|\mathbf{r}|}\Omega(|\mathbf{r}|) g_i(u_A(\mathbf{x}+\mathbf{r}, t), u_B(\mathbf{x}+\mathbf{r}, t))\, d\mathbf{r}\right],$$
$$i = A, B \tag{12.10}$$

This system is considered on the 2-D spatial domain $(0, 10)^2 \subset \mathbb{R}^2$ and complemented with periodic boundary conditions for both species. We consider two sets of initial conditions $u_A(\mathbf{x}, 0)$ and $u_B(\mathbf{x}, 0)$, corresponding to the left- and right-most frames of Figure 12.1(a): a single pellet of randomly mixed cell types, Figure 12.4 (center row, left), and a pellet of cell type A juxtaposed to a pellet of cell type B, Figure 12.4 (bottom row, left). The initial masses of cell types A and B are approximately equal for the initial condition shown in Figure 12.4 (center row, left), whereas there is a larger initial mass of cell type B in Figure 12.4 (bottom row, left).

The functions g_i in the cell adhesion term are parameterized by the self- and cross-adhesion parameters, $S_{AA}, S_{BB}, C_{AB} = S_{AB} = S_{BA}$ of the two cell types and we employ a logistic functional form (cf. Section 12.3.1, [6,64]):

$$g_i(u_A, u_B) := (S_{iA}u_A + S_{iB}u_B)\max\left\{0, 1 - \frac{u_A + u_B}{U_{\max}}\right\}, \quad i = A, B \tag{12.11}$$

The contributions $S_{ii}u_i$ account for self-adhesion whereas $S_{ij}u_j$ with $i \neq j$ account for cross-adhesion. The factor $\max\{0, 1 - \frac{u_A+u_B}{U_{\max}}\}$ is employed to limit the density to which an aggregate can reach; see the effect of the various forms for g in Figure 12.2. Under this form, the adhesive pull of a region increases at lower cell densities before decreasing at higher densities. The sensing region V is a circle with radius 1 and we use

$$D = \phi = R = 1, \quad \Omega(|\mathbf{r}|) = 1 \text{ for } |\mathbf{r}| \in [0, 1], \quad U_{\max} = 1 \tag{12.12}$$

Our first test is to demonstrate the capacity of Equation (12.10) to predict various final configurations according to the self- and cross-adhesion parameters, as illustrated in Figure 12.1. Accordingly, we start with a random mixture of cells of the two types in a pellet centered in the domain. The self-adhesion coefficients are fixed at $S_{AA} = 30$ and $S_{BB} = 15$ (i.e., population A has stronger self-adhesion), and we consider the impact of variation in the cross-adhesion strength C_{AB}. The results of the simulations are represented by plotting the differences of the cell densities u_A and u_B at large times (i.e.,

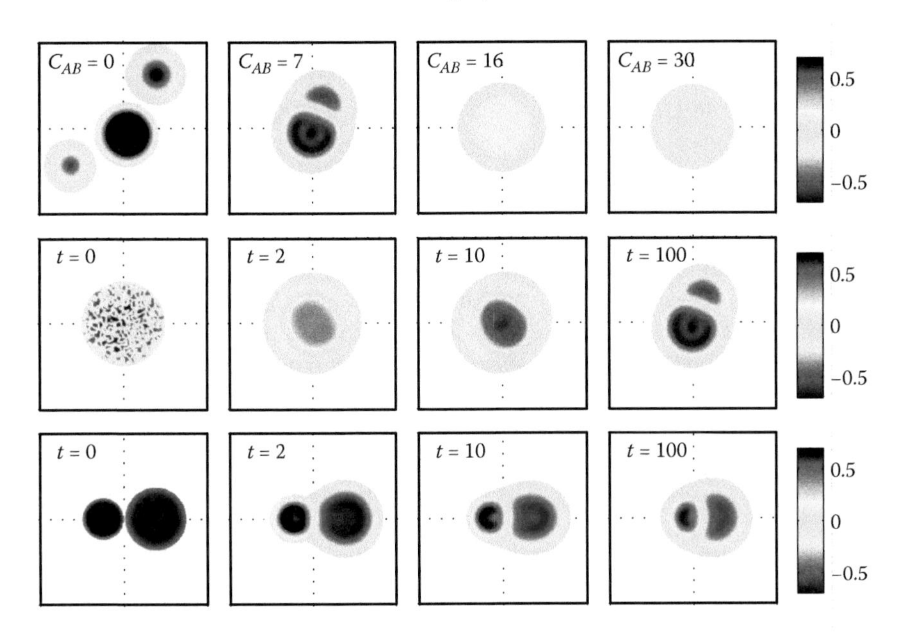

FIGURE 12.4 (See color insert following Page 398.) The plots show numerical approximations to the cell density differences $u_A(\mathbf{x}, t) - u_B(\mathbf{x}, t)$ of the cell–cell sorting model (Equation (12.10)). In regions with cell densities $u_A < 0.05$ and $u_B < 0.05$, the difference value is suppressed in the plots. Top row: The difference is plotted at time $t = 100$, when a (numerical) steady state is reached, for four different values of the cross-adhesion parameter C_{AB} and starting with the initial condition consisting of a single pellet with randomly mixed cell types (described below); all other parameters are as detailed in the main text. Depending on the choice of C_{AB}, from left to right, the four final configurations "complete sorting," "partial engulfment," "engulfment," and "mixing" (cf. Figure 12.1(c)) are attained. Middle and bottom rows: The plots show the time courses as solutions evolve to the steady-state distribution for the fixed cross-adhesion parameter $C_{AB} = 7$ but for the two different initial cell distributions; all other parameters are as detailed in the main text. For the middle row we initially consider a single pellet of radius 2.5 in the center of the domain with a random mixture of cells of type A and B such that $u_A(\mathbf{x}, 0) + u_B(\mathbf{x}, 0) = 0.8$ in the pellet's center and slightly decreasing toward the periphery; densities are zero outside the pellet. For the bottom row we initially consider two adjacent cell pellets, of radii 1.25 (type A) and ≈ 1.87 (type B), containing one cell type each at a density of ≈ 0.8; densities are zero outside the pellets. (This figure together with movies can be found in the online supplementary material.)

at numerical steady states), shown in Figure 12.4 (top row). Depending on the value of C_{AB}, the steady-state distribution attained corresponds to that predicted by the DAH based on the relative size of the adhesion coefficients (cf. Figure 12.1(c)). In particular, it is noted that the more strongly adhesive cell type A tends to accumulate in the center of the pellet, except for the case without any cross-adhesion, $C_{AB} = 0$, that is, total separation.

As a second exploration, we investigate the relatively insensitive nature of the final configuration with respect to the initial distribution of the populations (cf. Figure 12.1(a)). Here we choose $S_{AA} = 30$, $S_{BB} = 15$, and $C_{AB} = 7$; according to the DAH this predicts the partial engulfment of A by B at the steady state. Starting from the two sets of initial conditions described above, the time courses as solutions evolve to the steady-state distribution are plotted in the middle and bottom rows of Figure 12.4. Clearly, we observe evolution to the same pattern phenotype at the steady state; the differences in the rightmost configurations stem from the smaller proportion of A used in the bottom row.

As a final test of the continuous cell sorting model, we explore whether Equation (12.10), when extended to three cell populations, can predict the hierarchical relationship of adhesive populations, similar to Figure 12.1(b). For three populations A, B, and C obeying the self-adhesion hierarchy $S_{AA} > S_{BB} > S_{CC}$, simulations predict that population A becomes engulfed at the center, population C is confined to the periphery, and population B is sandwiched between A and C; see Figure 12.5.

12.5 Modeling Adhesion during Cancer Invasion

In this section we demonstrate the applicability of the continuous framework for cellular adhesion by considering a simple and minimalist model of cancer cell invasion into healthy tissue (cf. [33]). The model consists of three equations describing the cancer cell density (c), the extracellular matrix (ECM) density (v), and the concentration of a (generic) matrix degrading enzyme (MDE) (m). The model equations are given by:

$$\frac{\partial c(\mathbf{x},t)}{\partial t} = D_1 \nabla^2 c - \nabla \cdot \left[\frac{\phi c}{R} \int_V \frac{\mathbf{r}}{|\mathbf{r}|} \Omega(|\mathbf{r}|) g(c(\mathbf{x}+\mathbf{r},t), v(\mathbf{x}+\mathbf{r},t))\, d\mathbf{r} \right] + \mu_1 c(1 - c - v) \tag{12.13a}$$

$$\frac{\partial v(\mathbf{x},t)}{\partial t} = -\gamma m v + \mu_2 (1 - c - v) \tag{12.13b}$$

$$\frac{\partial m(\mathbf{x},t)}{\partial t} = D_3 \nabla^2 m + \alpha c - \lambda m \tag{12.13c}$$

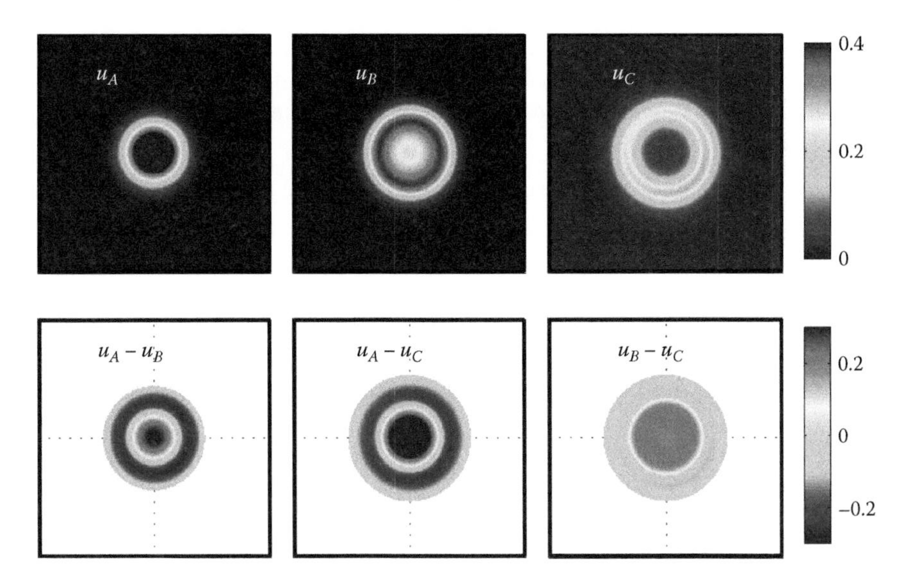

FIGURE 12.5 (See color insert following Page 398.) The plots show numerical solutions of Equation (12.10) extended to three cell types A, B, and C. Top row: Solutions u_A, u_B, and u_C at time $t = 50$. Bottom row: Pairwise solution differences at time $t = 50$. The initial condition is a randomly mixed pellet of the three cell types. The adhesion parameters are $S_{AA} = 45$, $S_{BB} = 30$, $S_{CC} = 15$, $C_{AB} = 31$, $C_{AC} = 0$, $C_{BC} = 16$. (This figure together with a movie can be found in the online supplementary material.)

For simplicity, we restrict our attention to a 2-D geometry and consider the above equations on the spatial domain $(-1.5, 1.5)^2 \subset \mathbb{R}^2$, subject to periodic boundary conditions. In this model, both the cancer cells and ECM occupy physical space while the volume occupied by MDE is assumed negligible. The cancer and ECM density equations above have been normalized such that the total density $c + v = 1$ characterizes fully occupied physical space.

In this simple model, cancer cell migration is assumed to arise from (1) random motility (the corresponding coefficient D_1 is rather small) and (2) a directed movement due to adhesive effects of cancer cells with themselves and the surrounding ECM. The sensing region V for the adhesion term is a circle of radius 0.1 and for the function $\Omega(|\mathbf{r}|)$ we select a linearly decaying function

$$\Omega(|\mathbf{r}|) = \frac{3}{\pi R^2}\left(1 - \frac{|\mathbf{r}|}{R}\right) \quad \text{for} \quad |\mathbf{r}| \in [0, R] \tag{12.14}$$

The linear decay of $\Omega(|\mathbf{r}|)$ models a diminishing influence of adhesive bonds toward the periphery of the sensing region. The leading factor in $\Omega(|\mathbf{r}|)$ follows from a normalization ensuring that the integral of $\Omega(|\mathbf{r}|)$ over the sensing region V is equal to 1, stipulating a fixed maximum capacity of cells to form adhesive bonds within the sensing region independent of its actual size. The magnitude of that capacity is captured in the (parameters of the) function g, which takes the form

$$g(c, v) = (S_{cc}c + S_{cv}v) \max\{0, 1 - (c + v)\} \tag{12.15}$$

This functional form implies that cancer cells adhere to themselves (self-adhesion parameter S_{cc}) and to the matrix (cross-adhesion parameter S_{cv}). Again, we include the limiting term such that g becomes 0 if the total density $c + v$ approaches the value of 1. In addition to cell migration, cancer cell proliferation is also incorporated through the employment of a logistic growth type law with growth rate μ_1 and "carrying capacity" dependent on the locally available space, $1 - c - v$.

The ECM is assumed to be nonmotile and degraded upon contact by MDEs at a rate γ. In the general formulation, a simple ECM production term permits regeneration of the ECM with rate μ_2 (note that in the simulations below, it is assumed that $\mu_2 = 0$). Finally, MDEs diffuse throughout the tissue (with diffusion constant D_3), are produced by the cells at rate α, and decay at rate λ. The following set of parameters is used in the simulation

$$\begin{aligned} &D_1 = 10^{-3}, \quad D_3 = 10^{-3}, \quad \mu_1 = 0.1, \quad \mu_2 = 0, \qquad \alpha = 0.1, \\ &\gamma = 10, \qquad \lambda = 0.5, \qquad R = 0.1, \quad S_{cc} = 0.05, \quad S_{cv} = 0.1 \end{aligned} \tag{12.16}$$

Clearly, the model in Equations (12.13) through (12.16) is highly simplified in its nature and excludes many pertinent biochemical interactions. However, the focus here is on the incorporation and effect of the adhesion term in a model of tumor invasion and, consequently, we wish to retain the simplicity of the model. Crucial questions for any model of cancer invasion are whether it permits the breakage of cancer cells from a central tumor mass and how cancer cell migration is affected by a heterogeneous tumor environment. To address these issues, we consider an initial tumor population concentrated at the center of the domain (representing the central tumor mass) and lying within a spatially structured ECM matrix. The initial MDE concentration is chosen to be proportional to the cell density. Simulations for a striped distribution in the initial ECM densities are shown in Figures 12.6 and 12.7.

In Figure 12.6 we observe the preferential accumulation and invasion of cancer cells along stripes of higher ECM density, in concert with degradation of that ECM. Cell migration obeys the restriction of physical space, that is, cells do not move into densely packed tissue. Cells at the tumor periphery do not accumulate in regions of low ECM density, but rather quickly cross these areas to concentrate at the front of the next ECM barrier. The variation in ECM density also leads to the formation of protrusions that stretch from the cancer mass into the healthy tissue. Due to the regular structure of the ECM, these protrusions are also regular. Similar results apply when cell proliferation is excluded; however, the protrusions now take the form of high-density tumor clumps extending along the ECM stripes; see Figure 12.7.

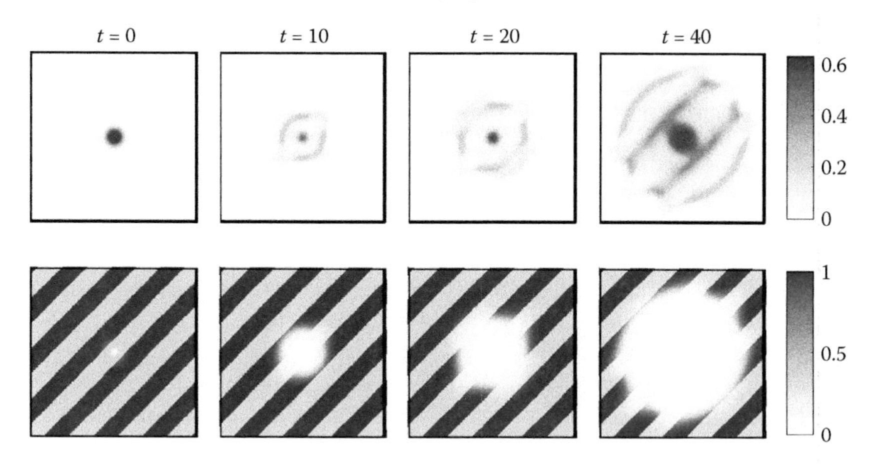

FIGURE 12.6 Simulation results for the model given by Equations (12.13) through (12.16) with a diagonally striped initial ECM distribution. Shown are the tumor cell density (top row) and the ECM density (bottom row) in the central part $(-1, 1)^2$ of the spatial domain at four time points. The MDE concentration displays similar features as the cell density and is not shown. (A color version of this figure together with a movie can be found in the online supplementary material.)

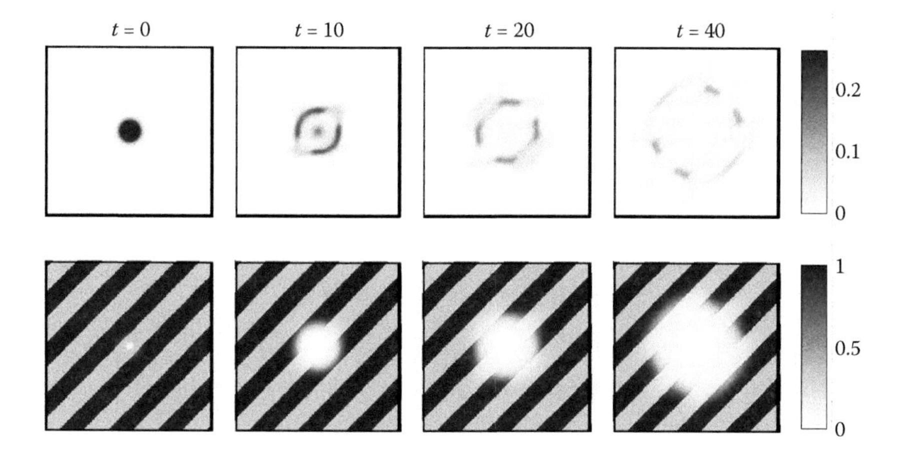

FIGURE 12.7 Simulation results for the model given by Equations (12.13) through (12.16) as in Figure 12.6 but without cancer cell proliferation (i.e., $\mu_1 = 0$). (A color version of this figure together with a movie can be found in the online supplementary material.)

12.6 Discussion and Open Questions

In this chapter we discussed the critical role played by cellular adhesion during a wide spectrum of biological processes, highlighting the need for mathematical models to capture this fundamental phenomenon. A brief review of existing models, discrete and continuous, has demonstrated their individual strengths and weaknesses. Of particular note is the lack of continuous models that can replicate the sorting behavior of multiple adhesive populations. As far as we are aware, the only continuous model that has been demonstrated to capture this property is that developed in [6]. Here we expanded the derivation of this model and corroborated its suitability through an extended numerical analysis that replicates the wide variety of cell sorting experiments, as predicted by the differential adhesion hypothesis (DAH). The ultimate success of this approach lies in its capacity for integration with existing continuous models; as a demonstration of its suitability, we considered its extension into a model for cancer invasion as originally studied in [33] (see also [64]). A second application, considered in [7], has been to the process of chick somitogenesis. There it was shown that upregulation of cellular adhesion, regulated through an underlying chemical signaling network, could drive both the epithelialization and subsequent sorting of pre-somitic cells into somites. The ubiquity of cellular adhesion would allow a catalog of potential applications to be listed: some typical examples, based on the history of modeling in these areas, include angiogenesis, wound healing, development of the slime mold *Dictyostelium*, and skeletal patterning.

In the continuous adhesion model, the microscopic processes (e.g., receptor binding) can be accounted for by a suitable choice of cell adhesion parameters. A crucial extension of this work lies in the development of truly multiscale models of cell adhesion, in which the sizes of adhesion parameters in the continuous model can be determined from the processes occurring at a microscopic scale. To achieve this, it will be necessary to derive models for cell adhesion from a realistic underlying description of individual cell behavior; as discussed in Section 12.2, a number of attempts have been made at exploring some of these issues (e.g., see [52,76]).

Mathematically, a striking feature in the modeling approach is that cellular adhesion is accounted for via a nonlocal (integral) advective type term. This fits coherently into the typical taxis–diffusion–reaction frameworks frequently employed in the modeling of pattern-formation type phenomena. Similarly, existing simulation packages for diffusion–reaction systems can be extended in modular fashion to allow the incorporation of such nonlocal terms. One difficulty, however, is the additional computational effort required to evaluate the nonlocal term. A suitable solution to this problem is outlined in the Appendix to this chapter, and results in a scheme that provides high-resolution simulations within reasonable computing times (at least for the case of a rectangular spatial domain with periodic boundary conditions). However, extra work will be required to extend the numerical techniques to more general situations, such as irregular geometries or three dimensions.

Analytically, a number of results are available on the properties of solutions. In [64], the boundedness of solutions was addressed under particular forms of the model. Specifically, it was shown (in one dimension) that for $g(u) = \alpha u \max\{0, 2 - u\}$, boundedness of u below 2 is possible under specific restrictions for the size of α and the form of $\Omega(|\mathbf{r}|)$. In [38] a related nonlocal model for chemotaxis was derived and global existence of solutions was proven for all finite sampling radii. A considerable number of questions, however, remain unanswered. Of particular interest are an extension of the boundedness results to spatially 2-D settings, the proper incorporation of other boundary conditions from both an analytical as well as a modeling point of view, and an analysis of the limiting scenario as the sampling radius $R \to 0$.

The formulation of the model presented here clearly simplifies many crucial components regarding the behavior of adhesive populations *in vivo*. For example, the dynamics of adhesive binding are assumed to correlate to overall cell–matrix densities rather than the concentrations of adhesive molecules at the cell membrane, the dynamics of which can vary both spatially and temporally according to intra- and extracellular signals. Further, while adhesion is assumed here only to generate forces resulting in cell migration, the signaling initiated through binding interplays with many facets of cell behavior, including division and apoptosis. Extending the model to include some of these complexities will further advance its relevance to understanding the role of adhesion in a wide variety of biological processes.

12.7 Appendix: Numerical Method

The models in this chapter are all solved following the method of lines (MOL). The rectangular spatial domain is covered with a uniform grid where each grid cell, or finite volume, is a square of side length h. In a first step of the MOL, the spatial derivatives are discretized on that grid and we employ a finite volume method (FVM) of order 2 (see, e.g., [32]). This transforms the PDE model into a large and, in general, stiff system of ordinary differential equations (ODEs), the MOL-ODE system:

$$\frac{d\mathbf{U}(t)}{dt} = \mathcal{F}(t, \mathbf{U}(t)), \quad \mathbf{U}(0) = \mathbf{U_0} \tag{12.17}$$

As is customary when using the FVM, the components of this ODE system represent approximations to the averages of the PDE solution in each finite volume. The numerical solution of the MOL-ODE in Equation (12.17) constitutes the second step of the MOL and an appropriate time integration scheme must be selected. Implicit time integration schemes can deal efficiently with the inherent stiffness of the MOL-ODE. We favor the linearly implicit, fourth-order Runge-Kutta method ROWMAP [78]. The multiple Arnoldi process used within this method for the solution of the linear equation systems in each time step makes this scheme particularly suited for the large ODE system at hand. Furthermore, the method does not require any computation of the Jacobian of the MOL-ODE by the user; the required Jacobian-times-vector products are computed automatically by a suitable finite difference approximation using the right-hand side $\mathcal{F}$ of the MOL-ODE.

The FVM described in [32] has been applied to taxis–diffusion–reaction systems. There, difficulties arise in regions of strong variation of the solution of the PDE, for example, near moving fronts. These are due to the taxis term of the model and special attention was given to ensure that the discretization of that term does not introduce oscillations or negative solution values in the solution of the MOL-ODE. This goal can be achieved, while maintaining the order 2 of the FVM as much as possible, using a second-order upwind discretization together with a nonlinear limiter function. The models of this chapter have a nonlocal adhesion term that is similar to the taxis terms in [32]. So we apply the same discretization to that term, with the added difficulty of the approximation of the integral. The adhesion term, in general, takes the following form for the adhesive species u_i of a vector $\mathbf{u}$ of concentrations or densities in the models

$$-\nabla \cdot [u_i(\mathbf{x}, t) \underbrace{\frac{\phi}{R} \int_V \frac{\mathbf{r}}{|\mathbf{r}|} \Omega(|\mathbf{r}|) g_i(\mathbf{u}(\mathbf{x} + \mathbf{r}, t)) \, d\mathbf{r}]}_{\text{the adhesive velocity in } (\mathbf{x}, t)} \tag{12.18}$$

The adhesive velocity, and consequently the integral, must be approximated on each edge of the spatial grid for each evaluation of the right-hand side of the MOL-ODE. This task constitutes the computational bottleneck of the whole numerical solution process. If we assume that (1) the sensing region V at each $\mathbf{x}$ is the same and time independent, and (2) we solve the PDE system on a rectangular domain with periodic boundary conditions, then the adhesive velocity in Equation (12.18) can be approximated on all vertical (or all horizontal) edges of the spatial grid simultaneously by evaluating a matrix–vector product $M\mathbf{G}$. Here, the matrix $M \in \mathbb{R}^{N,N}$, N the number of grid cells, and each row of $\boldsymbol{M}$ corresponds to the approximation of the nonlocal term on one edge. We arrive at this by first evaluating function g_i at the approximation $\mathbf{U}(t)$ yielding $\mathbf{G} \in \mathbb{R}^N$; second by reconstructing a function $\tilde{g}_i(\mathbf{x})$ from that data using bilinear interpolation; and third, by approximating the integral with $\tilde{g}_i$ instead of g_i. Thanks to a suitable basis representation of $\tilde{g}_i$, the matrix $\boldsymbol{M}$ will be independent of the data $\mathbf{G}$ and can be precomputed before the time integration of Equation (12.17) commences. Furthermore, the third step can be performed to any desired accuracy so that the overall accuracy of the nonlocal term evaluation hinges solely on the quality of the reconstruction of function g_i by $\tilde{g}_i$, that is, can be controlled by the spatial grid width h. Typically, the sensing region V is much smaller than the spatial domain of the PDE model. Consequently, the matrix $\boldsymbol{M}$ contains many zeros. However, in contrast to the approximation of derivatives, the fraction of non-zero elements of $\boldsymbol{M}$ remains constant with decreasing spatial grid width h. In that sense, sparse matrix techniques can only have a limited impact for the efficient evaluation of the matrix–vector product $\boldsymbol{M}\mathbf{G}$. At this point the periodic boundary conditions become important. These give rise to a matrix $\boldsymbol{M}$ having the structure of a block–circulant matrix with circulant blocks. Matrix–vector products with such matrices can be evaluated efficiently with fast Fourier transform (FFT) techniques. This substantially reduces the computational complexity and hence CPU time requirements for the evaluation of $\boldsymbol{M}\mathbf{G}$. More details of the integral approximation and evaluation can be found in [31].

In the discussion above we have assumed periodic boundary conditions for the PDE problem. This is not always suitable from the point of view of modeling. No-flux boundary conditions are frequently encountered and in the following we describe how they can be included in the computational framework. For the nonlocal term, the boundary conditions become only important in points $\mathbf{x}$ where the set V, centered at $\mathbf{x}$, intersects the boundary of the spatial domain of the PDE. For such $\mathbf{x}$ we follow the approach taken in [7]: that the integral of the nonlocal term is taken only over those $\mathbf{r} \in V$ such that $\mathbf{x} + \mathbf{r}$ is within the spatial domain. This modification implies that the matrix $\boldsymbol{M}$ changes from a block–circulant matrix with circulant blocks to a block–Toeplitz matrix with Toeplitz blocks. FFT techniques cannot be applied directly to such matrices but any such matrix can be embedded into a block–circulant matrix $\tilde{\boldsymbol{M}}$ with circulant blocks. The size of $\tilde{\boldsymbol{M}}$ will be larger than the size of $\boldsymbol{M}$ but for our application the increase will be modest. The vector $\mathbf{G}$ must also be padded with zeros in appropriate places to yield the extended vector $\tilde{\mathbf{G}}$. Now, the result vector of the matrix–vector product $\boldsymbol{M}\mathbf{G}$, which we want to compute, can be extracted from the result vector of the efficiently to evaluate extended matrix–vector product $\tilde{\boldsymbol{M}}\tilde{\mathbf{G}}$. We illustrate this for the Toeplitz to circulant case (i.e., no block structure), which is applicable for the simulation of spatially 1-D models. In that case, the matrix $\boldsymbol{M}$ is a banded Toeplitz matrix with, say, upper bandwidth m and lower bandwidth n

$$M = \begin{pmatrix} l_0 & l_1 & \cdots & l_m & 0 & \cdots & 0 \\ l_{-1} & \ddots & \ddots & & \ddots & \ddots & \\ \vdots & \ddots & & & & & \\ l_{-n} & & & & & & \\ 0 & \ddots & & & & & \\ \vdots & \ddots & & & & & \\ 0 & & & & & & \end{pmatrix} \in \mathbb{R}^{N,N} \qquad (12.19)$$

The corresponding circulant matrix $\tilde{M}$ then has $N + \max\{n, m\}$ rows and columns and is defined by its first column

$$(l_0, l_{-1}, \ldots, l_{-n}, 0 \ldots, l_m, l_{m-1}, \ldots, l_1)^\mathsf{T} \in \mathbb{R}^{N+\max\{n,m\}} \tag{12.20}$$

The extended vector $\tilde{\mathbf{G}}$ is given by $\tilde{\mathbf{G}} = (\mathbf{G}, \mathbf{0})^\mathsf{T} \in \mathbb{R}^{N+\max\{n,m\}}$ and then holds

$$M\mathbf{G} = \left[\tilde{M}\tilde{\mathbf{G}}\right]_{1,\ldots,N} \tag{12.21}$$

That is, only the first N entries of $\tilde{M}\tilde{\mathbf{G}}$ are used.

References

[1] B. Alberts, A. Johnson, J. Lewis, M. Raff, K. Roberts, and P. Walter (2002). *Molecular Biology of the Cell*, 4th edition. Garland Science, Taylor & Francis Group.

[2] A.R.A. Anderson (2005). A hybrid mathematical model of solid tumour invasion: the importance of cell adhesion. *IMA J. Math. Med. Biol.* 22:163–186.

[3] A.R.A. Anderson, M.A.J. Chaplain, E.L. Newman, R.J.C. Steele, and A.M. Thompson (2000). Mathematical modelling of tumour invasion and metastasis. *Comput. Math. Meth. Med. (*formerly *J. Theor. Med.)* 2:129–154.

[4] A.R.A. Anderson, A.M. Weaver, P.T. Cummings, and V. Quaranta (2006). Tumor morphology and phenotypic evolution driven by selective pressure from the microenvironment. *Cell* 127:905–915.

[5] K. Anguige and C. Schmeiser (2009). A one-dimensional model of cell diffusion and aggregation, incorporating volume filling and cell-to-cell adhesion. *J. Math. Biol.* 58(3):395–427.

[6] N.J. Armstrong, K.J. Painter, and J.A. Sherratt (2006). A continuum approach to modelling cell-cell adhesion. *J. Theor. Biol.* 243:98–113.

[7] N.J. Armstrong, K.J. Painter, and J.A. Sherratt (2009). Adding adhesion to a chemical signaling model for somite formation. *Bull. Math. Biol.* 71:1–24.

[8] K.M. Bagnall, S.J. Higgins, and E.J. Sanders (1989). The contribution made by cells from a single somite to tissues within a body segment and assessment of their integration with similar cells from adjacent segments. *Development* 107:931–943.

[9] A.L. Bauer, T.L. Jackson, and Y. Jiang (2007). A cell-based model exhibiting branching and anastomosis during tumor-induced angiogenesis. *Biophys. J.* 92:3105–3121.

[10] A.L. Berrier and K.M. Yamada (2007). Cell-matrix adhesion. *J. Cell. Physiol.* 213:565–573.

[11] W. Birchmeier (1994). Molecular aspects of the loss of cell adhesion and gain of invasiveness in carcinomas. *Int. Symp. Princess Takamatsu Cancer Res. Fund* 24:214–232.

[12] H.M. Byrne and M.A.J. Chaplain (1996). Modelling the role of cell-cell adhesion in the growth and development of carcinomas. *Math. Comput. Modelling* 24:1–17.

[13] R.W. Carthew (2007). Pattern formation in the *Drosophila* eye. *Curr. Opin. Genet. Dev.* 17:309–313.

[14] U. Cavallaro and G. Christofori (2004). Cell adhesion and signalling by cadherins and Ig-CAMs in cancer. *Nat. Rev. Cancer* 4:118–132.

[15] U. Cavallaro and G. Christofori (2004). Multitasking in tumor progression: signaling functions of cell adhesion molecules. *Ann. N.Y. Acad. Sci.* 1014:58–66.

[16] C.M. Cheney and J.W. Lash (1984). An increase in cell-cell adhesion in the chick segmental plate results in a meristic pattern. *J. Embryol. Exp. Morphol.* 79:1–10.

[17] G. Christofori (2003). Changing neighbours, changing behaviour: cell adhesion molecule-mediated signalling during tumour progression. *EMBO J.* 22:2318–2323.

[18] V. Cristini, J. Lowengrub, and Q. Nie (2003). Nonlinear simulation of tumor growth. *J. Math. Biol.* 46:191–224.

[19] J.C. Dallon and H.G. Othmer (2004). How cellular movement determines the collective force generated by the *Dictyostelium discoideum* slug. *J. Theor. Biol.* 231:203–222.

[20] A. Deutsch (1999). Cellular Automata and Biological Pattern Formation. Habilitation thesis, University of Bonn.

[21] A. Deutsch and S. Dormann (2005). *Cellular Automaton Modeling of Biological Pattern Formation: Characterization, Applications, and Analysis.* Boston, Birkhäuser.

[22] R. Dillon, K.J. Painter, and M.R. Owen (2008). A single-cell based model of cellular growth using the immersed boundary method. In B. Cheong Koo, Z. Li, and P. Li, Eds., *Moving Interface Problems and Applications in Fluid Dynamics*, Contemporary Mathematics. AMS, 1–16.

[23] D. Drasdo, R. Kree, and J.S. McCaskill (1995). Monte Carlo approach to tissue-cell populations. *Phys. Rev. E* 52:6635–6657.

[24] J.L. Duband, S. Dufour, K. Hatta, M. Takeichi, G.M. Edelman, and J.P. Thiery (1987). Adhesion molecules during somitogenesis in the avian embryo. *J. Cell. Biol.* 104:1361–1374.

[25] D. Duguay, R.A. Foty, and M.S. Steinberg (2003). Cadherin-mediated cell adhesion and tissue segregation: qualitative and quantitative determinants. *Dev. Biol.* 253:309–323.

[26] G.A. Dunn and J.P. Heath (1976). A new hypothesis of contact guidance in tissue cells. *Exp. Cell Res.* 101:1–14.

[27] R.A. Foty and M.S. Steinberg (2004). Cadherin-mediated cell-cell adhesion and tissue segregation in relation to malignancy. *Int. J. Dev. Biol.* 48:397–409.

[28] R.A. Foty and M.S. Steinberg (2005). The differential adhesion hypothesis: a direct evaluation. *Dev. Biol.* 278:255–263.

[29] P. Friedl and K. Wolf (2003). Tumour-cell invasion and migration: diversity and escape mechanisms. *Nat. Rev. Cancer* 3:362–374.

[30] J. Galle, G. Aust, G. Schaller, T. Beyer, and D. Drasdo (2006). Individual cell-based models of the spatial-temporal organization of multicellular systems: achievements and limitations. *Cytometry Part A* 69A:704–710.

[31] A. Gerisch (2009). On the approximation and efficient evaluation of integral terms in PDE models of cell adhesion. In press. DOI 10.1093/imanum. *IMA Journal of Numerical Analysis.* drp027.

[32] A. Gerisch and M.A.J. Chaplain (2006). Robust numerical methods for taxis–diffusion–reaction systems: applications to biomedical problems. *Math. Comput. Modelling* 43:49–75.

[33] A. Gerisch and M.A.J. Chaplain (2008). Mathematical modelling of cancer cell invasion of tissue: local and non-local models and the effect of adhesion. *J. Theor. Biol.* 250:684–704.

[34] J.A. Glazier and F. Graner (1993). Simulation of the differential adhesion driven rearrangement of biological cells. *Phys. Rev. E* 47:2128–2154.

[35] F. Graner and J.A. Glazier (1992). Simulation of biological cell sorting using a two-dimensional extended Potts model. *Phys. Rev. Lett.* 69:2013–2016.

[36] B.M. Gumbiner (2005). Regulation of cadherin-mediated adhesion in morphogenesis. *Nat. Rev. Mol. Cell Biol.* 6:622–634.

[37] T. Hillen (2002). Hyperbolic models for chemosensitive movement. *Math. Mod. Meth. Appl. Sci.* 12:1007–1034.

[38] T. Hillen, K. Painter, and C. Schmeiser (2007). Global existence for chemotaxis with finite sampling radius. *Disc. Cont. Dyn. Syst. B (DCDS-B)* 7:125–144.

[39] T. Hillen and K.J. Painter (2009). A user's guide to PDE models for chemotaxis. *J. Math. Biol.* 58:183–217.

[40] G.S. Hillis and A.D. Flapan (1998). Cell adhesion molecules in cardiovascular disease: a clinical perspective. *Heart* 79:429–431.

[41] T. Hofer, J.A. Sherratt, and P.K. Maini (1995). *Dictyostelium discoideum:* cellular self-organization in an excitable biological medium. *Proc. Biol. Sci.* 259:249–257.

[42] K. Horikawa, G. Radice, M. Takeichi, and O. Chisaka (1999). Adhesive subdivisions intrinsic to the epithelial somites. *Dev. Biol.* 215:182–189.

[43] E.F. Keller and L.A. Segel (1970). Initiation of slime mold aggregation viewed as an instability. *J. Theor. Biol.* 26:399–415.

[44] R. Keller (2002). Shaping the vertebrate body plan by polarized embryonic cell movements. *Science* 298:1950–1954.

[45] K.K. Linask, C. Ludwig, M.D. Han, X. Liu, G.L. Radice, and K.A. Knudsen (1998). N-cadherin/catenin-mediated morphoregulation of somite formation. *Dev. Biol.* 202:85–102.

[46] C.M. Lo, H.B. Wang, M. Dembo, and Y.L. Wang (2000). Cell movement is guided by the rigidity of the substrate. *Biophys. J.* 79:144–152.

[47] P. Macklin and J. Lowengrub (2007). Nonlinear simulation of the effect of microenvironment on tumor growth. *J. Theor. Biol.* 245:677–704.

[48] A.F. Marée and P. Hogeweg (2001). How amoeboids self-organize into a fruiting body: multicellular coordination in *Dictyostelium discoideum. Proc. Natl. Acad. Sci. USA* 98:3879–3883.

[49] M. Mareel and A. Leroy (2003). Clinical, cellular, and molecular aspects of cancer invasion. *Physiol. Rev.* 83:337–376.

[50] J. Moreira and A. Deutsch (2005). Pigment pattern formation in zebrafish during late larval stages: a model based on local interactions. *Dev. Dyn.* 232:33–42.

[51] T.J. Newman (2005). Modeling multicellular systems using subcellular elements. *Math. Biosci. Eng.* 2:613–624.

[52] T.J. Newman and R. Grima (2004). Many-body theory of chemotactic cell-cell interactions. *Phys. Rev. E* 70:051916.

[53] D. Ostrovsky, C.M. Cheney, A.W. Seitz, and J.W. Lash (1983). Fibronectin distribution during somitogenesis in the chick embryo. *Cell. Differ.* 13:217–223.

[54] K.J. Painter, D. Horstmann, and H.G. Othmer (2003). Localization in lattice and continuum models of reinforced random walks. *Appl. Math. Lett.* 16:375–381.

[55] E. Palsson (2008). A 3-D model used to explore how cell adhesion and stiffness affect cell sorting and movement in multicellular systems. *J. Theor. Biol.* 254:1–13.

[56] E. Palsson and H.G. Othmer (2000). A model for individual and collective cell movement in *Dictyostelium discoideum*. *Proc. Natl. Acad. Sci. USA* 97:10448–10453.

[57] S.D. Patel, C.P. Chen, F. Bahna, B. Honig, and L. Shapiro (2003). Cadherin-mediated cell-cell adhesion: sticking together as a family. *Curr. Opin. Struct. Biol.* 13:690–698.

[58] N.J. Popawski, M. Swat, J.S. Gens, and J.A. Glazier (2007). Adhesion between cells, diffusion of growth factors, and elasticity of the AER produce the paddle shape of the chick limb. *Physica A* 373:521–532.

[59] O. Pourquié (2001). Vertebrate somitogenesis. *Annu. Rev. Cell. Dev. Biol.* 17:311–350.

[60] I. Ramis-Conde, D. Drasdo, A.R.A. Anderson, and M.A.J. Chaplain (2008). Modeling the influence of the E-cadherin-beta-catenin pathway in cancer cell invasion: a multiscale approach. *Biophys. J.* 95:155–165.

[61] K.A. Rejniak (2007). An immersed boundary framework for modelling the growth of individual cells: an application to the early tumour development. *J. Theor. Biol.* 247:186–204.

[62] N.J. Savill and J.A. Sherratt (2003). Control of epidermal stem cell clusters by Notch-mediated lateral induction. *Dev. Biol.* 258:141–153.

[63] G. Schaller and M. Meyer-Hermann (2005). Multicellular tumor spheroid in an off-lattice Voronoi-Delaunay cell model. *Phys. Rev. E* 71:051910.

[64] J.A. Sherratt, S. Gourley, N.A. Armstrong, and K.J. Painter (2009). Boundedness of solutions of a nonlocal reaction-diffusion model for cellular adhesion. *Eur. J. Appl. Math.* 20:123–144.

[65] M.S. Steinberg (1962). Mechanism of tissue reconstruction by dissociated cells. II. Time course of events. *Science* 137:762–763.

[66] M.S. Steinberg (1962). On the mechanism of tissue reconstruction by dissociated cells, I. Population kinetics, differential adhesiveness, and the absence of directed migration. *Proc. Natl. Acad. Sci. USA* 48:1577–1582.

[67] M.S. Steinberg (1962). On the mechanism of tissue reconstruction by dissociated cells. III. Free energy relations and the reorganization of fused, heteronomic tissue fragments. *Proc. Natl. Acad. Sci. USA* 48:1769–1776.

[68] M.S. Steinberg (2007). Differential adhesion in morphogenesis: a modern view. *Curr. Opin. Genet. Dev.* 17:281–286.

[69] M.S. Steinberg and M. Takeichi (1994). Experimental specification of cell sorting, tissue spreading, and specific spatial patterning by quantitative differences in cadherin expression. *Proc. Natl. Acad. Sci. USA* 91:206–209.

[70] C.D. Stern and R.J. Keynes (1987). Interactions between somite cells: the formation and maintenance of segment boundaries in the chick embryo. *Development* 99:261–272.

[71] M. Takeichi (1988). The cadherins: cell-cell adhesion molecules controlling animal morphogenesis. *Development* 102:639–655.

[72] M. Takeichi (1995). Morphogenetic roles of classic cadherins. *Curr. Opin. Cell Biol.* 7:619–627.

[73] U. Tepass, D. Godt, and R. Winklbauer (2002). Cell sorting in animal development: signalling and adhesive mechanisms in the formation of tissue boundaries. *Curr. Opin. Genet. Dev.* 12:572–582.

[74] P.L. Townes and J. Holtfreter (1955). Directed movements and selective adhesion of embryonic amphibian cells. *J. Exp. Zool.* 128:53–120.

[75] S. Turner and J.A. Sherratt (2002). Intercellular adhesion and cancer invasion: a discrete simulation using the extended Potts model. *J. Theor. Biol.* 216:85–100.

[76] S. Turner, J.A. Sherratt, and D. Cameron (2004). Tamoxifen treatment failure in cancer and the nonlinear dynamics of TGFβ. *J. Theor. Biol.* 229:101–111.

[77] S. Turner, J.A. Sherratt, K.J. Painter, and N.J. Savill (2004). From a discrete to a continuous model of biological cell movement. *Phys. Rev. E* 69:021910.

[78] R. Weiner, B.A. Schmitt and H. Podhaisky (1997). ROWMAP: a ROW-code with Krylov techniques for large stiff ODEs. *Appl. Numer. Math.* 25:303–319.

[79] M.J. Wheelock, Y. Shintani, M. Maeda, Y. Fukumoto, and K. Johnson (2008). Cadherin switching. *J. Cell. Sci.* 121:727–735.

[80] S.M. Wise, J.S. Lowengrub, H.B. Frieboes, and V. Cristini (2008). Three-dimensional multispecies nonlinear tumor growth: I. Model and numerical method. *J. Theor. Biol.* 253:524–543.

[81] W. Wood and P. Martin (2002). Structures in focus: filopodia. *Int. J. Biochem. Cell Biol.* 34:726–730.

Chapter 13

Bridging Cell and Tissue Behavior in Embryo Development

Alexandre J. Kabla, Guy B. Blanchard, Richard J. Adams and L. Mahadevan

Contents

The tissues of animal embryos utilize large-scale morphological transformations to bring about highly sophisticated body plans. In contrast with plant tissues, where differential growth and change in cell shapes are the main morphogenetic mechanisms, animal cells have in addition the ability to move with respect to their neighboring cells. Movement can be either active (cell motility) or passive, for instance as a response to an imposed strain. Despite the progress in genetics and molecular biology, our understanding of developmental biology still suffers a lack of experimental description and mechanistic interpretation of how individual cell behaviors lead to well-organized collective movements at the tissue scale. One of the outstanding issues concerns the role of mechanical forces, for instance as a driving force for passive morphological changes, or as involved in signaling pathways regulating active cell behavior. In this chapter we summarize a framework specifically designed to quantify the kinematics of embryo development. By analyzing clusters of neighboring cells, we developed a multiscale geometrical description that decomposes tissue strains into two contributions: one associated with changes in cell shapes and the other with cell–cell slippage or motility. The emphasis on cell shapes and cell–cell slippage provides, in particular, a fully continuous framework especially suitable to capture temporal and spatial heterogeneities regardless of discrete events such as neighbor exchanges. We also show here explicitly how the statistics of cell shape changes depend on a microscopic assumption regarding cell–cell slippage and propose a simple geometrical principle that can be used to deploy a consistent and robust approach.

13.1 Quantifying Embryo Morphogenesis

13.1.1 Morphogenesis of the Animal Embryo

Modern views of morphogenetic mechanisms have been built from a multidisciplinary approach to the study of the problem. Models must span from subcellular mechanisms of cellular mechanics, adhesion, and polarized cell behaviors to a mechanical understanding of substantial portions of the embryo and its environment. Morphogenetic mechanisms appear to encompass many classes of cell behavior, ranging from individual cell shape changes and movements to the collective reshaping of sheets of cells during such movements as invagination during sea urchin gastrulation [15], the convergence and extension of the vertebrate body axis [14], or the neurulation of the brain [8]. This class of collective reorganization is considered here, and key questions remain about identifying the patterns of passive and active cell behavior that shape the change in tissue morphology.

Our current understanding has closely followed the availability of tools, principally for microscopy, that have provided a means to address the dynamics of morphogenetic change in a noninvasive way. Variations among the trajectories of cells is the first indication of deformation of the tissue. For instance, early studies by Jacobson on the neural plate of the newt pointed to a way of combining microscopic analyses and experimental manipulation to find a link between patterned regional variation in cell behavior—in this case, cell shape changes suggesting external mechanical forces responsible for the shaping of the neural plate [5].

While these early studies have been superseded, the basis of this approach is still recapitulated in current studies, including our own. To collect high-resolution and comprehensive 3-D data from which cell behavior and tissue morphogenesis can be measured has required the development of labeling and imaging methods capable of resolving cells in living embryos [17]. Software tools have been developed from which the trajectories and shapes of cells can be followed over time. Trajectories of cell centroids reveal the patterns of cell movements [7,16] but this is in itself difficult to interpret in terms of morphogenetic mechanisms until used to calculate tissue deformation [8,10] in the form of strain rates. Microscopic analysis of passive [13] and active tissue reshaping [20] has also provided evidence for cell movements or rearrangements underlying the morphogenesis of tissues.

13.1.2 Cell Intercalation and Rearrangements

Measures of cell area [5], cell neighbor number [22], and the change in cell neighbor topology [3] have been used to investigate the cellular mechanisms of morphogenesis. Interpretation of observations of cell rearrangements as passive or active events is highly dependent on additional biological or experimental insight into the problem. In recent years the power of molecular and genetic manipulation to investigate the relationship between cell behavior and morphogenesis has flourished, for instance in the role of planar cell polarity [23]. Deficiencies in cell behavior can be directly correlated with gross abnormalities of tissue morphogenesis [21]. However, to make full use of such methods, we require means of assaying the local morphogenetic characteristics of cells such as changes in their shapes and neighbor rearrangements. Independently, these measures give only their correlation but do not provide a direct link between events at the cell and tissue level. Linking them in a quantitative way is therefore necessary to explore causality and eventually identify underlying mechanisms.

Along that line, several novel approaches have been introduced to extract statistical and tensorial representations of cell behavior at a mesoscopic scale. Graner et al. [12] recently generalized tools primarily developed for foam mechanics into a generic framework for quantifying the strain and reorganization of cellular or granular materials based on the movement of cell centers and the evolution of their network of contacts. Their approach succeeds in relating material strains with the dynamics of neighbor exchanges, which represents one of the most visible characteristics of the long-term evolution of tissues.

On short time scales, however, neighbor exchange is a rare event and the local evolution of a piece of tissue is more appropriately described by a combination of cell shape change and minute relative movements of cells that is referred to as cell–cell slippage. To account for this continuous reorganization, one can develop a kinematic approach by quantifying cell shape change rather than the dynamics of the cell contact network. In the following sections, we present in detail the construction of such a framework recently developed to analyze strains and cell behaviors during embryo morphogenesis [4]. It has been validated on a number of experimental situations in 2-D, such as the Drosophila germband extension [6], Drosophila dorsal closure [11], or zebrafish neural plate formation [19]. Another novelty in the framework, introduced in the final section, is the explicit link to a microscopic model of relative movement at the interface of neighboring cells. Although in practice reasonable choices can be made at that level, this opens the possibility of more subtle adjustments of the kinematic model in order to better account for the biological variety of cell behavior at the microscale.

13.2 Strain Measurements

The experimental measure of strain and strain rates during tissue morphogenesis requires us to monitor the motion of material points over time. In heterogeneous and composite materials, such as biological tissues, internal displacement fields can be highly complex and one needs to consider at which length scale and time scale a coarse-grained description is relevant. In the context of embryo morphogenesis, time scales are typically on the order of tens of minutes up to a few hours, and movements typically involve a large number of cells. Quantifying cytoskeletal and cytoplasmic movements inside cells is largely irrelevant as such flows occur on much shorter time scales and length scales. Although these might be important to understand the biological origin of collective movements, it is *a priori* enough, in order to characterize the deformation at the tissue scale, to average internal motion at the cell scale. This can be done, for instance, by tracking the cell center of mass (determined from the cell contour) or the locations of nuclei if the latter are labeled [8][1]. As discussed in Section 13.3, movements within the cell will also be considered, encompassed in a single tensorial quantity, the cell shape strain rate.

13.2.1 Introduction of a Mesoscopic Scale

From 3-D movies of developing embryos where cell membranes are fluorescently labeled [17], the contour of each cell is detected by image analysis and

[1]Only a marginal amount of extracellular matrix is present in embryo tissue at the early stages of development. One can therefore assume that cells are contiguous in the tissue and occupy all its volume.

the cell center of mass is computed [4]. Individual cells can be followed over time and their trajectories recorded, as well as their full shape, at any time. This can be implemented either for a 2-D epithelium or a 3-D tissue. A common situation corresponds to planar processes along curved surfaces, where the local strain is suitably accounted for by a 2-D description. In this chapter, the methods are illustrated using 2-D examples, although each step can be generalized in 3-D.

Kinematic quantities in the tissue are then defined and measured at a mesoscopic length scale at which we linearize the cell displacement field. We introduce a time scale δt and length scale n_c expressed as a number of cell diameters. The latter can either represent a topological distance (first neighbors, second neighbors, ...) or a physical distance. These two quantities are used to select a domain surrounding each cell at each time (Figure 13.1). More precisely, such a neighborhood, denoted by $N(i,t)$, is defined as the collection of cells located around the cell i, at time t, at a distance at most n_c; all these cells are followed during a time interval $[t-\delta t/2, t+\delta t/2]$. Although the connectivity of the cells can be used to define the neighborhood, it should be stressed that it is not a requirement; it only serves the purpose of conveniently defining a cohort of cells whose relative motion is tracked for a certain time.

13.2.2 Tissue Strain Rates

Defining absolute strains in embryos is often inconvenient due to the lack of a meaningful reference state and the complexity of handling the large finite strains that accumulate over time. Rather than using such quantities, we therefore focus on strain rates that can be easily mapped to reveal temporal and spatial patterns and conveniently integrated if necessary. In the following, we describe the generic approach used to estimate tissue strain rates from cell trajectories. Such methods are broadly used in hydrodynamics [18] and solid mechanics [9] to quantify intrinsic strains and strain rates, and have been successfully applied to geophysical measurements, in particular in the context of plate tectonics [1].

13.2.2.1 Velocity Gradient

By convention in this chapter the index i represents the identity of a cell. Vector symbols are underlined, and tensors are written in **bold** characters. The position of a cell over time, or trajectory, is denoted by $\underline{r}_i(t)$. The cell velocity in the reference frame of the microscope is $\underline{v}_i(t) = d\underline{r}_i/dt$, and is calculated by linearizing the cell displacement over a time interval $t \pm \delta t/2$. Within a neighborhood $N(i,t)$, one can define a number of averaged quantities. The position of the cohort is defined by:

$$\underline{R}_i(t) = \left\langle \underline{r}_{i'}(t) \right\rangle_{i' \in N(i,t)} \tag{13.1}$$

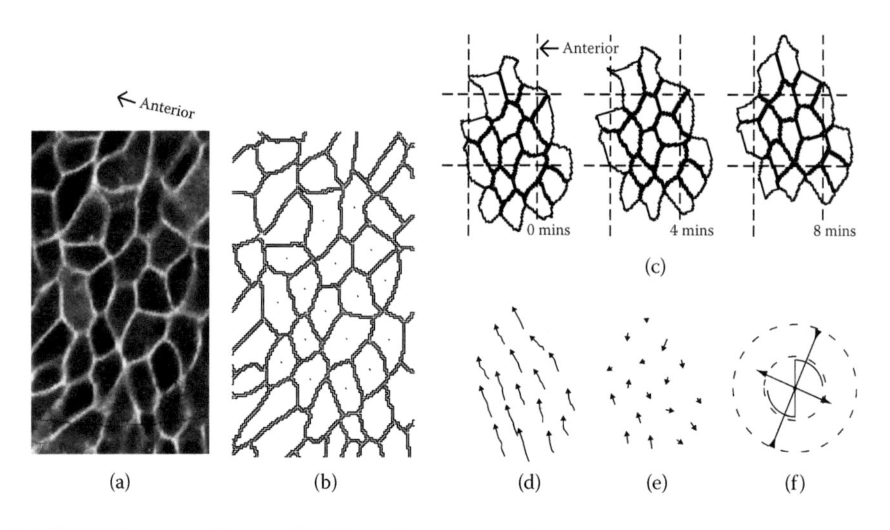

FIGURE 13.1 Example of tracking for a cell cohort in the zebrafish trunk. (a) Image of the first cellular layer flattened on a plane from a 3-D confocal stack. (b) Reconstructed cell membranes. (c) Evolution of a cell cohort, for the group radius corresponding to $n_c = 2$, and $\delta t = 8$ min. (d) Displacement field calculated from the above images. (e) Residual displacements once the mean velocity is subtracted. (f) Velocity gradient tensor $\mathbf{L_t}$ resulting from the linearization of the velocity field. The cross indicates the direction of eigenvectors of the symmetric part $\mathbf{D_t}$. The corresponding eigenvalues are coded in the length of the bars, with an arrow pointing outward for a positive eigenvalue (extension) or inward for a negative eigenvalue (convergence). The scythe motif represents the spin tensor $\mathbf{W_t}$, with a diameter in proportion to the magnitude of the antisymmetrical element. The radii of the dashed circles correspond to 0.8% and 1.6% per minute. (Data from [4].)

The average velocity of the cohort is:

$$\underline{V_i}(t) = \langle \underline{v_{i'}} \rangle_{i' \in N(i,t)} \tag{13.2}$$

The velocity field is then linearized within the neighborhood to calculate the velocity gradient tensor $\mathbf{L_t}$:

$$\underline{v_{i'}}(t) = \underline{V_i}(t) + \mathbf{L_t}\left(\underline{r_{i'}}(t) - \underline{R_i}(t)\right) + \underline{\delta v_{i'}} \tag{13.3}$$

The tensor $\mathbf{L_t}$ is, for instance, obtained from a least-squares fit of the experimental data. The residuals $\delta\underline{v}_{i'}$ can be used to assess the fit quality. The variation of the residuals with the neighborhood size n_c and integration time δt is in itself an interesting quantity that informs about the sources of the deviation, such as spatial heterogeneities, uniform noise in the position, etc. Nonaffinity in the displacement field is discussed later in this chapter.

13.2.2.2 Stretch and Rotation Rates

As strain rates intrinsically characterize infinitesimal strain increments, they can be appropriately decomposed into a sum of a stretch rate tensor $\mathbf{D_t}$ and a spin tensor $\mathbf{W_t}$, that is, a tensor accounting for the rate of rotation of the cohort domain. The tissue stretch rate tensor is defined as:

$$\mathbf{D_t} = \left(\mathbf{L_t} + \mathbf{L_t}^T\right)/2 \tag{13.4}$$

It is a symmetric tensor; its eigenvectors indicate the main axes of deformation, while the corresponding eigenvalues indicate the rate of elongation along these axes. The tissue spin tensor corresponds to the antisymmetrical part of the velocity gradient tensor:

$$\mathbf{W_t} = \left(\mathbf{L_t} - \mathbf{L_t}^T\right)/2 \tag{13.5}$$

13.2.3 Applications of Velocity Gradient Strain Tensor in Biology

The measurement of the velocity gradient tensor field has been applied to diverse embryological tissues, containing hundreds or thousands of cells, and for periods up to 3 hours [6,11,19]. For example, the fruit-fly germband converges to the ventral side of the embryo as it extends in the anterior–posterior axis in movements analogous to the convergence and extension of the vertebrate trunk. Analysis revealed that there was a gradient of increasing extension to the posterior, and a gradient of increasing convergence toward the ventral midline [6]. The latter gradient was correlated with tissue rotation in the flanking regions both anterior and posterior. The total accumulated strain of the tissue was comparable to published "shoelace" methods, and the time evolution of the process was found to be biphasic with a fast early phase followed by a slower phase. One direct advantage of mapping these quantities is to allow comparison between individuals to estimate inter-individual variability. The morphogenesis of the wild-type flies was then compared to various mutant fly strains whose morphogenesis is known to be abnormal during the convergence and extension process.

13.3 Cell Shapes and Intercalation

As mentioned in Section 13.1, changes in tissue morphology are accounted for at the cell scale by two main classes of evolution: cells can change shape and cells can rearrange (i.e., move relative to their closest neighbors). Both have a direct geometrical signature, and a proper kinematic description of

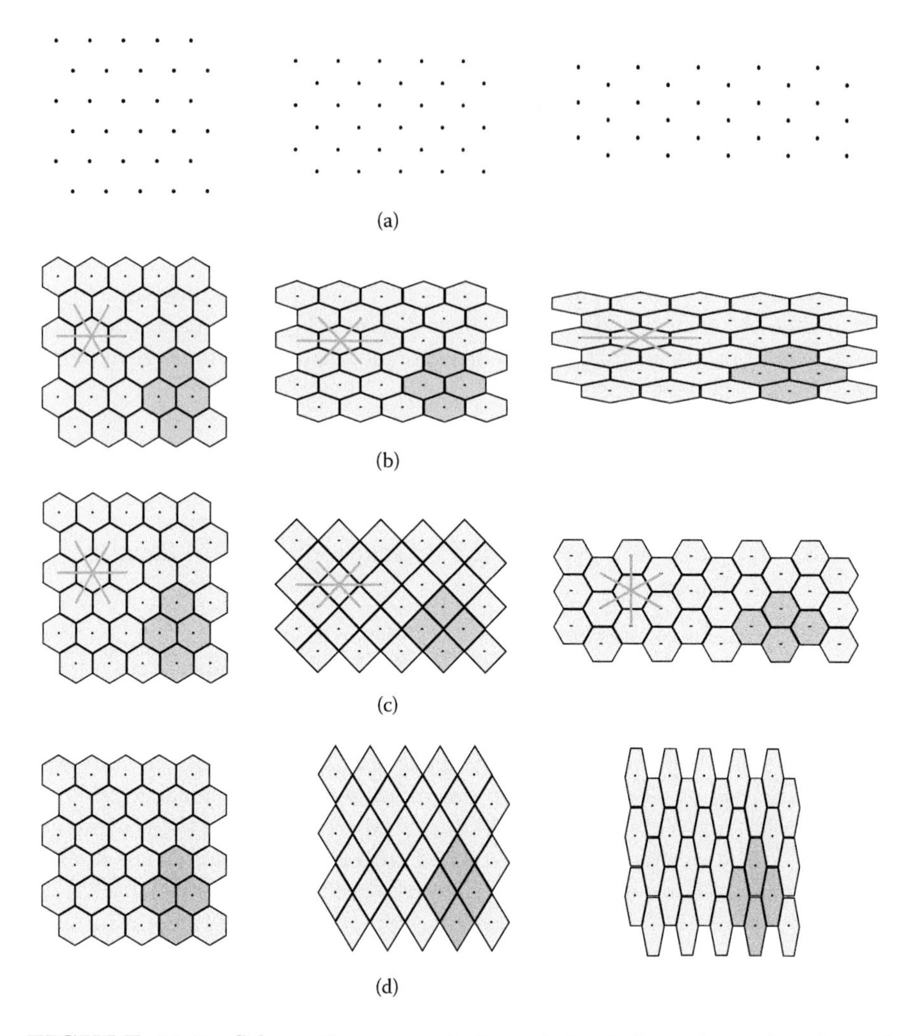

FIGURE 13.2 Schematic representation of the deformation of a piece of tissue; time increases from left to right. (a) Evolution of the cell centers, from which we can extract the tissue strain tensor. (b) and (c) Two opposite modes of shape deformations consistent with the tissue strain field shown in (a). The evolution of a ring of first neighbors is highlighted by line segments. (d) Shows another example where there is no tissue deformation, only a transfer between shape deformation and intercalation.

tissue morphogenesis must account for this. First, it should be highlighted that such information cannot be extracted from the motion of cell centers only. Figure 13.2 illustrates, for a uniform stretching of a regular hexagonal lattice, two different types of cellular dynamics that are consistent with the same overall tissue strain. In the first case (Figure 13.2(b)), the shape of each cell experiences exactly the same strain as the piece of tissue. In the second situation (Figure 13.2(c)), although cell shapes slightly evolve over time, cells mostly compensate the tissue strain by moving past each other while remaining, on average, isotropic. This movement corresponds to a purely intercalating group of cells where all the deformation is associated with cell slippage, eventually leading to topological reorganizations in the sample, that is, neighbor exchanges. These two examples therefore demonstrate that cell shape evolves independently of the tissue strain according to their ability to intercalate. As shown in Figure 13.2(d), even if the tissue is not changing shape, there can be an interplay between cell shapes and cell slippage leading to a reorganization of the cells in the tissue. In addition to the motion of the centers, information must therefore be extracted from the dynamics of cell reorganization or cell shape evolution in order to discriminate between cell shape and intercalation-driven tissue reshaping processes.

The approach introduced by Graner et al. [12] addresses that issue; it uses the dynamics of the network of neighboring cells to track rearrangements in addition to the tissue strain. The authors build a number of statistical quantities from the distribution of links between first or second neighbors. Each cell rearrangement causes the network of first neighbors to evolve since links are gained where cells come closer, and lost in directions where cells move away. This allows the building of tensorial quantities to quantify both the tissue strain and the reorganization of the cell within the tissue. This method has a number of advantages, including a simple implementation in 2-D and 3-D, and a broad range of applications, in particular in the field of granular systems and colloidal suspensions where the contact network is more relevant than the individual shape of the particles. However, in the context of biological tissues, it starts capturing intercalation only when neighborhood relationships evolve. For instance, in the first two examples in Figures 13.2(b) and 13.2(c), such a method would see a difference between the shape- and intercalation-driven processes only after cells rearrange (second half) because until then, both the center of mass locations and contact networks are exactly identical in the two examples.

To capture the continuous nature of cell shapes and cell motility independently of cell rearrangements, we introduce in this section a general approach based on cell shape and its statistical evolution that allows us to quantify the respective contributions of both shape variations and intercalation movements to the total tissue extension previously studied. The definition of the intercalation tensor is then discussed in light of its microscopical interpretation.

13.3.1 Cell Shape Evolution

We aim to determine the strain tensor that accounts for the evolution in shape of the cells contained in a given neighborhood during the time interval δt, that is, that transforms the initial collection of cell shapes into a collection of shapes that is statistically equivalent to the final cell shapes, in terms of

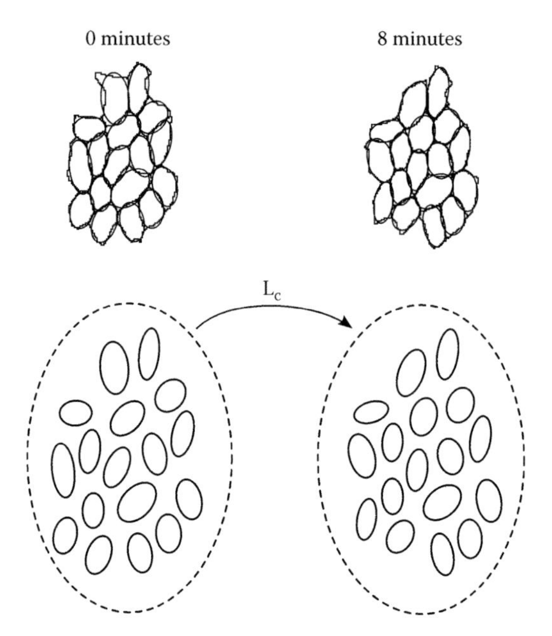

FIGURE 13.3 Illustration of the shape measurements. Top images show the evolution of a cohort of cells. Elliptical fits of individual cells are overlayed. The shape strain rate tensor is the transformation that, when applied to the initial collection of ellipses, matches the distribution of shapes in the final configuration.

orientation and elongation. Information concerning the relative placement of cells (as used in the previous section) and their neighborhood relationship is here irrelevant and discarded (see Figure 13.3).

Several approaches can be used to tackle this question. One is to consider, for each cell individually, the strain rate that best matches the shape evolution. The question of identifying the strain that transforms one shape into another has been extensively studied in image processing. Most registration methods (optical flow, digital image correlation, etc.) estimate the physical displacement of points in images and use it to deduce the local deformation. Unfortunately, such methods are not suitable for identifying the deformation of the cell shape because the movements of details of the membrane geometry (which is the only morphological feature available here) is locally non-affine and does not necessarily represent the average strain of the cell bulk. This is illustrated on Figure 13.4 where a regular arrangement of hexagon centers follows a simple shear deformation and cells accommodate it mostly by sliding on each other. We used a full registration method to estimate the strain tensor of individual shapes. The resulting tensor captures primarily the movement of membranes, which results in typically three solutions where the spin of the shape is either null or twice as big as the rotation component of the tissue velocity gradient. It is difficult to

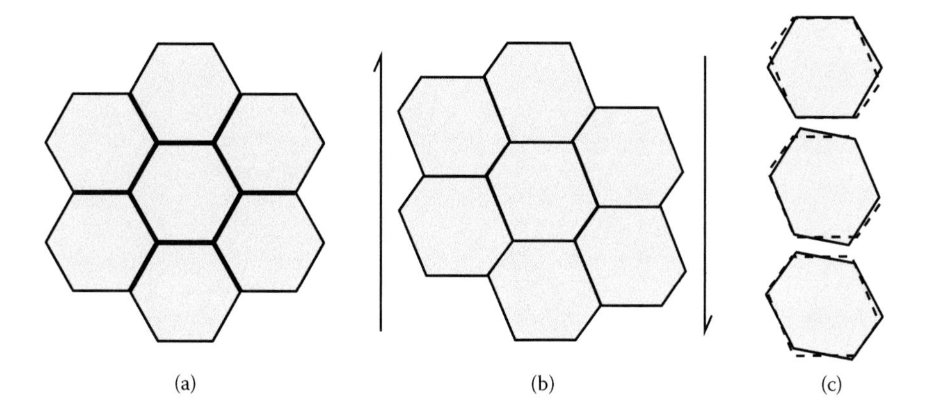

FIGURE 13.4 Typical issues arising with full registration methods applied to situations where non-affine shape deformations are present. (a) Regular hexagonal pattern. (b) Its evolution after a pure shear deformation. Shapes are recalculated in this example from the Voronoi tesselation of the cell centers. (c) Three local solutions for the shape strain typically obtained when trying to find the strain tensor that minimizes the nonoverlapping area. Dashed lines show the target shape displayed in (b), and the solid line, the calculated transformation of the initial hexagon in (a).

choose between these options, and there is no reason to believe *a priori* that the bulk of the cell tends to rotate on itself with respect to the tissue. This simple situation shows that, in general, registration methods on the membrane contour cannot be used to extract the cell shape strain. The only reliable information we can use at this stage corresponds to the global anisotropy of the cell shape, represented, for instance, by an elliptical fit to its shape. Another approach, introduced by Aubouy et al. [2] and not developed here, uses a (symmetric) texture tensor to characterize the average cell shape at the neighborhood scale. In all cases, we end up with an "elliptical" representation of the cell shapes, and we basically need to know what linear application transforms a given ellipse into another. Unfortunately, there is not a unique solution to the problem.

Let us define two elliptical shapes C_1 and C_2. Each of them is fully characterized by a pair of orthogonal vectors $(\underline{a}_1, \underline{b}_1)$ and $(\underline{a}_2, \underline{b}_2)$ representing, respectively, their minor and major axis directions and length (see Figure 13.5a). To transform C_1 into C_2, one can, for instance, first rotate C_1 until $\underline{a}_1$ is along $\underline{a}_2$, and then stretch the shape accordingly; or equivalently, first stretch the shape and then rotate it (Figure 13.5c). In both cases, points located along the main axis of C_1 end on the main axis of C_2. One could also directly search for a symmetric tensor that transforms C_1 into C_2 (Figure 13.5d). The eigenvectors of this tensor are not, in general, along the axis of either ellipse and, in contrast to the previous situation, the vectors $\underline{a}_1$ and $\underline{b}_1$ would not be mapped into the vectors $\underline{a}_2$ and $\underline{b}_2$. In fact, a continuum

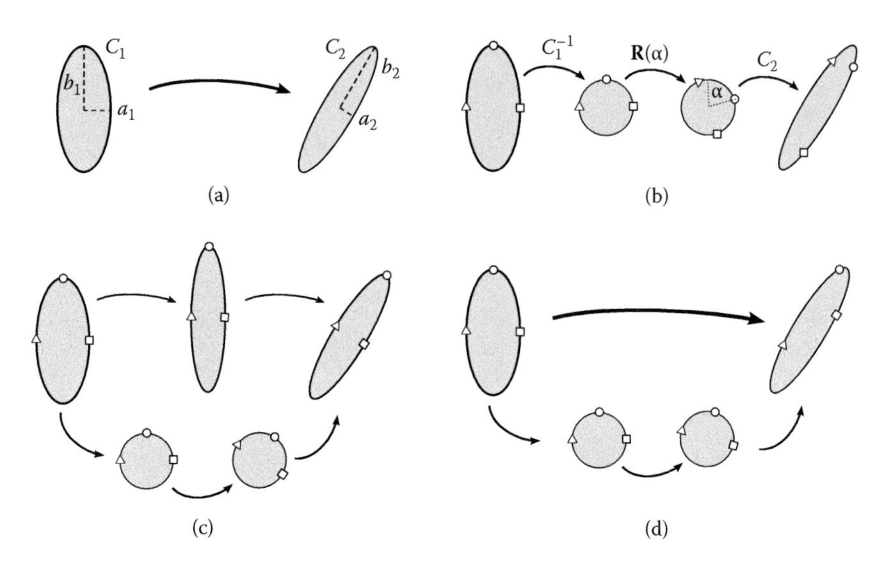

FIGURE 13.5 Family of deformations that transform one ellipse into another. (a) Description of the two ellipses C_1 and C_2. (b) Generic decomposition of the transformation into two stretches and an arbitrary rotation. Symbols indicate how a few specific points move with the deformation. (c) Example of a deformation that preserves the position of points with respect to the main axis of the ellipse. (d) Example of a deformation constrained to be symmetrical.

of solutions can easily be found. Let $\mathbf{C_1}$ and $\mathbf{C_2}$ be the (symmetric) stretch matrices that transform a circle of radius one onto, respectively, C_1 and C_2:

$$\mathbf{C_1} = \begin{bmatrix} a_1 & 0 \\ 0 & b_1 \end{bmatrix}, \mathbf{C_2} = \begin{bmatrix} a_2 & 0 \\ 0 & b_2 \end{bmatrix} \tag{13.6}$$

expressed on the basis of unit vectors along the principal axis of C_1 and C_2, respectively. Let $\mathbf{R}(\alpha)$ be a rotation of angle α. For all values of α, the transformation $\mathbf{A}(\alpha) = \mathbf{C_2}\mathbf{R}(\alpha)\mathbf{C_1}^{-1}$ transforms C_1 into C_2 (see Figure 13.5b). Depending on the value of the angle α, the image of a given point along C_1 is going to move along C_2, as illustrated on Figure 13.5. This simple argument shows that building a strain tensor based on an elliptic representation of a shape is fundamentally undetermined.

A natural hypothesis to fully specify the shape deformation is to set the shape rotation component to match the tissue rotation, that is, so that the cells are not rotating with respect to the tissue. This ensures, in particular, that if a piece of tissue is uniformly rotated, the cell shape strain matches the tissue strain. In the case of a pure stretch of a tissue without intercalation, the shape deformation is also identical to the tissue strain, as expected. If we denote $\mathbf{L_c}$ as the cell shape strain rate, and use the same decomposition into

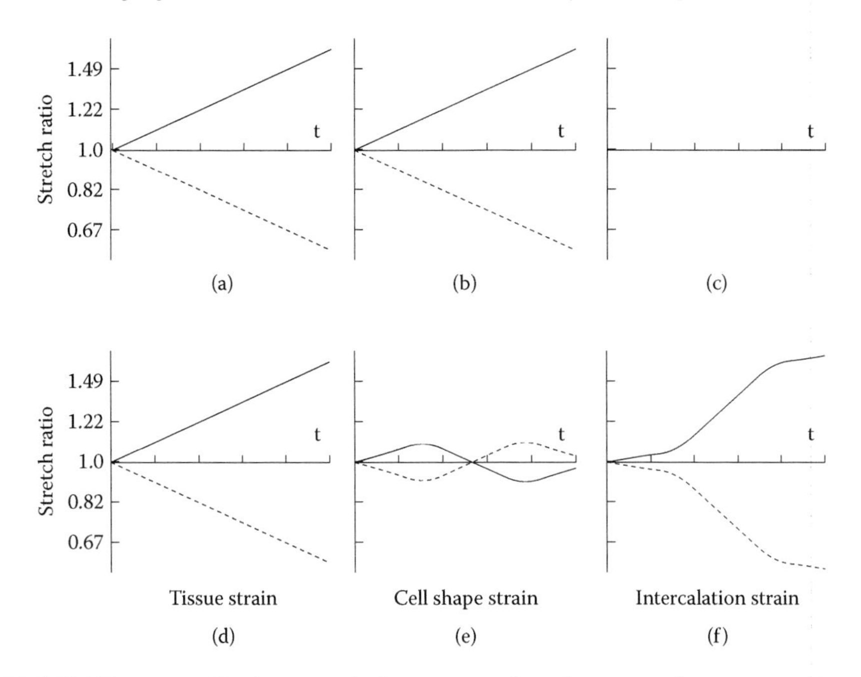

FIGURE 13.6 Evolution of the strains for the examples presented on Figures 13.2(b) (a to c) and 13.2(c) (d to f). The stretch ratio represents the ratio between the current tissue length and the initial length. Solid lines correspond to the direction of extension (horizontal), and dashed lines to the direction of convergence (vertical). From G.B. Blanchard, A.J. Kabla, N.L. Schultz, L.C. Butler, B. Sanson, N. Gornkiel, L. Mahadevan, and R.J. Adams (2009) Tissue tectonics: morphogenetic strain rates, cell shape change and intercalation. *Nature Methods*, in press. (With permission).

a symmetrical and anti-symmetrical part, we obtain:

$$\mathbf{L_c} = \mathbf{D_c} + \mathbf{W_c} \text{ with } \mathbf{W_c} = \mathbf{W_t} \tag{13.7}$$

A general method used to calculate the symmetrical component of the cell shape strain rate for a local domain is introduced in [4]. Figures 13.6(a) and b show the evolution of the total tissue strain for the hexagonal pattern described on Figure 13.2(b). In this example we find, as expected, that the cell shape follows precisely the tissue stretch. The same kind of plot is presented in Figures 13.6(d) and 13.6(e) for the intercalating tissue (Figure 13.2(c)). In that case, the shape strain slightly varies about zero, indicating indeed that cells do not deform on average.

As all the volume of the local group of cells is simply the sum of cell volumes (there is no free space between cells), the cell shape strain rate tensor must also account for any variation in tissue volume. This implies that the trace of $\mathbf{L_c}$ is identical with the trace of $\mathbf{L_t}$.

$$\text{Tr } \mathbf{L_t} = \text{Tr } \mathbf{L_c} \tag{13.8}$$

13.3.2 Tensorial Representation of Cell Intercalation

As suggested by the examples above (Figure 13.2), identical shape and tissue strain rates are a signature of nonintercalating tissues. On the other hand, the presence of collective slippage between cells leads to a mismatch between tissue strain rate and cell shape strain rate. More generally, the part of the tissue strain that is not accounted for by the shape evolution reflects collective movements of cells past each other. This suggests a very simple definition for a strain rate tensor $\mathbf{L_i}$ characterizing intercalation based on the shape change and tissue strain rate tensors, defined as:

$$\mathbf{L_t} = \mathbf{L_c} + \mathbf{L_i} \tag{13.9}$$

The physical meaning of this tensor is explored further in the next section. We calculated that quantity for the examples introduced on Figures 13.2(b) and 13.2(c), as plotted on the graphs in Figures 13.6(c) and 13.6(f) showing the evolution of the accumulated intercalation strain for both examples. As expected, there is no intercalation for the first case. More interestingly, there is a monotonic and continuous increase in the total intercalation for the second example. We can identify a few important properties of the intercalation rate tensor $\mathbf{L_i}$. First, the trace of $\mathbf{L_i}$ is zero, as all volume variations in the tissue are accounted for by the shapes. Second, assuming that the shape rotation is identical to the tissue rotation, $\mathbf{L_i}$ is symmetrical. It therefore has a diagonal form with two eigenvalues of opposite sign.

13.3.3 Handling Cell Division

While we monitor the evolution of a neighborhood, cell division might happen and rules regarding the handling of such events should be defined in the light of their influence on tissue deformation. If one considers the situation of a local neighborhood in which a cell division event occurs during the time of observation δt, the mere fact that a new membrane now splits a large cell into two smaller units does not influence directly the surrounding tissue (see Figure 13.7).What matters in practice is the change in cell shape before and after the division. As a consequence, to measure kinematic quantities that are relevant for a neighborhood, cells that divide during the time δt of the measurement are maintained artificially linked: the location and shape of the composite entity are obtained by merging the two cells. In practice, δt is small enough that cells remain in contact during the strain measurement.

13.3.4 Applications

The breakdown of the tissue strain rate tensor into cell shape and cell intercalation strain rate tensors allows us to map these two cell behaviors in space and over time in real tissues, just as can be done for the tissue strain rate tensor. Measured intercalation strain rates emerged as predominantly pure shear

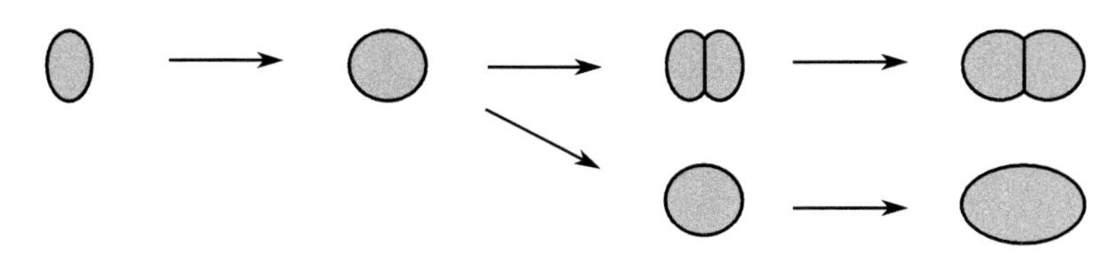

FIGURE 13.7 Illustration of the evolution of a cell elongating in a nonintercalating tissue. On the top path, one cell is followed through a sequence of change in shape, division event, and change in shape. On the bottom path, we represent a cell changing shape with a strain that is identical to the top situation with the daughter cells merged together. Both scenarios are expected to have the same effect on the surrounding tissue at a mesoscopic scale.

deformations, as predicted. In the fruit fly example [6] introduced in Section 13.2.3, the relative contributions of two cellular mechanisms were very revealing. Cell shape change and intercalation were equally strong during the early fast phase of germband extension, but cell shape change then reduced quite quickly, leaving only cell intercalation during the slow phase. Interestingly, in a mutant where the cell intercalation machinery was significantly compromised, the total tissue deformation was unaffected during the fast phase, but the relative contribution of the cellular mechanisms was different, with cell shape stretch significantly increased, compared to the wild-type flies, at the expense of cell intercalation. During the slow phase, cell shapes rounded up, suggesting that cell shape elongation was an elastic response to an external pull. Further insights have been gained into the formation of the zebrafish forebrain [8], convergence and extension in the zebrafish trunk [19], and the amnioserosa tissue of the fruit fly [11] using these methods.

13.4 Intercalation and Slippage

Intercalation is defined as the mismatch between the tissue strain rate and the cell shape strain rate. Its definition has been postulated from the fact that only cell intercalation can explain the difference between shape deformation and tissue strain. In this section we develop a microscopic interpretation of this tensor and show that it is intimately related to cell–cell slippage.

13.4.1 Cell–Cell Slippage

Our approach is based on the simultaneous quantification of movements within the cells (cell shape tensor) and at the cell cohort scale. The cell shape tensor is a very coarse description of material movement in the cell, which is known to be highly complex; it provides, however, the minimal description needed to compare cell and tissue strains. Figures 13.8(a) and 13.8(c) reuse the examples introduced on Figures 13.2(b) and 13.2(c) to illustrate the link between cell

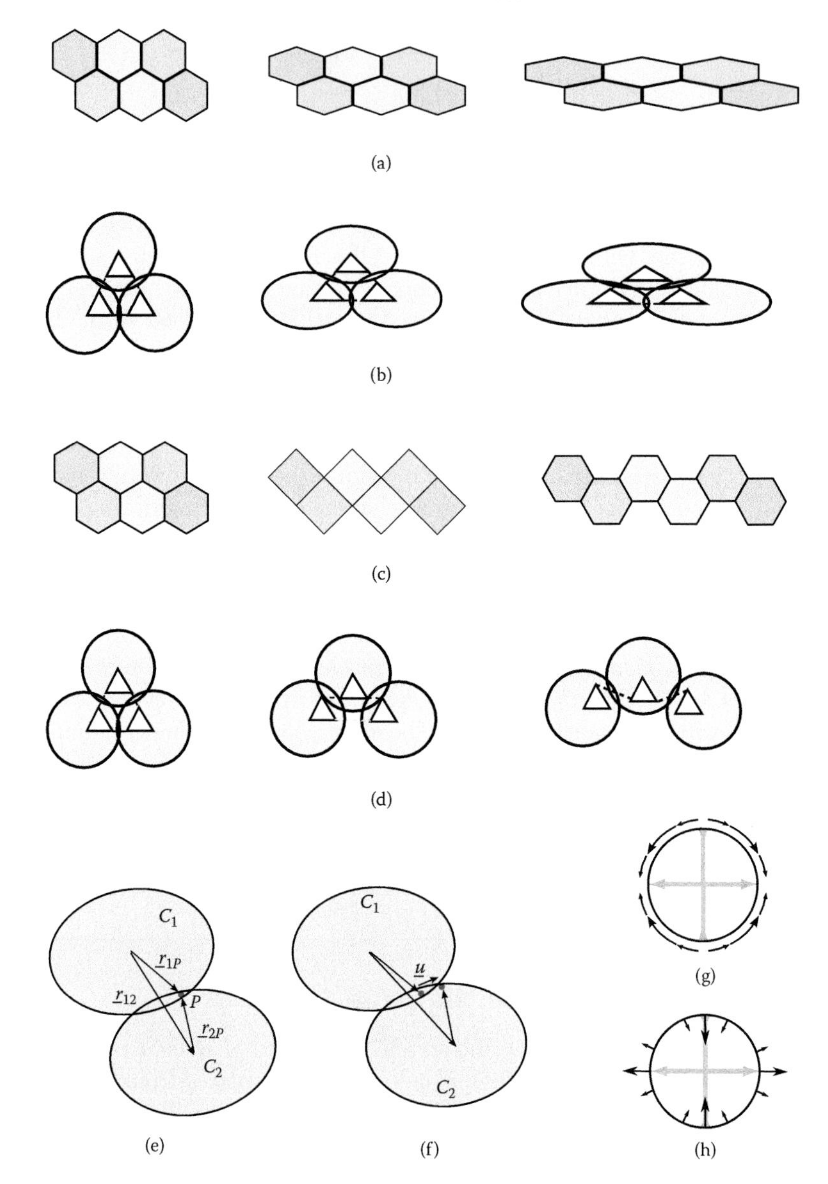

FIGURE 13.8 Illustration of the link between intercalation and slippage. (a) and (c) show a few cells extracted from Figures 13.2(b) (cell shape change) and 13.2(c) (intercalation), respectively. The ellipses in (b) and (d) represent the average shapes of cells in (a) and (c). White triangles within each ellipse are deformed accordingly to the cell shape strain tensor. (e) and (f) illustrate the construction of the slippage velocity (see text). (g) and (h) show the representation of the intercalation tensor in terms of slippage velocity. Vectors plotted along the circle correspond to the tangential (g) and normal (h) components of the slippage vector.

slippage and intercalation. The white triangles inside cells in Figures 13.8(b) and 13.8(d) evolve according to the cell shape strain rate $\mathbf{L_c}$ and allow us to visualize the strain within each cell. The shifts appearing among the three facing triangles during intercalation highlight, at the microscopic scale, the consequences of a mismatch of tissue and cell shape strain rates: in intercalating tissues, slippage occurs between neighboring cells, meaning that a domain about the cell interface necessarily experiences a local strain that does not correspond to the cell shape strain. Because we decided to ignore the complexity of the movements within cells, we represent here this complex flow near the cell membrane by a discontinuity at the cell interface, which can then be estimated geometrically.

We consider two neighboring cells C_1 and C_2. A point P is chosen along their interface. The velocity of P can be calculated in two different ways. Assuming it belongs to C_1, its velocity relative to the center of mass of C_1 is $\underline{u}_1 = \mathbf{L_c}\underline{r}_{1P}$. Assuming now it belongs to the cell C_2, its velocity would be $\underline{u}_2 = \mathbf{L_t}\underline{r}_{12} + \mathbf{L_c}\underline{r}_{2P}$. The discontinuity $\underline{u}$ at the interface can therefore be deduced:

$$\underline{u} = \underline{u_2} - \underline{u_1} = \mathbf{L_t}\underline{r}_{12} + \mathbf{L_c}\underline{r}_{2P} - \mathbf{L_c}\underline{r}_{1P} = (\mathbf{L_t} - \mathbf{L_c})\underline{r}_{12} = \mathbf{L_i}\underline{r}_{12} \tag{13.10}$$

This construction demonstrates that the intercalation rate tensor is indeed a natural physical quantity for characterizing the movement of cells past each other. The velocity $\underline{u}$ has *a priori* components along and perpendicular to the cell membrane. In the case of the intercalating hexagons, the corresponding intercalation tensor as well as the tangential and normal components of the velocity $\underline{u}$ are shown in Figures 13.8(g) and 13.8(h).

The component u_s of $\underline{u}$ along the cell membrane is a direct measurement of slippage velocity. It has a maximum at an angle of about 45° of the main axis of the intercalation tensor (see Figure 13.5(g)). However, along the eigendirections of the latter, the slippage velocity vanishes, and most of the movement occurs normal to the interface. This component indicates the evolution of the ellipses overlap, corresponding qualitatively to the surface area of the cell–cell interfaces. Where $\underline{u}$ points inward, the contact area between the cells increases; where $\underline{u}$ points outward, the contact area decreases. Ultimately, this means that new neighbors are, respectively, gained or lost along these directions.

The description above sets the microscopic picture underlying the definition of the intercalation tensor introduced in the previous section. The framework provides us with three different scales: (1) the cell cohort, which is a mesoscopic scale to describe local tissue strain; (2) the single cell, for which we measure an internal strain rate; and finally (3) the cell–cell interface, along which slippage can be estimated.

13.4.2 Total Slippage

We assumed in the previous section that the rotation rate of cell shapes was the same as the tissue rotation rate. One argument for such a choice is that

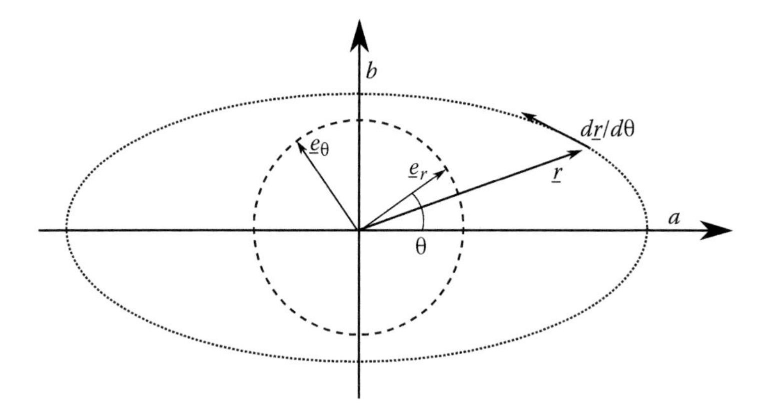

FIGURE 13.9 Geometrical symbols used in the text. The ellipse is parametrized by the angle θ: $\underline{r}(\theta) = (a\cos(\theta); b\sin(\theta))$.

it guaranties that cells of non-intercalating tissues do not rotate against each other. We need, however, to analyze further this choice as it is unclear how the same argument applies to fully or partially intercalating tissues.

We showed on Figure 13.5 that the choice of the shape rotation has a direct influence on the movements of points at the surface of an ellipse. Because these movements control, in part, cell slippage in the context of tissues, one can study how any particular assumption regarding cell rotation influences cell slippage at the microscale. A first step is to construct a measure of the total slippage required to transform the cell cohort and then explore ways to use such a physical quantity to build a self-consistent kinematic description.

The amount of cell–cell slippage S locally in the tissue can be quantified by integrating the squared slippage velocity along an elliptical contour C representing the average cell elongation in a neighborhood:

$$S = \int_C \left(\underline{u} \cdot \underline{e}_s\right)^2 ds \tag{13.11}$$

where s corresponds to the curvilinear abscissa along the ellipse. The ellipse contour C is parameterized by an angle θ so that $\underline{r}(\theta) = (a\cos(\theta); b\sin(\theta))$ (see Figure 13.9). Defining the matrix:

$$\mathbf{C} = \begin{bmatrix} a & 0 \\ 0 & b \end{bmatrix} \tag{13.12}$$

the position along the contour can be rewritten as $\underline{r}(\theta) = \mathbf{C}\underline{e}_r(\theta)$. The tangent to the ellipse is along $d\underline{r}/d\theta = \mathbf{C}\underline{e}_\theta$. The unit vector tangent to the ellipse is given by $\underline{e}_s = \mathbf{C}\underline{e}_\theta/(ds/d\theta)$. The total slippage can therefore be rewritten as:

$$S = \int_C \left(\mathbf{L_i}\mathbf{C}\underline{e}_r(\theta) \cdot \mathbf{C}\underline{e}_\theta\right)^2 \left(\frac{ds}{d\theta}\right)^{-1} d\theta \tag{13.13}$$

In Section 13.3.1, we characterized the family of deformations that transforms a particular ellipse into another. For each value of the angle α, we obtain a different shape strain tensor $\mathbf{A}(\alpha)$ that can be written for small strains as $\mathbf{A}(\alpha) = \mathbf{Id} + \mathbf{L_c}(\alpha)\delta t$. From this, the decomposition of the shape strain rate directly allows us to determine the tensors $\mathbf{D_c}(\alpha)$ and $\mathbf{W_c}(\alpha)$, the latter associated with the angular velocity $\omega_c(\alpha)$. For simplicity, we directly express the different tensors as a function of ω_c rather than α in the following.

For each value of ω_c, we therefore have a corresponding intercalation rate tensor $\mathbf{L_i}(\omega_c)$. Unless ω_c is identical to the tissue angular velocity, denoted by ω_t, $\mathbf{L_i}$ is not symmetrical, and has an antisymmetrical component corresponding to an angular velocity $\omega_i = \omega_t - \omega_c$. For a given cell elongation and a given tissue velocity gradient and shape evolution, we can now study how the total slippage S varies as a function of the shape angular velocity ω_c or equivalently ω_i. It remains true that Tr $\mathbf{L_i} = 0$ for all ω_i, although all elements of the tensor change with ω_i, taking the general form below:

$$\mathbf{L_i}(\omega_i) = \begin{bmatrix} \beta(\omega_i) & \gamma(\omega_i) - \omega_i \\ \gamma(\omega_i) + \omega_i & -\beta(\omega_i) \end{bmatrix} \tag{13.14}$$

For small anisotropy of the cell shapes, the total slippage becomes (see Appendix at end of chapter for details):

$$S(\omega_i) = 2\pi(ab)^{3/2}\left(\omega_i^2 - \frac{1}{2}\left(\frac{a}{b} - 1\right)\omega_i\gamma + \frac{\beta^2 + \gamma^2}{2}\right) \tag{13.15}$$

The expression above shows that the total slippage $S(\omega_i)$ depends on both the cell shape aspect ratio and orientation with respect to the deformation. This leads to the identification of several cases of interest, as discussed below.

13.4.3 Typical Situations

13.4.3.1 Tissue Strain without Intercalation

If the tissue deforms with the velocity gradient $\mathbf{L_t}$, and if the same tensor $\mathbf{L_t}$ also deforms the cell shapes, this implies that $\mathbf{L_i}(\omega_i = 0) = \mathbf{0}$. Therefore, slippage is minimal for $\omega_i = 0$, as $S(\omega_i = 0)$ is strictly equal to zero. In the simple case of nonintercalating tissues, the shape strain rate that provides minimal slippage gives a meaningful output. Figure 13.10a shows the evolution of $S(\omega_i)$ in the context of a simple shear of both the tissue and the shape for a cell elongating along the shear direction ($a/b = 1.21$).

13.4.3.2 Tissue Intercalation along the Orientation of Cell Elongation

In the case where the tissue intercalates along one of the main orientations of the cell shape, $\mathbf{L_i}(0)$ is diagonal in the basis of the ellipse axes. It results in

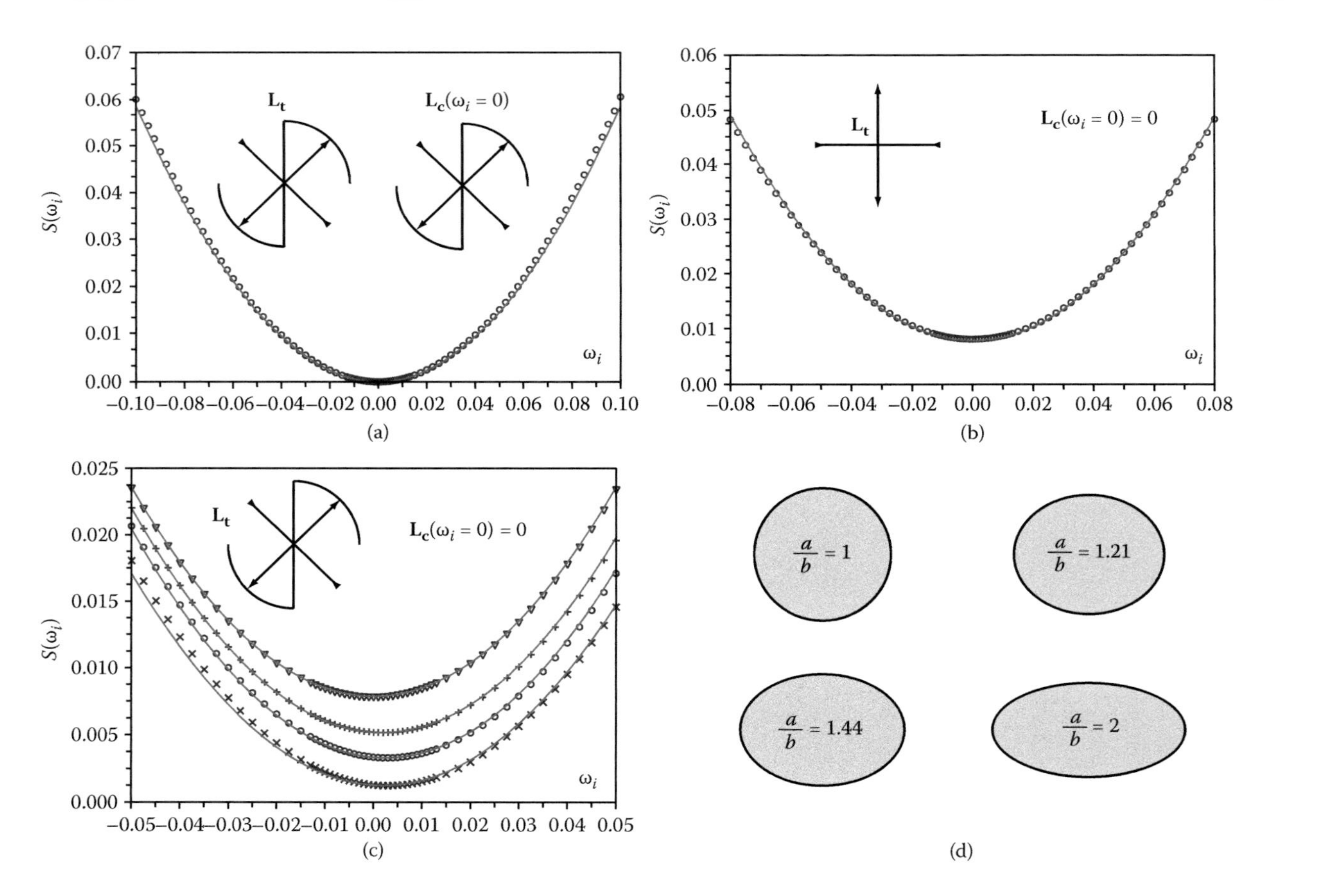

FIGURE 13.10 Total slippage calculated for a range of strain situations and cell shapes. Symbols correspond to the direct calculation of Equation (13.13). Solid lines show the small anisotropy approximation calculated from Equation (13.15). (a) Slippage for a simple shear deformation; $\mathbf{L_t} = \mathbf{L_c} = [0, 0.1; 0, 0]$, aspect ratio $a/b = 1.21$. (b) Pure intercalation along the cell principal axis of extension; $\mathbf{L_t} = [0.05, 0; 0, -0.05]$, $\mathbf{L_c} = \mathbf{0}$, $a/b = 2$. (c) Pure intercalation during a simple shear deformation, $\mathbf{L_t} = [0, 0.1; 0, 0]$, $\mathbf{L_c} = \mathbf{0}$, $a/b = 1, 1.21, 1.44$, and 2, respectively, from top to bottom. (d) Elliptical shapes for

$\gamma(\omega_i = 0) = 0$. Moreover, by symmetry, $\gamma(\omega_i)$ is an odd function, implying that $\gamma(\omega_i) \sim \omega_i$. This means, in particular, that $S(\omega_i) = S(-\omega_i)$. Figure 13.10b displays the total slippage as a function of ω_i for such a situation, with highly elongated cells ($a/b = 2$). We observe indeed that the slippage is minimal when the shape rotates exactly with the tissue. This situation is quite frequent in reshaping tissues as the cell elongation is usually caused by the strain history itself, either as a passive adaptation or as an active contribution. In that situation, a symmetry argument is, in practice, enough to rule out any cell shape rotation other than the tissue rotation itself. It is also consistent with the minimal slippage idea.

13.4.3.3 Tissue Intercalation not along the Orientation of Cell Elongation

This corresponds to the most generic type of deformation. A common example is a simple shear deformation of a tissue with cells elongated along or normal to the shear direction. Such a case is addressed on Figure 13.10(c) for four values of the cell anisotropy. We observe that the shape spin that provides the smallest slippage is therefore different from the tissue spin. Consistent with Equation (13.15), this shift increases with the cell anisotropy. However, even for highly deformed cells ($a/b \geq 2$), the difference between tissue and shape rotation would be less than 10%.

In the three situations described above, our initial choice for the cell shape rotation rate seems justified *a posteriori* based on a minimal slippage argument. Assuming that the cells do not rotate in a reference frame attached to the tissue (i.e., $\omega_t = \omega_c$, $\omega_i = 0$) provides an excellent approximation, if not an exact answer. In many situations the result could come from symmetry arguments. However, in cases where the cell shape is not aligned with the eigenvectors of the intercalation strain rate tensor, there is no trivial answer and the robustness of the kinematic approach had to be verified. Although we provide here a workable and generic approach to quantify tissue morphogenesis, the kinematic description of cellular movements remains empirically linked with a microscopic assumption that sets the cell rotation and the amount of cell–cell slippage.

The main advantage of developing an approach based on slippage is to clearly highlight the microscopic origin and consequences of choices made at the tensorial level. Although it seems natural to penalize choices of rotation that induce unnecessary cell slippage, one should question how this penalty is practically determined, and in particular if slippage should be penalized the same way for all directions as we did above. The case of steady simple shear deformation is, for instance, a situation that remains to be explored. If cells move along layers, one can reasonably imagine that the slippage is mostly localized between layers, and not so much between cells of the same layer, as depicted on Figure 13.11. In such a situation, the kinematic description should

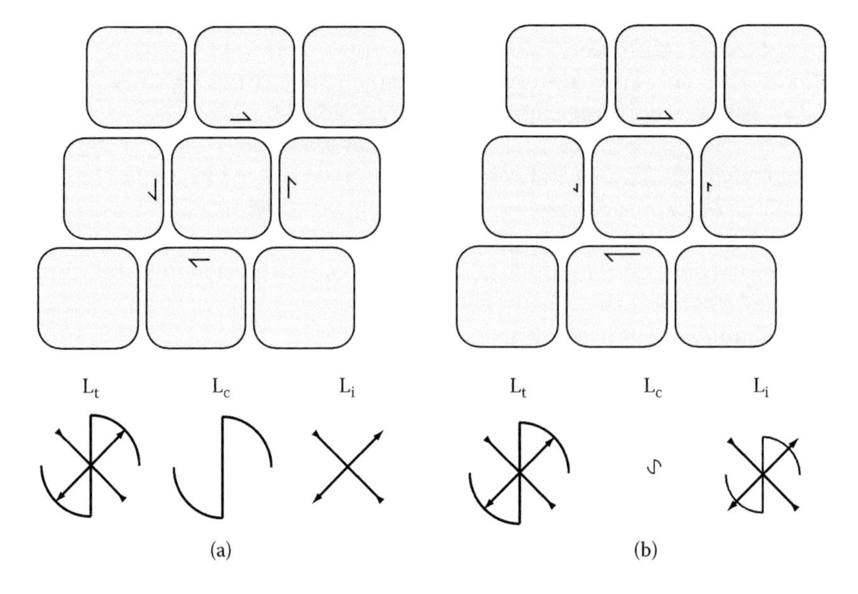

FIGURE 13.11 Two different shear situations: (a) Typical approach developed here, where cells rotate according to the tissue spin and slippage is shared among all neighbors; (b) Alternative situation in which slippage is more localized. To account for such a case, the minimal slippage argument must be adapted.

therefore account for the peculiarity of cell–cell interactions and be aware of the existence of planes of sliding arising either from anisotropic adhesion or active cell crawling. Allowing the intercalation tensor to have, in general, a rotational component then provides a natural way to express, within the same kinematic framework, that slippage inside cell layers is low compared to its value between layers.

13.4.4 Intercalation and Tissue Microstructure

Figure 13.8 provides a microscopic interpretation of the intercalation strain rate tensor. We first characterized in detail the slippage behavior of cells past each other, which mostly concerns cells contacting at about 45° to the eigenvectors of the intercalation tensors, where relative cell slippage is non-negligible (see Figure 13.8(g)). Cell movements along the direction of extension or convergence of the tissue are more subtle. Along these directions, slippage is low on average and the orientation of the relative velocity $\underline{u}$ defined in Equation (13.10) is mostly normal to the cell membrane, indicating that neighboring cells tend in practice to decrease or increase their contact area. However, these simple dynamics can only exist for a finite amount of time. The example of intercalating hexagons provides a good illustration of

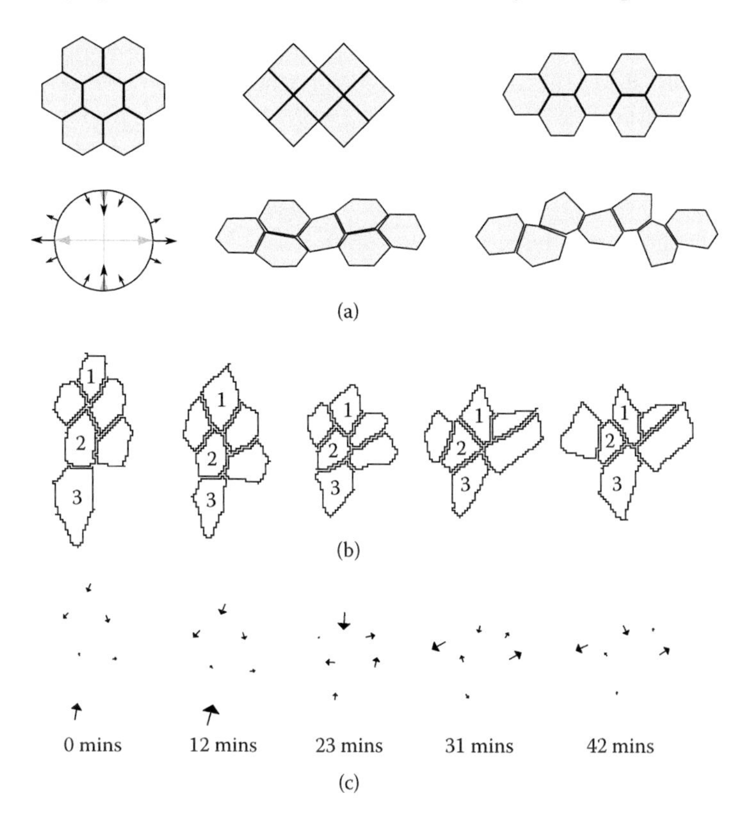

FIGURE 13.12 (a) Evolution of a purely intercalating tissue. During the first three steps, the hexagon network can deform in an affine way. Beyond that, non-affine displacement is required (images are only qualitative illustrations). (b) and (c) Details of the evolution of a few neighboring cells extracted from experimental data (zebrafish, [4]). As cells 1 and 2 come closer, the field is nearly affine. Once cells 1, 2, and 3 form a chain along the converging axis, highly non-affine displacements are required to satisfy the mean tissue and shape strains.

the process (Figure 13.12(a)), showing that even if it is well ordered, a purely intercalating tissue can reorganize in a homogeneous or affine manner only up to deformations on the order of 60%.

As cells move past each other, the detail of the distribution of membrane orientation evolves, creating, in particular, new contacts along the direction of convergence, and decreasing their number in the extension direction. Once cells come into contact along the direction of convergence, the affine displacement of cells becomes impossible and cells have to move in a non-affine way to accommodate the strain by intercalation. The last two steps of Figure 13.12(a) show a schematic of that process. Evidence of such behavior can be readily found in experimental records. Figures 13.12(b) and 13.12(c) show the evolution over time of an intercalating cluster of cells. During the first 15 minutes, the displacement field is globally affine, and a pair of cells can be seen converging and forming a new interface. However, after 20 minutes, the cells 1, 2, and 3 form a chain that cannot intercalate further in an affine way without overlapping. From the resulting displacement fields, the non-affinity can be quantified by the value of $< \delta v_i^2 >$ at each time, $\underline{\delta v}_i$ being the residual velocity in Equation (13.3). For the five time steps shown on Figure 13.12(c), we measure $< \delta v_i^2 >= 0.10, 0.19, 0.87, 0.68, 0.52$, respectively, showing as expected a dramatic rise between 12 and 23 minutes.

Non-affine cell displacements cause significant heterogeneities in the velocity field (quantified above from the residuals). This also leads to variations in the cell shape strain rates within a neighborhood. The precise amount of slippage in a tissue has been therefore underestimated in our previous first-order approach, which neglected heterogeneities in the relative displacement field, in slippage, and in cell shapes. Non-affinity is probably not relevant at the tissue scale but, because it reflects the detailed structure and dynamics of the tissue and membrane orientation, it may highlight intrinsic mechanisms used by cells to produce or respond to macroscopic strains. In the case of living tissues, experimental studies must be carried out to quantify non-affinity and relate it to intrinsic material properties and in particular to its geometrical organization at the cell scale.

We developed a formalism here to quantify strains in reshaping tissues. In addition to strain rate measurements based on individual cell trajectories in the tissue, we introduced a quantification of cell shape changes and intercalation. The kinematic approach leads to a set of tensorial quantities defined for a local group of cells, approximating the local complexity by an affine description of the tissue deformation, and an elliptical representation of the cell shapes. Such a level of description represents a strong advantage to account for the global tissue morphogenesis; it allows us to define quantities that are continuous and easy to map with high resolution in space and time. We showed, in particular, that these quantities also reflect intrinsic cell behavior, such as shape change or slippage.

This framework introduces an explicit dependence on a microscopic assumption for cell–cell movements that sets the amount of rotation of the cell shapes. We propose to choose the rotation component so that cell–cell slippage is minimal; this corresponds in most practical cases to assuming that cells do not rotate in the tissue frame. However, this leads to nontrivial situations when, for instance, slippage occurs during a simple shear flow. Further development in quantifying morphogenetic movements will therefore be concerned with higher-order descriptions, accounting for the common and as yet unresolved situation of sheared layers and non-affine displacement fields.

13.5 Appendix: Total Slippage for Nearly Isotropic Cell Shapes

The measure S of the total slippage along the membrane of an ellipse can be calculated analytically if the intercalation rate tensor is known.

The geometrical notations used below are defined on Figure 13.9. The quantity S is calculated from Equations (13.13) and (13.14). We define the mean ellipse radius R and anisotropy λ by $R^2 = ab$ and $\lambda = a/R = R/b$. Equation (13.13) writes:

$$\tilde{S} = \frac{S}{R^3} = \oint_C \left(\mathbf{L_i}\tilde{\mathbf{C}}\underline{e}_r(\theta) \cdot \tilde{\mathbf{C}}\underline{e}_\theta\right)^2 \left(\frac{d\tilde{s}}{d\theta}\right)^{-1} d\theta \quad \text{with} \quad \tilde{\mathbf{C}} = \mathbf{C}/R = \begin{bmatrix} \lambda & 0 \\ 0 & 1/\lambda \end{bmatrix} \tag{13.16}$$

and $\tilde{s} = s/R$. This leads to:

$$\frac{d\tilde{s}}{d\theta} = \sqrt{\frac{1}{2}\left(\lambda^2 + \frac{1}{\lambda^2} - \left(\lambda^2 - \frac{1}{\lambda^2}\right)\cos(2\theta)\right)} \tag{13.17}$$

and

$$\mathbf{L_i}\tilde{\mathbf{C}}\underline{e}_r(\theta) \cdot \tilde{\mathbf{C}}\underline{e}_\theta = \frac{d\underline{\tilde{r}}}{d\theta} \cdot \underline{\tilde{u}} = \gamma\cos(2\theta) - \frac{\beta}{2}\left(\lambda^2 + \frac{1}{\lambda^2}\right)\sin(2\theta) + \omega_i \tag{13.18}$$

Assuming small anisotropy for the cell shape, we define $\epsilon = \frac{1}{2}(\frac{a}{b} - 1)$ such as $\lambda \approx 1 + \epsilon$ and $\lambda^{-1} \approx 1 - \epsilon$. The two expressions above can be simplified into:

$$\frac{d\tilde{s}}{d\theta} = 1 - \epsilon\cos(2\theta) \tag{13.19}$$

and

$$\mathbf{L_i}\tilde{\mathbf{C}}\underline{e}_r(\theta) \cdot \tilde{\mathbf{C}}\underline{e}_\theta = \gamma\cos(2\theta) - \beta\sin(2\theta) + \omega_i \tag{13.20}$$

By substituting the slippage integral, we get:

$$\tilde{S} = \frac{S}{R^3} = \oint_C \left(\omega_i^2 - 2\epsilon\omega_i\gamma\cos^2(2\theta) + (\gamma\cos(2\theta) - \beta\sin(2\theta))^2\right) d\theta \tag{13.21}$$

providing after integration a relationship corresponding to Equation (13.15):

$$\tilde{S} = \frac{S}{R^3} = 2\pi\left(\omega_i^2 - \epsilon\omega_i\gamma + \frac{\beta^2 + \gamma^2}{2}\right) \tag{13.22}$$

References

[1] R.W. Allmendinger, R. Relinger, and J. Loveless (2007). Strain and rotation rate from GPS in Tibet, Anatolia, and the Altiplano. *Tectonics* 26:TC3013.

[2] M. Aubouy, Y. Jiang, J.A. Glazier, and F. Graner (2003). A texture tensor to quantify deformations. *Granular Matter* 5:67–70.

[3] C. Bertet, L. Sulak, and T. Lecuit (2004). Myosin-dependent junction remodelling controls planar cell intercalation and axis elongation. *Nature* 429:667–671.

[4] G.B. Blanchard, A.J. Kabla, N.L. Schultz, L.C. Butler, B. Sanson, N. Gorfinkiel, L. Mahadevan, and R.J. Adams (2009). Tissue tectonics: morphogenetic strain rates, cell shape change and intercalation. *Nat. Meth.*, 6(6):458–464.

[5] M.B. Burnside and A.G. Jacobson (1968). Analysis of morphogenetic movements in the neural plate of the newt *Taricha torosa. Dev. Biol.* 18:537–552.

[6] L.C. Butler, G.B. Blanchard, A.J. Kabla, N. J. Lawrence, D.P. Welchman, L. Mahadevan, R.J. Adams, and B. Sanson (2009). Cell shape changes indicate a role for extrinsic tensile forces in *Drosophila* germband extension. *Nature Cell Biology*, 11(7):859–865.

[7] M.L. Concha and R.J. Adams (1998). Oriented cell divisions and cellular morphogenesis in the zebrafish gastrula and neurula: a time-lapse analysis. *Development* 125:963–994.

[8] S.J. England, G.B. Blanchard, L. Mahadevan, and R.J. Adams (2006). A dynamic fate map of the forebrain shows how vertebrate eyes form and explains two causes of cyclopia. *Development* 133:4613–4617.

[9] Y.C. Fung and P. Tong (2001). *Classical and Computational Solid Mechanics*, Singapore, World Scientific, Advanced Series in Engineering Science, Vol. 1.

[10] N.S. Glickman, C.B. Kimmel, M.A. Jones, and R.J. Adams (2003). Shaping the zebrafish notochord. *Development* 130:873–887.

[11] N. Gorfinkiel, G.B. Blanchard, R.J. Adams, and A. Martinez Arias (2009). Mechanical control of global cell behaviour during dorsal closure in *Drosophila. Development* 136:1889–1898.

[12] F. Graner, B. Dollet, C. Raufaste, and P. Marmottant (2008). Discrete rearranging disordered patterns. I. Robust statistical tools in two or three dimensions. *Eur. Phys. J. E.* 25:349–369.

[13] R. Keller and J.P. Trinkaus (1987). Rearrangement of enveloping layer cells without disruption of the epithelial permeability barrier as a factor in *Fundulus epiboly. Dev. Biol.* 120:12–24.

[14] R. Keller (2006). Mechanisms of elongation in embryogenesis. *Development* 133:2291–2302.

[15] T. Kominami and H. Takata (2004). Gastrulation in the sea urchin embryo: a model system for analyzing the morphogenesis of a monolayered epithelium. *Dev. Growth. Differ.* 46:309–326.

[16] T. Langenberg, T. Dracz, A.C. Oates, C. Heisenberg, and M. Brand (2006). Analysis and visualization of cell movement in the developing zebrafish brain. *Dev. Dynamics* 235:928–933.

[17] S.G. Megason and S.E. (2007). Fraser imaging in systems biology. *Cell* 130:784–795.

[18] L.M. Milne-Thomson (1996). *Theoretical Hydrodynamics*, 5th edition. New York, Dover Publications.

[19] N.L. Schultz, G.B. Blanchard, A.J. Kabla, L. Mahadevan, and R.J. Adams, Morphogenesis of the zebrafish neural plate, in preparation.

[20] J. Shih and R. Keller (1992). Cell motility driving mediolateral intercalation in explants of *Xenopus laevis. Development* 116:901–914.

[21] J.B. Wallingford, B.A. Rowning, K.M. Vogeli, U. Rothbcher, S.E. Fraser, and R.M. Harland (2000). Dishevelled controls cell polarity during *Xenopus* gastrulation. *Nature* 405:81–85.

[22] J. Zallen and R. Zallen (2004). Cell-pattern disordering during convergent extension in *Drosophila. J. Phys. Cond. Mat.* 16:S5073–S5080.

[23] J. Zallen (2007). Planar polarity and tissue morphogenesis. *Cell* 129:1051–1063.

Chapter 14

Modeling Steps from Benign Tumor to Invasive Cancer: Examples of Intrinsically Multiscale Problems

Dirk Drasdo, Nick Jagiella, Ignacio Ramis-Conde, Irene E. Vignon-Clementel, and William Weens

Contents

14.1 Introduction

Cancer arises from a series of mutations manifested in phenotypic changes of both the cells and the local tissue structure. At the stage of an *in situ* tumor or neoplasia, a pre-state on the path to invasive cancer, cells divide out of control but still form a compact colony well separated from its environment. The transition from an *in situ* tumor to invasive cancer is marked by a number of steps. This includes angiogenesis, the formation of new blood vessels to supply the growing tumor with oxygen and nutrients, and detachment of cells from the tumor that subsequently invade the tissue and the blood vessels to be transported into distant organs where they can lead to the formation of secondary tumors called metastases. Angiogenesis is the process during which endothelial cells divide and generate new vessels sprouting toward the tumor as a response to angiogenesis factors secreted directly and indirectly by the tumor cells. Many angiogenesis factors have been identified [77]. The most prominent one is probably VEGF, which is related to platelet-derived growth factors (PDGF). A shortage of oxygen triggers an increase in the intracellular concentration of an active form of the protein hypoxia-inducible factor 1 (HIF-1), which then stimulates transcription of the VEGF gene. The protein VEGF is secreted into the extracellular space acting on the nearby endothelial cells as described above. A lack of oxygen favors cells that can survive at a lower oxygen concentration, which explains why in hypoxic regions cells with a small oxygen demand can be found.

Significant lack of oxygen and nutrients such as glucose triggers the death of cells by a process called necrosis. Multicellular spheroids grown in homogeneous isotropic liquid suspensions, often used as an *in vitro* model to study tumors in the avascular phase, show the formation of a central spheroidal necrotic core above a certain size at which nutrients become limiting. In multicellular spheroids, the oxygen and glucose enter the tumor border equally from all sides. Comparisons of experiments with models suggest that the concentration of oxygen and glucose controls the size of the necrotic core while the speed at which the tumor diameter grows seems to be constant (i.e., the tumor diameter grows linearly) and controlled by a biomechanical form of contact inhibition [27]. The same linear growth of the diameter can be observed in monolayers [14], where cell colonies grow as an approximately one-cell-thick layer. It can also be found *in vivo*, for example in xenografts of human NIH3T3 cells in the mouse model [66], indicating a generic character of this growth law. However, different from monolayers and multicellular spheroids, which grow only up to about a millimeter in diameter, the NIH3T3 tumors can grow up to several centimeters. As the multicellular spheroids, the xenografts have a largely spherical shape but in contrast to multicellular spheroids they are well vascularized and show only decent necrotic and apoptotic figures. Hence the induction of new vessels permits growth of the tumor cell population up to about three to four magnitudes more than multicellular spheroids and, as in the case of NIH3T3 cells, is capable of avoiding the formation of a central necrotic core.

Metastasis emerges from cancer cells invading distant organs and creating new tumors. At this stage, the prognosis of the individual becomes poorer and surgical methods are insufficient to eradicate cancer. To achieve metastasis, a malignant cell needs first to reach the vascular or lymph system, penetrate into a vessel, and extravasate in a new distant position where it can generate a new colony. The processes of intravasation and extravasation are similar: the cancer cell opens a gap in the wall of the *tunica intima* after disrupting the cell–cell bonds between endothelial cells and deforms its cytoplasm to cross into the stream. Angiogenesis, as a process that stimulates the growth of vessels into the tumor, facilitates metastasis.

We present in this chapter mathematical models of angiogenesis, cell detachment, and intravasation (Figure 14.1). Many aspects of tumor growth have been studied using mathematical models. In most cases deterministic models of the reaction–diffusion type or continuum mechanical models have been used (for comprehensive reviews, see [1,10,50,53,70]). Continuum models addressing this issue assume that growth is mechanically regulated [7,16] or nutrient limited (e.g., [19,51,76]). These are well suited to the description of large-scale phenomena where the cell and tissue properties vary smoothly over a length scale of several cell diameters.

To study small-scale phenomena or situations in which the properties of the cells vary over distances comparable to the size of a cell, single-cell-based models permit a higher degree of spatial resolution than models in which subcellular properties are replaced by locally averaged quantities. Most individual-based models can be characterized as either lattice-based or lattice-free models (for reviews, see [4,9,22,25,56,58]). In some lattice models, each lattice site can be occupied by at most one cell (e.g., [3,8,13,24,26,48]) while in others one cell may span many lattice sites (e.g., [39,42,74]). Within the

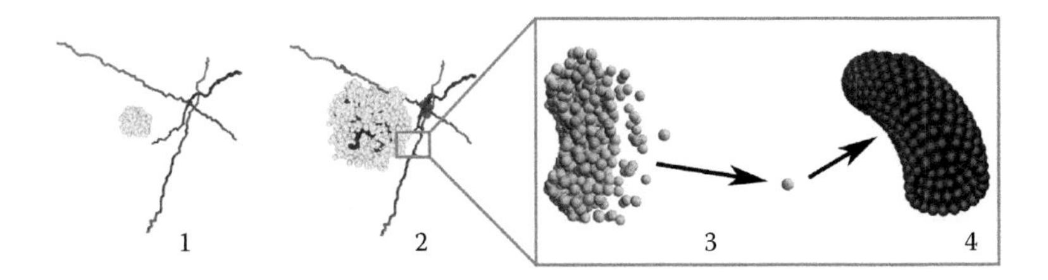

FIGURE 14.1 Stages of the transition from an *in situ* tumor to invasive cancer. (1) Loss of contact inhibition leads to uncontrolled growth; (2) necrotic core of tumor induces angiogenesis; (3) invasive tumor cells gain ability to migrate into surrounding tissue and (4) blood vessels that results in metastasis. (The pictures represent results of our 3-D model simulations explained in this chapter.)

class of lattice-free models, cells have been parameterized by measurable biophysical and biological parameters and approximated by deformable spheres or ellipses (e.g., [23,27–30,36,37,62,68]) and, more recently, in a multi-center approach by aggregates of spheres [60].

In our models the tumor cells will be represented by individual agents. In the section on angiogenesis we compare the growth kinetics of cell populations in 3-D from a single precursor cell up to a population of several thousands of cells in cases where nutrients or oxygen are not limiting, a tumor grows in a static vascular environment, and where tumors can induce the formation of new vessels. The models consider the competition between contact inhibition- and nutrient/oxygen-limited growth. For these questions it turns out to be sufficient to model individual cells within a cellular automaton model where the dynamics are rule based. The vessels are modeled explicitly as discrete objects with a simplified lumped model relating flow and pressure inside them while the diffusion of oxygen, nutrients, and growth factors is represented by continuum equations. In the subsequent section we study cell detachment and intravasation in a multiscale model where the cells are modeled as homogeneous, isotropic elastic sticky objects, and where the cell–cell adhesion forces are controlled by the intra-regulatory processes, represented by coupled systems of ordinary differential equations [67,68].

14.2 Tumor Growth and Angiogenesis

At small population sizes where oxygen, nutrients, and contact inhibition are not limiting, a tumor grows exponentially fast. In liquid suspension, EMT6/Ro mouse mammary carcinoma cells form multicellular spheroids; they show a transition from exponential to linear growth at a population size of about 1000 cells at nutrient conditions similar to those *in vivo.* Experiments in agarosis gel mimicking a tissue-like biomechanical environment show that the growth kinetics of multicellular spheroids is almost unaffected at moderate agarosis gel concentrations [41] so the transition between the exponential and linear growth phase would be expected to occur at about the same size. The transition size also varies with the cell type. For example, for those fibroblasts for which cells migrate almost freely at small population sizes, a later transition is expected but the growth of the population is still expected to become eventually linear [13,26]. This is observed for NIH3T3 cells, which show a significant migratory activity until several days of growth in monolayer culture. However, NIH3T3 cells also eventually form linearly growing cell assemblies in monolayer culture (Hoehme, Drasdo, Hengstler, unpublished). *In vitro* growing multicellular spheroids form an approximately central spheroidal necrosis at about 400 μm [59]. However, the linear expansion is almost unaffected by the formation of the necrotic core except in very unfavorable nutrient and oxygen conditions [34]. This may be explained by a biomechanical form of contact inhibition known as the key growth-limiting factor [27–29,71]. Freyer and Sutherland observed a significant slow-down in tumor expansion only if both oxygen and glucose concentrations were low [33,34].

The *in vivo* situation is extensively described in a recent book [54] on tumor angiogenesis. A tumor grows up to a size of 1 to 2 mm in diameter, nourished by the diffusion of oxygen and nutrients of the existing vasculature. This phase is often called avascular growth; many aspects of this growth phase are studied *in vitro* by growing multicellular spheroids that are described above. A tumor can remain stable at that size. However, if as a consequence of their proliferation, tumor cells lack oxygen and nutrients, they can trigger several mechanisms to generate a higher density of blood vessels and thus increase their needed supply [12,69]. The main mechanism is the formation of new capillaries (angiogenesis), sprouting from existing vessels, as a consequence of two phenomena. Hypoxic and underfed tumor cells and other cells from the micro-environment secrete soluble factors that target the nearby blood vessels, make their endothelial cells proliferate, migrate, sprout, or form a tube. These endothelial cells also produce soluble factors such as VEGF [32] that amplify the preceding effect. In parallel with these angiogenic factors, anti-angiogenic factors are also produced. The angiogenic switch occurs if growth factor promoters overbalance anti-angiogenic factors [40]. New blood vessels are created toward the hypoxic regions of the tumor [18]. Cancer tissues have indeed extensive regions of hypoxia compared to normal tissue [75], usually associated with necrotic regions. This is due to the rapid growth of tumor mass that increases the distance between some cells and blood vessels, and to the nonfunctioning of some of the generated blood vessels (structural malformation, fluctuation in blood flow). A constant vessel network remodeling thus occurs, as the blood vessels inside the tumor may not be very functional, may collapse due to the high pressure generated by the surrounding proliferating tumor cells that cannot be counterbalanced by a too unstable flow, or may die due to hypoxia or anti-angiogenic factors (see in addition the references to biological articles in [11] and [53]).

The dynamics and the heterogeneous 3-D spatial structure of a tumor are thus governed by a complex interplay of different components and scales (from gene expression changes to population competition at the tissue level). In this section we model the key players of the first stages of tumor growth: tumor cells, factors influencing them or secreted by them, and the influence of and on the vascularization.

There are mainly three types of approaches for which representative recent references are given: (1) continuum models (PDEs in time and space) that describe how the solid tumor front grows (see, e.g., [15]); (2) more precise continuum models where the density of the different components evolves in time and space (see, e.g., [63,79]); and (3) agent-based models that describe each entity individually (for example, at the cellular level and individually how tumor and endothelial cells grow, divide, move, and die). The different approaches have also been combined in hybrid or multiscale models, mostly in two dimensions of space. Examples include agent-based models for cells and continuum models [2,45,47,71,78] or simplified assumed profiles [11,48] for diffusion of oxygen, nutrients, and/or growth factors or inhibitors. Note that in [47] the different tumor zones were represented with either agent-based models or continuum models. The effect of the vasculature is explored with a given network [2] or includes angiogenesis and remodeling [11,48,51,61]; angiogenesis alone is an active subject of modeling (see, e.g., [17,53,55]).

In this section we model the cell population by a cellular automaton model. The advantage of this model type is that it permits efficient simulations at moderate computation time while the dynamics have been shown to be as for more detailed biophysically related individual-based models [13,26]. The subcellular scale of oxygen, nutrients, and growth factors induces the choice of a continuum description of their respective conservations of mass. The diameter of a capillary is of the same order of magnitude as the size of tumor cells. Therefore blood vessels are represented individually. Constitutive laws and threshold-based rules express the interplay of the different components of the system. This chapter aims to describe, through modeling, the key mechanisms of tumor growth rather than targeting a specific system for which more precise information would be needed to go beyond qualitative results, although parameters were chosen to be as realistic as possible to determine relevant macroscopic behaviors (see Table 14.1).

14.2.1 The Model

14.2.1.1 Cellular Automaton Model

The cellular automaton model is an extension of a similar introduced in previous work [13,26] and is defined by the following set of rules:

R1: *Cell:* One cell can either occupy one or two sites on a Voronoi diagram. (In the following we refer to a Voronoi cell as a Voronoi lattice site to distinguish it clearly from a biological cell.) At the beginning of a cycle, a cell always occupies only one Voronoi lattice site.

R2: *Cell cycle/replication:* The cell cycle is subdivided into m intervals that reflect how far a cell is advanced in the cycle. At the beginning of the cycle the internal counter M is set to $M = 0$. If the oxygen and glucose conditions are permissive for proliferation, then the cell enters the cell cycle. This is mimicked by the transition $M = 0 \rightarrow M = 1$ that takes place only if the product of glucose and oxygen concentrations satisfies $[G] \cdot [O_2] \geq P^{crit}$ [71]. A proliferating cell successively increases its internal counter starting from $M = 0$ with a rate $m\mu$ by one until $M = m$. Here $\mu = \tau^{-1}$, where τ is the expectation value of the cycle time.

TABLE 14.1 Parameters for Tumor Growth and Its Environment

Parameter	Value	Unit	Ref.
V_{vc}	1200	μm^3	[45]
$\mu_{\max}$	0.032	cell·h^{-1}	Fitted from [34]
k	5	—	Fitted from [34]
m	10	—	Fitted from [34]
m_g	2	—	Fitted from [34]
m_d	8	—	Fitted from [34]
μ_n	0.0072	cell·h^{-1}	[71]
p^{crit}	0.025	mM^2	[71]
μ_l	0.01	cell·h^{-1}	this work
K_G^G	0.07	mM	Fitted from [33]
K_O^G	0.05	mM	Fitted from [33]
$q_{G,\max}$	$52.8 \cdot 10^{-17}$	mol·cell^{-1}·s^{-1}	Fitted from [33]
$q_{G,\min}$	$11.2 \cdot 10^{-17}$	mol·cell^{-1}·s^{-1}	Fitted from [33]
K_O^O	0.005	mM	Fitted from [33]
K_G^O	0.39	mM	Fitted from [33]
$q_{O,\max}$	$16.6 \cdot 10^{-17}$	mol·cell^{-1}·s^{-1}	Fitted from [33]
$q_{O,\min}$	$8 \cdot 10^{-17}$	mol·cell^{-1}·s^{-1}	Fitted from [33]
D_O	6300000	$\mu m^2 \cdot h^{-1}$	[71]
D_G	378000	$\mu m^2 \cdot h^{-1}$	[71]
D_{GF}	100	$\mu m^2 \cdot h^{-1}$	[45]
$[O_2]^{bv}$	0.07	mM	[33,34,45], order of magnitude of [11]
$[G]^{bv}$	5.5	mM	[33,45]
$[GF]^{bv}$	1	mM	[48]
d^{bv}	150	μm	Order of magnitude of [48]
$r_{ij}{}^0$	10	μm	[11]
μ^{bv}	0.1	Pa·s	Order of magnitude from [35]
P_{in}	100	Pa	Defined up to a multiplicative constant
P_{out}	100	Pa	Defined up to a multiplicative constant
$l_{\max}$	100	μm	[48]
θ^{GF}	0.01	mM	[11]
θ^{O_2}	0.01	mM	[11]
t_{EC}	40	h	[11]
$d_{\max}$	35	μm	[11]
θ^{ss}	0.5	—	[48]
p_c^{TC}	80	%	[48]
t_r	50	h	[48]
$\theta_{bv}^{O_2}$	0.01	mM	[48]

R2CG: *Cell growth:* At $M = m_g \leq m$, a cell tries to grow (i.e., to increase its volume). It can grow only if there is an empty Voronoi lattice site within a distance $\Delta L = kl (k \geq 1, l$: cell diameter) and if the oxygen and glucose conditions are still permissive. As a consequence of cell growth, this Voronoi lattice site is filled. Hence, for $m_g < m$, the doubling of the mass takes place in the cell cycle and not in the moment of cell division as in the case of the classical Eden model [31] or more recent work [26].

R2CD: *Cell division:* After a cell doubles its volume, it increases its internal counter M until $M = m$ by increments of one, again with rate $m\mu$. If $M = m$ the cell splits into two daughter cells where each daughter occupies only one Voronoi lattice site.

R3: *Cell necrosis:* Glucose and oxygen concentrations control necrosis. If the product of glucose and oxygen concentrations $[G] \cdot [O_2] < P^{crit}$, then cells become necrotic with the rate μ_n [71].

R4: *Cell lysis:* Necrotic cells decompose with the rate μ_l and consequently free the Voronoi sites they were occupying.

We consider early stages of vascularization where detachment and invasion of cells into their environment do not occur yet. For this reason we here neglect cell migration as it could be shown to not affect the growth kinetics as long as migration occurs only along the tumor surface [13]. The time evolution of the system is computed using the Gillespie algorithm [38] assuming that the underlying system dynamics of the multi-cellular system can be modeled by a master equation for the multivariate probability distribution to find the configuration $\underline{X}$ at time t. Here, $\underline{X} = (x_1, x_2, x_3, ...)$ where x_i denotes the number of cells in compartment i. In other words, the space is subdivided into compartments that here are the Voronoi lattice sites. In our model each Voronoi lattice site can only be occupied at most by one cell, $x_i \in [0, 1]$ (0: unoccupied, 1: occupied). The master equation applies to the conditional probability distribution so it needs the specification of an initial condition. In our simulations we start with a single cell $N(t = 0) = 1$ centered in the middle of our 3-D Voronoi lattice. The possible transitions from this into another configuration are denoted by rates for each process (cell growth, division, and death). This implies that each is assumed to be a Poisson process. However, for $M > 1$ the duration τ of the cell cycle emerges from the sum of M Poisson processes and thus is Erlang distributed. Within the Gillespie algorithm, the time until the next event is calculated by $\Delta t = -\log(\xi)/\sum_{\sigma'} W(\sigma \rightarrow \sigma')$ where $\sum_{\sigma'} W(\sigma \rightarrow \sigma')$ is the total rate by which the current state σ can be left. $W(\sigma \rightarrow \sigma')$ denotes the individual rates that lead from the current state σ to the accessible state σ' characterized by any of the processes explained above. $\xi \in (0, 1]$ is a uniformly distributed random variable. After time Δt a transition is chosen with regard to its relative weight $W(\sigma \rightarrow \sigma')/\sum_{\sigma'} W(\sigma \rightarrow \sigma')$. To eliminate fluctuation effects that emerge from individual time evolution paths in the configuration space we average parameters that we use to quantify our observations (observables) over many realizations.

14.2.1.2 Oxygen and Nutrients

The nearby vasculature releases oxygen and glucose, which diffuse into the local environment and nourish the tumor. This behavior is described by the following diffusion equation:

$$\begin{aligned} \frac{\partial[c]}{\partial t} &= D_c \Delta[c] - \omega_c \\ [c] &= [c]^{bv} \text{ at EC nodes} \\ [c] &= [c]^{bv} \text{ on the external boundary} \end{aligned} \tag{14.1}$$

$c = O$ or G reads for oxygen or glucose. The stationary boundary sources $[c]^{bv} = [G]^{bv}$ and $[O_2]^{bv}$ represent capillaries that provide glucose and oxygen to the nearby tissue, respectively. D_c is the diffusion coefficient. ω_G and ω_O are the cancer cell consumption rates of glucose and oxygen defined by:

$$\omega_c(i, j, k) = \begin{cases} q_c & \text{If there is a cancer cell at position } (i, j, k) \\ 0 & \text{Otherwise} \end{cases} \tag{14.2}$$

where q_c are the functions

$$q_G = q_{G,\max} \cdot \frac{[G]}{K_G^G + [G]} \cdot \left[1 - \left(1 - \frac{q_{G,\min}}{q_{G,\max}}\right) \cdot \frac{[O_2]}{K_O^G + [O_2]}\right] \tag{14.3}$$

$$q_O = q_{O,\max} \cdot \frac{[O_2]}{K_O^O + [O_2]} \cdot \left[1 - \left(1 - \frac{q_{O,\min}}{q_{O,\max}}\right) \cdot \frac{[G]}{K_G^O + [G]}\right] \tag{14.4}$$

14.2.1.3 Growth Factors

In this model the endothelial growth factors are released by the (hypoxic) necrotic cells and diffuse into the tumor environment following the equation

$$\begin{aligned} \frac{\partial[GF]}{\partial t} &= D_{GF} \Delta[GF] \\ [GF] &= [GF]^{bv} \text{ at necrotic nodes} \\ [GF] &= 0 \text{ on the external boundary} \end{aligned} \tag{14.5}$$

where D_{GF} is the diffusion constant and $[GF]^{bv}$ is the boundary source of growth factors released by the necrotic cells.

For a list of all the parameters used in the simulations, see Table 14.1.

14.2.1.4 Vascularization, Angiogenesis, and Remodeling

The models for the vascularization and its adaptation to the micro-environment are largely inspired by the 2-D model of [11] and the 3-D model in [48]. We also refer to [11] for biological references of every mechanism that these models describe.

- *Preexisting network:* The preexisting network of vessels is generated on the random lattice, in common with the tumor cells. The vessel orientation follows the three spatial directions, with an average distance between vessels of d^{bv}. Each node of the lattice is thus either free, a tumor cell (TC), or an endothelial cell (EC). Two neighboring EC nodes are linked by an edge e_{ij} of length $|e_{ij}|$ that represents a blood vessel of radius r_{ij}, initially at the homogeneous value of ${r_{ij}}^0$. Flow through a vessel Q_{ij} and pressure at the nodes P_i and P_j are computed in all the vessels based on the simplest resistance law (Poiseuille law) that linearly relates the pressure gradient in a segment to the flow through it. Poiseuille law reads as follows:

$$P_i - P_j = \frac{8\mu^{bv}|e_{ij}|}{\pi {r_{ij}}^4} Q_{ij} \tag{14.6}$$

  μ^{bv} is the dynamic viscosity of the blood, taken constant as a first approximation. Pressure is prescribed as a boundary condition at the entrances P_{in} and exits P_{out} of the network, and solved at nodes using Kirchoff's law. A measure of the shear stress f_{ij} in the vessel is calculated as a linear function of the pressure gradient and the radius: $f_{ij} = r_{ij}(P_i - P_j)$. We denote by f_{ij}^0 its value in the initial network.
- *Angiogenesis:* A sprout can form from a blood vessel with a probability proportional to the time step divided by the endothelial proliferation time t_{EC}, if certain criteria are met. From a given EC node i, a new blood vessel that goes until the sprouting node j can emanate if the distance between the two existing EC nodes is smaller than a maximum length $l_{\max}$. In addition, along the possible new vessel path all nodes must be free and the growth factor concentration higher than the threshold θ^{GF} that characterizes the angiogenic switch.
- *Remodeling:* Within the living tumor zones (proliferating and quiescent zones), blood vessels cannot sprout but they can dilate due to proliferation induced by growth factors if the local growth factor concentration is above the threshold θ^{gf}. This occurs with a certain probability proportional to the time step divided by the endothelial proliferation time t_{EC}, up to a maximum diameter $d_{\max}$, with an increment of the radius r_{ij} of $1/2\pi(|e_{ij}| + 1)$. In contrast, under-perfused vessels can also collapse due to the high pressure generated by the proliferation of tumor cells or disappear because they are not functional enough and thus experience hypoxia or are sensitive to the anti-angiogenic factors. This is modeled by the collapse of a vessel if its shear force is too low (f_{ij}/f_{ij}^0 is below a critical value θ^{ss}) and the density of tumor cells is too large (percentage of TC nodes above p_c^{TC}), with a probability proportional

to the time step divided by the collapsing time constant t_r. The vessel can also be removed with probability 0.5 if the flow is zero and the local concentration of oxygen is below a critical threshold $\theta_{bv}^{O_2}$.

The vascular network responds to changes in the local micro-environment by angiogenesis or remodeling. The local radius, pressure, flow, and shear values are thus continuously updated. In turn, the changing vascular network influences the growth of the tumor as explained in the cellular model above.

14.2.2 Results

We compared the growth of a tumor not constrained by nutrient and oxygen limitation with a tumor in a static blood vessel environment and a tumor that is able to modify the static blood vessel environment by triggering the formation and remodeling of blood vessels. In Figure 14.2, the radius of the tumor is plotted versus time for three cases: (1) without oxygen or nutrient limitation, (2) with nutrient limitation but without the angiogenic switch, and (3) with nutrient limitation and angiogenesis.

In the "no limitation" scenario, oxygen and nutrient concentrations are set to be high enough to meet the tumor demands, both in space and time. All cells can divide and the tumor thus first expands exponentially (zone a of Figure 14.2). After some time, the cells in the center cannot divide further due to contact inhibition and they become quiescent. When the proliferating rim reaches a constant thickness, the radius becomes a linear function of time (zone c of Figure 14.2).

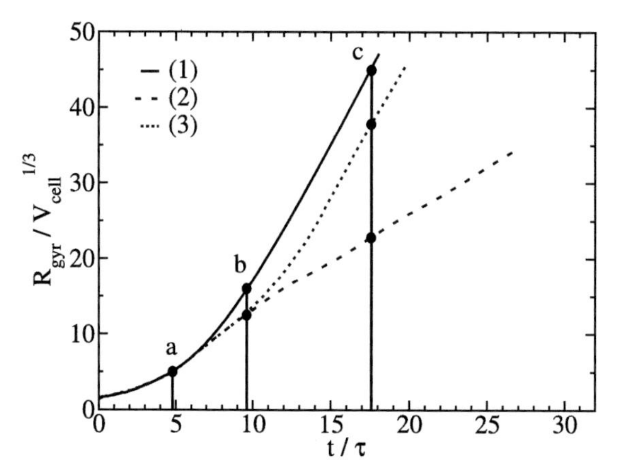

FIGURE 14.2 Time evolution of the radius of a tumor cell population for three different scenarios: (1) without any nutrient limitation (solid line), (2) nutrient limited growth in vascularized tissue (dashed line), and (3) nutrient limited growth inducing angiogenesis (dotted line).

For the two other simulated scenarios there is nutrient limitation: oxygen and nutrients are supplied by sources (blood vessels) and diffuse out of them in the interstitial space but they are also locally consumed by the cells. As the tumor mass expands, there is a first period where demands are lower than supplies (as seen by the superposition of the three curves in zone a of Figure 14.2). Then, supplies cannot balance demands due to an increasing consumption: this is the nutrient limitation phase. This slows the growth of the tumor, as can be seen by the decreasing slope in zone b of curves (2) and (3) of Figure 14.2.

After some time, the angiogenic switch occurs and enables the tumor to expand with a higher speed as indicated by the higher slope of the curve (3) compared to curve (2) in zone c of Figure 14.2: its demands of oxygen and nutrients are better fulfilled. Indeed, it reaches back the speed of the no limitation case (slopes of curve (1) and (3) in the zone c).

In addition to the tumor size, its structure varies significantly with the different environmental conditions. When there are neither oxygen nor nutrient limitations (Figure 14.3 (1a),(1b),(1c), (2a), (3a)), cells are either proliferating (yellow) or quiescent (green), but none of the cells are necrotic. In contrast, when oxygen or nutrients are lacking because their diffusion from blood vessels is not fast enough and their local concentrations are too low, necrotic cells (blue) appear in the center (Figure 14.3 2b and 3b).

As a response to hypoxia and hyponutrition, cells produce growth factors that diffuse through the tissue, reach the existing blood vessels, and finally trigger sprouting from them to create new blood vessels (Figure 14.3 (3c)). If no new blood vessels are created, the necrotic zone increases (blue region in Figure 14.3 (2c) larger than in Figure 14.3 (3c), where little necrosis can be observed). Note the quiescent zones around the blood vessels inside the tumor (Figure 14.3 (2c) and (3c)): in these regions, cells have enough oxygen and nutrients but they cannot divide due to contact inhibition of growth. As time passes, the case without limitation continues to grow with a spherical shape and without any necrosis (Figure 14.3 (1c)). In the limited case, the tumor continues to grow and tries to grow toward or along blood vessels (Figure 14.3 (2c)). In the angiogenic case, new vessels are forming toward and inside the tumor (Figure 14.3 (3c)) as it continues to grow with a speed closer to the no-limitation case (Figure 14.2).

14.2.3 Discussion

In this section we studied the interplay of biomechanically induced contact inhibition and oxygen/nutrient limitation on the growth kinetics of tumors in three cases: (1) tumor growth not constrained by the lack of oxygen or nutrients as it can be partly observed in monolayers and the early avascular phase of tumors or multicellular spheroids growing *in vitro*; (2) tumor growth in a *static* vascular network, where neovascularization does not occur; and (3) tumor cell growth can induce the formation of new vessels.

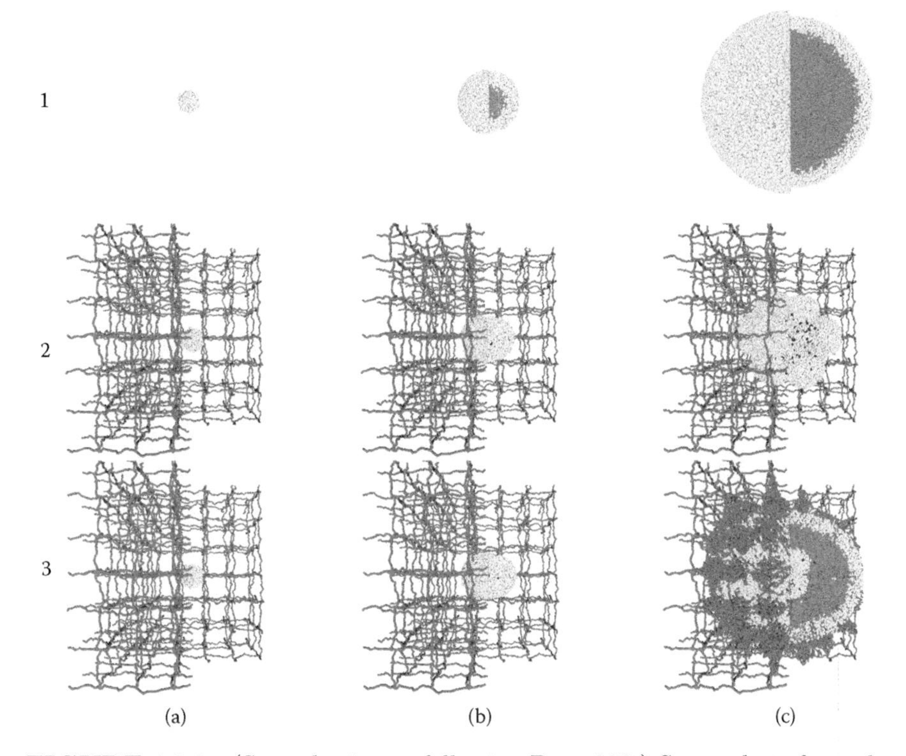

FIGURE 14.3 (See color insert following Page 398.) Screenshots from the simulations at times $t/\tau = 4.8$ (*left*), $t/\tau = 9.6$ (center) and $t/\tau = 17.6$ (right). Each figure consists of an exterior view on the left-hand side and a central-cropped view of the simulated domain on the right-hand side. The colors indicate proliferating (yellow), quiescent (green), and necrotic cells (blue), as well as blood vessels (red). The upper sequence shows the reference simulation of growth without any nutrient limitation in contrast to the lower two sequences showing the scenarios of nutrient-limited tumor growth in vascularized tissue without (center) and with angiogenesis (bottom).

For the angiogenesis we used a cellular automaton approach for the tumor cells and for the blood vessels and modeled the blood flow by Poiseuille technique and allowed for leakage of oxygen and nutrients in the extracellular space, which we mimicked by reaction–diffusion equations. The rules of angiogenesis were largely motivated by the work of Rieger and co-workers [11,48,78].

In all cases we found the tumors to grow approximately spherical for the parameter values we had chosen. So the expansion can be quantified by studying the time development of the tumor radius. We find that the radius grows exponentially fast for small times, changing to linear growth later on. This is precisely what could be observed in monolayers, multicellular spheroids, and in xenografts (see Section 14.1). The existence of linear growth is closely related to a proliferating rim of constant size. If the spatial density of proliferating cells is constant (as we find), then the size of the proliferating rim determines the speed of growth. In case (1) the speed of growth is controlled by a biomechanical form of contact inhibition (e.g., [27] and references therein). In this and other references [15,37] we have shown that this form of contact inhibition of growth can be explained by the degree of compression of cells within the tumor caused by the local pressure within the tumor tissue. During tumor expansion, this pressure profile increases from a small value at the tumor border (marking the outer border of the proliferating rim) toward a value at which cell proliferation is contact inhibited (marking the interior border of the proliferating rim). In case (2) the speed of growth is controlled by the penetration depth at which the concentrations of the oxygen and nutrients fall below the value necessary to permit cell proliferation. Because this penetration depth is below the size of the proliferating rim dictated by contact inhibition (case (1)), the growth speed in case (2) is below that in case (1). In case (3), angiogenesis after its onset generates sufficient new blood vessels so that oxygen and nutrient supply is—after a short transient—not limiting as in case (2). Accordingly, the tumor grows with the same speed as in the oxygen/nutrient-unlimited case (1). Moreover, in case (3) we find only decent mitotic figures. Both the linear growth and the decent mitotic figures have been observed in xenografts of NIH3T3 cells, which are well vascularized [72].

14.3 Later Stages of Cancer: Invasion and Intravasation

In this section we give an introduction to two versions of a multiscale force-based model. The first version studies the importance of cell–cell adhesion in cancer cell invasion. The second version approaches the problem of cancer cell intravasation. They are both structured as follows. At the intracellular scale, the protein concentrations are governed by a system of ordinary differential equations derived from the reaction systems (14.11) and (14.13) (for the exact formulation of the equations, see [67, 68]). At the cellular scale, the cell–cell forces are based on a modified Hertz model (Equations (14.7) to (14.9)) where the intensity of the adhesive forces depends on the intracellular cadherins available to travel to the cell surface to form bonds. Finally, a cell moves according to a (Langevin) equation of motion, which accounts for the main biophysical characteristics of the multicellular system (Equations (14.10), (14.12), (14.14), and (14.15)). For a more detailed explanation of the equations and methods, a detailed list of parameter values, and further results, see [67,68].

14.3.1 Biophysical Model of a Single Cell

The cell is characterized as an isotropic elastic object capable of migration and division and parameterized by kinetic, biophysical, and biological parameters that can be experimentally measured. We describe below the key features of this modeling approach.

Cell–cell shape: We assume that an individual cell in isolation is spherical and characterized by a radius R.

Cell–cell interaction: With decreasing distance between the centers of two cells (e.g., upon compression), both their contact area and the number of adhesive contacts increase, resulting in an attractive interaction. On the other hand, if cells are spheroidal in isolation, a large contact area between them significantly stresses their cytoskeleton and membranes. Furthermore, experiments suggest that cells only have a small compressibility (the Poisson numbers are close to 0.5, [6,52]). In this instance, both the limited deformability and the limited compressibility give rise to a repulsive interaction. We model the combination of the repulsive and attractive energy contributions by a modified Hertz model [37,71] where the potential V_{ij} between two cells of radius R_i and R_j is given by:

$$V_{ij} = \underbrace{\sqrt{(R_i + R_j - d_{ij})^5}\frac{1}{5\tilde{E}_{ij}}\sqrt{\frac{R_i R_j}{R_i + R_j}}}_{\text{repulsive contribution}} + \underbrace{\epsilon_s}_{\text{adhesive contribution}} \tag{14.7}$$

The first term of the equation models the repulsive interaction, the second term the adhesive interaction, and $\tilde{E}_{ij}$ is defined by

$$\tilde{E}_{ij}^{-1} = \frac{3}{4}\left(\frac{1-\sigma_i^2}{E_i} + \frac{1-\sigma_j^2}{E_j}\right) \tag{14.8}$$

Here, E_i, E_j are the elastic moduli of the cells i, j, σ_i, σ_j the Poisson ratios of the spheres. $\epsilon_s \approx -\rho_{ij} A_{ij} W_s$, where $W_s \approx 25 k_B T$ (T: temperature, k_B: Boltzmann constant) is the energy of a single bond, $A_{ij}(d_{ij})$ the contact area between cells i, j, and ϱ_m the density of surface adhesion molecules in the contact zone, in our case the density of E-cadherin [21]. The interaction force results from deriving the potential function

$$\underline{F}_{ij} = -\left(\frac{\partial V_{ij}}{\partial d_{ij}}\right)\left(\frac{d(d_{ij})}{dx}, \frac{d(d_{ij})}{dy}, \frac{d(d_{ij})}{dz}\right) \tag{14.9}$$

The modified Hertz model approximates a cell as an elastic sphere. It superimposes the repulsive force that emerges in case of a deformation or compression of the sphere with an attractive contribution due to cell–cell adhesion.

14.3.2 Coupling Intracellular and Extracellular Scales

Adhesion forces between cells are controlled by the density of the β-catenin–E-cadherin complexes ($[E/\beta]$) positioned at the cell membrane within the cell–cell contact zone. Following our previous work [68] we take as the adhesion energy $W_s \varrho_m = 200\ \mu N m^{-1}$, so that the surface receptor density is $\varrho_m = 200\ \mu N m^{-1} W_s^{-1}$. We use this value as a maximum density of the cadherin–β-catenin complex in the membrane and define the actual density by:

$$\rho_{ij} = \min(\rho_i, \rho_j), \quad \rho_i = \frac{[E/\beta]_i}{E_T} \varrho_m$$

Figure 14.4 shows the resulting force function depending on the different ϱ_m^{ij} values. By modifying the intracellular concentration of β-catenin, the cells can control the concentration of $[E/\beta]$ complexes and thereby the strength of the intercellular adhesion force.

14.3.3 Cell Movement

Cells move under the influence of forces and a random contribution to the locomotion that results from the local exploration of space. A simple form of the force equation that accounts for the basic elements of our biological systems can be written in the form:

$$\underbrace{\underline{\underline{\Gamma}}_{is}^{f} \underline{v}_i}_{\text{substrate friction}} + \sum_{i\,nn\,j} \underbrace{\underline{\underline{\Gamma}}_{ij}^{f} \left(\underline{v}_i - \underline{v}_j\right)}_{\text{cell-cell friction}} = \underbrace{\sum_{i\,nn\,j} \underline{F}_{ij}}_{\text{forces}} + \underbrace{\underline{f}_i(t)}_{\text{noise}} \tag{14.10}$$

Inertia terms have been neglected due to the high friction of cells with their environment. $\underline{v}_i$ is the velocity of the cell i at time t, and the sums are over the nearest neighbors in contact with cell i. The substrate friction term denotes the friction force with the substrate and the cell–cell friction denotes the friction forces with the nearest neighbor cells. The tensors $\underline{\underline{\Gamma}}_{ij}^{f}$ and $\underline{\underline{\Gamma}}_{is}^{f}$ denote cell–cell friction and cell–substrate friction, respectively. The forces term denotes the force that cell i exerts on the other cells in contact with it, previously calculated in Equations (14.7) and (14.8). The noise term models the random component in the cell movement (the micro-motility) and is chosen to be uncorrelated as explained in [28,68].

14.3.4 Cell–Cell Detachment and Invasion of Local Tissue

Cadherins are the main proteins involved in preserving cell–cell adhesion and tissue structure. They are integral membrane calcium-dependent proteins. Historically they were named after the spatial distributions where they were first discovered: N-cadherin (neural), E-cadherin (epithelial), P-cadherin (placental), etc. Lately they have been found present in other different tissues and

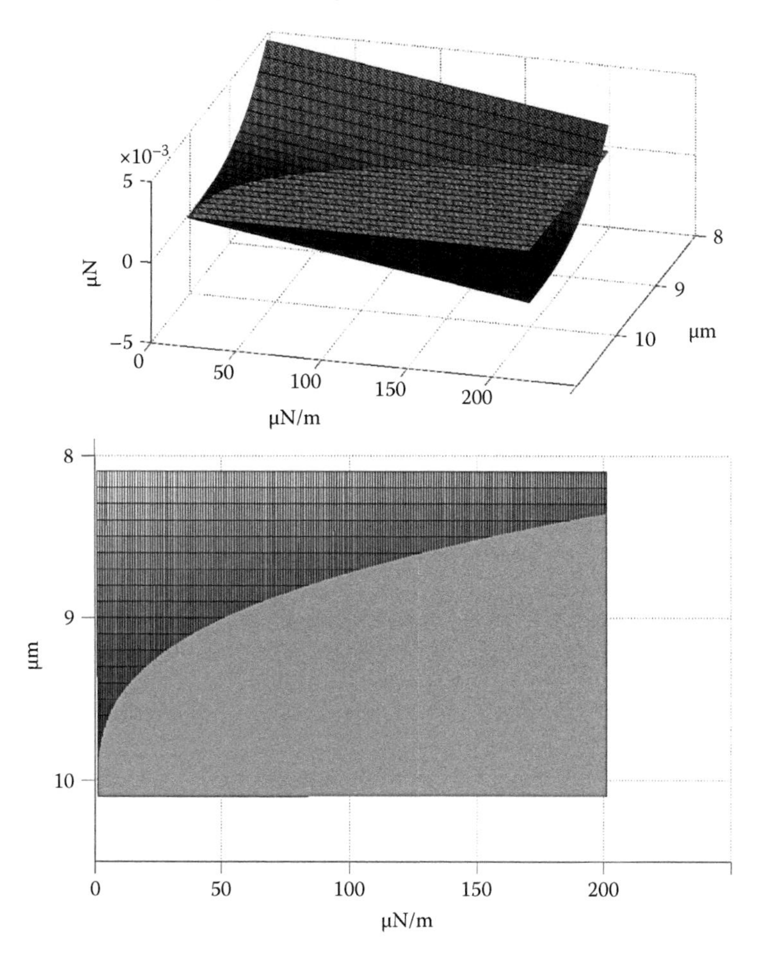

FIGURE 14.4 The top plot shows the force function between two cells. Variables are cell–cell distance (in μm) and adhesion energy per unit of area in contact (in μN/m). The bottom plot shows the vertical view of the same graph for better observation of the adhesive interaction between cells depending on the E-cadherin concentration forming bonds. The gridded part determines the zone where the net force is adhesive.

organs. Among them, the most studied molecule is E-cadherin, which binds at the extracellular domain to the E-cadherin molecules of other cells and, in the intracellular domain, to a multi-protein complex formed by catenins (p120, β- and α-catenin).

In the E-cadherin adhesive system, β-catenin joins α-catenin in a complex that links E-cadherin and the actin filaments of the cytoskeleton. When β-catenin is not forming part of the adhesive complex, it is rapidly phosphorylated and degraded within the ubiquitin proteasome pathway. If the degradation of β-catenin is not performed efficiently, it can eventually bind to transport proteins such as T cell factor (Tcf) and be translocated into the cell nucleus where it interacts with transcription factors. As a result of this interaction, the cells expressing high levels of nuclear β-catenin become more invasive [49,80]. The implications of β-catenin deregulation are essential in cancer invasion.

Here we model a simplified version of the cadherin–β-catenin pathway as follows. When the cells come into contact, the cadherin travels from the cytosol to a position at the membrane where it links the neighboring cell with the cytoskeleton via the scaffolding proteins of the catenin family. In the case where the cells separate from each other, the β-catenin and the E-cadherin molecules are internalized into the cytosol so that β-catenin can interact with proteasome units and be degraded. If the degradation of β-catenin does not occur fast enough, it is transported into the cell nucleus where it interacts with transcription factors. As a result of this interaction, the cell may change its adhesive state to a nonadhesive state by sequestering the rest of the cadherins forming bonds with other cells into the cytosol [44]. Some experimentalists have proposed the existence of a soluble β-catenin threshold over which the concentration is high enough to be transported into the nucleus and interact with transcription factors. We assume that a cell will undertake the decision to detach from its neighbors if the soluble β-catenin levels overcome the threshold of $c_T \approx 50\ nM$. This mimics the observations of Kemler et al. [46] when studying epithelial bud development. They showed that an upregulation of soluble β-catenin was followed by a downregulation of the E-cadherin-mediated bonds with the local neighbors.

There is evidence that cytoplasmic E-cadherin translocates to the membrane after binding to β-catenin at the endoplasmic reticulum [20]. For simplicity we assume that the complexes are formed in the cell membrane. We consider three possible different states of the E-cadherins: catenin free in the cytoplasm ($[E_c]$), catenin free in the cell membrane ($[E_m]$), and the complex E-cadherin–β-catenin forming bonds in the cell membrane ($[E/\beta]$). As one cell comes into contact with another cell, the cadherin in the cytoplasm moves to the cell surface. These interactions can be described by the following reaction scheme:

$$\begin{aligned}
[E_c] &\rightarrow_{\{contact\}} [E_m] \\
[\beta] + [E_m] &\xrightarrow{\nu} [E/\beta] \\
[E/\beta] &\rightarrow_{\{detachment\}} [E_c] + [\beta]
\end{aligned} \tag{14.11}$$

The β-catenin degradation process takes place after forming a complex with the proteasome. In the framework of our model, this proteasome variable should be understood as a complex of proteins that, after different biochemical interactions, degrades soluble β-catenin; that is,

$$[\beta] + [P] \underset{k^+}{\overset{k^-}{\rightleftharpoons}} [C] \xrightarrow{k_2} [P] + \omega$$

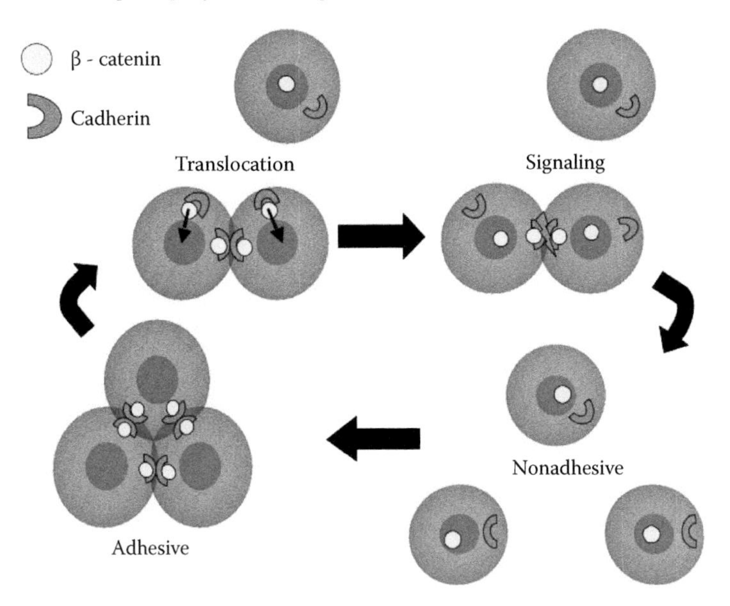

FIGURE 14.5 Schematic diagram showing the modeling of cell–cell adhesion dynamics. From an initial set of three cells in an adhesive state, one cell detaches and this induces the translocation of β-catenin from the membrane to the nucleus of the other two binding neighbors. This in turn stimulates intracellular signaling pathways [43,44] that promote a temporal change in the adhesive state of the neighboring cells to become nonadhesive. Eventually the levels of soluble β-catenin become downregulated and all the cells become adhesive again. From O. Huber, C. Bierkamp, and R. Kemler (1996). Cadherins and catenins in development. *Curr. Opin. Cell Biol.* 8:685–691, and C. Jamora, R. DasGupta, P. Koclenlewski, and E. Fuchs (2003). Links between signal transduction, transcription and adhesion in epithelial bud development. *Nature* 422:317–322. (With permission).

where ω denotes the product of the degradation process. Using the law of mass action, this scheme is converted into a system of four ODEs governing the intracellular dynamics of the concentrations of the molecules. Figure 14.5 shows how this system of reactions can trigger cell–cell detachment. If the concentration of soluble β-catenin is high enough to be transported into the nucleus and interact with transcription factors, this may trigger a decision process in the cell to detach from the neighbors. As a consequence of the detachment, the neighbors in contact will suffer an upregulation of soluble β-catenin, starting the detachment cycle again.

The active decision of a cell to migrate can be triggered in different ways, all of them involving an upregulation of the soluble β-catenin, which overcomes the critical threshold c_T. One case where this happens is if the cytoplasmic concentration of β-catenin is upregulated due to a failure in the proteasome system. A further case would be if the detachment of local neighbors upregulates the soluble β-catenin concentration. In both cases, β-catenin enters the nucleus and triggers cell migration (the precise model of the intracellular β-catenin regulation is explained in the next subsection). In our simulations we extended Equation (14.10) with a chemotaxis term as in [73] modeling a morphogen gradient:

$$\underbrace{\underline{\underline{\Gamma}}^f_{is}\underline{v}_i}_{\text{substrate friction}} + \sum_{i\,nn\,j} \underbrace{\underline{\underline{\Gamma}}^f_{ij}\left(\underline{v}_i - \underline{v}_j\right)}_{\text{cell-cell friction}} = \underbrace{\sum_{i\,nn\,j} \underline{F}_{ij}}_{\text{forces}} + \underbrace{\underline{f}_i(t)}_{\text{noise}} + \underbrace{\chi\underline{\nabla}Q(t)}_{\text{chemotaxis}} \tag{14.12}$$

We have chosen the parameters of the chemotaxis term such that the chemotaxis force remained below the threshold at which the cells can detach from the tumor surface by the pulling force generated by the chemotaxis force. Hence, the cells would need to downregulate the cadherin–β-catenin adhesion complexes to be able to detach from the neighbors.

14.3.5 Protein Dynamics

To illustrate the response of possible malfunctions in the intracellular control of the β-catenin concentration, we study simulations for different attachment/detachment scenarios. If the cell remains adhered to its neighbors, almost all of the β-catenin remains bound to the E-cadherin complexes at the cell membrane. Alternatively, if a cell detaches, β-catenin is released into the cytosol so that the concentration of soluble β-catenin (i.e., β-catenin in the cytosol) increases.

Figure 14.6 shows the protein dynamics of a single cell that attaches to other cells. As can be seen from the figure, soluble β-catenin is rapidly sequestered from the cytoplasm by the cadherins to form the $[E/\beta]$ complex. As long as the contacts are maintained, the soluble β-catenin concentration remains at a low level. If some of the neighboring cells detach, then the concentration of E-cadherin forming bonds will be partially reduced.

Figure 14.7 shows concentrations of the intracellular variables for two different detachment scenarios. In the plot at the top we assume that a cell that had contacts to many neighbor cells loses all its bonds with all its neighbors at $t \approx 0.4$, which triggers a dramatic increase in the β-catenin concentration in the cytoplasm. This soluble β-catenin enters the nucleus in excess of the threshold concentration necessary to initiate migration and promote cell movement via transcription. In the plot at the bottom, at time $t \approx 0.4$, the cell has lost only about 1/4 of its bonds with the neighbors and the soluble β-catenin concentration is insufficient to cause cell migration.

We implement the intracellular dynamics model explained above in every single cell of the individual-force-based model. The advantage of using this type of modeling approach is that it not only allows us to explicitly include the influence of intracellular pathways, but also provides a realistic approach to model the biophysical properties of individual cells, which cannot be neglected

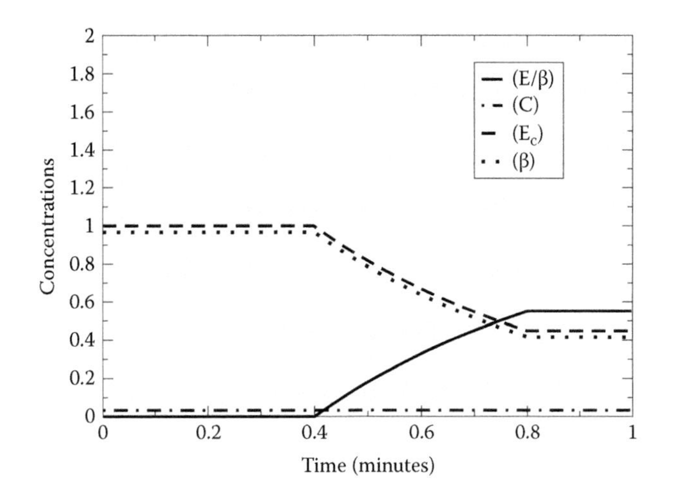

FIGURE 14.6 Plot showing the intracellular concentrations of different biochemicals (see legend) over time when a cell attaches to a group of cells at $t \approx 0.4$ min. β-Catenin is rapidly sequestered by the cadherins that travel to the cell surface to form bonds.

when studying tissue organization. We reproduce *in silico* different scenarios of relevance in cancer growth and invasion, and study the behavior of detachment waves in epithelial layers and how it can produce the epithelial–mesenchymal transition (EMT). We also study the β-catenin distribution in small tumors and how its upregulation can induce invasion.

14.3.6 Results: Invasion of Local Tissue

The epithelial-to-mesenchymal transition is a process in which a well-polarized layer of cells becomes diffuse and loses the initial structure and compactness. This transition occurs in a similar way at the tumor surface when invasion occurs: the mass of outer cells loses contact and invades further into the tissue. To model the epithelial–mesenchymal transition, we set as initial conditions a layer of cells with fixed boundaries (i.e., unable to move). We include a constant force term as if it were a constant source of chemoattractant (*chemotaxis* term in Equation (14.9)) that diffuses toward the tissue in the form of the equation

$$\frac{\partial Q(t)}{\partial t} = D_Q \triangle Q(t)$$

We assume that in one of the cells the β-catenin concentration is upregulated, as discussed in Section 14.3.4. Figure 14.8 shows how the cells migrate and the configuration of the epithelial layer is lost. The color of the cells determines the intracellular concentrations of soluble β-catenin. The lighter color cells (in yellow) have higher concentration of soluble β-catenin and will tend to

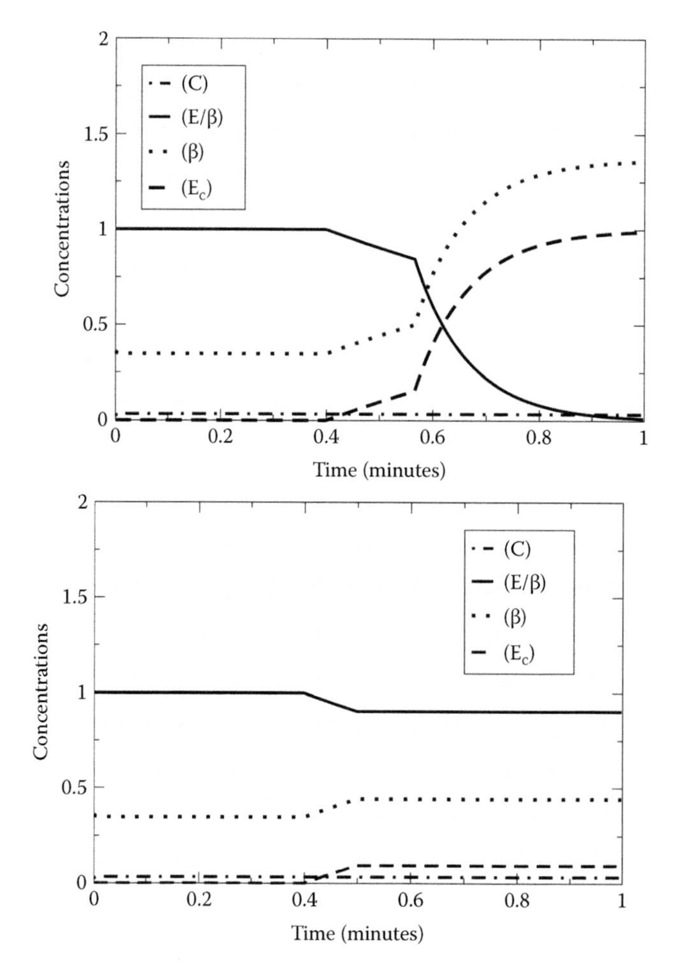

FIGURE 14.7 Plots showing the concentrations over time of the intracellular variables under two different scenarios. In the top plot, a cell which previously had many neighbor cells has lost contact with each of its neighbors (for details see [67]). The β-catenin concentration increases dramatically and it enhances mechanisms that promote invasion. In the bottom plot, the total contact area has been only moderately reduced (for details see [67]); soluble β-catenin is maintained under the threshold levels ($c_T = 0.5$) that enhance migration.

detach from their neighbors. The darker cells (in red) have low concentrations of soluble β-catenin, and most of the β-catenin is found at the membrane forming a complex with E-cadherin. If the proteasome system is downregulated and one cell detaches, it induces the same behavior in neighboring cells.

The same migration mechanism can be observed in a tumor: cells detach from the outer rim and migrate toward a source of attractants. Figure 14.9

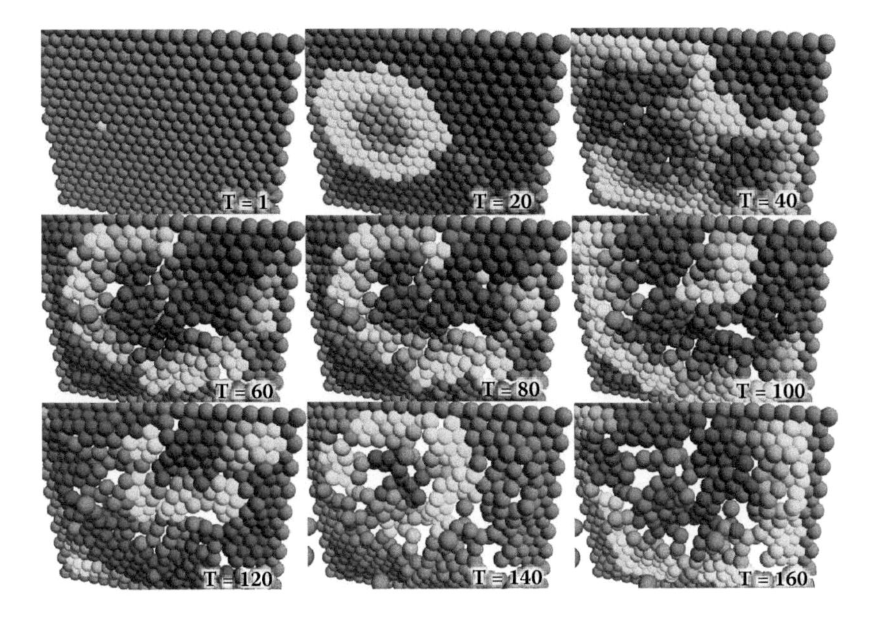

FIGURE 14.8 (See color insert following Page 398.) Plots showing how malfunctions in the proteasome system can alter the layer configuration producing the epithelial–mesenchymal transition. In this figure, the cells migrate toward a source of attractants escaping from the initial epithelial configuration. Migration can occur only when the catenin levels are above a determined threshold (yellow/lighter color cells). Time is measured in minutes.

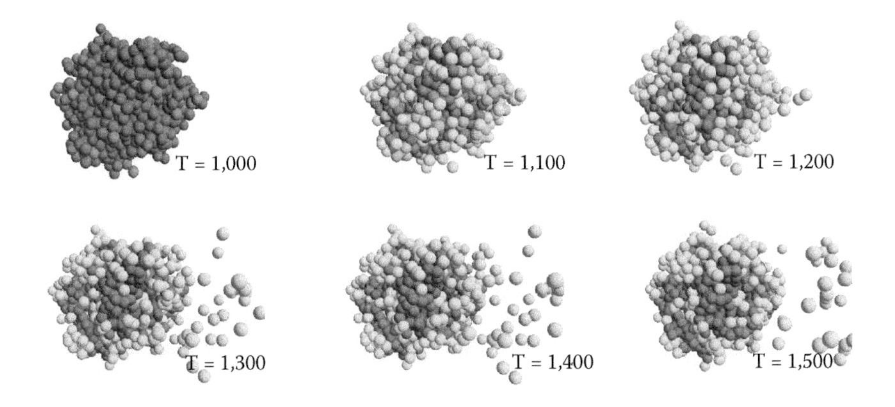

FIGURE 14.9 (See color insert following Page 398.) Plot showing how a small tumor invades further tissue stimulated by a source of morphogen located on the right-hand side of the tumor. Cells decide to detach gradually when the intracellular concentration of β-catenin is upregulated (yellow/ lighter color cells). Time is measured in minutes.

shows the spatio-temporal dynamics of an invasive tumor undergoing EMT. The proteasome functionality has been downregulated to the point where migration occurs. It can be observed that the outer cells migrate and detach from the main tumor mass, the new cells at the outer rim lose part of their E-cadherin bonds, upregulate soluble β-catenin as well, and also enhance their migration and invasion. These findings suggest how invasion can be a gradual process produced by activity of cells that detach from the tumor surface.

14.3.7 Cell Intravasation: Hallmark of Metastasis

Metastasis is a crucial process in the growth of a cancer leading to the formation of secondary tumors at sites distant from the primary tumor in the host. To colonize distant organs, the malignant cells need to get into a blood or lymph vessel, be transported in the vascular system and eventually attach to the inner wall of the vessel, and escape from the vasculature. In this new location, the malignant cell proliferates to form a secondary tumor. Intravasation and extravasation are defined as the processes of a cell entering and leaving the vascular network, respectively. These are essential natural mechanisms used by specialized cells to travel to distant organs. However, the same mechanisms are used by cancer cells to create colonies and secondary tumors [57]. Both intravasation and extravasation occur in a similar way. During intravasation the cancer cell attaches to the endothelial wall of a vessel, and pushes apart the endothelial cells to squash in and enter the vascular network. During extravasation the cell leaves the vascular network by the same process. This migration through the endothelial tissue is also known as transendothelial migration (TEM). The physical properties of the cell combined with the intra- and intercellular protein interactions that govern cell–cell adhesion are the driving forces of TEM. Formation and detachment of bonds involve the interactions of, among other molecules, N-cadherins and VE-cadherins (vascular endothelial cadherins), and the activation of related protein pathways.

The blood vessel wall consists of three distinct layers of tunics [5] from the inner wall to the outer wall. The *tunica intima* is composed of endothelium and rests on a connective tissue membrane rich in elastic and collagenous fibers. The *tunica media* makes up the bulk of the vessel wall and includes smooth muscle fibers and a thick layer of elastic connective tissue. Finally, the *tunica adventitia* attaches the vessel to the surrounding tissue. It is a thin layer that consists of connective tissue, elastic collagenous fibers, and minute vessels that are the beginnings of small capillaries which will help to irrigate the surrounding tissue. The *tunica intima* is the last cell layer that a cancer cell needs to cross to reach the inside of the vasculature. To achieve TEM, the metastatic cell needs to break, among other cellular adhesion molecules, the VE-cadherin bonds that hold the endothelial cells of the *tunica intima*. Whether these bonds are broken by the mechanical pressure exerted by the malignant cell on the endothelial wall, or whether there are other biological mechanisms involved is not yet completely elucidated.

The cells in our model can attach to each other via VE-cadherin or N-cadherin molecules. Precisely which of these two binding proteins is used by an endothelial cell to bind to another cell will depend on the adhesion molecule that the adhering cell is expressing at that precise moment. If the malignant cell is attaching to the endothelial cell, it will create an N-cadherin-mediated bond. If on the other hand, the endothelial cell comes into contact with another endothelial cell, they will try to preserve the architecture of the *tunica intima*, creating VE-cadherin-mediated bonds [64,65].

When two cells come into contact, the cadherins change into a stimulated state prior to forming cell–cell bonds. Here we refine the intracellular dynamics used in the invasion version of the multiscale model by including that cadherins are transported to the intermembrane region to form bonds with the neighbors after forming a complex with β-catenin [20]. This is translated into our model by considering the following possible states for VE- and N-cadherin: (a) free in the cytoplasm ($[VE]$, $[N]$); (b) in the cytoplasm in a stimulated state before forming a complex with β-catenin ($[VE_s]$, $[N_s]$), and; (c) forming bonds at the intermembrane position ($[VE/\beta]$, $[N/\beta]$).

The expression of N-cadherin molecules forming bonds between the malignant cell and the endothelial cells activates the Src kinase activity [64,65]. In our model, these enzymes can target both the N-cadherin mediated bonds between the cancer and the endothelial cells, and the VE-cadherin mediated bonds formed between two endothelial cells. Figure 14.10 shows the action of the Src kinases in the bonds of the cells forming the endothelial wall. The cancer cell comes into contact with the *tunica intima* and activates the kinases after forming N-cadherin-mediated bonds (Figure 14.10.1). The kinases do not

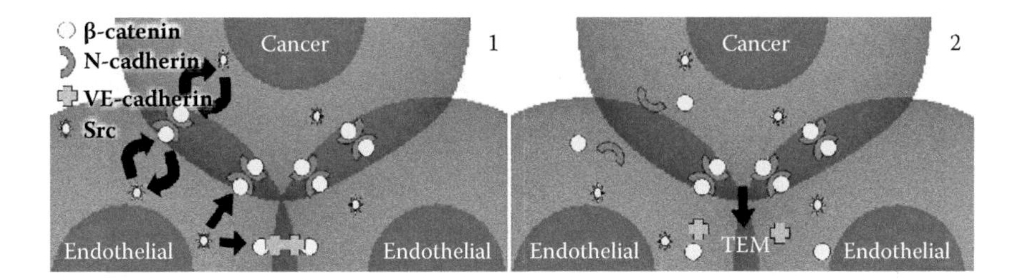

FIGURE 14.10 Diagram showing the intra- and intercellular protein pathways considered in the model. 1: The cancer cell comes into contact with the endothelial cells forming the wall and creates N-cadherin-mediated bonds. As a consequence of the activation of the N-cadherin pathway, the Scr kinase activity is upregulated. 2: The β-catenin linked to the VE-cadherin and N-cadherin molecules is phosphorylated by the Src kinases. After the β-catenin molecules bound to VE-cadherin are phosphorylated, the VE-cadherin-mediated bonds holding the endothelial cells are downregulated and TEM migration can be achieved.

only target the N-cadherin–β-catenin complexes but also the VE-cadherin–β-catenin complexes and disrupt the bonds between endothelial cells thereby facilitating TEM (Figure 14.10.2). As we did for modeling the degradation of soluble β-catenin, we include a generic proteasome (variable $[P]$) that degrades β-catenin after forming a complex with it ($[C]$). We also assume that both types of cadherins can be recruited to form bonds again. The main biochemical reactions affecting the cadherin adhesion pathways are the following (z denotes either VE or N):

$$[z] \rightharpoonup_{\{\text{contact}\}} [z_s]$$
$$[\beta] + [z_s] \rightharpoonup [z/\beta]$$

Cell–cell detachment can happen by a combination of physical stress, that is:

$$[z/\beta] \rightarrow_{\{\text{detachment}\}} [z] + [\beta] \tag{14.13}$$

and the action of tyrosine kinases:

$$[z/\beta] + [Src] \underset{k_z^-}{\overset{k_z^+}{\rightleftharpoons}} [S_z] \rightarrow [z] + [\beta] + [Src]$$

where $[S_z]$ denotes the complex Src-z-cadherin–β-catenin. Finally, the generic proteasome variable accounts for the degradation of β-catenin:

$$[\beta] + [P] \underset{k^+}{\overset{k^-}{\rightleftharpoons}} [C] \overset{k_2}{\rightarrow} [P] + \omega$$

where again ω denotes the product of the degradation process. These reactions will produce different adhesion dynamics, depending on the type of cell observed. Following the observations of Qi et al. [65], we assume that the malignant cell does not express VE-cadherin. Hence for the set of intracellular equations for the cancer cell, $z = N$. In the case of the endothelial cells there will be a set of equations where $z = N$ to explain the heterotypic bonds with the malignant cell and a set of equations where $z = VE$ to explain the homotypic bonds between endothelial neighbors.

14.3.7.1 Equations of Motion

We model the cell movement by two sets of equations of motion, being slightly different for the endothelial and for cancer cells. In the equation for the malignant cell, we take into account friction terms with the environment and with cells, adhesive–repulsive forces between cells, random movement, and directed movement.

$$\underbrace{\underline{\underline{\Gamma}}_{cs}^f \underline{v}_c}_{\text{substrate friction}} + \sum_{c\,nn\,j} \underbrace{\underline{\underline{\Gamma}}_{cj}^f \left(\underline{v}_c - \underline{v}_j\right)}_{\text{cell-cell friction}} = \underbrace{\sum_{c\,nn\,j} \underline{F}_{cj}}_{\text{ad.-rep. forces}} + \underbrace{\underline{f}_c(t)}_{\text{r. movement}} - \underbrace{\hat{\underline{F}}_c}_{\text{directed movement}} \tag{14.14}$$

The directed movement term models the active decision of the malignant cell to move toward the vessel. Endothelial cells forming the *tunica intima* move according to the force balance equation:

$$\underbrace{\underline{\underline{\Gamma}}^{f}_{is}\underline{v}_i}_{\text{substrate friction}} + \sum_{i\,nn\,j}\underbrace{\underline{\underline{\Gamma}}^{f}_{ij}\left(\underline{v}_i - \underline{v}_j\right)}_{\text{cell-cell friction}} = \underbrace{\sum_{i\,nn\,j}\underline{F}_{ij}}_{\text{ad.-rep. forces}} + \underbrace{\underline{f}_i(t)}_{\text{r. movement}} + \underbrace{\underline{\hat{F}}_{c,i}}_{\text{response forces}} \tag{14.15}$$

On the right-hand side, the third term (response forces) is only active when the malignant cell comes into contact with the endothelial wall and is the response to the force exerted by the malignant cell in the endothelial cells (directed movement in Equation (14.14)). We set $\underline{\hat{F}}_{c,i} = -\frac{[N/\beta]_{c,i}}{\sum_{c\,nn\,i}[N/\beta]_{c,i}}\underline{\hat{F}}_c$ where the subscripts denote the pair of cells sharing the N-cadherin bonds to ensure the total force is zero if the cancer cell squashes in the vessel.

14.3.8 Results: Cell Intravasation

Figure 14.11 shows the spatio-temporal dynamics of the multiscale simulations. The coloring of the cells denotes the scale of the type of protein used

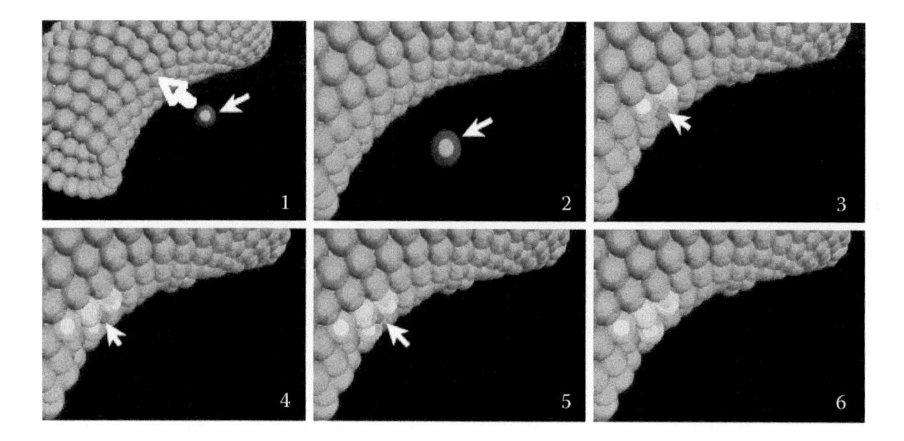

FIGURE 14.11 (See color insert following Page 398.) Plot showing the spatio-temporal evolution dynamics of malignant cell (solid arrow pointing at red colored nucleus cell) approaching a blood vessel to undergo TEM. The green (dark) initial color of the endothelial cells forming the blood vessel denotes the VE/β concentration. When the malignant cell attaches the vessel, the VE–cadherin bonds are disrupted and new N-cadherin bonds are formed (yellow/light gray). After some time, the malignant cell manages to disrupt the endothelial bonds enough to open a gap in the vessel and undergo TEM. Time scales in the frames correspond to the intracellular simulations of Figure 14.12 at time 1 = 0 min, 2 = 50 min, 3 = 100 min, 4 = 200 min, 5 = 300 min, 6 = 400 min.

when forming bonds, that is, green (dark gray) is used for VE-cadherin bonds and yellow for N-cadherin bonds. The malignant cell approaches the *tunica intima* and attaches to the endothelial cells. At the moment when contact occurs (frame 3, $t \approx 100$ min), the malignant cell and the endothelial cells start forming heterotypic N-cadherin bonds (light gray/yellow). As time evolves it can be observed how the VE-cadherin bonds (dark gray/green) between the endothelial cells are disrupted, and the malignant cell is able to open a gap in the endothelial wall and undergo TEM (frame 6, $t \approx 400$ min).

From the assumptions, a malignant cell that is only able to express N-cadherin therefore only induces the formation of N-cadherin bonds with the endothelial cells. At this stage within the endothelial cells, the two types of cadherins start competing for the soluble β-catenin. Figure 14.12 shows the intracellular dynamics of the cadherin and β-catenin proteins corresponding to the simulations shown in Figure 14.11. The plot at the top shows the intracellular protein dynamics of an endothelial cell to which the cancer cell attaches. At the moment when the malignant cell comes into contact with the endothelial cell ($t \approx 100$ min), the VE-cadherin concentration forming bonds between the endothelial neighbors decreases. This is initially caused by the degradation of the VE-cadherin–β-catenin complexes by the Src enzymes. At time $t \approx 450$ min, due to a combination of the degradation of the VE-cadherin bonds by the Src activity and the physical forces exerted by the malignant cell on the wall, the bonds between endothelial cells are totally disrupted and the VE-cadherins are internalized in the cytosol. As a consequence, soluble β-catenin is upregulated. The plot at the bottom shows the intracellular protein dynamics of the cancer cell. When it attaches to the endothelial wall at time $t \approx 100$ min, the N-cadherin in the cytosol is stimulated and binds to β-catenin. This complex is transported to the membrane to form heterotypic bonds with the endothelial cells. At time $t \approx 450$ min, the cancer cell succeeds in opening a gap in the *tunica intima* and undergoes TEM. As a consequence of this intravasation and losing contact with the *tunica intima*, the N-cadherin molecules forming bonds in the intermembrane position are internalized into the cytosol.

Figure 14.13 shows the time evolution dynamics of the adhesive forces between the endothelial cell to which the cancer cell attaches at time $t \approx 100$ min, and its endothelial neighbors forming the *tunica intima*. It can be observed that the strength of the adhesive forces decreases until cell detachment (at $t \approx 450$ min), as far as the VE-cadherin concentration is downregulated (see top plot of Figure 14.12). The slow reduction of the adhesive forces is caused by the intracellular action of the Src enzymes. At time $t \approx 450$ min the cancer cell undergoes TEM, thereby breaking the bonds between the endothelial cells and the adhesive forces disappear. In this simulation, detachment occurs by a combination of physical and biological causes.

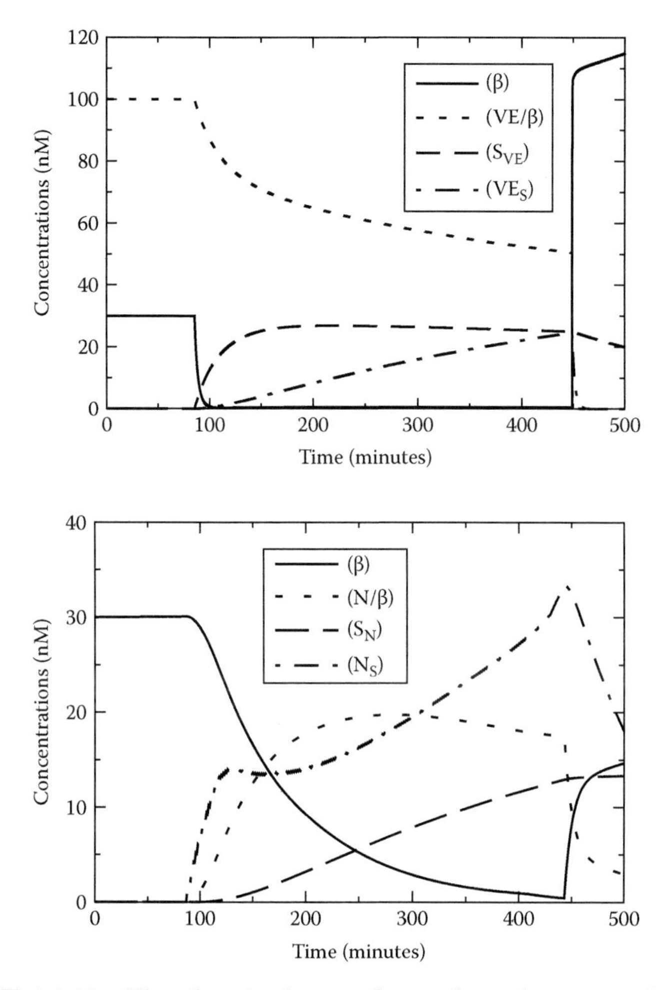

FIGURE 14.12 The plots in the top figure show the temporal evolution of the intracellular protein concentrations of an endothelial cell to which the cancer cell attaches at time $t = 100$ min. As time evolves, the VE-cadherin concentration at the surface of the cell is partially decreased by the action of the enzymes. At $t \approx 450$ min the cancer cell disrupts the bonds formed by the endothelial cells and, as a consequence, the VE-cadherin at the surface of the endothelial cell is dramatically decreased. The plot at the bottom shows the protein concentrations of the cancer cell when attaching to the endothelial wall. At time $t = 100$ min the cell comes into contact with the *tunica intima* and the N-cadherin travels to the membrane to form heterotypic bonds. As a combination of the biochemical pathways and the physical forces, the cancer cell undergoes TEM at time $t \approx 450$ min.

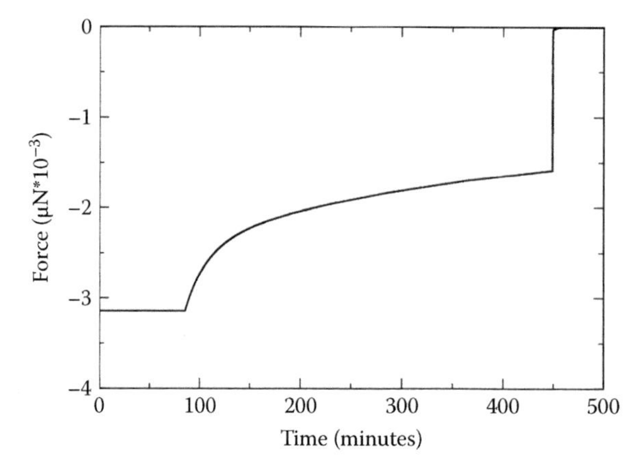

FIGURE 14.13 Plot showing the time evolution of the adhesion force of the same endothelial cell as the one in Figure 14.12 with the neighbors forming the vessel. It can be observed that the intensity of the adhesive forces decreases over time.

14.3.9 Discussion

In this section we studied the epithelial–mesenchymal transition that leads to cell detachment and invasion, and the subsequent intravasation, the process by which a malignant cell enters a blood vessel. We find that the interplay of biomechanical forces and the biochemical reactions that trigger their strength are able to explain the detachment of cells leading to both invasion and intravasation.

In TEM we assumed a morphogen release from the neighboring blood vessels that directs the movement of cells after they detached. We studied the case where chemotactic forces alone were insufficient to trigger a mechanical rupture of the cell–cell contacts necessary for cell detachment. We considered this case because the morphogen gradient is sensed by all cells at the tumor surface up to a certain penetration depth, so the forces would have to be very strong to pull out individual cells. In our model the strength of adhesion is controlled by the concentration of E-cadherin–β-catenin complexes in the cell membrane. We represented the β-catenin core module explicitly in our model to realistically mimic the possible key functions to which β-catenin contributes in the EMT (epithelial–mesenchymal transition). We find that an upregulation of β-catenin in the cytosol of a single cell (e.g., due to a defect in the degrading apparatus of β-catenin in the cytosol) can be sufficient to trigger a detachment cascade from other cells at the surface as a consequence of the inter- and intracellular dynamics of β-catenin.

In intravasation, a malignant cell arriving at the wall of a blood vessel under the influence of a chemotaxis force pushes against the endothelial wall and thereby induces the formation of N-cadherin-mediated bonds with the neighboring endothelial cells. As a consequence VE-cadherin-mediated bonds among endothelial cells are locally downregulated by a competition process that involves the Src enzyme. This competition weakens the strength of the contacts between the endothelial cells, which enables the cancer cell to squash in between endothelial cells—again under the influence of the chemotaxis force—and eventually enter the blood vessel. So here the interplay of the biochemical control of the strength of cell–cell contacts and the chemotaxis force permit the cancer cell to deploy its malignant potential.

14.4 Conclusion

In this chapter we presented model results of different steps from an *in situ* tumor to invasive cancer. These are angiogenesis, the process of formation of new blood vessels stimulated by tumor cells to ensure their oxygen and nutrient supply, and the detachment of cells from the tumor mass and their intravasation, the process by which they invade into blood vessels. In our models, which are all in three dimensions, the tumor cells were presented individually within agent-based models.

In the first part of this chapter we compared the growth kinetics of tumors for (1) oxygen and nutrient unlimited growth, (2) growth in a static vascular environment, and (3) growth if the vascular environment is remodeled by angiogenesis. Our model is able to explain the growth characteristics found in approximately spherically growing tumors such as multicellular spheroids, an experimental system used to mimic tumors in their avascular phase, and xenografts of NIH3T3 mouse fibroblast cells, which also have spherical shapes.

In the second part we studied the influence of intracellular biochemical pathways involved in cell–cell contact formation on the strength of cell–cell adhesion forces. We find that it is crucial to include the intracellular machinery that controls the cell–cell adhesion to mimic the experimental observations both during cell detachment from a tumor leading to tissue invasion and during intravasation. Hence our results stress the importance of taking into account both biomechanical and biochemical factors at the relevant scales of the different processes. Moreover, including the intracellular molecular regulation in the models permits the simulation of predictions that can be validated experimentally because cell parameters such as micromotility, strength of cell–cell adhesion or material parameters cannot be directly modified in the experiment while molecular concentrations can.

In all our simulations we used multiscale models. In the first part on angiogenesis we coupled individual-cell-based models with a continuous description of the concentration of oxygen, glucose, and growth factor. In the second part on the epithelial–mesenchymal transition and intravasation, we explicitly represented the key functions of the intracellular biochemical pathways that control the strength of cell–cell adhesion by a coupled system of ordinary differential equations for the concentration of the involved intracellular molecules, and coupled them directly to the strength of cell–cell forces, that is, a parameter on the cellular scale. In all our simulations we tried to choose realistic parameter values inferred from the literature—except in our model in the first section where we shifted the process of angiogenesis toward smaller tumor sizes to reduce computational expense.

We believe that simulations that permit an analysis of the development of cancer from a single precursor cell up to its size at the stage of clinical manifestation may soon become possible, even in quantitative agreement with experimental and clinical observations, provided that the necessary data on all respective scales will be made available. This might enable us to mimic the effect of therapies *in silico*, and guide experiments and tests to improve therapy schedules and therapies in the not so distant future.

Acknowledgments

DD's work is supported by EU grants CANCERSYS and PASSPORT, and BMBF grants LungSys under 0315415F and Hepatosys under 313074D. NJ acknowledges support by the EU network DEASE. DD, NJ, IVC, and WW contributed to modeling the avascular and vascularized tumor growth, DD and IRC to modeling the invasion and metastases.

References

[1] J.A. Adam and N. Bellomo (1997). *A Survey of Models for Tumor-Immune System Dynamics*, Boston, Birkhäuser.

[2] T. Alarćon, H.M. Byrne, and P.K. Maini (2005). A multiple scale model for tumor growth. *Multiscale Model. Simul.* 3(2):440–475.

[3] T. Alarćon, H.M. Byrne, and P.K. Maini (2004). A mathematical model of the effects of hypoxia on the cell cycle of normal and cancer cells. *J. Theor. Biol.* 229:395–411.

[4] M.S. Alber, M.A. Kiskowski, J.A. Glazier, and Y. Jiang (2002). On cellular automaton approaches to modeling biological cells. In *Mathematical Systems Theory in Biology, Communication, and Finance*, Eds. J. Rosenthal and D.S. Gilliam, New York, Springer.

[5] B. Alberts, A. Johnson, J. Lewis, M. Raff, K. Roberts, and P. Walker (2007). *Molecular Biology of the Cell*, New York, Garland Science Textbooks.

[6] J. Alcaraz, L. Buscemi, M. Grabulosa, X. Trepat, B. Fabry, R. Farre, and D. Navajas (2003). Microrheology of human lung epithelial cells measured by atomic force. *Biophys. J.* 84:2071–2079.

[7] D. Ambrosi and F. Mollica (2002). On the mechanics of a growing tumor. *Ing. J. Eng. Sci.* 40(12):1297–1316.

[8] A. Anderson (2000). Mathematical modeling of tumor invasion and metastasis. *J. Theor. Med.* 2:129–154.

[9] A.R.A. Anderson, M.A.J. Chaplain, and K.A. Rejniak (2007). *Single-Cell-Based Models in Biology and Medicine*, Basel, Birkhäuser.

[10] R.P. Araujo and D.L. McElwain (2004). A history of the study of solid tumour growth: the contribution of mathematical models. *Bull. Math. Biol.* 66:1039–1091.

[11] K. Bartha and H. Rieger (2006). Vascular network remodeling via vessel cooption, regression and growth in tumors. *J. Theor. Biol.* 241(4):903–918.

[12] G. Bergers and L.E. Benjamin (2003). Tumorigenesis and the angiogenic switch. *Nat. Rev. Cancer* 3(6):401–410.

[13] M. Block, E. Schoell, and D. Drasdo (2007). Classifying the growth kinetics and surface dynamics in growing cell populations. *Phys. Rev. Lett.* 99:248101.

[14] A. Bru, S. Albertos, J.L. Subiza, J.L. Garcia-Arsenio, and I. Bru (2003). The universal dynamics of tumor growth. *Biophys. J.* 85:2948–2961.

[15] H. Byrne and D. Drasdo (2009). Individual-based and continuum models of growing cell populations: a comparison. *J. Math. Biol.* 58(4–5):657–687.

[16] H. Byrne and L. Preziosi (2003). Modelling solid tumour growth using the theory of mixtures. *Math. Med. Biol.* 20:341–366.

[17] V. Capasso and D. Morale (2009). Stochastic modelling of tumour-induced angiogenesis. *J. Math. Biol.* 58(1-2):219–233.

[18] P. Carmeliet (2003). Angiogenesis in health and disease. *Nat. Med.* 9(6):653–660.

[19] M.A.J. Chaplain (1996). Avascular growth, angiogenesis and vascular growth in solid tumours: the mathematical modelling of the stages of tumour development. *Math. Comput. Modelling* 23(6):47–87.

[20] Y.T. Chen, D.B. Stewart, and W. Nelson (1999). Coupling assembly of the E-cadherin/β-catenin complex to efficient endoplasmic reticulum exit and basal-lateral membrane targeting of E-cadherin in polarized MDCK cells. *J. Cell Biol.* 144:687–699.

[21] S.E. Chesla, P. Selvaraj, and C. Zhu. (1998). Measuring two-dimensional receptor-ligand binding kinetics by micropipette. *Biophys. J.* 75:1553–1557.

[22] T.C. Cickovski, C. Huang, R. Chaturvedi, T. Glimm, H.G.E. Hentschel, M. Alber, J.A. Glazier, S.A. Newman, and J. A. Izaguirre (2005). A framework for three-dimensional simulation of morphogenesis. *IEEE/ACM TCBB* 2(3):273–288.

[23] J.C. Dallon and H.G. Othmer (2004). How cellular movement determines the collective force generated by the *Dictyostelium discoideum* slug. *J. Theor. Biol.* 231:203–222.

[24] S. Dormann and A. Deutsch (2002). Modeling of self-organized avascular tumor growth with a hybrid cellular automaton. *In Silico Biol.* 2:393–406.

[25] D. Drasdo (2003). On selected individual-based approaches to the dynamics of multicellular systems. In *Multiscale Modelling and Numerical Simulations*, Eds. W. Alt, M. Chaplain, M. Griebel, and J. Lenz, Basel, Birkhäuser, pp. 169–203.

[26] D. Drasdo (2005). Coarse graining in simulated cell populations. *Adv. Complex Syst.* 8(2-3):319–363.

[27] D. Drasdo and S. Höhme (2005). A single-cell based model to tumor growth *in-vitro*: monolayers and spheroids. *J. Phys. Biol.* 2:133–147.

[28] D. Drasdo, S. Höhme, and M. Block (2007). On the role of physics in the growth and pattern formation of multi-cellular systems: what can we learn from individual-cell based models? *J. Stat. Phys.* 128:287–345.

[29] D. Drasdo and S. Höhme (2003). Individual-based approaches to birth and death in avascular tumors. *Math. and Comp. Modelling* 37:1163–1175.

[30] D. Drasdo, R. Kree, and J.S. McCaskill (1995). Monte Carlo approach to tissue-cell populations. *Phys. Rev. E* 52(6):6635–6657.

[31] M. Eden (1961). A two-dimensional growth process. In *Proc. 4th. Berkeley Symposium on Mathematics and Probability*, Ed. J. Neyman, Vol. IV, pp. 223–239, Berkeley, University of California Press.

[32] N. Ferrara, H.P. Gerber, and J. LeCouter (2003). The biology of VEGF and its receptors. *Nat. Med.* 9(6):669–676.

[33] J.P. Freyer and R.M. Sutherland (1985). A reduction in the *in situ* rates of oxygen and glucose consumption of cells in EMT6/Ro spheroids during growth. *J. Cell. Physiol.* 124:516–524.

[34] J.P. Freyer and R.M. Sutherland (1986). Regulation of growth saturation and development of necrosis in EMT6/Ro multicellular spheroids by the glucose and oxygen supply. *Cancer Res.* 46:3504–3512.

[35] Y.C. Fung (1990). *Biomechanics: Motion, Flow, Stress, and Growth*, New York, Springer.

[36] J. Galle, G. Aust, G. Schaller, T. Beyer, and D. Drasdo (2006). Single-cell based mathematical models to the spatio-temporal pattern formation in multi-cellular systems. *Cytometry Teil A* 69A:704–710.

[37] J. Galle, M. Loeffler, and D. Drasdo (2005). Modelling the effect of deregulated proliferation and apoptosis on the growth dynamics of epithelial cell populations *in vitro*. *Biophys. J.* 88:62–75.

[38] D.T. Gillespie (1977). Exact stochastical simulations of coupled chemical reactions. *J. Phys. Chem.* 81(25):2340–2361.

[39] F. Graner and J. Glazier (1992). Simulation of biological cell sorting using a two-dimensional extended potts model. *Phys. Rev. Lett.* 69(13): 2013–2016.

[40] D. Hanahan and J. Folkman (1996). Patterns and emerging mechanisms of the angiogenic switch during tumorigenesis. *Cell* 86(3):353–364.

[41] G. Helmlinger, P.A. Netti, H.C. Lichtenfeld, R.J. Melder, and R.K. Jain (1997). Solid stress inhibits the growth of multicellular tumor spheroids. *Nat. Biotech.* 15(8):778–783.

[42] P. Hogeweg (2000). Evolving mechanisms of morphogenesis: on the interplay between differential adhesion and cell differentiation. *J. Theor. Biol.* 203:317–333.

[43] O. Huber, C. Bierkamp, and R. Kemler (1996). Cadherins and catenins in development. *Curr. Opin. Cell Biol.* 8:685–691.

[44] C. Jamora, R. DasGupta, P. Koclenlewski, and E. Fuchs (2003). Links between signal transduction, transcription and adhesion in epithelial bud development. *Nature* 422:317–322.

[45] Y Jiang, J. Pjesivac-Grbovic, C. Cantrell, and J.P. Freyer (2005). A multiscale model for avascular tumor growth. *Biophys. J.* 89(6):3884–3894.

[46] R. Kemler, A. Hierholzer, B. Kanzler, S. Kuppig, K. Hansen, M.M. Taketo, W.N. de Vries, B.B. Knowles, and D. Solter (2004). Stabilization of β-catenin in the mouse zygote leads to premature epithelial mesenchymal transition in the epiblast. *Development* 131: 5817–5824.

[47] Y. Kim, M.A. Stolarska, and H.G. Othmer (2007). A hybrid model for tumor spheroid growth *in vitro*. I. Theoretical development and early results. *Math. Models Meth. Appl. Sci.* 17:1773–1798.

[48] D.-S. Lee and H. Rieger (2006). Flow correlated percolation during vascular remodeling in growing tumors. *Phys. Rev. Lett.* 96:058104.

[49] E. Lee, A. Salic, R. Kru, M.W. Kirschner, and R. Heinrich (2003). The roles of APC and axin derived from experimental and theoretical analysis of the Wnt pathway. *PLoS Biol.* 1:116–132.

[50] L. Preziosi (2003). *Cancer Modelling and Simulation.* Boca Raton, FL, Chapman & Hall/CRC.

[51] P. Macklin, S. McDougall, A.R. Anderson, M.A. Chaplain, V. Cristini, and J. Lowengrub (2009). Multiscale modelling and nonlinear simulation of vascular tumour growth. *J. Math. Biol.* 58(4-5):765–798.

[52] R.E. Mahaffy, C.K. Shih, F.C. McKintosh, and J. Käs (2000). Scanning probe-based frequency-dependent microrheology of polymer gels and biological cells. *Phys. Rev. Lett.* 85:880–883.

[53] N.V. Mantzaris, S. Webb, and H.G. Othmer (2004). Mathematical modeling of tumor-induced angiogenesis. *J. Math. Biol.* 49(2):111–187.

[54] D. Marmé and N. Fusenig Adam (2008). *Tumor Angiogenesis*, Berlin, Springer.

[55] S.R. McDougall, A.R. Anderson, and M.A. Chaplain (2006). Mathematical modelling of dynamic adaptive tumour-induced angiogenesis: clinical implications and therapeutic targeting strategies. *J. Theor. Biol.* 241(3): 564–589.

[56] R.M.H. Merks and J.A. Glazier (2005). A cell-centered approach to developmental biology. *Physica A* 352:113–130.

[57] E. Mira, R.A. Lacalle, C. Gomez-Mouton, E. Leonardo, and S. Manes (2004). Quantitative determination of tumor cell intravasation in a real-time polymerase chain reaction-based assay. *Clin. Exp. Metastasis* 19:313–318.

[58] J. Moreira and A. Deutsch (2002). Cellular automata models of tumour development—a critical review. *Adv. Complex Syst.* 5:247–267.

[59] W.J. Mueller-Klieser (1987). A review on cellular aggregates in cancer research. *Cancer Res. Clin. Oncol.* 113(2):101–122.

[60] T. Newman (2005). Modeling multi-cellular systems using sub-cellular elements. *Math. Biosci. Eng.* 2:613–624.

[61] M.R. Owen, T. Alarćon, P.K. Maini, and H.M. Byrne (2009). Angiogenesis and vascular remodelling in normal and cancerous tissues. *J. Math. Biol.* 58(4-5):689–721.

[62] E. Palsson and H.G. Othmer (2000). A model for individual and collective cell movement in dictyostelium discoideum. *Proc. Natl. Acad. Sci. USA* 12(18):10448–10453.

[63] L. Preziosi and A. Tosin (2009). Multiphase modelling of tumour growth and extracellular matrix interaction: mathematical tools and applications. *J. Math. Biol.* 58(4-5):625–656.

[64] J. Qi, N. Chen, J. Wang, and C.-H. Siu (2005). Transendothelial migration of melanoma cells involves N-cadherin-mediated adhesion and activation of the β-catenin signalling pathway. *Mol. Biol. Cell* 16:4386–4397.

[65] J. Qi, J. Wang, O. Romanyuk, and C. Siu (2006). Involment of Src family kinases in N-cadherin phosphorylation and β-catenin dissociation during transendothelial migration of melanoma cells. *Mol. Biol. Cell* 17:1261–1272.

[66] M. Radszuweit, M. Block, J.G. Hengstler, E. Schoell, and D. Drasdo (2009). Comparing the growth kinetics of cell populations in two and three dimensions. *Phys. Rev. E.* 79(5): 051907.

[67] I. Ramis-Conde, M.A.J. Chaplain, A.R.A. Anderson, and D. Drasdo (2009). Multi-scale modelling of cancer cell intravasation: the role of cadherins in metastasis. *J. Phys. Biol.* 6(1):16008–16020.

[68] I. Ramis-Conde, D. Drasdo, M.A.J. Chaplain, and A.R.A. Anderson (2008). Modelling the influence of the E-cadherin–β-catenin pathway in cancer cell invasion: a multi-scale approach. *Biophys. J.* 95:155–165.

[69] D. Ribatti, A. Vacca, and F. Dammacco (2003). New non-angiogenesis dependent pathways for tumour growth. *Eur. J. Cancer* 39(13):1835–1841.

[70] T. Roose, S.J. Chapman, and P.K. Maini (2007). Mathematical models of avascular tumor growth. *SIAM Rev.* 49:179–208.

[71] G. Schaller and M. Meyer-Hermann (2005). Multicellular tumor spheroid in an off-lattice Voronoi-Delaunay cell model. *Phys. Rev. E* 71:051910.

[72] I.B. Schiffer, S. Gebhard, C.K. Heimerdinger, A. Heling, J. Hast, U. Wollscheid, B. Seliger, B. Tanner, S. Gilbert, T. Beckers, S. Baasner, W. Brenner, C. Spangenberg, D. Prawitt, T. Trost, W.G. Schreiber, B. Zabel, M. Thelen, H.A. Lehr, F. Oesch, and J.G. Hengstler (2003). Switching off HER-2/neu in a tetracycline-controlled mouse tumor model leads to apoptosis and tumor-size-dependent remission. *Cancer Res.* 63:7221–7231.

[73] A. Stevens (2000). The derivation of chemotaxis equations a limit dynamics of moderately interacting stochastic many particles systems. *SIAM J. Appl. Math.* 61(1):172–182.

[74] E.L. Stott, N.F. Britton, J.A. Glazier, and M. Zajac (1999). Stochastic simulation of benign avascular tumor growth using the potts model. *Math. Comput. Model.* 30:183–198.

[75] P. Vaupel (2004). Tumor microenvironmental physiology and its implications for radiation oncology. *Semin. Radiat. Oncol.* 14(3):198–206.

[76] J.P. Ward and J.R. King (1997). Mathematical modelling of avascular-tumor growth. *IMA J. Math. App. Med. Biol.* 14:39–69.

[77] R.A. Weinberg (2007). *The Biology of Cancer*, New York, Garland Science.

[78] M. Welter, K. Bartha, and H. Rieger (2008). Emergent vascular network inhomogeneities and resulting blood flow patterns in a growing tumor. *J. Theor. Biol.* 250(2):257–280.

[79] S.M. Wise, J.S. Lowengrub, H.B. Frieboes, and V. Cristini (2008). Three-dimensional multispecies nonlinear tumor growth model and numerical method (2008). *J. Theor. Biol.* 253(3):524–543.

[80] A.S.T. Wong and B.M. Gumbiner (2003). Adhesion independent mechanism for suppression of tumour cell invasion by E-cadherin. *J. Cell Biol.* 161:1191–1203.

Chapter 15

Delaunay Object Dynamics for Tissues Involving Highly Motile Cells

Tilo Beyer and Michael Meyer-Hermann

Contents

15.1 Introduction

At present a variety of methods exist to investigate and simulate the organization of cellular compounds with different levels of spatial and temporal resolution. Delaunay object dynamics (DOD) provides a platform to model different systems on a common physical basis [7,34]. The biophysics of multicellular systems, although influenced by details of the biological system in consideration, is the guiding principle common to different species or tissues.

Within the DOD platform, only the specific properties of cells in a given system have to be adapted. It allows us to separate the phenotype of a cell seen in experiments (surface markers, gene expression patterns, signaling cascades) from more universal biophysical phenotypes (such as speed, cell division, surface area, adhesion). The latter phenotypes can be mapped to physical forces acting on cells. In analogy with molecular dynamics methods, a reference "force field" representing classes of biophysical interactions forms the basis of a DOD simulation. Assuming that, on the qualitative level, biophysical properties of cells are universal, the DOD platform can be refined independently of the considered biological system. The biophysical parameters will differ on the quantitative level only, which are easily adapted.

The DOD simulation is applied to secondary lymphoid tissues (SLTs) [14,62], which only recently gained modelers' interest [8]. SLTs develop shortly after birth and are the sites of interaction of immune effectors with pathogens. Several immune responses are triggered in this tissue, including the generation of specific antibodies. In contrast with the majority of other multicellular systems, SLTs are dominated by fast migrating cells. Stable structures emerge despite the cell population exchanged by cell influx and efflux on a daily basis. A simulation of SLT ontogenesis is presented here and the effects of the biophysical cell features on pattern formation are highlighted.

15.2 Delaunay Object Dynamics

The DOD framework is designed to cover contact-dependent forces acting between cells [7,34]. It is based on a weighted Delaunay triangulation, an extension of the normal Delaunay triangulation [17,41]. The triangulation provides the cell neighborhood topology, which might change rapidly due to fast cell migration. The topology determines the interaction partners—or field of view—of a cell, permitting us to determine the local environment of every cell independent of its position, size, and the cells surrounding it. The dual Voronoi tessellation [2,41] can provide information about the shape of a cell as defined by cell contact surfaces and volume [51]. The contact area determines the strength of the adhesion and friction forces between cells and allows us to describe contact-dependent signals in terms of receptor–ligand pairs. Immunological tissue is a prime example where nonadhesive contact-dependent signals also play a major role [22].

15.2.1 Weighted Delaunay Triangulation

In DOD each cell i is represented as a sphere at position $\mathbf{x}_i$ with radius R_i. For the purpose of the weighted Delaunay triangulation, each cell corresponds to a vertex that is defined by the pair $X_i = (\mathbf{x}_i, R_i)$. The weighted Delaunay triangulation is defined using the empty orthosphere criterion [2,17,41]. In three dimensions, four vertices X_k, X_l, X_m, X_n forming a tetrahedron uniquely define an orthosphere. The orthosphere is empty if for any vertex $X_i \neq X_k, X_l, X_m, X_n$

$$\begin{vmatrix} x_{k,1} - x_{i,1} & x_{k,2} - x_{i,2} & x_{k,3} - x_{i,3} & \|\mathbf{x}_k - \mathbf{x}_i\|^2 - R_k^2 + R_i^2 \\ x_{l,1} - x_{i,1} & x_{l,2} - x_{i,2} & x_{l,3} - x_{i,3} & \|\mathbf{x}_l - \mathbf{x}_i\|^2 - R_l^2 + R_i^2 \\ x_{m,1} - x_{i,1} & x_{m,2} - x_{i,2} & x_{m,3} - x_{i,3} & \|\mathbf{x}_m - \mathbf{x}_i\|^2 - R_m^2 + R_i^2 \\ x_{n,1} - x_{i,1} & x_{n,2} - x_{i,2} & x_{n,3} - x_{i,3} & \|\mathbf{x}_n - \mathbf{x}_i\|^2 - R_n^2 + R_i^2 \end{vmatrix} > 0 \quad (15.1)$$

holds, provided that the four vertices are oriented positively, that is, if

$$\begin{vmatrix} x_{k,1} & x_{k,2} & x_{k,3} & 1 \\ x_{l,1} & x_{l,2} & x_{l,3} & 1 \\ x_{m,1} & x_{m,2} & x_{m,3} & 1 \\ x_{n,1} & x_{n,2} & x_{n,3} & 1 \end{vmatrix} > 0 \quad (15.2)$$

$\mathbf{x}_i = (x_{i,1}, x_{i,2}, x_{i,3})$ are the position coordinates of the vertex X_i.

A set of nonoverlapping tetrahedrons covering a set of vertices forms a Delaunay triangulation if all orthospheres attributed to these tetrahedrons do not contain any further vertices. The Delaunay triangulation, and with it the neighborhood topology, change when cells are moving, and when they are added or removed from the system due to flux, death, or division. The algorithms that allow a local adaptation of the triangulation in response to these processes have been developed and published previously [9,52]. Thus, a total re-triangulation after a simulation time step is not necessary.

15.2.2 Geometry of the Cell

The mechanical interaction of cells relies on geometric variables such as distance, contact surface, and volume. These quantities can be drawn from the weighted Delaunay triangulation using its dual graph, the Voronoi tessellation [2,41,52]. The calculation is based on the decomposition of the Voronoi polyhedra into triangles. The corners of the polyhedra are identical to the centers of the orthosphere (Figure 15.1). Herefrom the center-contact distances, the contact areas, and the volume of a cell follow.

The Voronoi tessellation is not an optimal approximation for the cell shape under all circumstances. If cells are not densely packed—thus, not in physical contact—the Voronoi contact surface and volume would grow to artificially large values with increasing distance between the cells. A better approximation is to compute the contact area as the minimum of the sphere overlap and the Voronoi face (Figure 15.2). Both are located in the same plane, thus not changing the position or orientation of the cell contact. The Voronoi contact area is the better description for high density of cells because it takes into

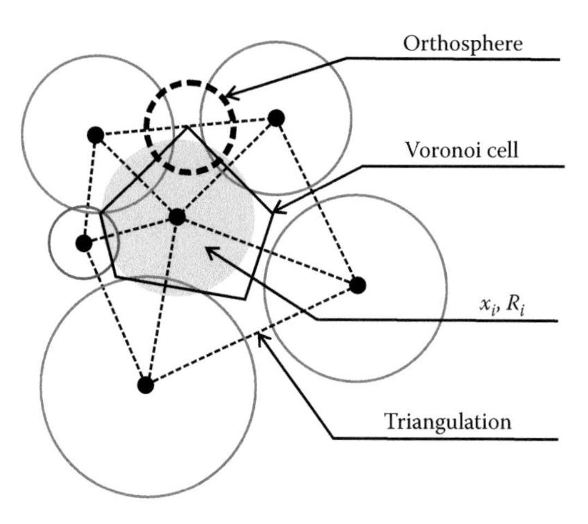

FIGURE 15.1 Duality of the Delaunay triangulation and the Voronoi tessellation. The Voronoi cell (black polygon) of a sphere representing the cell at position $\mathbf{x}_i$ with radius R_i (gray disk) is determined by the orthospheres (dashed circle) of the triangulation (dashed lines).

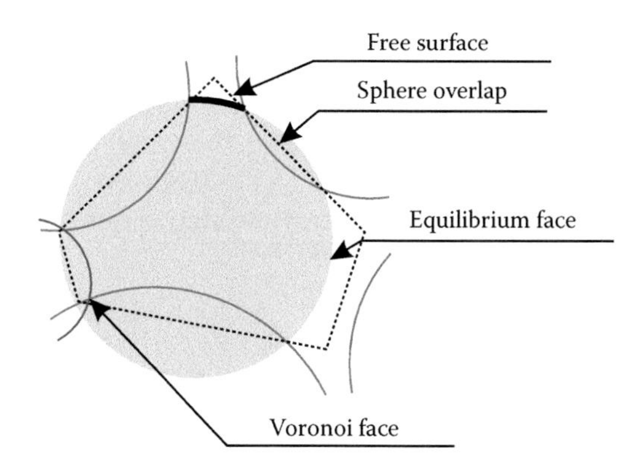

FIGURE 15.2 Different surface contributions of a cell. For physical cell contacts, the minimum between Voronoi surface area and the sphere overlap is used. This choice can result in small parts of the cellular surface being in contact with the medium (thick black arc, "free surface"). The integral of these parts of the cell membrane is taken into account to model the part of the surface in contact with the medium. For parts of the cell lacking physical contact with a distant cellular neighbor, the overlap of a sphere within equilibrium distance (the equilibrium face) is used to calculate the contact surface.

account many-body configurations. The virtual overlap of the spheres is the better approximation for low-density systems and becomes zero in the limit of cells losing physical contact to each other [51,53]. Thus, the minimum of the two contact areas turns out to be close to the real contact of two elastically deforming spherical cells.

If the sphere overlap becomes zero and cells lose contact, the cell is considered in contact with the medium. The contact surface with the medium is calculated as the minimum of the Voronoi face and the sphere overlap with an equal-sized virtual cell of the same kind at equilibrium distance (Figure 15.2). The equilibrium distance is defined with respect to the Johnson-Kendall-Roberts forces [26] (introduced below). It is assumed that the cell's contact surface with the medium corresponds to a relaxed cell configuration. The intermediate state where cells are in contact with other cells but a part of the surface is still in contact with the medium (Figure 15.2) is not resolved in the model. Only the total surface in contact with the medium can be estimated by calculating the difference between the ideal sphere surface and the sum of all contact areas.

A similar argument is applied to the volume of cells, which is defined as the minimum of the sphere volume and the Voronoi cell volume.

It is not required to use the Voronoi tessellation for the geometry of the cell. It is also possible to use the Delaunay triangulation only for neighborhood relationships and use an independent description of the cell shape. Note that the neighborhood topology limits the complexity of cell shapes. For example, dendritic or strongly elongated shapes cannot be described with the Delaunay triangulation.

15.2.3 Equations of Motion

The dynamics of cells in the DOD platform are determined by the forces acting on and generated by cells in physical contact. The method is as flexible in the choice of forces as are molecular dynamics methods. A generic decomposition of force components is provided in this section.

The underlying equations are Newtonian equations of motion in the over-damped approximation [15,53]. In the approximation, acceleration of cells and conservation of moment are neglected and therefore the inertial term can be omitted. The equation of motion for the cell i at position $\mathbf{x}_i$ then reads

$$\begin{aligned} 0 &\approx m_i \ddot{\mathbf{x}}_i \\ &= \mathbf{F}_i^{act}(\phi_i) + \mathbf{F}_i^{drag}\left(\dot{\mathbf{x}}_i, \{\dot{\mathbf{x}}_j\}_{\mathcal{N}_i^c}\right) + \sum_{j \in \mathcal{N}_i} \mathbf{F}_{ij}^{act}(\phi_i) \\ &\quad + \sum_{j \in \mathcal{N}_i^c} \left[-\mathbf{F}_{ji}^{act}(\phi_j) + \mathbf{F}_{ij}^{pass}(\mathbf{x}_i, \mathbf{x}_j)\right] \end{aligned} \tag{15.3}$$

The various contributions to the force acting on cell i are listed below and further explained in the subsequent sections (Figure 15.3).

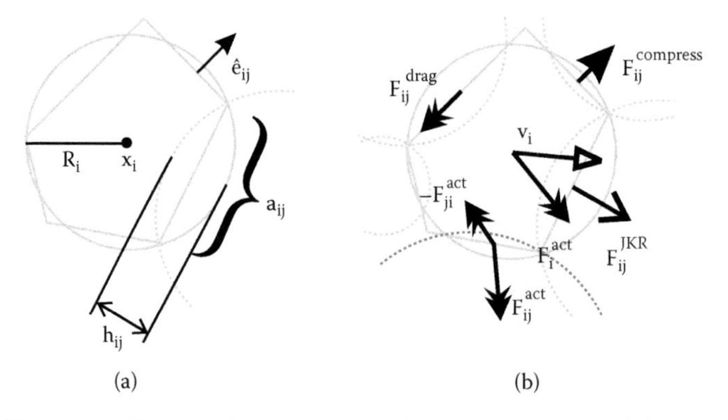

FIGURE 15.3 Essential parameters of the DOD model. (a) The geometry of the cell is determined by its position $\mathbf{x}_i$ and radius R_i. Depending on the position of the surrounding cells j, the contact area a_{ij} and virtual cell/sphere overlap h_{ij} are calculated. In addition the contact area is characterized by the normal vector $\hat{\mathbf{e}}_{ij}$. (b) The normal vector $\hat{\mathbf{e}}_{ij}$ determines the direction of the passive forces between cells $\mathbf{F}_{ij}^{compress}$ and $\mathbf{F}_{ij}^{JKR}$. The forces $\mathbf{F}_{ij}^{act}$ generated by the cytoskeleton of the cell i act either on neighbor cells or the medium. Forces $\mathbf{F}_{ji}^{act}$ generated by cell j act on cell i only when both cells are in physical contact. Cytoskeletal forces acting on the medium are summarized into one vector $\mathbf{F}_i^{act}$. The speed $\mathbf{v}_i$ of the cell results from the balance of all the aforementioned forces with the drag forces $\mathbf{F}_{ij}^{drag}$, which act tangentially on each contact surface, and from the resistance $\mathbf{F}_{i,med}^{drag}$ of the medium (not shown).

1. The active forces $\mathbf{F}_i^{act}$ and $\mathbf{F}_{ij}^{act}$ on a cell i depend only on the phenotype ϕ_i of cell i and the phenotypes, ϕ_j of the interaction partners. ϕ_i is a state vector of the cell's internal degrees of freedom. The details of ϕ_i define the biological system under consideration. The degrees of freedom may include surface molecule concentrations, cell cycle status, or cell-type or cell-specific parameters. Variables of the state vector ϕ_i can be dynamic or static. The dynamics of some internal degrees of freedom may, for example, depend on cytoskeletal changes. All forces resulting from cytoskeletal rearrangements are summarized as active forces. The forces can be exerted on neighbor cells or to the surrounding medium. Therefore the set $\mathcal{N}_i$ includes all cells j that are next neighbors in the Delaunay triangulation, independent of whether they are in physical contact with the cell i. This accounts for forces acting on the noncellular environment. In addition, the active forces $\mathbf{F}_{ji}^{act}(\phi_j)$ of neighbor cells act on cell i. But in this case, only the set of cells $j \in \mathcal{N}_i^c$ in physical contact to cell i are included.
2. Passive forces $\mathbf{F}^{pass}$ depend on the positions of the cell i and those neighbor cells $j \in \mathcal{N}_i^c$ in physical contact with cell i. There are two contributions to this force resulting from the cell's deformability

and compressibility:

$$\mathbf{F}_{ij}^{pass}(\mathbf{x}_i, \mathbf{x}_j) = \mathbf{F}_{ij}^{JKR}(\mathbf{x}_i, \mathbf{x}_j) + \mathbf{F}_{ij}^{compress}(\mathbf{x}_i, \mathbf{x}_j) \quad (15.4)$$

where $\mathbf{F}_{ij}^{JKR}$ contains elastic interactions and adhesion (see Equation (15.6)).

3. Passive and active forces are counterbalanced by drag forces $\mathbf{F}^{drag}$. They depend on the velocity $\dot{\mathbf{x}}_i$ of the cell i and the velocities $\{\dot{\mathbf{x}}_j\}_{\mathcal{N}_i^c}$ of the neighbors cells j being in physical contact with cell i.

15.2.4 Active Force of Migrating Cells

Cells are not only subject to forces exerted on them by their environment, but are themselves major sources for mechanical stress due to the dynamics of their cytoskeletons. One cause for cell-generated mechanical stress is migration. Lymphocyte migration is most relevant for the application of DOD to immunological tissue, and thus a model is needed to show how cells exert forces on their surroundings.

It is known that during movement, cells do exert strong forces perpendicular to their direction of motion [43,58]; these forces can even exceed the net force in the direction of cell migration [43]. Immune cells use a migration mode that depends on the so-called constriction ring [29,39,46,65]. Cells use a ring formed by their cytoskeleton as an anchor to the extracellular matrix (ECM). The contraction of the rear of the cell generates outward-directed pressure conveyed to the environment.The cell squeezes itself through the ring, which remains fixed with respect to the ECM. In a simplified modeling approach [7], a cell forms only one ring that is traveling to the end of the cell before a new ring forms at the front. The force acting on cell i by exerting active forces on a neighbor cell j is given by

$$\mathbf{F}_{ij}^{act} = a_{ij} p_i^* \, sign[(\mathbf{x}_{ij}^* - \mathbf{x}_i^*) \cdot \mathbf{o}_i] \, \frac{\mathbf{x}_{ij}^* - \mathbf{x}_i^*}{\|\mathbf{x}_{ij}^* - \mathbf{x}_i^*\|} \quad (15.5)$$

with the cell polarization vector $\mathbf{o}_i$, cell surface contact point $\mathbf{x}_{ij}^*$, constriction ring center $\mathbf{x}_i^*$ (which generally differs from the cell's center), interaction area a_{ij}, and the pressure p_i^* generated by the cell i cytoskeleton.

In addition to the forces resulting from the ring, a constant active force $-\mathbf{F}_i^{act}(\phi_i)$ is directly exerted on the ECM opposite to the direction of the cell's orientation $\mathbf{o}_i$, that is, the cell is pushed forward against the ECM by the force $\mathbf{F}_i^{act}(\phi_i)$ in the direction $\mathbf{o}_i$. This additional force may be generated either by the integrated pressure of the constriction ring model directly exerted to the ECM or by the filopodia dynamics of the classical multistep migration model (reviewed in [28,36]). Recent data demonstrate that both migration modes are active in immune cells [68] although the contribution of the constriction ring seems to be greater during chemotactic responses of migrating lymphocytes [20,68].

The polarization $\mathbf{o}_i$ of the cell is part of its phenotype ϕ_i. It is dynamic and therefore the direction of the migration and the active forces are also dynamic. In the model the persistence time T_p, which describes how long the cell maintains its orientation [1,35], is used. The cell reorients according to a sufficiently strong gradient of a chemotactic stimulus [47] only if the persistence time has past since the last reorientation event. Without sufficient chemotactic stimuli, cells orient randomly and perform a persistent random walk. This model could be extended by including the texture of the ECM and coupling the direction of motion to the direction of fibers that guide lymphocyte movement [3].

15.2.5 Adhesive and Elastic Cell–Cell Interaction

Mechanical parameters of cells are typically measured either in suspension in which cells adopt a spherical shape or—in the case of fibroblast-like cells—spread on substrates [5,13,37]. In most cases both conditions deviate strongly from the *in situ* situation. The Johnson-Kendall-Roberts (JKR) model [26] is applicable to the spherical cells in suspension [13] and therefore may also apply to immune cells, which belong to the few cells that show roughly spherical shapes in the tissue [35]. Cells in the present simulation are treated as soft elastic adhesive spheres according to the JKR model. Other cell types may require different mechanical models for an adequate description [60].

The JKR force is typically formulated independent of the contact area between two spheres [26]. However, the contact area derived in the DOD framework is a result of a multibody configuration of cells and is not directly related to the forces acting between two cells. In particular, when cells are densely packed and get closer to each other, the contact area decreases such that the repulsive force by the JKR model would decrease instead of the required increase of repulsion. Thus, the JKR model is rewritten based on the virtual cell overlap $h_{ij} = R_i + R_j - \|\mathbf{x}_i - \mathbf{x}_j\|$ where R_i and R_j are the cell radii:

$$\begin{aligned} F_{ij}^{JKR}(\mathbf{x}_i, \mathbf{x}_j) &= E_{ij}^* \sqrt{R_{ij}^*}\, h_{ij}^{3/2} - \sqrt{6\pi\sigma_{ij} E_{ij}^* {R_{ij}^*}^{3/2} h_{ij}^{3/2}} \\ \frac{1}{E_{ij}^*} &= \frac{3}{4}\left[\frac{1-\nu_i^2}{E_i} + \frac{1-\nu_j^2}{E_j}\right] \\ \frac{1}{R_{ij}^*} &= \frac{1}{R_i} + \frac{1}{R_j} \end{aligned} \tag{15.6}$$

with the cell's elastic moduli E_i and E_j, Poisson numbers ν_i and ν_j, and the surface energy σ_{ij}. The force acts in the direction of the normal $\hat{\mathbf{e}}_{ij}$ on the contact face:

$$\mathbf{F}_{ij}^{JKR}(\mathbf{x}_i, \mathbf{x}_j) = F_{ij}^{JKR}(\mathbf{x}_i, \mathbf{x}_j)\hat{\mathbf{e}}_{ij} \tag{15.7}$$

The approximation using the virtual cell overlap treats attachment and detachment symmetrically, while in reality one would expect different relations

[51,67]. To account for this difference would require the calculation of the relative movement between each cell pair in every time step. It is not clear, *a priori*, which neighbor cells will detach or attach. This depends on the movement of all cells calculated from the equations of motion (Equation (15.3)). An iterative procedure is needed to determine whether cells are attaching or detaching, and the equations of motion have to be solved again until a consistent solution is found [44]. Solving the equations under these conditions is a nontrivial task. Concerning the experimental situation (i.e., the available data and the little contribution of adhesion molecules to immunological tissue organization [27]), this seems to be an unnecessary effort.

The JKR model treats adhesion differently from the common biological terminology [10,18]. In physical terms it is a reversible process while in biology many phenomena imply active changes by the cell (e.g., affinity changes of integrins). The attachment of a cell to a substrate by means of adhesion requires additional forces. The forces generated by the adhesive contact are not sufficient to drive the required deformation of the cell [18]. The required additional forces can be active forces of the cell generated by the cytoskeleton or outside forces acting on the cell (e.g., the experimentalist pushing the cell). On the contrary, the forces required to detach the cell can be sufficiently high that the cell may break before contact with the substrate is lost. Thus, the JKR model can only be applied in situations in which adhesion forces do not dominate the system behavior. For tissues that strongly depend on adhesion (such as endothelial layers), the JKR force will be replaced with more appropriate descriptions within the DOD framework.

15.2.6 Many-Cell Interactions

As pointed out in the previous section the JKR model cannot properly account for many-body interactions. In tissue, however, this is the normal situation. Therefore using the JKR force alone can impose a problem in certain situations. The JKR model defines an equilibrium distance of two cells induced by the balance of surface energy and elastic repulsion. For two cells, this leads to a negligible deviation of the volume compared to the (conserved) real volume of a cell. However, in the case that a cell is strongly compressed by its surrounding cells, a corresponding large relaxing force is not generated by the JKR model. Thus, cells might remain in a highly compressed state for too long because many-body interactions are neglected. To account for this and approximately ensure volume conservation, a cell pressure concept is included [15,53]. The pressure of cell i is calculated as deviation of the actual (simulated) cell volume V_i from the target (real cell in relaxed state) volume V_i^*

$$P_i = K_i \left(1 - \frac{V_i}{V_i^*}\right) \tag{15.8}$$

$$K_i = \frac{E_i}{3(1 - 2\nu_i)} \tag{15.9}$$

where a linear compression model with compressibility K_i is used. The forces resulting from this pressure are exerted between cells by adding the term

$$\mathbf{F}_{ij}^{compress} = a_{ij}(P_i - P_j)\hat{\mathbf{e}}_{ij} \tag{15.10}$$

to the passive cell forces $\mathbf{F}_{ij}^{pass}(\mathbf{x}_i, \mathbf{x}_j)$ (Equation (15.4)). a_{ij} is the contact area of the cell with neighbor cells or the medium. This effective many-cell interaction force vanishes in the limit of low-density cells.

15.2.7 Friction Forces

The motion of cells is damped by friction forces that arise from several sources. Adhesion molecules impose resistance to cellular movement when dragged laterally along the plasma membrane. Also the unbinding and rebinding of adhesion molecules to their ligands on other cells and the ECM effectively generate a friction force. In addition to specific adhesion, a number of cell membrane components can also interact and contribute to friction [10].

Similar to active forces, the friction forces are split into two components. One component represents the interaction with the environment (e.g., the ECM):

$$\mathbf{F}_{med,i}^{drag} = -\gamma_{med}\,\dot{\mathbf{x}}_i \tag{15.11}$$

The linear velocity dependence is justified by the low Reynolds numbers of cell migration and the overdamped approximation [15,45,53]. The second component describes the friction between two cells i and j in contact:

$$\mathbf{F}_{ij}^{drag} = \gamma_{ij}[\dot{\mathbf{x}}_{ij} - \hat{\mathbf{e}}_{ij}(\hat{\mathbf{e}}_{ij} \cdot \dot{\mathbf{x}}_{ij})] \tag{15.12}$$

which depends on the relative cell velocity $\dot{\mathbf{x}}_{ij} = \dot{\mathbf{x}}_j - \dot{\mathbf{x}}_i$. Only the component of the relative velocity tangential to the contact surface is taken into account (Figure 15.3). The restriction on the tangential part is justified because a traversed part would contradict the energy-conserving nature of the JKR model (Equation (15.6)).

The friction coefficient γ_{ij} has the dimension of viscosity times length scale and is symmetric ($\gamma_{ij} = \gamma_{ji}$). γ_{ij} is chosen proportional to the contact area a_{ij} between the cells. The overall drag force in Equation (15.3) is then given by

$$\begin{aligned}\mathbf{F}_i^{drag} &= \mathbf{F}_{med,i}^{drag} + \sum_{j\in\mathcal{N}_i} \mathbf{F}_{ij}^{drag} \\ &= -\eta_{med} R_i \left(1 - \frac{A_i}{A_i^{tot}}\right)\dot{\mathbf{x}}_i + \sum_{j\in\mathcal{N}_i^c} (\eta_i R_i + \eta_j R_j)\frac{a_{ij}}{A_i^{tot}}[\dot{\mathbf{x}}_{ij} - \hat{\mathbf{e}}_{ij}(\hat{\mathbf{e}}_{ij} \cdot \dot{\mathbf{x}}_{ij})]\end{aligned} \tag{15.13}$$

with medium viscosity η_{med} and the cell-specific viscosities η_i. $A_i = \sum_{j\in\mathcal{N}_i^c} a_{ij}$ is the surface in contact with other cells. $A_i^{tot} = \sum_{j\in\mathcal{N}_i} a_{ij}$ is the total surface of a cell. The specific form of the friction coefficients is motivated by the Stokes relation for the friction of a sphere at velocity $\dot{\mathbf{x}}$ in a medium with viscosity η: $\mathbf{F}^{Stokes} = 6\pi\eta R\dot{\mathbf{x}}$.

15.2.8 Three-Level Delaunay Object Dynamics

The concrete phenotype ϕ_i and its dynamics depend on the multicellular system under consideration. In general they will be influenced either by the contact interaction with neighbor cells, which can be described using the contact face a_{ij} introduced above, or by long-range communication using secreted molecules. The simplest way of communication via molecules is diffusion. The DOD simulation is completed with an underlying grid to solve reaction–diffusion equations. In particular, such a grid is used for chemotactic factors that influence the polarization $\mathbf{o}_{ij}$ of a cell. Thus, in total, DOD covers three scales to simulate biological systems:

1. The level of highest resolution is the representation of the single cell phenotype ϕ_i, which specifies how cells adapt their behavior to the environment.
2. The second level represents biophysical and contact-dependent interactions of cells based on the weighted Delaunay triangulation.
3. The third level describes long-distance communication via diffusive substances.

15.3 DOD Simulation of Secondary Lymphoid Tissue

SLTs (secondary lymphoid tissues) are important sites for hosting immune cells and provide a platform for immune reactions [21]. To understand their function, the development of the unique organization of SLT has been studied using DOD simulations [7,8]. The striking difference between SLTs and the majority of other tissues that have gained attention is their more fluid-like nature. Most of the volume of lymphoid tissue is occupied by highly motile and fast cells, which are predominantly lymphocytes along with other immune cells. The motility of these cells is comparable to the migration pattern seen in the slug stage of *Dictyostelium discoideum* [15,45]. However, in contrast to slime mold, there is still a network of sessile cell populations present and, moreover, lymphocyte numbers are only kept constant on average by a equilibrium of influx and efflux of cells into and out of the tissue.

Secondary lymphoid tissues have a specific structure in mammals in which B and T cells are kept in separate compartments [14,21,38,62]. B cells are attracted by the chemokine CXCL13 produced by follicular dendritic cells (FDCs). Together these two cell types form several spherical or ovoid primary lymphoid follicles (PLFs) in each SLT, which altogether form the B zone.

T cells occupy the area between the follicles and are supported by stromal cells called fibroblastic reticular cells (FRCs). The FRCs produce the chemokine CCL21, which is supposed to keep the T cells in the T zone. While FDCs and FRCs are sessile populations, B and T cells enter the tissue via blood vessels that are exclusively located in the T zone. After several hours of travel through their respective compartments, lymphocytes leave the tissue again via lymphatic vessels that are also located in the T zone.

15.3.1 Speed Distribution of B and T Cells

The model for active cell migration (Equation (15.5)) requires knowledge of several parameters for each cell type in SLTs: the force on the ECM F_i^{act}, the cell pressure p_i^*, and the friction parameters η. Unfortunately, corresponding experimental data for the biophysics of lymphocytes are not available. To reduce the parameter space and for the purpose of illustration, the friction is assumed to be equal for all cell types and the medium ($\eta_{med} = \eta_i, \forall i$). Although the values for the friction are not known explicitly, the order of magnitude for the "strength" F^{act} of a cell is provided by experiments [4,12], thus confining the physiological range for η to values around 500 nN μm^{-2}s. This value is consistent with cytoplasmatic viscosities [5,6].

As a reference parameter set for lymphocyte migration, the remaining force parameters for B and T cells have been determined by a fit to measured speed distributions [7,35]: $F_{Bcell}^{act} = 18 \pm 3\,\text{nN}$, $p_{Bcell}^* = 0.04 \pm 0.01\,\text{nN}\,\mu\text{m}^{-2}$, $F_{Tcell}^{act} = 22 \pm 3\,\text{nN}$, $p_{Tcell}^* = 0.06 \pm 0.02\,\text{nN}\,\mu\text{m}^{-2}$. The full parameter set used for the simulation of the SLT ontogenesis is given in Tables 15.1, 15.2, 15.3, and 15.4 in the Appendix (Section 15.5).

15.3.2 Essential Cellular Interactions

The formation of the SLT organization has been discussed in detail previously [7,8]. Briefly, the simulations start from a background composed of chemokine-negative stromal cells and few blood vessels. The emerging organization of the SLT is a result of the following minimal set of assumptions (Figure 15.4):

- There is constant immigration of lymphocytes via blood vessels.
- Lymphocytes leave SLT via lymphatic vessels.
- Stromal cells differentiate into FRCs upon sufficient lymphocyte contact.
- FRCs are replaced by FDCs upon sufficient B cell contact.
- FDCs are replaced by FRCs upon loss of sufficient B cell contact.
- FDCs secrete a diffusing substance that inhibits FDC generation.
- Vessels are dynamic and positively correlated with FRC.
- Vessels are negatively correlated with FDC.
- FRCs secrete CCL21, which attracts B and T cells.
- FDCs secrete CXCL13, which attracts B cells.

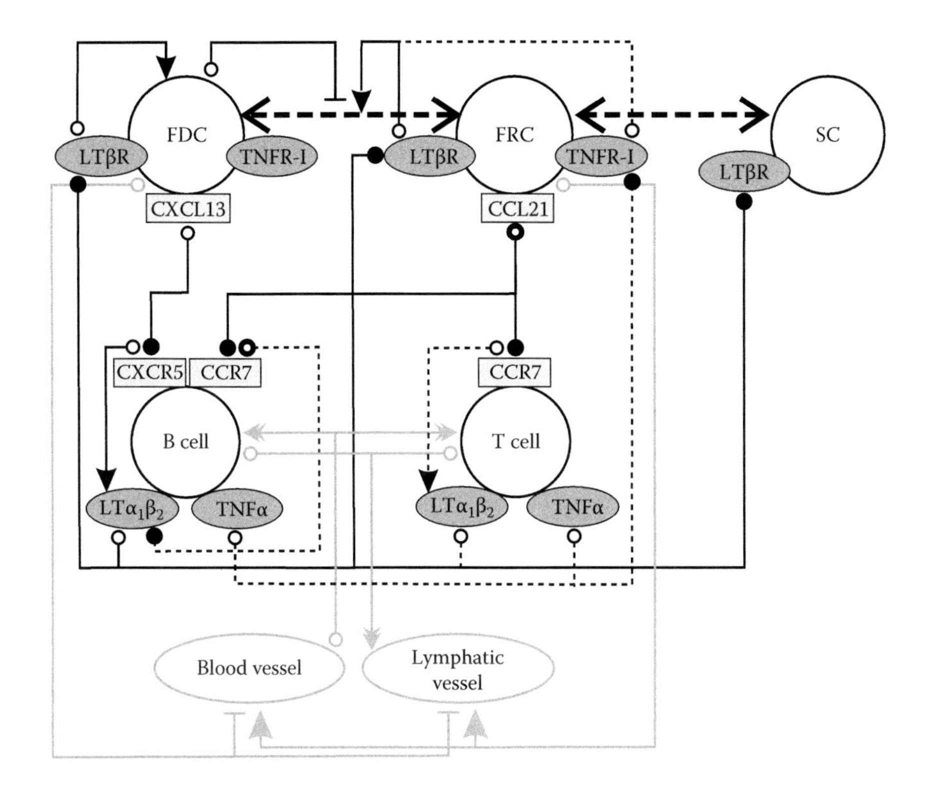

FIGURE 15.4 Essential interactions in SLT ontogenesis. Interactions are directed from open circles to arrows for positive feedback and with lines for negative feedback. Receptor–ligand pairs are indicated with closed and open circles, respectively. Stromal cell differentiation is indicated with dashed lines. Gray lines show effects of the vessel structure with migration of cells represented by double arrows. Interactions shown with dotted lines are not taken into account in the model. They represent possible interactions with experimental evidence that they are not mandatory. From Y.X. Fu, and D.D. Chaplin (1999). Development and maturation of secondary lymphoid tissues. *Ann. Rev. Immunol.* 17:399–433; G. Muller, U.E. Hopken, and M. Lipp (2003). The impact of CCR7 and CXCR5 on lymphoid organ development and systemic immunity. *Immunol. Rev.* 195:117–135; and A.V. Tumanov, S.I. Grivennikov, A.N. Shakhov, S.A. Rybtov, E.P. Koroleva, J. Takeda, S.A. Nedospasov, and D.V. Kuprash (2003). Dissecting the role of lymphotoxin in lymphoid organs by conditional targeting. *Immunol. Rev.* 195:106–116. (With permission.)

Thus, the stromal cells together with FRCs and FDCs represent a twice excitable medium that receives positive feedback from T and B cells and self-inhibition of the highest excited state (i.e., FDCs).

15.3.3 Primary Lymphoid Follicle Formation

Based on the above assumptions, the typical structure of secondary lymphoid tissues can be generated (Figure 15.5). The resulting structures are stable while B and T cells constantly flow in through blood vessels and leave the tissue through lymphatic vessels. The flow equilibrium of the immune cells is accompanied by equilibrated dynamics of FRCs, FDCs, and vessel generation. These components are in steady state with respect to their number and position.

According to the simulation, the SLT forms via a sequence of the following steps (Figure 15.5). With the first lymphocytes entering the tissue, FRCs are induced that produce CCL21 and attract both T and B lymphocytes. The presence of FRCs drives the development of vessels, leading to a homogeneous distribution of blood and lymphatic vessels inside the T zone. The presence of B cells is sufficient to replace some FRCs by FDCs. FDCs secrete CXCL13, attracting further B cells and leading to the formation of an aggregate. The attraction of B cells by CXCL13 and the induction of FDCs by B cell contact form a positive feedback loop. The growth of the developing PLF is limited by the negative feedback in which FDCs suppress the formation of new FDCs. Due to the anti-correlation of vessels and FDCs, B cells have to migrate through the T zone to reach the PLF. B cells can only leave the follicle and the SLT by migration to the surface of the follicle where they come into close contact with the lymphatic vessels in the T zone. This migration pattern is facilitated by a rather flat CXCL13 distribution in follicles. In addition, the B cells in a follicle extend beyond the FDC network such that a proportion of B cells migrates at the border of the follicle where the exit vessels are located. This implies that the PLF can be subdivided into the FDC area (which is smaller than the total B cell area) and a follicular border that is devoid of FDCs. Therefore vessels can develop in the follicular border, permitting the exit of B cells. B cells from the center of the follicle reach the periphery by random motion and can also leave the PLF.

15.3.4 Chemokinesis versus Chemotaxis

In the simulation, CCL21 and CXCL13 are described as chemotactic factors. However, in particular for CCL21, experimental data indicate a predominantly chemokinetic effect [55,66]. The simulation framework provides a hint as to why chemotactic effects cannot be seen easily by two-photon microscopy [36,65]. The distribution of each chemokine is rather flat within each compartment and chemotactic gradients are virtually absent (Figure 15.6). Thus, cells will not respond chemotactically in these zones and perform a random

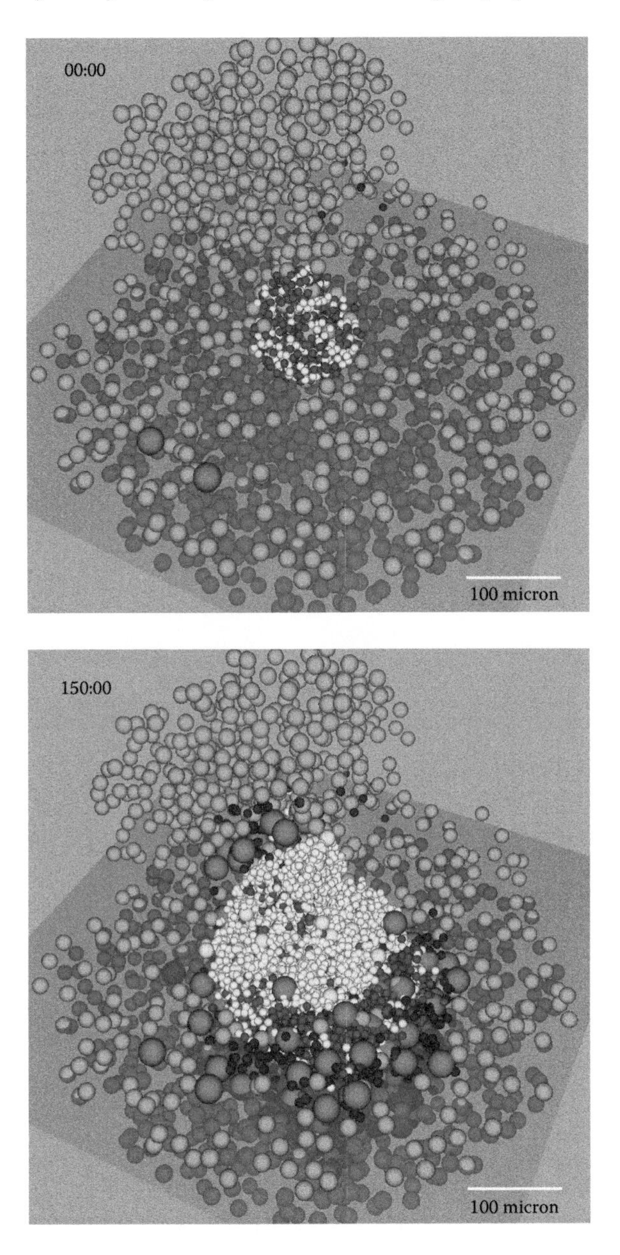

FIGURE 15.5 (See color insert following Page 398.) PLF formation. Time is shown as hh:mm. The simulation starts with a stromal background, few blood vessels, and some lymphocytes (00:00). The induction of CCL21 production by FRCs is correlated with the presence of vessels in the T zone (150:00). B cells induce and colocalize with FDCs to form a PLF adjacent to the T zone. The presence of FDCs largely suppresses vessel formation in the PLF (white: B cells, dark blue: T cells, green: FRC, yellow: FDC, red: blood vessels, dark gray: lymphatic vessels).

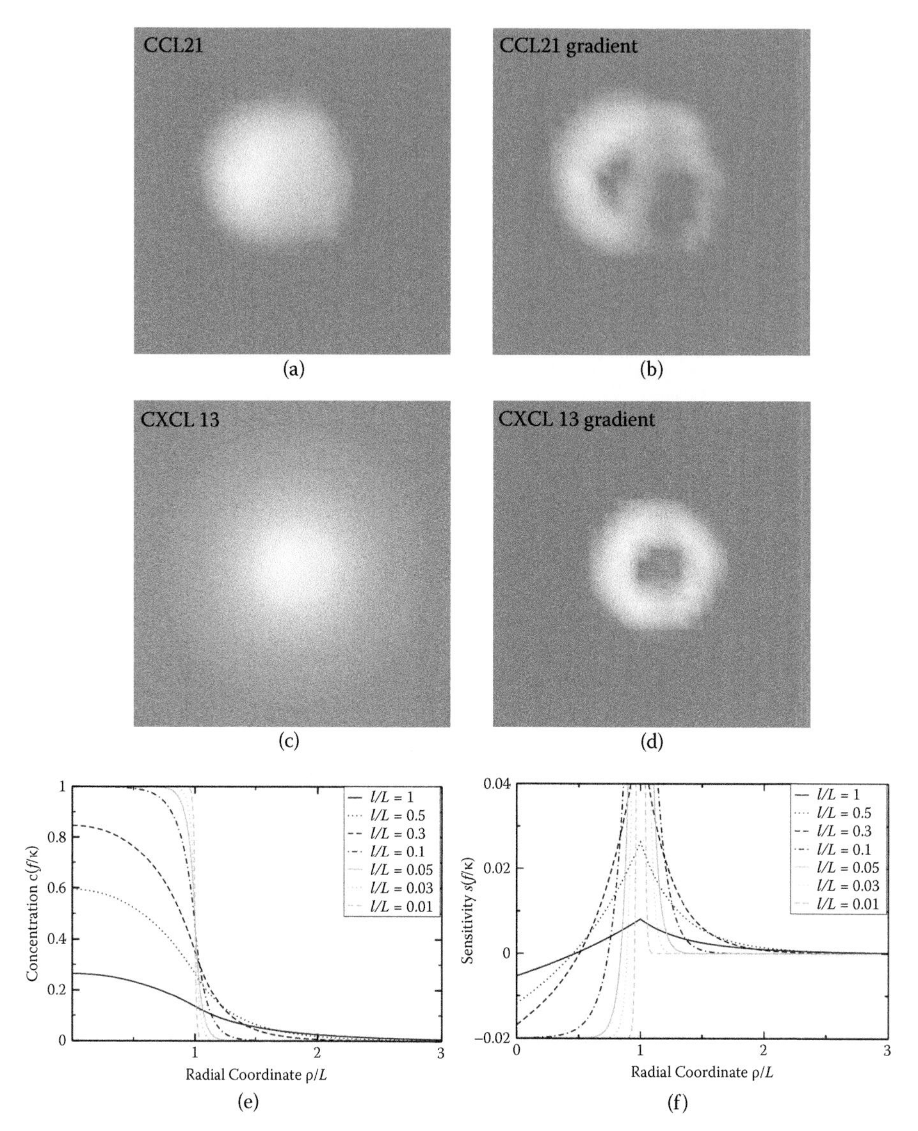

FIGURE 15.6 (a) through (d): chemokine distributions of the follicle. Values increase from gray to white. For both chemokines the gradients are significant at the periphery while the central regions show a rather flat profile. (e) and (f): chemokine profiles (concentration c) and sensitivity for various chemokine ranges l/L (Equation (15.15)) in response to a spherical homogeneous source of radius L. The concentration c is plotted against the radial coordinate ρ scaled by the radius of the source L. Cells respond to the chemokine gradient when the sensitivity $s = 2R\,|dc/d\rho| - \alpha c > 0$ with $\alpha = 0.02$ following the observation that the cell can respond to a 2% difference in the chemokine concentration across its length $2R$ [47]. The sensitivities are plotted for $2R = L/25$, which corresponds to a homogeneous chemokine source of $2L = 300$ μm diameter and a cell diameter of $2R = 6$ μm [7].

motion. Only at the boundary do cells show a preference for their direction of motion, in agreement with previous experimental work [63]. This effect also becomes obvious when looking for the analytical solution of a homogeneous spherical source. Solving the stationary diffusion equation with an unspecific decay of the chemokine κ and a homogeneous spherical source f of size L (spherical coordinates with radius ρ)

$$0 = D\left(\frac{\partial^2 c}{\partial \rho^2} + 2\rho \frac{\partial c}{\partial \rho}\right) - \kappa c - f\,\Theta(L - \rho) \tag{15.14}$$

yields the solution

$$c = \frac{f}{\kappa\rho}\left\{\rho\Theta(L-\rho) + \Theta(\rho-L)\left[L\cosh\frac{\rho-L}{l} + l\sinh\frac{\rho-L}{l}\right] - (L+l)e^{-\frac{L}{l}}\sinh\frac{\rho}{l}\right\} \tag{15.15}$$

with $l = \sqrt{D/\kappa}$ the typical length scale of the system. Even for relatively long-ranged chemokines (e.g., $l = 0.3$), the central part ($\rho < 0.5$) does not permit chemotaxis responses of cells that occur only at the periphery ($0.5 < \rho < 1$) (Figure 15.6).

15.4 Summary

The Delaunay object dynamics (DOD) method centers the description of multicellular systems around the geometry and physics of cell–cell contacts [7,34]. The architecture of the simulation tool is agent based and relies on a weighted Delaunay triangulation to deal with a continuous representation of cell positions and sizes. The Delaunay triangulation provides information about the next neighbors of a cell. The dual Voronoi tessellation can serve to extract information about geometry parameters such as cell volume and cell contact areas. In line with other models [15,16], DOD concentrates on the impact of a physical description of cells onto pattern formation of multicellular systems. Similar to molecular dynamics, DOD relies on equations describing elementary cellular interactions that are thought to be universal for large classes of tissues. Thus the different terms underlying the dynamics (Equation (15.3)) can be adapted and refined as necessary without the need to change the remaining platform architecture. This permits a straightforward extension to other specific biological systems with specific cell behavior and properties. In the multiscale agent-based approach of DOD, only the internal dynamics of cells need to be adapted while the equations describing contact interaction can remain unchanged. Improved experimental and/or theoretical data on the mechanical cell interactions can be a easily included in the model without touching the specific properties of an already-implemented specific biological system like SLT. Thus, the DOD framework combines the discrete and specific nature of cell phenotypes with their universal nature as physical objects and allows for an incremental improvement in biomechanical and biological features. In particular, this makes it possible to disentangle the effects of specific pathways and biophysical parameters on the system's behavior.

The pattern formation and homeostasis of SLT were presented to exemplify the DOD approach. The set of known local (molecular and cellular) interactions leads to the formation of the final structure of the SLT with the two compartments and is sufficient to explain the emergence of the global tissue pattern in all stages of its development. A hallmark of SLT organization is that, in contrast to classical pattern formation, these patterns are the result of a flow equilibrium of cells. Lymphocytes are exchanged on a daily basis and still self-organize into the observed stable pattern.

This example clearly demonstrates that the DOD method has the potential to uncover subtle effects of the cell biophysics on tissue organization. DOD is a quantitative method in the sense that each parameter of the model corresponds to a measurable quantity. This, together with the modular architecture, improves the predictive power of this method. The predictive power is a fundamental requirement for using computational approaches to identify relevant mechanisms in complex biological systems and for the design and guidance of novel experiments that verify these predictions.

15.5 Appendix: Simulation Parameters

The parameters used in the simulation of the SLT are listed in the tables. The references support the used value. If no reference or comment is given, a systemic model parameter was chosen to achieve sufficient accuracy in the simulation. SLT-specific parameters are discussed in detail in [7,8].

TABLE 15.1 Biophysical Parameters for Lymphocytes.

Parameter	Value	Remarks/Ref.
B/T cell diameter	6 μm	[11,23,56,59]
E_i	1 kPa	[5,6,13,19]
ν_i	0.4	[24,32]
σ_{ij}	0–0.3 nN μm^{-1}	[37,60]
F^{act}_{Bcell}	18 nN	[7,35]
F^{act}_{Tcell}	22 nN	[7,35]
p^*_{Bcell}	0.04 nN μm^{-2}	[7,35]
p^*_{Tcell}	0.06 nN μm^{-2}	[7,35]
T_p	120–180 s	[35,63]
η_i, η_{med}	500 nN μm^{-1} s	[5–7,28,35,63]

TABLE 15.2 Threshold-Model for FDC Induction by B Cell Contact

Parameter	Value	Remarks/Ref.
$LT\alpha_1\beta_2$ threshold	1	Arbitrary units
B cell $LT\alpha_1\beta_2$ expression	$0.025\,\mu m^{-2}$	Unknown
FRC/FDC differentiation time	3 h	[31,33,59]

TABLE 15.3 Reaction of Chemokines CXCL13 and CCL21 with Their Receptors CXCR5 and CCR7

Parameter	Value	Remarks/Ref.
diffusion coefficient D	$10\text{–}100\ \mu m^2\,s^{-1}$	[49]
receptor internalization rate k_i	$5 \cdot 10^{-5} \ldots 3 \cdot 10^{-2}\,s^{-1}$	[40]
receptor recycling rate k_r	$1 \cdot 10^{-4} \ldots 7 \cdot 10^{-3}\,s^{-1}$	[40]
dissociation constant K_d	$0.2 \ldots 5\,nM$	[30,48,54,64,69]
association rate k_{on}	$2.5 \cdot 10^5 \ldots 10^8\,M^{-1}s^{-1}$	[48]
dissociation rate k_{off}	$10^{-4} \ldots 1\,s^{-1}$	(from K_d and k_{on})
number of receptors per lymphocyte R_{tot}	$10^4\text{–}10^5$	[42,64]
chemokine production rate per stromal cell Q	$2.5 \cdot 10^1 \ldots 10^4\,s^{-1}$	[25,61]

TABLE 15.4 System Parameters and Numerical Tolerances

Parameter	Value	Remarks/Ref.
Size of diffusion grid	1200 μm	
Grid resolution	22 μm	
Max. cell displacement Δx	0.5 μm	
Min. time resolution Δt	10 s	
Lymphocyte influx	1 s^{-1}	[50,70]
B:T cell ratio	0.4:0.6	[50,70]
Size of simulation area	600 μm	
Number of FRC	3500	

References

[1] E. Albrecht and H.R. Petty (1998). Cellular memory: neutrophil orientation reverses during temporally decreasing chemoattractant concentrations. *Proc. Natl. Acad. Sci. USA* 95(9):5039–5044.

[2] F. Aurenhammer (1987). Power diagram: properties, algorithms and applications. *SIAM J.* 16(1):78–96.

[3] M. Bajenoff, J.G. Egen, L.Y. Koo, J.P. Laugier, F. Brau, N. Glaichenhaus, and R.N. Germain (2006). Stromal cell networks regulate lymphocyte entry, migration, and territoriality in lymph nodes. *Immunity* 25(6):989–1001.

[4] N.Q. Balaban, U.S. Schwarz, D. Riveline, P. Goichberg, G. Tzur, I. Sabanay, D. Mahalu, S. Safran, A. Bershadsky, L. Addadi, and B. Geiger (2001). Force and focal adhesion assembly: a close relationship studied using elastic micropatterned substrates. *Nature Cell Biol.* 3(5):466–472.

[5] A.R. Bausch, W. Moller, and E. Sackmann (1999). Measurement of local viscoelasticity and forces in living cells by magnetic tweezers. *Biophys. J.* 76(1):573–579.

[6] A.R. Bausch, F. Ziemann, A.A. Boulbitch, K. Jacobson, and E. Sackmann (1998). Local measurements of viscoelastic parameters of adherent cell surfaces by magnetic bead microrheometry. *Biophys. J.* 75(4):2038–2049.

[7] T. Beyer and M. Meyer-Hermann (2007). Modeling emergent tissue organization involving high-speed migrating cells in a flow equilibrium. *Phys. Rev. E* 76(2):021929.

[8] T. Beyer and M. Meyer-Hermann (2008). Mechanisms of organogenesis of primary lymphoid follicles. *Int. Immunol.* 20(4):615–623.

[9] T. Beyer, G. Schaller, A. Deutsch, and M. Meyer-Hermann (2005). Parallel dynamic and kinetic regular triangulation in three dimensions. *Comput. Phys. Commun.* 172(2):86–108.

[10] P. Bongrand (1995). Adhesion of cells. In *Handbook of Biological Physics*, Ed. R. Lipowsky and E. Sackmann, Vol. I, Amsterdam, Elsevier Science, pp. 755–803.

[11] M. Bornens, M. Paintrand, and C. Celati (1989). The cortical microfilament system of lymphoblasts displays a periodic oscillatory activity in the absence of microtubules: implications for cell polarity. *J. Cell Biol.* 109(3):1071–1083.

[12] K. Burton, J.H. Park, and D.L. Taylor (1999). Keratocytes generate traction forces in two phases. *Mol. Biol. Cell* 10(11):3745–3769.

[13] Y.S. Chu, S. Dufour, J.P. Thiery, E. Perez, and F. Pincet (2005). Johnson-Kendall-Roberts theory applied to living cells. *Phys. Rev. Lett.* 94(2):028102.

[14] J.G. Cyster (2005). Chemokines, sphingosine-1-phosphate, and cell migration in secondary lymphoid organs. *Annu. Rev. Immunol.* 23:127–159.

[15] J.C. Dallon and H.G. Othmer (2004). How cellular movement determines the collective force generated by the *Dictyostelium discoideum* slug. *J. Theor. Biol.* 231(2):203–222.

[16] D. Drasdo (2003). On selected individual-based approaches to the dynamics in multicellular systems. In *Polymer and Cell Dynamics: Multiscale Modeling and Numerical Simulations*, Eds. W. Alt, M. Chaplain, M. Griebel, J. Lenz, Basel, Birkhäuser, pp. 169–204.

[17] H. Edelsbrunner and N.R. Shah (1996). Incremental topological flipping works for regular triangulations. *Algorithmica* 15(3):223–241.

[18] E. Evans (1995). Physical actions in biological adhesion. In *Handbook of Biological Physics. Vol. I*, Eds. R. Lipowsky and E. Sackmann, Amsterdam, Elsevier Science, pp. 723–753.

[19] G. Forgacs, R.A. Foty, Y. Shafrir, and M.S. Steinberg (1998). Viscoelastic properties of living embryonic tissues: a quantitative study. *Biophys. J.* 74(5):2227–2234.

[20] P. Friedl, S. Borgmann, and E.B. Brocker (2001). Amoeboid leukocyte crawling through extracellular matrix: lessons from the *Dictyostelium* paradigm of cell movement. *J. Leukoc. Biol.* 70(4):491–509.

[21] Y.X. Fu and D.D. Chaplin (1999). Development and maturation of secondary lymphoid tissues. *Annu. Rev. Immunol.* 17:399–433.

[22] M. Gunzer, C. Weishaupt, A. Hillmer, Y. Basoglu, P. Friedl, K.E. Dittmar, W. Kolanus, G. Varga, and S. Grabbe (2004). A spectrum of biophysical interaction modes between T cells and different antigen-presenting cells during priming in 3-D collagen and *in vivo*. *Blood* 104(9):2801–2809.

[23] W.S. Haston and J.M. Shields (1984). Contraction waves in lymphocyte locomotion. *J. Cell Sci.* 68:227–241.

[24] A. Hategan, R. Law, S. Kahn, and D.E. Discher (2003). Adhesively-tensed cell membranes: lysis kinetics and atomic force microscopy probing. *Biophys. J.* 85(4):2746–2759.

[25] W.-S. Hu (2004). Stoichiometry and kinetics of cell growth and product formation. *Cell. Bioprocess Technol.*, University of Minnesota, pp. 1–18.

[26] K. Johnson, K. Kendall, and A. Roberts (1971). Surface energy and the contact of elastic solids. *Proc. Roy. Soc. London A* 324(1558):303–313.

[27] T. Lammermann, B.L. Bader, S.J. Monkley, T. Worbs, R. Wedlich-Soldner, K. Hirsch, M. Keller, R. Forster, D.R. Critchley, R. Fassler, and M. Sixt (2008). Rapid leukocyte migration by integrin-independent owing and squeezing. *Nature* 453(7191):51–55.

[28] D.A. Lauffenburger and A.F. Horwitz (1996). Cell migration: a physically integrated molecular process. *Cell* 84(3):359–366.

[29] W. Lewis (1931). Locomotion of lymphocytes. *Bull. Johns Hopkins Hosp.* 49:29–36.

[30] F. Lin, C.M. Nguyen, S.J. Wang, W. Saadi, S.P. Gross, and N.L. Jeon (2004). Effective neutrophil chemotaxis is strongly influenced by mean IL-8 concentration. *Biochem. Biophys. Res. Commun.* 319(2):576–581.

[31] F. Mackay and J.L. Browning (1998). Turning off follicular dendritic cells. *Nature* 395(6697):26–27.

[32] A.J. Maniotis, C.S. Chen, and D.E. Ingber (1997). Demonstration of mechanical connections between integrins, cytoskeletal filaments, and nucleoplasm that stabilize nuclear structure. *Proc. Natl. Acad. Sci. USA* 94(3):849–854.

[33] R.E. Mebius, P. Rennert, and I.L. Weissman (1997). Developing lymph nodes collect $CD4^+CD3^-$ $LT\beta^+$ cells that can differentiate to APC, NK cells, and follicular cells but not T or B cells. *Immunity* 7(4):493–504.

[34] M. Meyer-Hermann (2008). Delaunay object dynamics: cell mechanics with a 3D kinetic and dynamic weighted Delaunay triangulation. *Curr. Top. Dev. Biol.* 81:373–399.

[35] M.J. Miller, S.H. Wei, I. Parker, and M.D. Cahalan (2002). Two-photon imaging of lymphocyte motility and antigen response in intact lymph node. *Science* 296(5574):1869–1873.

[36] T.J. Mitchison and L.P. Cramer (1996). Actin-based cell motility and cell locomotion. *Cell* 84(3):371–379.

[37] V.T. Moy, Y. Jiao, T. Hillmann, H. Lehmann, and T. Sano (1999). Adhesion energy of receptor-mediated interaction measured by elastic deformation. *Biophys. J.* 76(3):1632–1638.

[38] G. Muller, U.E. Hopken, and M. Lipp (2003). The impact of CCR7 and CXCR5 on lymphoid organ development and systemic immunity. *Immunol. Rev.* 195:117–135.

[39] J. Murray, H. Vawter-Hugart, E. Voss, and D.R. Soll (1992). Three-dimensional motility cycle in leukocytes. *Cell. Motil. Cytoskeleton* 22(3):211–223.

[40] N.F. Neel, E. Schutyser, J. Sai, G.H. Fan, and A. Richmond (2005). Chemokine receptor internalization and intracellular trafficking. *Cytokine Growth Factor Rev.* 16(6):637–658.

[41] A. Okabe, B. Boots, K. Sugihara, and S.N. Chiu (2000). *Spatial Tessellations: Concepts and Applications of Voronoi Diagrams, 2nd ed.* New York, John Wiley & Sons.

[42] T. Okada, V.N. Ngo, E.H. Ekland, R. Forster, M. Lipp, D.R. Littman, and J. G. Cyster (2002). Chemokine requirements for B cell entry to lymph nodes and Peyer's patches. *J. Exp. Med.* 196(1):65–75.

[43] T. Oliver, M. Dembo, and K. Jacobson (1999). Separation of propulsive and adhesive traction stresses in locomoting keratocytes. *J. Cell Biol.* 145(3):589–604.

[44] G. Oron and H. Herrmann (1998). Exact calculation of force networks in granular piles. *Phys. Rev. E* 58:2079–2089.

[45] E. Palsson and H.G. Othmer (2000). A model for individual and collective cell movement in *Dictyostelium discoideum*. *Proc. Natl. Acad. Sci. USA* 97(19):10448–10453.

[46] E. Paluch, M. Piel, J. Prost, M. Bornens, and C. Sykes (2005). Cortical actomyosin breakage triggers shape oscillations in cells and cell fragments. *Biophys. J.* 89(1):724–733.

[47] C.A. Parent and P.N. Devreotes (1999). A cell's sense of direction. *Science* 284(5415):765–770.

[48] A.J. Pelletier, L.J. van der Laan, P. Hildbrand, M.A. Siani, D.A. Thompson, P.E. Dawson, B.E. Torbett, and D.R. Salomon (2000). Presentation of chemokine SDF-1-alpha by fibronectin mediates directed migration of T cells. *Blood* 96(8):2682–2690.

[49] G.J. Randolph, V. Angeli, and M.A. Swartz (2005). Dendritic-cell trafficking to lymph nodes through lymphatic vessels. *Nat. Rev. Immunol.* 5(8):617–628.

[50] R. Sacca, C.A. Cuff, W. Lesslauer, and N.H. Ruddle (1998). Differential activities of secreted lymphotoxin-α3 and membrane lymphotoxin-α1-β2 in lymphotoxin-induced inflammation: critical role of TNF receptor 1 signaling. *J. Immunol.* 160(1):485–491.

[51] G. Schaller (2006). On selected numerical approaches to cellular tissue. Ph.D. thesis, Johann Wolfgang Goethe University, Frankfurt am Main.

[52] G. Schaller and M. Meyer-Hermann (2004). Kinetic and dynamic Delaunay tetrahedralizations in three dimensions. *Comput. Phys. Commun.* 162(11):9–23.

[53] G. Schaller and M. Meyer-Hermann (2005). Multicellular tumor spheroid in an off-lattice Voronoi-Delaunay cell model. *Phys. Rev. E.* 71(5):051910.

[54] H. Slimani, N. Charnaux, E. Mbemba, L. Saffar, R. Vassy, C. Vita, and L. Gattegno (2003). Binding of the CC-chemokine RANTES to syndecan-1 and syndecan-4 expressed on HeLa cells. *Glycobiology* 13(9):623–634.

[55] A.N. Stachowiak, Y. Wang, Y.C. Huang, and D.J. Irvine (2006). Homeostatic lymphoid chemokines synergize with adhesion ligands to trigger T and B lymphocyte chemokinesis. *J. Immunol.* 177(4):2340–2348.

[56] C.B. Thompson, I. Scher, M.E. Schaefer, T. Lindsten, F.D. Finkelman, and J.J. Mond (1984). Size-dependent B lymphocyte subpopulations: relationship of cell volume to surface phenotype, cell cycle, proliferative response, and requirements for antibody production to TNP-Ficoll and TNP-BA. *J. Immunol.* 133(5):2333–2342.

[57] A.V. Tumanov, S.I. Grivennikov, A.N. Shakhov, S.A. Rybtsov, E.P. Koroleva, J. Takeda, S.A. Nedospasov, and D. V. Kuprash (2003). Dissecting the role of lymphotoxin in lymphoid organs by conditional targeting. *Immunol. Rev.* 195:106–116.

[58] K.S. Uchida, T. Kitanishi-Yumura, and S. Yumura (2003). Myosin II contributes to the posterior contraction and the anterior extension during the retraction phase in migrating Dictyostelium cells. *J. Cell Sci.* 116(1):51–60.

[59] W. van Ewijk and T.H. van der Kwast (1980). Migration of B lymphocytes in lymphoid organs of lethally irradiated, thymocyte-reconstituted mice. *Cell. Tissue Res.* 212(3):497–508.

[60] C. Verdier (2003). Rheological properties of living materials: from cells to tissues. *J. Theor. Med.* 5(2):67–91.

[61] J.L. Vissers, F.C. Hartgers, E. Lindhout, C.G. Figdor, and G.J. Adema (2001). BLC (CXCL13) is expressed by different dendritic cell subsets *in vitro* and *in vivo*. *Eur. J. Immunol.* 31(5):1544–1549.

[62] U.H. von Andrian and T.R. Mempel (2003). Homing and cellular traffic in lymph nodes. *Nat. Rev. Immunol.* 3(11):867–878.

[63] S.H. Wei, I. Parker, M.J. Miller, and M.D. Cahalan (2003). A stochastic view of lymphocyte motility and trafficking within the lymph node. *Immunol. Rev.* 195:136–159.

[64] K. Willimann, D.F. Legler, M. Loetscher, R.S. Roos, M.B. Delgado, I. Clark-Lewis, M. Baggiolini, and B. Moser (1998). The chemokine SLC is expressed in T cell areas of lymph nodes and mucosal lymphoid tissues and attracts activated T cells via CCR7. *Eur. J. Immunol.* 28(6):2025–2034.

[65] K. Wolf, R. Muller, S. Borgmann, E.B. Brocker, and P. Friedl (2003). Amoeboid shape change and contact guidance: T-lymphocyte crawling through fibrillar collagen is independent of matrix remodeling by MMPs and other proteases. *Blood* 102(9):3262–3269.

[66] T. Worbs, T.R. Mempel, J. Bolter, U.H. von Andrian, and R. Forster (2007). CCR7 ligands stimulate the intranodal motility of T lymphocytes *in vivo*. *J. Exp. Med.* 204(3):489–495.

[67] F. Yang (2003). Load-displacement relation in adhesion measurement. *J. Phys. D* 36(19):2417–2420.

[68] K. Yoshida and T. Soldati (2006). Dissection of amoeboid movement into two mechanically distinct modes. *J. Cell. Sci.* 119(18):3833–3844.

[69] R. Yoshida, M. Nagira, M. Kitaura, N. Imagawa, T. Imai, and O. Yoshie (1998). Secondary lymphoid-tissue chemokine is a functional ligand for the CC chemokine receptor CCR7. *J. Biol. Chem.* 273(12):7118–7122.

[70] A.J. Young (1999). The physiology of lymphocyte migration through the single lymph node *in vivo*. *Semin. Immunol.* 11(2):73–83.

Index

D

E

F

N

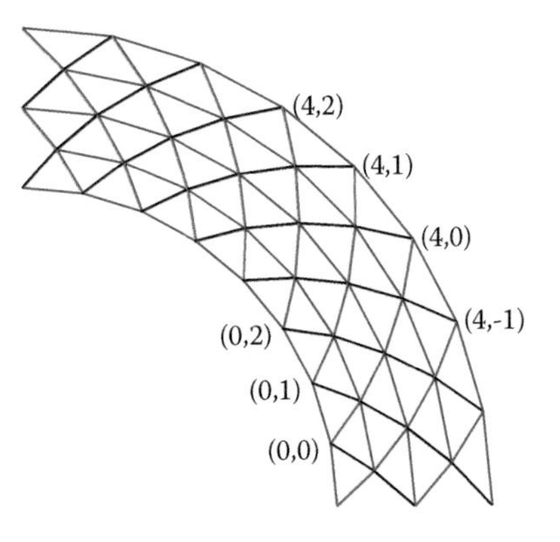

COLOR FIGURE 2.10

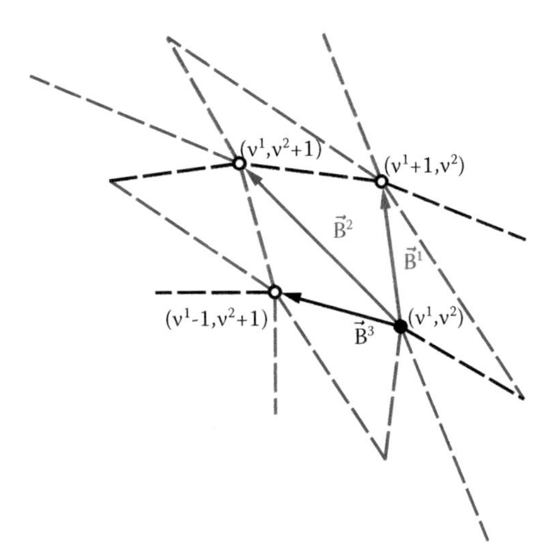

COLOR FIGURE 2.11

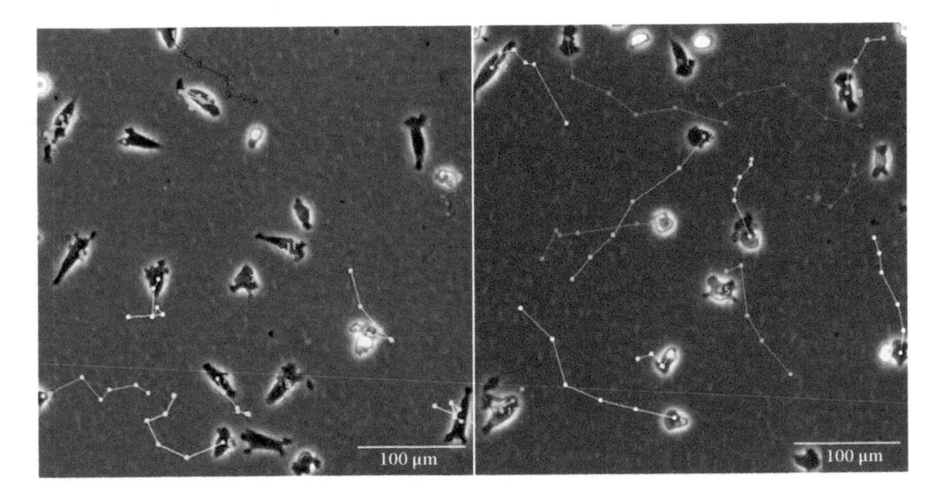

COLOR FIGURE 3.3

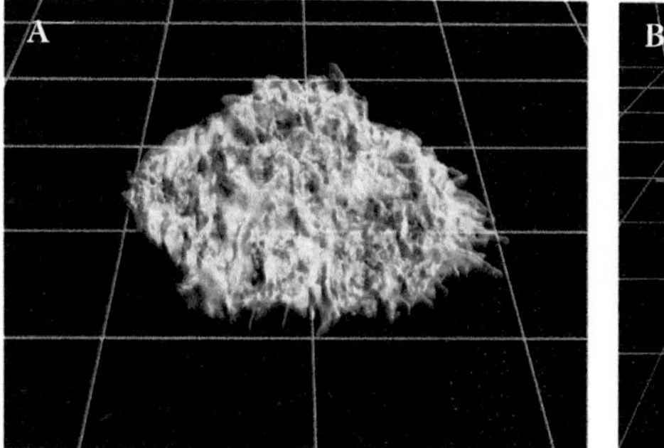

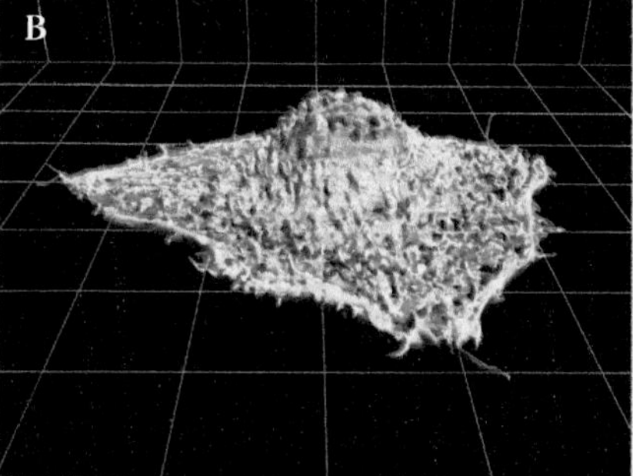

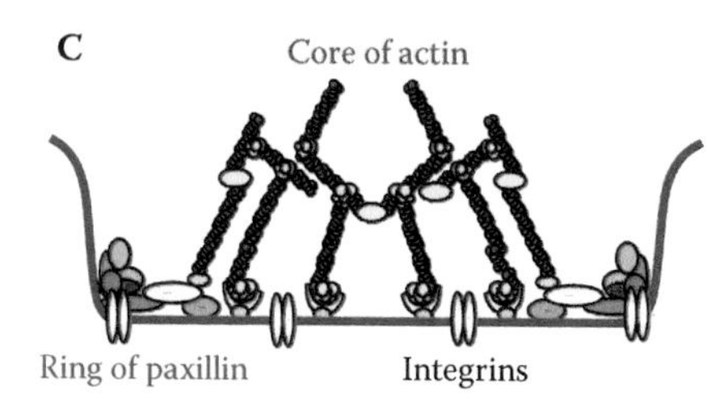

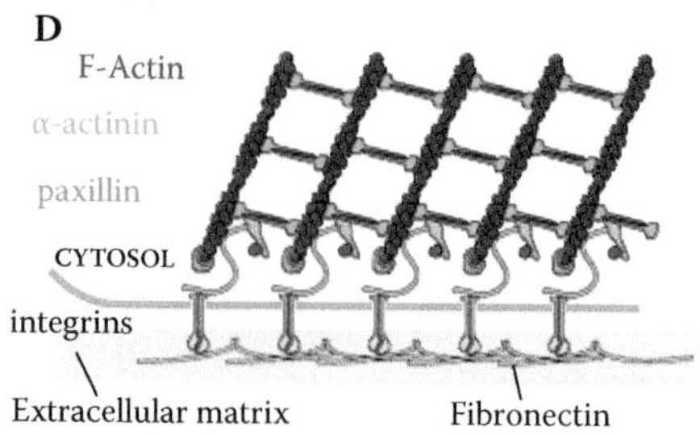

COLOR FIGURE 8.8

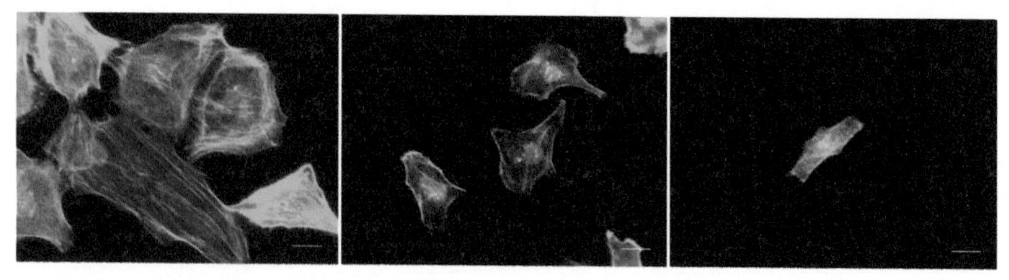

COLOR FIGURE 9.10

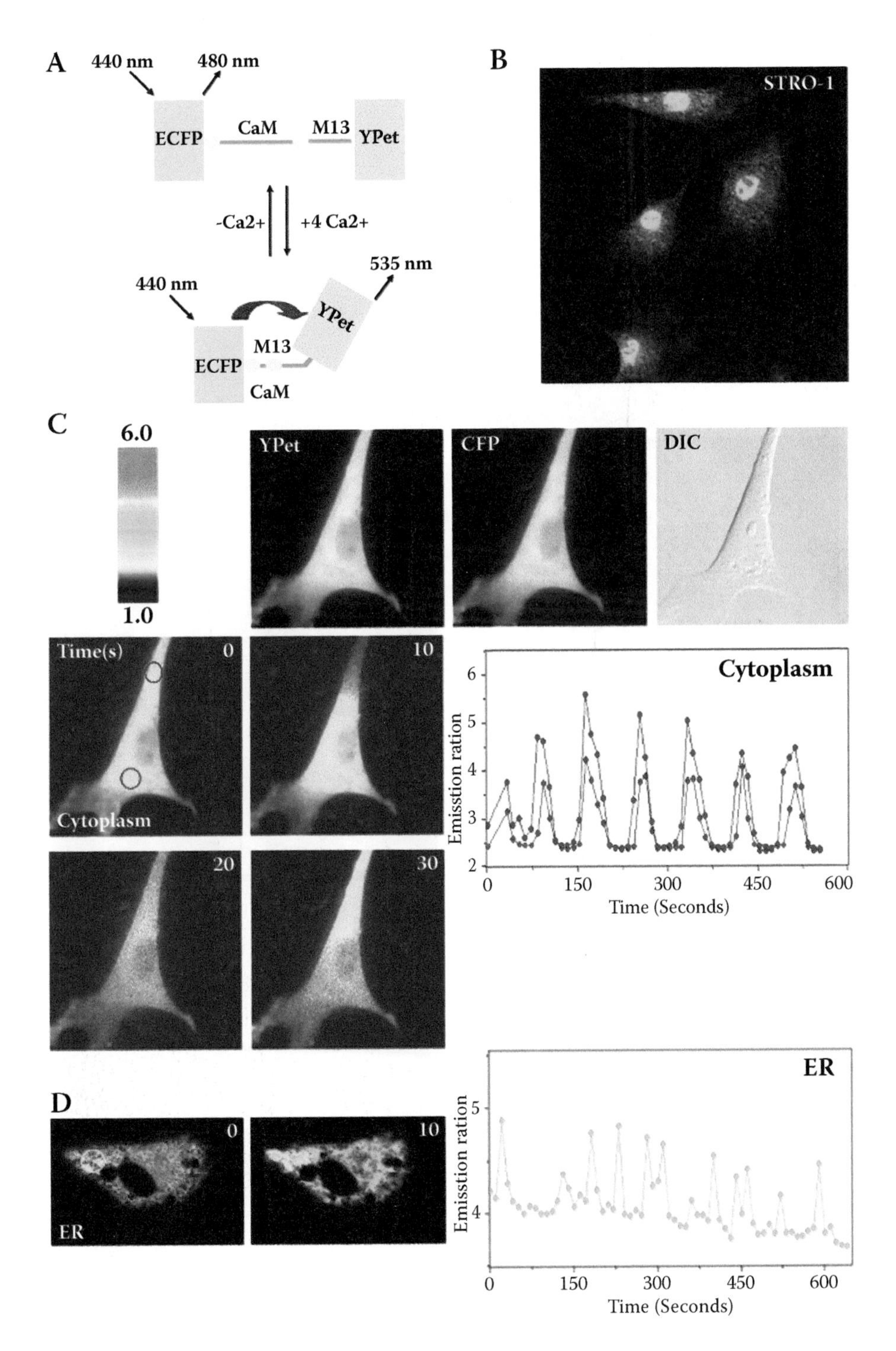

COLOR FIGURE 10.1

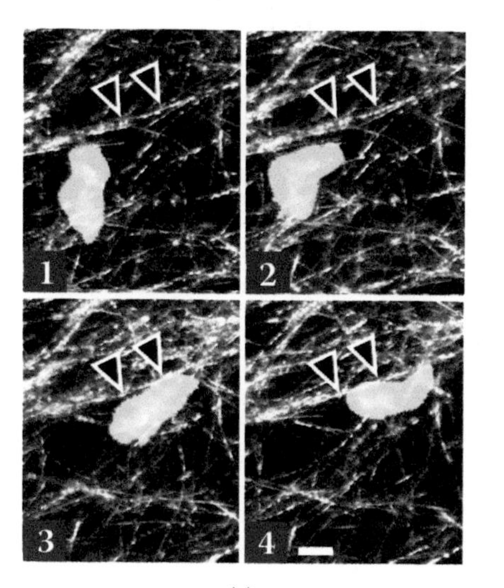

(a)

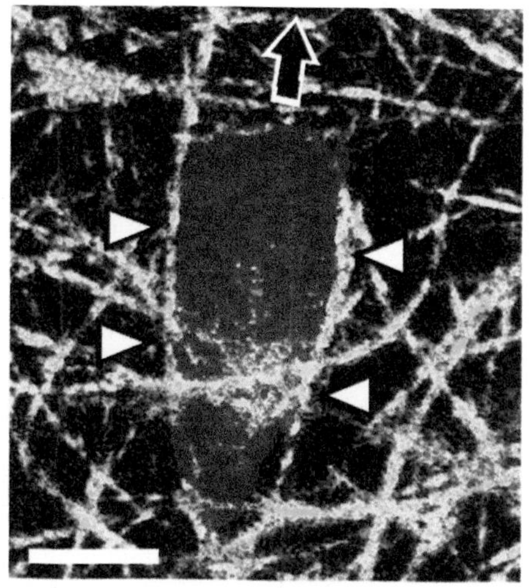

(b)

COLOR FIGURE 11.1

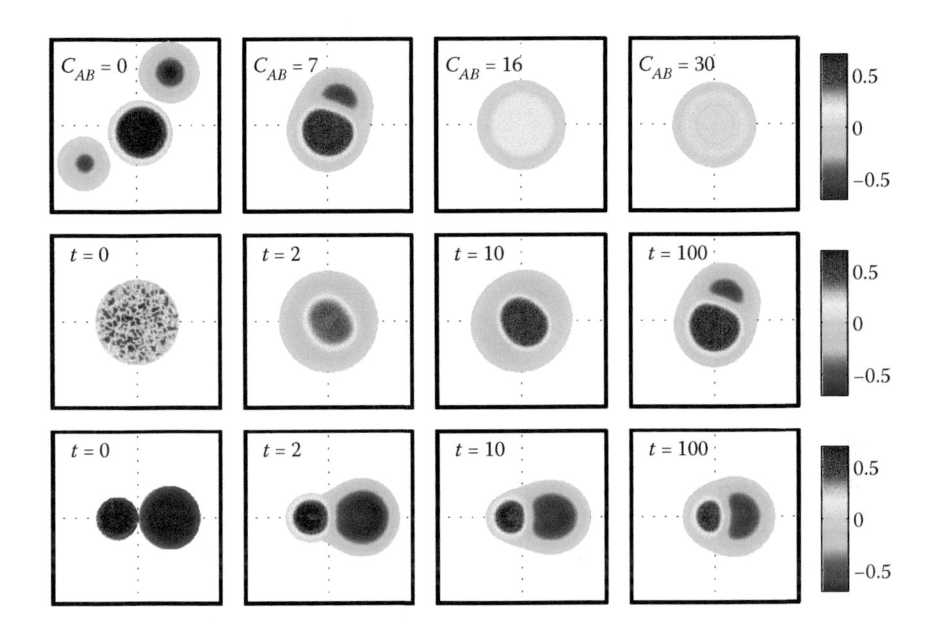

COLOR FIGURE 12.4

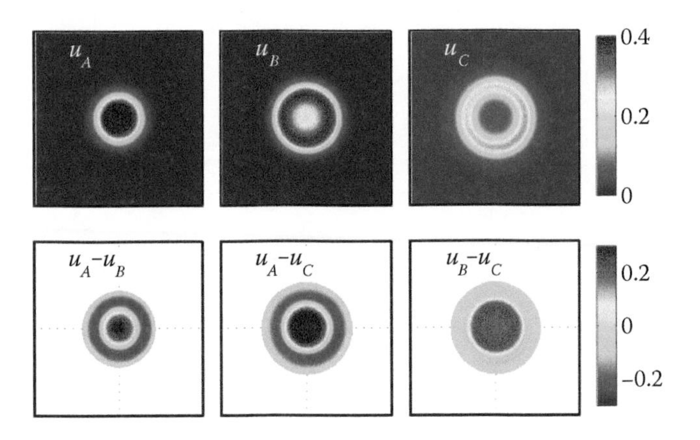

COLOR FIGURE 12.5

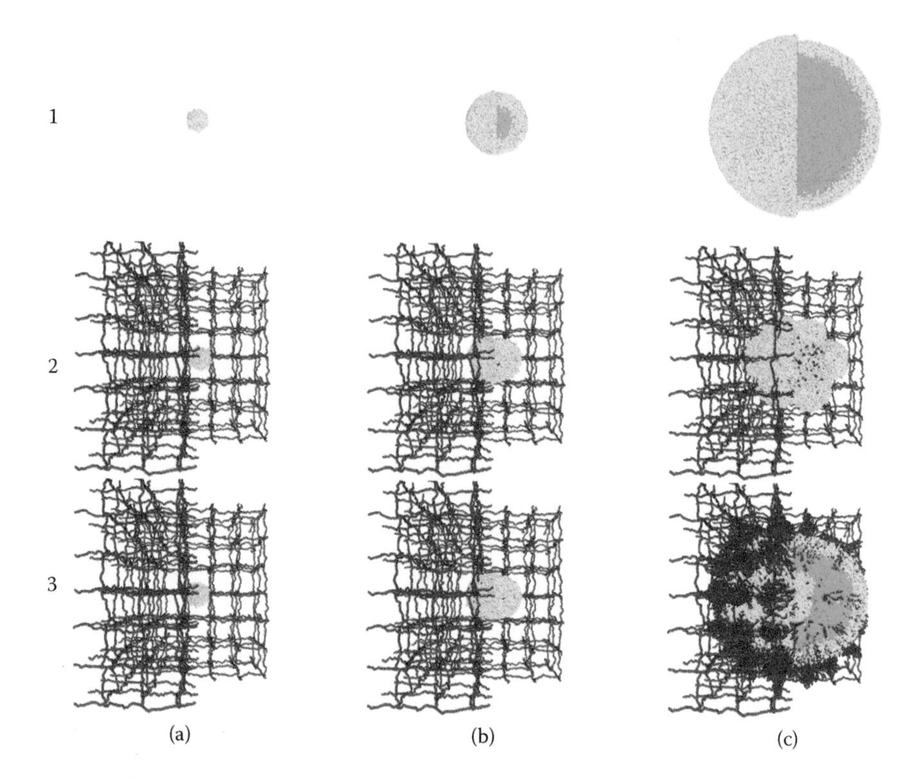

COLOR FIGURE 14.3

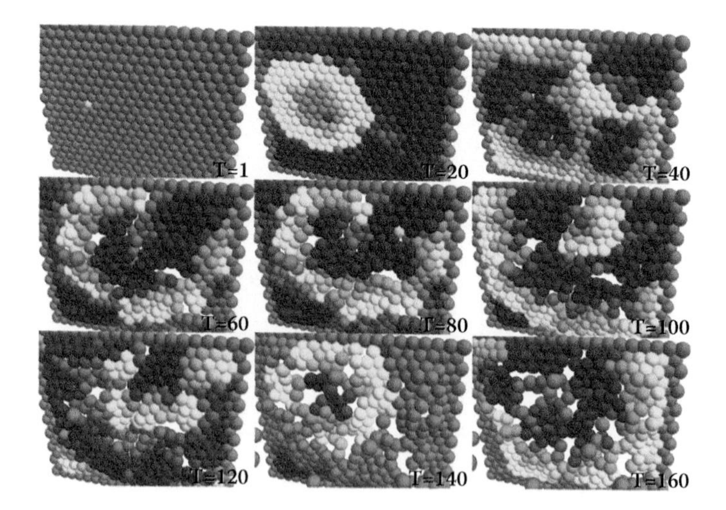

COLOR FIGURE 14.8

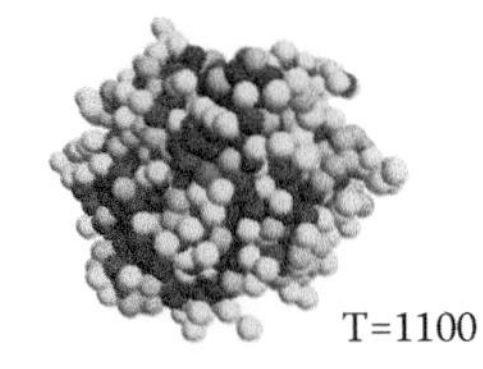

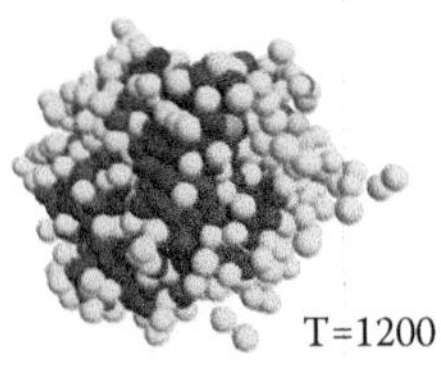

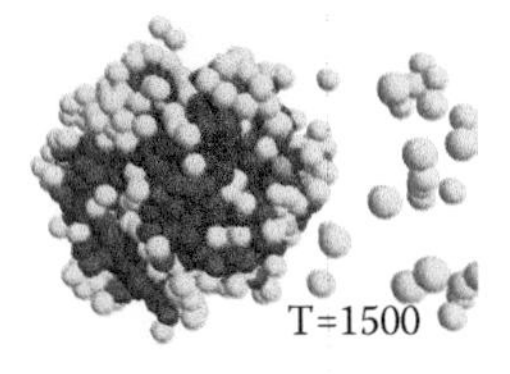

COLOR FIGURE 14.9

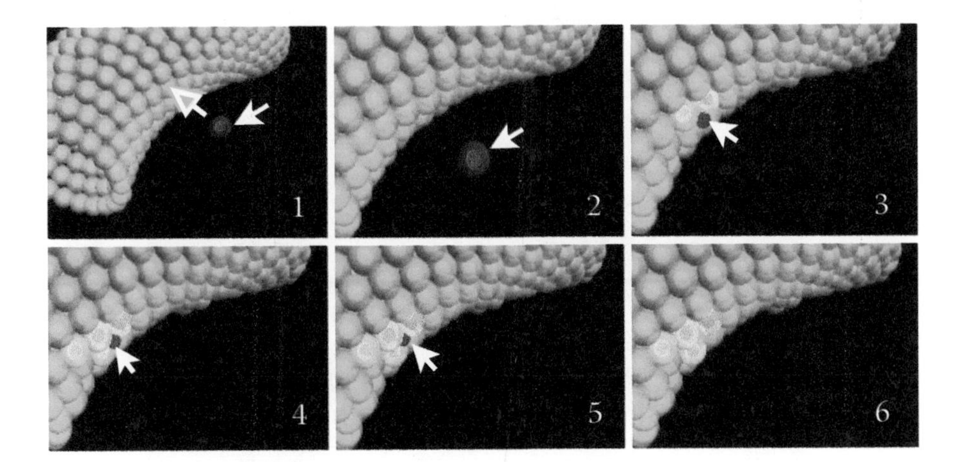

COLOR FIGURE 14.11

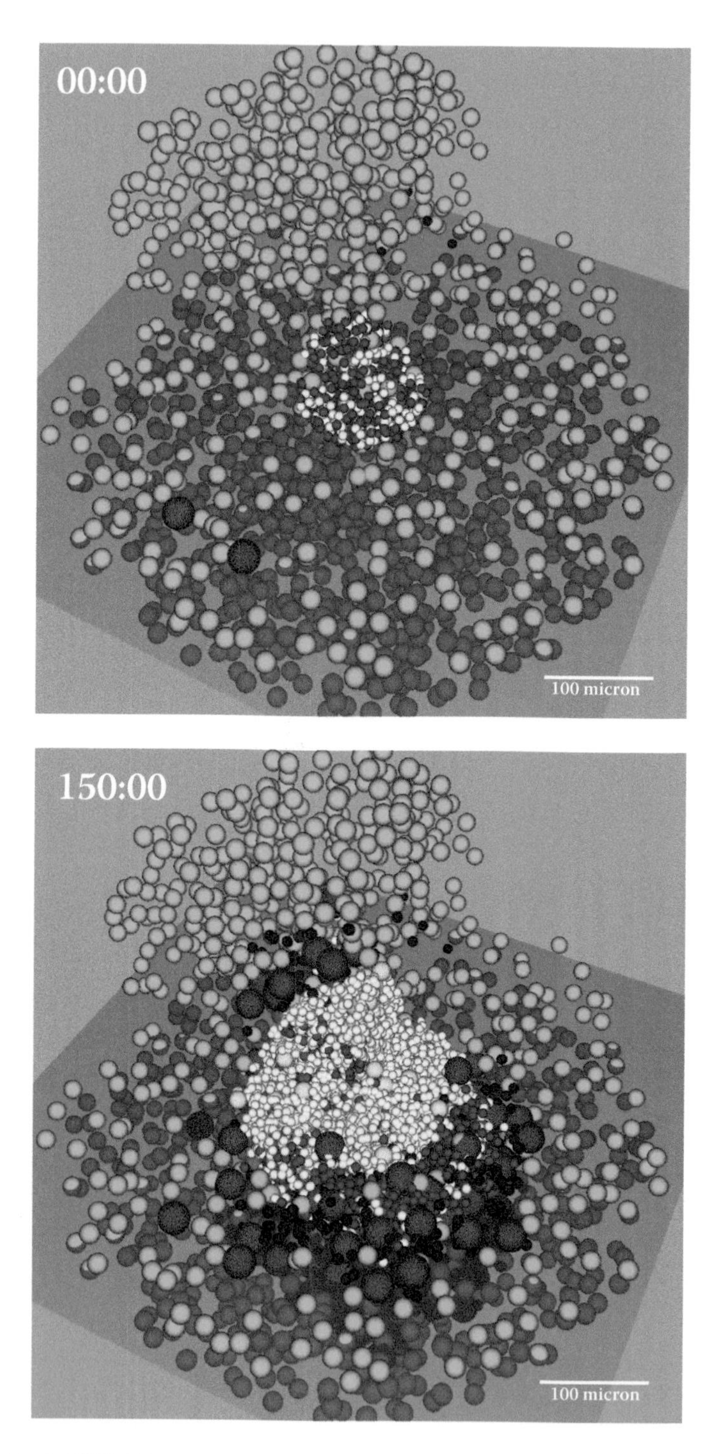

COLOR FIGURE 15.5

T - #0153 - 171019 - C8 - 234/156/21 - PB - 9780367384524